Geology of Pluton-related Gold Mineralization at Battle Mountain, Nevada

Ted G. Theodore

with a section on Potassium-Argon Chronology of Cretaceous and Cenozoic Igneous Activity, Hydrothermal Alteration, and Mineralization

by Edwin H. McKee

and a section on Lone Tree Gold Deposit

by Edward I. Bloomstein, Bruce L. Braginton, Roy W. Owen, Ronald L. Parratt, Kenneth C. Raabe, and Warren F. Thompson

and a section on Geology of The Marigold Mine Area

by Douglas H. McGibbon and Andy B. Wallace

and a section on Geology, Mineralization, and Exploration History of the Trenton Canyon Project

by Robert P. Felder

and other sections by Robert L. Oscarson and D. M. DeR. Channer

Monographs in Mineral Resource Science No. 2
Center for Mineral Resources, The University of Arizona®
and the U.S. Geological Survey

Center for Mineral Resources
Gould-Simpson Building
The University of Arizona
Tucson, Arizona 85721
Presentation copyright © by The Arizona Board of Regents
Published 2000
Printed in the United States of America

Library of Congress Catalog Catalog Card Number: 00-102928

ISBN 0-96612330-1

MONOGRAPHS IN MINERAL RESOURCE SCIENCE

Text format: Meg A. Watt

The Center for Mineral Resources is a joint venture of the Geoscience Department of The University of Arizona and the U.S. Geological Survey, for the purpose of furthering collaborative research in mineral resource science. To this end, its Monograph Series publishes extended original studies of topics in the scientific and societal aspects of mineral resources. The scope is international. The goal is to provide a venue for high-quality, timely, and affordable presentation, either of single-component studies or multi-component studies linked by a common theme, including color maps. Submissions should be entirely digitial and with documented previous reviews. More information can be obtained from the Center for Mineral Resources, The University of Arizona, Gould-Simpson Building, Tucson, Arizona, 85721.

Distributed by The University of Arizona Press, 1230 N. Park Avenue, Suite 102, Tucson, Arizona 85719.

Printed on acid-free paper.

Contents

Plates

Figures

Tables

Preface

This second volume from the Center for Mineral Resources presents the results of many years of work in the Battle Mountain district by one of the U.S. Geological Survey's eminent economic geologists and his 12 collaborators. With 33 years of experience in the district, Ted Theodore has synthesized pertinent information ranging from tectonics to microchemistry. Thus, he draws together numerous geochronologic, exploration, and other threads for this volume, also presented by his co-authors. Publication comes at a time when relations among Nevada's big gold districts are receiving great attention. Key to their understanding are integrated studies such as this, that document and synthesize features across one of the major districts in this important region.

Further monographs in line for the series derive from both USGS and university sources, and address themes not limited to North America. We intend to continue providing an outlet for regional mineral-resource monographs in a high-quality series with large-format maps.

Eric R. Force
Editor, Center for Mineral Resources

Frontispiece Oblique aerial photograph viewed to southwest in October, 1992, showing open pits and gold production facilities in the general area of the Marigold mining complex. Informally named Havallah Hills make up low hills in upper right part of photograph. Large open pit in central part of photograph is at site of Eight South deposit, and small open pit in upper left is site of Top Zone deposit. Old Marigold Mine is at head of arrow. Buffalo Mountain is at skyline on upper right, and Tobin Range is at upper left corner of photograph. Photograph courtesy of Valerie Parks, published with permission granted to Marigold Mining Co.

Geology of pluton-related gold mineralization at Battle Mountain, Nevada

Ted G. Theodore

ABSTRACT

The Battle Mountain Mining District contains many of the classic pluton-related metal deposits in the northern Great Basin of Nevada, including Au skarns and a number of porphyry Cu and stockwork Mo systems. Geologic relations of most base- and precious-metal deposits in the southern part of the mining district were well established prior to this investigation, but the geology of a significant number of Au deposits in the northern half continued to be contentious many years after their discovery. The northern part of the mining district—specifically the Valmy, North Peak, and Snow Gulch 7–1/2 minute quadrangles—has been re-examined in this report in order to clarify geologic relations of these enigmatic Au deposits. As a result, it has been determined that the three quadrangles contain a number of highly mineralized and economically important sites that show widespread introduction of structurally-controlled Tertiary Au whose orientations partly are inherited from much older fabrics. Significant Au ore bodies are present (1) at Lone Tree Hill, (2) in the general area of the Eight South, Top Zone, and East Hill/UNR deposits, and (3) at the Trenton Canyon-Valmy-North Peak cluster of deposits. Most ores in these deposits are oxidized. The first two areas were in large-scale production of Au in 1998, whereas the North Peak deposit began limited production late in 1996. Prior to mining, the deposits included approximately 228 tonnes Au (7.3 million troy oz Au) in minable reserves; the bulk of the Au was in the Lone Tree

area. Widespread exploration activity continued through middle 1998, and the likelihood for additional discoveries of economic Au resources is persuasive on the basis of preliminary evaluations. In addition, many Au deposits in the northern part of mining district were covered by varying thicknesses of Quaternary gravel and alluvium, and, for some deposits, Tertiary gravel and alluvium. The latter, in places, includes small amounts of Tertiary ash-flow tuff. Moreover, several exposed porphyry Cu and stockwork Mo systems are present in the southern parts of the quadrangles. However, most sites of extensively Au–mineralized rock are present north of the Oyarbide fault, whereas the porphyry systems crop out south of the Oyarbide fault. The Oyarbide fault is a post-mineral, northeast-striking Miocene fault, which has a vertical component of offset of approximately 700 m with its northwestern block faulted down.

Most of the 15 currently (1999) known Au deposits north of the Oyarbide fault are hosted almost entirely by altered, intensely fractured lower Paleozoic and upper Paleozoic sedimentary rocks, and the distribution of the deposits is controlled strongly by faults, many of which have north-south strikes. Such fault orientations may be related partly to deep-seated structures as well as north-trending folds associated with accretion of the Roberts Mountains and Golconda allochthons, respectively during the Antler and Sonoma orogenies. However, many of these Au deposits do not show unequivocal spatial and genetic ties to felsic intrusions. Nonetheless, the 15 deposits are herein classified as distal-disseminated Ag–Au deposits, and are considered to have formed on the upper and lateral fringes of genetically related porphyry- or pluton-related mineralized systems on the basis of some compelling geologic evidence. Because of their geologic setting, Au deposits north of the Oyarbide fault apparently represent tops of porphyry Cu systems, and they are envisioned to represent products of far-traveled, Au–enriched fluids of probable late Eocene and (or) early Oligocene age. In addition, sulfur isotopic ratios, mostly from hydrothermal barite at the Eight South Au deposit, cluster tightly near +10 to +11 per mil $\delta^{34}S$, which suggests derivation of sulfur primarily from a magma and a homogeneous sulfur source. Isotopic ratios of fluid-inclusion waters in barite from the Lone Tree deposit cluster near those of magmatic waters. The isotopic ratios, as well as the recent discovery of high temperature, highly saline fluid inclusions in mineralized veins at Lone Tree, and sericitically altered dikes associated with Au–mineralized rocks at Trenton Canyon, thus are all compatible with the proposed deposit-model designations. However, the Au occurrences show many differences when compared to other deposits elsewhere previously assigned to the distal-disseminated Ag–Au category. These differences possibly are the result of (1) an expansion of the model to include deposits spread vertically over a significant distance above buried porphyry Cu systems, as well as (2) oxidation of the deposits for a significant part of the Tertiary. Nonetheless, the large number of Au occurrences present north of the post-mineral Oyarbide fault, as well as the absence of well-exposed porphyry Cu systems, distinguishes the northern area from the southern one primarily on the basis of paleodepths. All Au deposits north of the Oyarbide fault have current levels of exposure that should be close to the uppermost parts of large porphyry Cu systems.

The mining district south of the Oyarbide fault represents a somewhat deeper geologic environment because of the presence of at least seven exposed porphyry Cu or stockwork Mo systems. These systems are present at Copper Canyon, Copper Basin-Buckingham, Trenton Canyon, Buffalo Valley Gold, Buffalo Valley Molybdenum, Elder Creek, and Paiute Gulch. All seven of these porphyry systems are surrounded by widespread polymetallic veins, and some of them also include prominent mineralized skarns, including the 2–million-oz Au skarn at Fortitude in the Copper Canyon area.

INTRODUCTION

The adjoining Valmy, North Peak, and Snow Gulch 7–1/2 minute, bounded by latitudes 40° 52' 30" and 40° 37' 30" and longitudes 117° 00' and 117° 15', contain the highly mineralized area around Lone Tree Hill, and mineralized areas near the Old Marigold Mine (fig. 1). As used in this report, the Battle Mountain Mining District is considered to include all of the Antler Peak 15–minute quadrangle as originally defined by Roberts (1964) and Roberts and Arnold (1965), as well as the Valmy quadrangle which includes the Lone Tree area (fig. 1). This differs from the usage of Tingley (1992) who restricted his use of the term "Battle Mountain Mining District" to essentially all bedrock areas of the Snow Gulch and Galena Canyon and the eastern parts of the North Peak and Antler Peak 7–1/2 minute quadrangles. Tingley (1992) also refers to an area around the Buffalo Valley gold mine in the northwest quadrant of the Antler Peak 7–1/2 minute quadrangle as the "Buffalo Valley Mining District," and he places the area of Lone Tree Hill near the northeast end of his "Buffalo Mountain Mining District," although the gold deposits and

geology at Lone Tree Hill are allied closely with the Battle Mountain Mining District.

The three quadrangles studied are in Humboldt and Lander Counties, in large part south of the Humboldt River—they are approximately 50 km southeast of the city of Winnemucca and 20 km west-northwest of the unincorporated town of Battle Mountain (fig. 1). The Valmy quadrangle encompasses both Lone Tree Hill and Treaty Hill, which are, respectively, south and north of the Humboldt River. These two small hills form isolated inselbergs rising from the surrounding pediment. The bulk of the bedrock area of the Valmy quadrangle is present in the southwest part of the quadrangle, which makes up a large part of the low-lying, informally named Havallah Hills (Frontispiece). The North Peak quadrangle includes North Peak, the highest point in the Battle Mountain Range at an 8,550–ft elevation, together with many rugged ridges leading up to this point. The Snow Gulch quadrangle is immediately east of the North Peak. A number of gold deposits are present in the pediment area in the general area of the Old Marigold Mine near the northern border of the North Peak quadrangle, and one of the large deposits in this cluster of deposits, the Eight South deposit, was in production in the early 1990s (Frontispiece).

Previous geologic studies

Previous geologic studies in the Battle Mountain Mining District include the landmark studies by Roberts (1964) and Roberts and Arnold (1965), who described the stratigraphy, structure, and ore deposits of the Antler Peak quadrangle and the regional implications of the geology therein. Several other reports that focus primarily on structural and stratigraphic relations in Paleozoic rocks of the upper plate of the Golconda thrust in the mining district also deserve mention because of the significant contributions they have made to our understanding of the complexities of these enigmatic rocks (fig. 2). Miller and others (1982) documented the imbricate style of tectonic overthrust relations within the Golconda allochthon in the Willow Creek area of the mining district, and Murchey (1990) established a biostratigraphic zonation in these rocks that provided the basis upon which geologic mapping (pls. 1, 2; Doebrich, 1992, 1995) could be conducted with confidence. Madrid (1987) remapped at large scale a small area including some critical geologic relations originally documented by Roberts among the Upper Cambrian Harmony Formation, the Ordovician Valmy Formation, and the Middle Pennsylvanian Battle Formation near the south border of the North Peak quadrangle. Madrid (1987) confirmed that the relations among the rocks there had been interpreted correctly by Roberts. Elsewhere in the

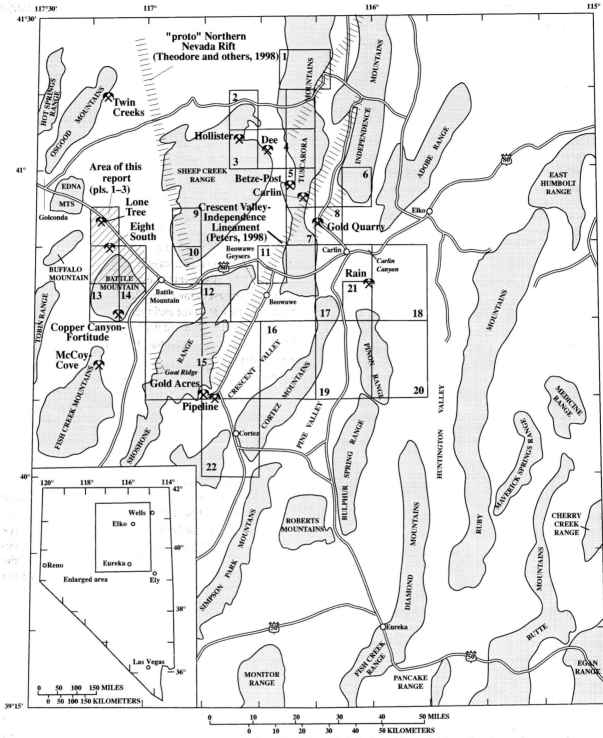

Figure 1 Index map of north-central Nevada showing location of the Valmy, North Peak, and Snow Gulch quadrangles, major mountain ranges, areas discussed, and locations of some nearby quadrangles, as well as locations of "proto" northern Nevada rift of Theodore and others (1998) and the Crescent Valley-Independence lineament of Peters (1998). 1, Henry and others (1998); 2, Willow Creek Reservoir (Wallace and John, 1998); 3, Willow Creek Reservoir SE (A.R. Wallace, unpub. data, 1999); 4, Santa Renia Fields and Beaver Peak (Theodore and others, 1998); 5, 7 Rodeo Creek NE and Welches Canyon (Evans, 1974a, b); 6, Swales Mountain (Evans and Ketner, 1971); 8, Schroeder Mountain (Evans and Cress, 1972); 9, Izzenhood Spring (Wallace and John, 1998); 10, Stony Point (D.A. John and others, unpub. data, 1999); 11, Bob's Flat (S.G. Peters, unpub. data, 1999); 12, Mule Canyon (D.A. John, unpub. data, 1999); 13, Antler Peak 7–1/2 minute (Doebrich, 1995); 14, Galena Canyon (Doebrich, 1994); 15, Mount Lewis and Crescent Valley (Gilluly and Gates, 1965); 16, Frenchie Creek (Muffler, 1964); 17–20, Carlin, Dixie Flats, Pine Valley, and Robinson Mountain (Smith and Ketner, 1978); 21, Ravens Nest (R.M. Tosdal, unpub. data, 1999); 22, Cortez (Gilluly and Masursky, 1965).

mining district, some recent studies include those by Seedorff and others (1991) who described the Buffalo Valley gold deposit, Theodore and others (1992) who mapped and described the Buckingham stockwork molybdenum system, and Theodore and Jones (1992) who described some of the relations among various types of jasperoid near the promontory at Elephant Head, immediately to the east of the Buckingham and Copper Basin areas. The Fortitude gold skarn deposit is described by Myers (1994), and ratios of initial melt-water concentration to saturation melt-water concentration for the Late Cretaceous stock at Trenton Canyon (fig. 2) is calculated by Ratajeski (1995). Doebrich (1995) and Doebrich and others (1995, 1996) respectively describe the geology of the Antler Peak 7-1/2 minute quadrangle and some of the latest discoveries of ore deposits in the mining district. The provenance of the Upper Cambrian Harmony Formation and the Ordovician Valmy Formation was inferred by Smith and Garrels (1994) from geochronologic studies of detrital zircons obtained from a small number of sample sites in the mining district. Peters and others (1996) include numerous references to relations of mineralized rocks in the Battle Mountain Mining District as part of their regional synthesis of metallogeny in the northwest quadrant of Nevada. In addition, many unpublished geologic, geochemical, and geophysical studies of the mining district, which have been prepared for various companies over the last few decades, unfortunately reside in company files essentially unheralded and essentially unfamiliar to the economic geologic community at large; the geophysical studies include some innovative applications of state-of-the-art remote sensing and controlled-source-audiomagnetotelluric technologies.

Prior to this study, only a small number of published geologic reports (Bloomstein and others, 1992, 1993; Braginton, 1993; Norman and others, 1996) have been devoted specifically to the immediate area of Lone Tree Hill, primarily because of the limited bedrock there and the small number of unproductive prospects known before discovery of the Lone Tree and Stonehouse deposits (pl. 1). In addition, published relatively small-scale regional geologic maps of the area (Ferguson and others, 1952; Willden, 1963) do not show presence of Middle Pennsylvanian Battle Formation in the northern part of Lone Tree Hill because of the small size of the outcrop. Nor do they show the presence of the Permian Edna Mountain Formation at Treaty Hill (pl. 1). If the outcrop area of the Battle Formation on Lone Tree Hill had been large enough to be shown on these geologic maps, then it is highly probable that far more attention by more exploration groups would have been devoted to the area, and discovery of the major deposits in all likelihood would have occurred sooner. This reasoning results from the fact that many major replacement ore bodies in the Battle Mountain Mining District are hosted by calcareous and hematitic conglomerate of

the Battle Formation (Roberts and Arnold, 1965; Theodore and Blake, 1975; Theodore and others, 1986). Malachite-stained fault breccia along several minor faults, which are not shown on plate 1, was noted in several prospects on the east and west sides of Lone Tree Hill in the late 1950s and early 1960s during a reconnaissance mineral survey (Southern Pacific Company, 1964, p. 80–81). The shallow underground workings at these prospects were driven in the search of copper by two brothers, the Messers. Planck, during the 1930s (Gene DiGrazia [deceased], oral commun. to D.H. McGibbon, 1990). Subsequently in the 1960s and 1970s, the general area of Lone Tree Hill was considered by several companies as an exploration target for a buried porphyry copper system (D.H. McGibbon, oral commun., 1992).

This report is the latest in a series of cooperative studies between the U.S. Geological Survey and private industry concerning various aspects of mineralized rocks of the Battle Mountain Mining District. Roberts (1964) and Roberts and Arnold (1965) previously had described the geology of the Antler Peak 15–minute quadrangle part of the mining district, regional setting of the mining district with respect to major allochthonous and autochthonous sequences of rock, and many small ore deposits then known throughout the area. The latter report includes an extensive accumulation of geologic data from underground workings accessible at that time. Theodore and others (1973) reported on the chemistry of many plutons in the Battle Mountain Mining District and the potassium-argon ages of their minerals. The section below by E.H. McKee adds a number of ages to the data base of radiometrically dated samples by describing the geochronologic implications of 20 newly determined potassium-argon ages of minerals and whole-rocks. Theodore and Blake (1975, 1978) described the middle Tertiary Cu–Au–Ag East and West ore bodies at Copper Canyon. Several other reports (Nash and Theodore, 1971; Theodore and Nash, 1973; Batchelder and others, 1976; Batchelder, 1973, 1977) discussed composition of fluids genetically associated with the deposits at Copper Canyon, and geochemical zonation of mineralizing fluids there. Metal zonation of the deposits is described by Blake and others (1978), Blake and Kretschmer (1983), Blake and others (1984), and Theodore and others (1986), and a description of the Buckingham system is included in both Blake and others (1979) and Theodore and others (1992). Mineral chemistry and fluid-inclusion relations of some of the barren Tertiary skarns in the northern part of the Buckingham area are characterized in detail by Theodore and Hammarstrom (1991), and Theodore and Jones (1992) describe the geology and geochemistry of jasperoids in the Elephant Head area, just to the east of Buckingham. Doebrich (1995) describes the geology and mineral deposits of the Antler Peak 7-1/2 minute quadrangle, and Doebrich and others (1995) include descrip-

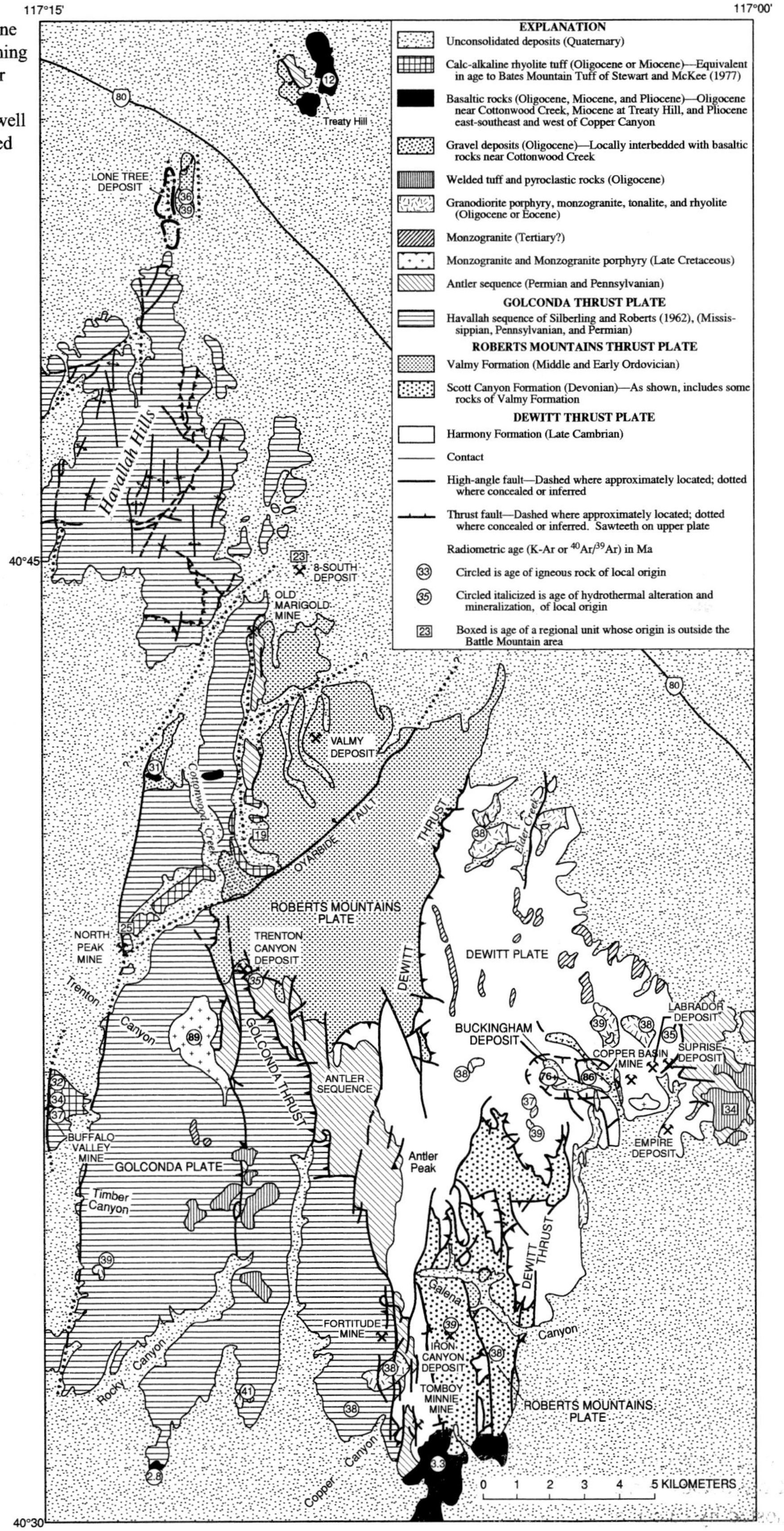

Figure 2 Geologic sketch map Lone Tree area and Battle Mountain Mining District showing locations of major mines and areas of base- and precious-metal mineralization, as well as locations of radiometrically dated samples.

tions of some recent discoveries ore deposits together with their sizes and grades. The geology of the Galena Canyon 7–1/2 minute quadrangle is described by Doebrich (1994). Kotlyar and others (1995) re-examined the implications of geochemical relations in surface rocks at Copper Canyon, and Gostyayeva and others (1996) studied the implications of fluid-inclusion relations in the Elder Creek porphyry Cu system, which is present in the northeast part of the mining district. Kotlyar and others (1998) synthesized enormous amounts of three-dimensional geochemical data surrounding the Fortitude Au skarn at Copper Canyon, and showed presence of a strongly geochemically anomalous zone well below the Fortitude. In addition, they concluded that the small laccolithic intrusive body at the center of the porphyry Cu sysytem at Copper Canyon could not have been the source of all metals introduced into its surrounding wallrocks. Most of the metals must have come from a much larger magmatic root to the laccolith.

In this report, we particularly focus geologically on the geometry of tectonically imbricated lithotectonic packages of rock, as well as their depositional environments, present in the Devonian, Mississippian, Pennsylvanian, and Permian Havallah sequence which makes up the upper plate of the Golconda thrust. Such emphasis is placed here on rocks of the Golconda allochthon because of (1) problems in discriminating properly rocks near the trace of the Golconda thrust and along some of the other tectonically higher thrusts in the allochthon, and because (2) calcareous content of many sequences of rock in the allochthon suggests that they should form excellent hosts for replacement-type ore. Furthermore, provenance studies of rocks in the Golconda allochthon shed light on questions concerning relations between the paleocontinental margin and depositional sites of rocks that constitute the allochthon. Nevertheless, our studies build upon well-founded prior studies of Roberts (1964) principally as a result of recent availability of large-scale topographic maps, and as a result of development in the 1970s of techniques for extraction of microfossils from fine-grained siliceous sedimentary rocks. Preliminary versions of the geology of the Valmy and North Peak quadrangles (Theodore, 1991a, b) have been refined substantially in this report (pls. 1, 2) because of numerous critical relations that have since been exposed and evaluated at depth by drilling, and because of the much greater density of sites from which fossil data have been obtained in the intervening years. An earlier preliminary description of the geology of the Snow Gulch quadrangle is contained in Theodore (1994).

Acknowledgments

This study of the geology of the Valmy, North Peak, and Snow Gulch quadrangles could not have been assembled as thoroughly as it has without cooperation and encouragement of the managements of Cordex Exploration Co., Marigold Mining Co., Santa Fe Pacific Gold Corp., and Santa Fe Pacific Mining Inc., the latter now (1999) a part of Newmont Gold Mining Co. A special gratitude must be accorded to Andy B. Wallace of Cordex Exploration Co., who provided geologic details of bedrock buried underneath gravels in the general area of the Eight South deposit, and to Robert P. Felder and Ronald L. Parratt, then of Santa Fe Pacific Mining, Inc., who provided partial support to the Nevada Bureau of Mines and Geology for the fieldwork undertaken by Theodore during early stages of this study. This fieldwork was completed partly under Cooperative Research and Development Agreement 9300–1–94. Theodore also benefited immensely from discussions in the field with Russell T. Ashinhurst, David W. Blake, Mike Broch, Edward I. Bloomstein, Peter E. Chapman, Jeff L. Doebrich, Robert P. Felder, Ronald D. Luethe, Joe Lamana, Bill Matlack, Raul J. Madrid, Linda B. McCollom, Douglas H. McGibbon, Benita L. Murchey, Ralph J. Roberts, Michael G. Sawlan, Trevor J. Thomas, Patrick R. Wotruba, and Chester T. Wrucke, and from able field assistance furnished by Robert L. Oscarson over a number of years. Robert W. Schafer, BHP Minerals International, Inc., made available results of drilling by Utah Mining and Construction Co. in the Trenton Canyon area, and Jim McGlasson, formerly of St. George Metals Inc., provided an internal company report that summarizes exploration activities up to 1993 in the Trenton Canyon area. Kent Ratajeski, University of Maryland, kindly provided some samples and chemical data on the pluton at Trenton Canyon. The preceding investigations of Ralph J. Roberts established the regional tectonostratigraphic framework for much of the studies reported here. Provenance of geologically critical calcareous sandstones in the Havallah sequence was determined by A.K. Armstrong, U.S. Geological Survey. Finally, I would like to express my gratitude to all current and former managers of various companies who granted access to proprietary data over several decades that allowed me to evolve scientifically with the truly fascinating unfolding by private industry of hidden mineral wealth in the general area of Battle Mountain.

Affiliations of many authors involved with this report have changed since early drafts of the report were first assembled, and in mid–1999 the affiliations were as follows: E.I. Bloomstein (Earth Search Resources, Ltd.), B.L. Braginton (Newmont Gold Co.), D.M.DeK. Channer (Guaniamo Mining Co.), R.W. Owen (Newmont Gold Co.),

R.L. Parratt (Homestake Mining Co.), K.C. Raabe (Newmont Gold Co.), W.F. Thompson (Barrick Gold Exploration Inc.), D.H. McGibbon (Glamis Gold, Ltd.), A.B. Wallace (Cordex Exploration Co.), and R.P. Felder (Pittston Nevada Gold Co.). McKee, Oscarson, and Theodore are still (2000) with the U.S. Geological Survey.

Initial digital files of the geologic maps of the quadrangles (pls. 1–3) were prepared by Nancy Wotruba, Battle Mountain, Nev., and final conversion of those digital files into color separates ready for offset printing was done by Barry C. Moring and Robert J. Miller. Many of the final figures were prepared by Boris B. Kotlyar and Judy Weathers.

REGIONAL STRUCTURAL AND STRATIGRAPHIC FEATURES OF THE BATTLE MOUNTAIN MINING DISTRICT

The locations of many ore deposits throughout north-central Nevada clearly are structurally controlled at their district-scale geologic settings, and this relation also applies to some of the major deposits in the Battle Mountain Mining District. However, the importance of apparent mineralized trends or belts at the regional scale in Nevada remained, even in the early 1990s, problematical when one considers that detailed studies, whose primary purpose is establishing fundamental geologic controls of regional-scale alignments of ore deposits, were then still in their early stages (Seedorff, 1991). By the middle 1990s, however, geology of the mineral belts—in particular the world class Carlin, Nev., trend of Au deposits (fig. 1)—was finally being assembled at scales adequate to provide some comprehension of how regional alignment of deposits and mining districts may have occurred. Some type of regional structural control appears to have been influential on distribution of ore deposits along the northwest-aligned Carlin trend (Roberts, 1966; Bagby and Berger, 1985; Madrid and Roberts, 1992), especially if one compares areal distribution of decalcified, silicified, or argillized rock, shown to be as much as 3 km wide, and location of the most productive Carlin-type deposits—also referred to as sediment-hosted Au deposits (SHGD) (Arehart, 1996)—for a distance of approximately 35 km from the Dee deposit on the northwest to the Gold Quarry deposit on the southeast (Seedorff, 1991, fig. 17b). The linear geometry of the

pinching and swelling of these altered rocks in this part of the Carlin trend strongly suggests that, whatever the underlying cause controlling distribution of this alteration, it must reflect, in part, a number of high-angle northwest-striking faults, possibly extending to crustal depths, and readily mappable at large map scales (Barrick Gold Exploration Inc., unpub. data, 1993; Theodore and others, 1998). Recently, Volk and Lauha (1993) ascribed these northwest-striking faults to Reidel structures associated with a period of transpressional shear. In addition, Peters (1998) suggested that Au–mineralized rocks along a significant part of the Carlin trend may be in large part reflecting influx of mineralizing fluids along the northeast-trending Crescent Valley-Independence Lineament (CVIL) which extends northeast for approximately 90 km from Gold Acres to the Independence Mining District (fig. 1). Theodore and others (1998) interpreted sinistral shear along major fault strands of the CVIL as having been initiated during the late Paleozoic Humboldt orogeny of Ketner (1977)—the fault strands probably reflect motion along transfer faults associated with approximately north-south compression that, in the area of the Carlin trend, continued up until the Jurassic. Theodore and others (1998) also suggest that (1) early deformation along site of the future Carlin trend of gold deposits represents a dilational jog between two of the major fault strands of the CVIL, and that (2) dextral shear may have occurred along a "proto" northern Nevada rift (fig. 1).

Another structural control on areal distribution of many deposits in Nevada involves relatively shallow-dipping thrust zones of regionally extensive tectonic plates and windows through these plates (Roberts, 1966), although a large number of recently discovered precious-metal deposits are now known not to be spatially associated with windows in the allochthon of the Roberts Mountains thrust (Seedorff, 1991; Madrid and Roberts, 1992). Nonetheless, the Roberts Mountains thrust, which crops out mainly east of Battle Mountain, has the Bootstrap, Dee, Betze, Carlin, Bullion Monarch, Blue Star, Genesis, and several other Carlin-type Au deposits near its trace (fig. 1; Bagby and Berger, 1985). Recent deep discoveries, however, along the Carlin trend at the Meikle Mine and Deep Post ore bodies differ significantly in terms of their elevated Au concentrations and overall sulfide mineral content from many of the other Au deposits elsewhere along the trend. In the Battle Mountain Mining District, numerous ore deposits in the Copper Canyon area (Roberts and Arnold, 1965; Theodore and Blake, 1975, 1978; Myers and Meinert, 1991; Doebrich, 1995; Doebrich and others, 1995, 1996) are confined largely to the Pennsylvanian and Permian Antler sequence in the lower plate of the nearby Golconda thrust, as are many gold deposits in the two quadrangles that are the focus of much of this report (pls. 1, 2). Thus, in the area of the Battle Mountain Mining District, some spatial and petrogenetic association exists between mineralized rocks, which probably formed during the Tertiary, and the much earlier emplaced regional-scale thrusts.

Antler orogeny

A large part of north-central Nevada is covered by siliciclastic and volcanic rocks in the upper plate of the Roberts Mountains thrust whose easternmost trace bounds an accretionary fold-thrust terrane that preserves but one small part of an apparently much more widespread contractional or shortening event (Antler orogeny) of middle Paleozoic age in the Cordillera (Nilsen and Stewart, 1980; Oldow and others, 1989; Miller and others, 1992; Poole and others, 1992; Smith and others, 1993). At the approximate latitude of the Battle Mountain Mining District, the present width of allochthonous rocks from the apparent leading edge of the Roberts Mountains thrust on the east to the westernmost known window in the East Range (Whitebread, 1994) is approximately 150 km. In the East Range, lower Paleozoic carbonate rocks are exposed through a window in the Willow Creek thrust, a thrust whose age of major displacement must be younger than the Triassic Rochester Rhyolite. It is possible that these lower Paleozoic carbonate rocks are themselves allochthonous. The regional trace of the leading edge of the Roberts Mountains thrust near the Carlin trend may have been deflected farther to the east during the late Paleozoic Humboldt orogeny and its largely north-south shortening (see also, Theodore and others, 1998). However, because of additional unknown amounts of post-Roberts-Mountains-thrust crustal extension that might be anywhere in the 70 to 200 percent range in this region of the Great Basin during the Tertiary (Stewart,1980), the minimum amount of transport along the sole thrust of the Roberts Mountains allochthon must be substantially less than 150 km. The present maximum tectonostratigraphic thickness of the Roberts Mountains allochthon presumably is about 5 km (Miller and others, 1992). The Roberts Mountains thrust carried oceanic volcanic rocks, chert, shale, and quartzarenite of early and middle Paleozoic age eastward over lower plate, continental shelf or platformal carbonate rocks of an equivalent age mostly during Mississippian thrusting in the Antler orogeny (Roberts and others, 1958; Roberts, 1964; Speed and Sleep, 1982; Oldow and others, 1989; Miller and others, 1992). Conodont collections from the Lower Mississippian Dale Canyon Formation in the Sulphur Spring Range, Nev., suggest that the Roberts Mountains allochthon was emplaced during late Kinderhookian or early Osagean time (Johnson and Visconti, 1992).

As defined by P.B. King in Nilsen and Stewart (1980), the Antler orogenic belt

*** is a belt of rocks orogenically deformed in mid-Paleozoic time that extends north-northeastward across central Nevada, southward into the Mojave Desert region of southeastern California, and northward into central Idaho as far as the Salmon River arch. It is bounded on the east by cratonic miogeosynclinal rocks, and on the west by upper Paleozoic rocks of the Golconda allochthon, as well as lower Mesozoic rocks. It is composed of *** rocks that were deformed and thrust eastward *** along the Roberts Mountains thrust. This orogenic event was not accompanied (so far as known) by any plutonism or severe metamorphism, but some sort of deep-seated effect is indicated by the persistence of a positive belt along the orogenic zone through the remainder of Paleozoic time, and even later. A thick sequence of flysch was deposited in a foreland basin to the east of the orogenic belt following uplift of the belt ***

This definition of the Antler orogeny was subsequently accepted by Roberts and Madrid (1991) who affirmed that temporally associated phases of compression and movement of the Roberts Mountains allochthon, together with their accompanying uplift and debris, all comprise the orogeny. Roberts and Madrid (1991) also present a concise summary of the historical evolution of ideas concerning a Paleozoic deformation in this part of the Cordillera. However, Miller and others (1992) emphasize that active-arc magmatism of Mississippian age present in rocks that comprise the Havallah basin, which at one time was offshore of the Antler orogenic welt, must be considered magmatism essentially coeval with the Antler orogeny.

Geologic evidence exists that some deformation within the Roberts Mountains allochthon may have occurred during Late Devonian time (Murphy and others, 1984), and additional evidence to support this conclusion is discussed below. The Antler orogeny seems to have been the first major tectonism to affect regional sedimentary patterns in the Cordilleran carbonate platform or miogeocline after rifting during the Proterozoic of the continental crystalline crust (Stewart, 1980). However, some rocks in the Cordilleran miogeocline also may have been affected by, and others owe their origins to, an early Paleozoic orogeny as proposed by Willden (1979). Some authors (including Speed and Sleep, 1982; Snyder and Brueckner, 1983, and others) have proposed that the allochthons of the Roberts Mountains thrust may compose a number of internally deformed tectonic packets of rock which were emplaced owing to collision of far-traveled, east-facing island arcs with North America. Others (Burchfiel and Davis, 1972, 1975; Miller and others, 1984, and many others) suggest that it is more likely that the arcs associated with the Antler orogeny are west facing. Reviews of these alternative hypotheses are also presented by Nilsen and Stewart (1980), Silberling (1986), and Madrid (1987). However, some recently acquired U–Pb isotopic data from zircons in the Upper Cambrian Harmony Formation, Ordovician Valmy Formation, Ordovician Palmetto Formation, and Devonian Slaven Chert, all in the allochthon of the Roberts Mountains thrust, apparently are consistent, on the one hand, with an interpretation that significant collision of a volcanic arc with its continental margin did not occur during the Antler orogeny (Wallin, 1990). On the other hand, as Wallin (1990) further points out as have many others, ample evidence exists in California that volcanic arcs were present along the continental margin during the early and late Paleozoic. Therefore, these isotopic data may reflect a sampling problem owing to possible presence of a paleogeographic barrier that prevented arc-derived detritus from reaching the sites sampled in the formations listed above for the isotopic studies. Further in-depth study of the geochronology of detrital zircons in the Harmony and Valmy Formations by Smith and Garrels (1992, 1994) suggest that the rocks of the Harmony Formation may have been derived from the Kootenay and Yukon-Tanana terranes and other minor terranes in Alaska and in northern Canada, and that the Valmy Formation may owe its ultimate origin to the Peace River Arch area of northern Alberta.

The most logical interpretation remains that allochthons of the Roberts Mountains thrust represent back-arc subduction-accretion wedges associated with west-facing magmatic arcs and their accompanying east-plunging subduction slabs. Rocks that now compose the allochthon of the Roberts Mountains thrust probably represent sedimentary deposits laid down in a series of deep off-shore basins that formed intermittently along the continental margin during lower Paleozoic time (Miller and others, 1992; Poole and others, 1992). Additional uranium-lead analyses of abraded zircons in lower Paleozoic formations in the Roberts Mountains allochthon further suggest both off-shore and cratonal sources and that the allochthon is not a far-travelled package of rocks (Gehrels and others, 1993).

The initial tectonic welt that was slated to become the Antler orogenic highlands apparently formed on the margin of the continent during the *Siphonodella sandbergi* conodont biofacies zone which is purported to have been extant in eastern Nevada and western Utah between approximately 3.0 and 4.5 m.y. following the Devonian and Carboniferous boundary (Sandberg and others, 1982). According to these same authors, the main highland phase

of the Antler orogeny then occurred during an immediately succeeding approximately 20–m.y. time interval at which time a major transgressive-regressive cycle occurred in the Mississippian Antler foreland basin primarily in a broad region well to the east of the mining district. Harbaugh and Dickinson (1981) and Trexler and Nitchman (1990) essentially propose a two-stage chronology for events associated with the Antler orogeny as reflected in the Mississippian foreland basin: (1) collision of western North America with allochthons of the Antler orogeny concomitant with downwarping of the continental margin resulting in the sedimentation of retrogradational submarine fans showing increasing water depths with time, and (2) subsequent uplift and folding of the basin, including some deposition of siliciclastic sedimentary rocks showing fining upward channel-floodplain cycles across a shallow marine shelf. These latter rocks commonly include mixtures of carbonate and siliciclastic detritus.

A continental-scale extensional event along the Cordilleran margin during a 14–m.y. time interval in the Late Devonian may have immediately preceded displacements along the Roberts Mountains thrust (Turner, 1985; Turner and others, 1989), a conclusion supported also by field and geochemical data derived from altered basalt of the Devonian Slaven Chert (Madrid, 1987). As Madrid noted, application of Hf, Th, and Ta discrimination diagrams of Wood and others (1979) to Middle Cambrian to Upper Devonian alkalic basalts of the Roberts Mountains allochthon results in a cluster of these data in the within-plate-basalt field and are most compatible with emplacement of the basalt in a rifted continental basement setting.

Although Ketner and Smith (1982), Ketner and others (1993a, 1993b), and Ketner (1994; written commun., 1996) have questioned the middle Paleozoic timing of crustal shortening along the Roberts Mountains thrust, Johnson (1983) presented a succinct summary of geologic relations supporting such a middle Paleozoic age. In addition, Jansma and Speed (1993) suggest that prelithification outcrop-scale structures in the Mississippian Chainman Shale, which is part of the Antler foreland basin, are penecontemporaneous with the Antler orogeny. It is difficult to envision how a post-Pennsylvanian regionally extensive tectonic event could affect the rocks of the upper plate of the Roberts Mountains thrust as intensely and as penetratively as the rocks show throughout much of north-central Nevada without deforming the Pennsylvanian and younger rocks of the overlap assemblage. Nonetheless, one can maintain that rocks of the overlap assemblage were carried "piggy back," and some even postulate that the Roberts Mountains allochthon had been emplaced initially during the Mesozoic (Ketner and Smith, 1982; Ketner and others, 1993a, 1993b). However, large-scale geologic mapping by Madrid (1987), described above, confirmed that the Dewitt thrust (Roberts, 1964), where it crops out in the southern part of the North Peak quadrangle (pl. 2), is a Paleozoic thrust fault that juxtaposed rocks of the Upper Cambrian Harmony Formation and the Ordovician Valmy Formation. These formations, separated by the Dewitt thrust, are projected as having been overlapped unconformably by essentially unfolded rocks of the Pennsylvanian and Permian Antler sequence. However, as first reported by Roberts (1949), the basal formation of the Antler sequence, the Middle Pennsylvanian Battle Formation, where it crops out less than 1 km south of the south boundary of the North Peak quadrangle in NE 1/4 sec. 29, T. 32 N., R. 43 E., appears to lap unconformably across a minor high-angle fault between the Harmony and Valmy Formations (see also, Doebrich, 1992, 1995). Thus, major tectonic displacements involving crustal shortening in this area must predate deposition of the Battle Formation and must have been concentrated between the Harmony and Valmy Formations along the Dewitt thrust because removal of apparent offsets on the minor high-angle faults there still results in a requisite projection of the basal contact of the Battle Formation across the projected plane of the Dewitt thrust (pl. 2). Near the head of Cow Canyon, in the Galena Canyon quadrangle, relatively undisturbed strata of the Battle Formation rest unconformably on folded rocks of the Scott Canyon Formation (Doebrich, 1994). Thus, a pre-Middle Pennsylvanian age of thrusting and shortening is indicated (fig. 3). The designation by Roberts (1964) of the Dewitt thrust as a major splay belonging to the Roberts Mountains allochthon apparently is correct. Madrid (1987) suggests that rocks of the Harmony Formation at the leading edge of the Dewitt thrust where they crop out in the Shoshone Range, approximately 40 km southeast of the Valmy and North Peak quadrangles, may have been displaced roughly 100 km from the original site of deposition of the formation (see also, Miller and others, 1992).

Evidence is present in the Galena Canyon area of the mining district that the site of deposition of the Harmony Formation may have been much closer (J.L. Doebrich, written commun., 1995). This evidence consists of large olistolithic blocks of the Harmony Formation included in some parts of the Scott Canyon Formation. In addition, as will be discussed below, the Battle Formation, where it crops out near the south edge of the North Peak quadrangle (pl. 2), apparently laps across another intraformational low-angle thrust fault that juxtaposes rocks belonging to unit 4, the highest tectonic unit in the vertical stacking pattern of the Valmy Formation. Displacements along this fault also must be pre-Pennsylvanian in age. Finally, Jones (1993) finds that a single west-vergent phase of pre-Pennsylvanian deformation has affected rocks of the Harmony Formation in the Hot Springs Range, approximately 70 km northwest of the study area.

Additional evidence for tectonic emplacement of rocks of the Harmony Formation during the Devonian is present in some outcrops near the mouth of Galena Canyon in the east-central part of the Battle Mountain Mining District. In this area, volcaniclastic rocks and, in places, basalts with poorly developed pillows of the presumably mostly Devonian Scott Canyon Formation contain numerous clasts and blocks of poorly sorted feldspathic arenite of the Harmony Formation (fig. 4A, B). Some rocks previously mapped by Roberts (1964) in the general area of Galena Canyon as part of the Scott Canyon Formation apparently should be included now with the Ordovician Valmy Formation. Ordovician-age conodonts have been recovered from a small outcrop of black thin-bedded chert near the confluence of Iron and Galena Canyons and also from an olistolithic block of limestone in greenstone near Cow Canyon, a minor tributary of Galena Canyon (McCollum and others, 1987; M.B. McCollum, oral commun., 1992). At least four localities are known from which Devonian-age radiolarians also have been recovered from the Scott Canyon Formation, one as described in Theodore and others (1992) and the others in the general area of Devonian-age radiolarians initially reported by Stewart and Suczek (1977). This latter site is in Scott Canyon, also a minor north-trending tributary of Galena Canyon, at a 5,750–ft elevation, almost on the section line between secs. 11 and 14, T. 31 N., R. 43 E. (C.A. Suczek, written commun., 1992). Devonian radiolarians from here were identified by Brian Holdsworth. Devonian radiolarians were later recollected from the same site and their regional implications then described by Jones and others (1978) and three additional Devonian-age radiolarian collections from the same thin sequence of chert were made by Susan Boundy-Sanders (written commun., 1993). In addition, some olistoliths of archaeocyathan-bearing Lower Cambrian limestone derived from an unknown source also are present in rocks of the Scott Canyon Formation near the locality shown in figure 4. Some olistolithic blocks of feldspathic arenite derived from the Harmony Formation are as much as 5 to 8 m wide and the largest of the limestone olistoliths in the same general area is as much as 100 m in length. Thus, rocks of the Harmony Formation, which at Galena Canyon and elsewhere in the mining district tectonically overlie rocks of the Scott Canyon Formation (Roberts, 1964), must be older than volcanic rocks in a debris-flow unit of the Scott Canyon Formation that host them (Doebrich, 1994). Moreover, the presence of an in-place, generally topographically high Harmony Formation close to the arenite-clast locality suggests that the leading edge of the allochthonous rocks of the Harmony Formation may also have been nearby that locality in Devonian time as a tectonically uplifted source for the clasts and blocks now present in basaltic rocks in a debris-flow unit of the Scott Canyon Formation. Some Ordovician-age limestone olistoliths also are now known to be present in basaltic rocks of the Scott Canyon Formation (L.B.

WEST   EAST

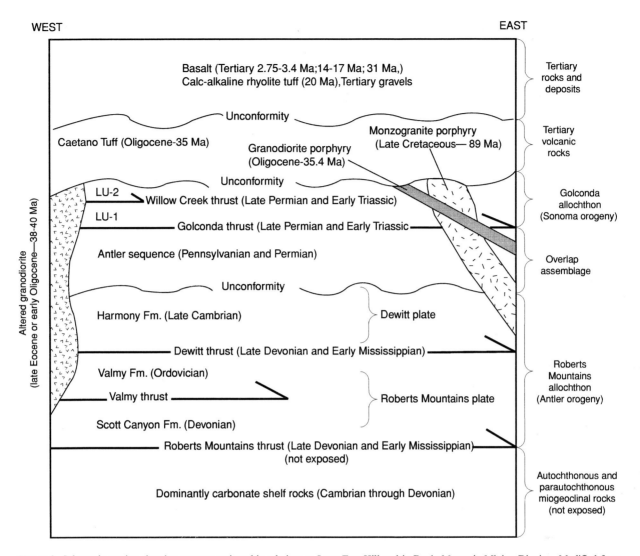

Figure 3 Schematic section showing tectonostratigraphic relations at Lone Tree Hill and in Battle Mountain Mining District. Modified from Roberts (1964) and Theodore and Jones (1992). Arrow, approximate direction of thrust event.

McCollum, oral commun., 1993) as described above. If the above sequence of geologic events is generally correct, then shortening during the Devonian in the foreland thrust-and-fold belt as inferred from intra-allochthon slip between rocks of the Harmony and Scott Canyon Formations may have been much more pronounced than previously thought. Nonetheless, continental-scale final emplacement of the Roberts Mountains allochthon seems to have taken place during Mississippian time.

The present complicated array of crustal and supracrustal rocks in north-central Nevada is a culmination of geologic events protracted over an extremely long timespan. Before the Antler orogeny, distribution of basinal-, slope-, and shelf-assemblage rocks in the Cordilleran belt was arrayed in a somewhat uniform fashion with basinal rocks on the west and carbonate shelfal rocks on the east (Stewart and Poole, 1974; Poole and others,1992). The North American continent in Late Devonian time was

probably amalgamated with Greenland, Eurasia west of the Urals, and the Iberian peninsula as one continental mass, referred to as Laurasia, and that the North American part of this continental mass was largely south of the equator (Smith and others, 1981). Speed and Sleep (1982) maintained that a large accretionary prism was underthrust westwards by the continental slope and outer shelf of the North American plate; the accretionary prism then became the Roberts Mountains allochthon. Alternately, Whiteford and others (1983) and Miller and others (1984;1992) concluded from geologic relations in uppermost Devonian to Lower Permian rocks exposed in the Independence Mountains and elsewhere in northern Nevada, that emplacement of the Roberts Mountains thrust resulted from thrusting in a backarc basin offshore of western North America associated with eastwards underthrusting. As further noted by Oldow and others (1989), oblique convergence of accretionary prisms associated with the Antler orogeny may

Figure 4 Photographs showing clasts and blocks of feldspathic arenite of the Upper Cambrian Harmony Formation (*A*), and Middle and Upper Cambrian, archaeocyathid-bearing limestone (*B*) hosted by volcaniclastic rocks of the Devonian Scott Canyon Formation. *A*, Knifepoint at contact between feldspathic arenite (fa) and volcaniclastic rock; *B*, small clot of limestone, at head of arrow, totally engulfed by volcaniclastic rock. Locality in SE 1/4 sec. 13, T. 31 N., R. 44 E., near mouth of Galena Canyon in Battle Mountain Mining District.

result in transmission of some strike-slip faults into the foreland thrust and fold belt of the orogen (see also, Avé Lallemant and Oldow, 1988). Two major east-striking dextral faults, apparently late Mesozoic in age, show approximately 45– to 80–km offsets of Proterozoic, Paleozoic, and Mesozoic rocks in the western Great Basin (Stewart, 1985), and these dextral faults may be following zones of weakness inherited from a previous episode of oblique convergence during the Antler orogeny. Burchfiel and Royden (1991) suggested that the Antler orogenic belt, with its lack of a clearly-defined, coeval magmatic arc and its association with a broad zone of subsidence and extension to the west of the allochthonous thrust plates, may be somewhat analogous with the geologically young thrust belts in the Mediterranean.

Composition of the crust inferred from isotopic studies

The composition of crust that underlay this part of Nevada covered largely by the allochthon of the Roberts Mountains thrust has been inferred from isotopic study of plutons that have been emplaced into the sedimentary accretionary rocks. Previous $^{87}Sr/^{86}Sr$ isotopic studies of granitic plutons by Kistler (1983, 1991) and Kistler and Fleck (1994) indicated that the crust underlying the area of the quadrangles may be characterized by initial isotopic signatures that are greater than 0.706 (fig. 5). However, Sr whole-rock isotopic analyses of 2,137 samples of Mesozoic granitoids in the region suggests that the 0.7050 isopleth for $^{87}Sr/^{86}Sr$ conceptually fixes the geographic position of the Late Proterozoic rifted continental margin (Robinson, 1993, 1994). From the distribution of these values and values of Nd and Sm isotopic ratios, Robinson (1993, 1994) then infers that the western sialic margin of North America extends as far as barely west of Winnemucca; the area of the Battle Mountain Mining District is thus located in a region wherein plutons apparently could be derived from several different sources. The most likely possible source of magmas in the area of the quadrangles is a Late Proterozoic rift-generated crust that apparently was underplated during rifting beneath uplifted and attenuated continental crust (Robinson, 1993, 1994). Furthermore, the western Idaho suture of Leeman and others (1992) that apparently juxtaposes oceanic and continental terranes is approximately coincident with their recognition of a 0.706 isopleth for $^{87}Sr/^{86}Sr$ isotopic ratios defined by data from Miocene and Pliocene volcanic rocks, and, as shown by them, this northerly-trending isopleth would project just to the east of the mining district. Somewhat earlier, Farmer and DePaolo (1983) suggested from neodymium and strontium isotopic data that the line marking those plutons showing initial $^{87}Sr/^{86}Sr$ ratios greater than 0.708 best indicates the western edge of continental crystalline crust. They then placed the 0.708 contour about approximately 85 km east of the area of the Valmy, North Peak, and Snow Gulch quadrangles (Farmer and DePaolo, 1984). Farmer and DePaolo (1983, 1984) also reported a systematic variation in the initial Nd isotopic compositions of granite, used in an unrestricted sense, in the western Great Basin that they interpreted to reflect derivation of granite, including that in the Battle Mountain Mining District, primarily from crustal components. Nonetheless, Wright and Wooden (1991) have further determined that Sr, Nd, and Pb isotopic data for Late Cretaceous and Cenozoic plutons that are present generally to the east of the Battle Mountain Mining District, apparently define a broad east-west crustal transition zone between Archean rocks on the north and Proterozoic rocks on the south. This east-

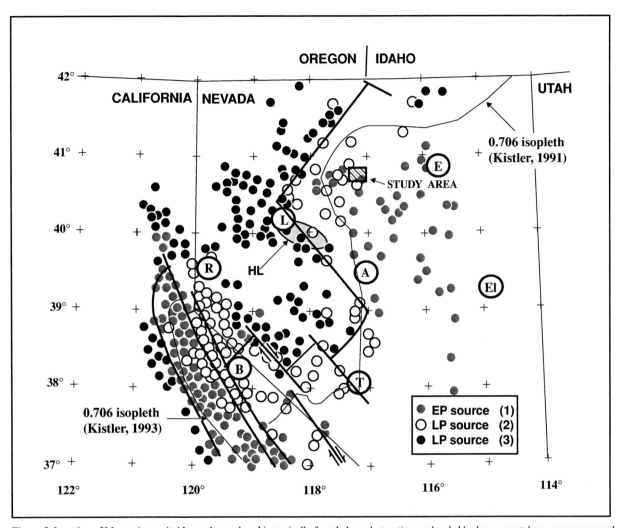

Figure 5 Location of Mesozoic granitoid samples analyzed isotopically for whole-rock strontium and coded by lower crustal magma sources and the 0.706 $^{87}Sr/^{86}Sr$ isopleths determined by Kistler (1991) and Kistler and Fleck (1994). Modified from A.C. Robinson (written communs., 1993, 1994). EP source (1): Early Proterozoic lower continental crust of basaltic composition. LP source (2): Late Proterozoic, slightly alkalic basaltic rift-generated lower crust. LP source (3): Late Proterozoic rift-generated lower crust, probably emplaced as oceanic crust as continental segments spread apart. W, Winnemucca; E, Elko; El, Ely; A, Austin; L, Lovelock; R, Reno; T, Tonopah; B, Bodie; HL, Humboldt lopolith.

west transition zone trends roughly towards the area of this report. In addition, Solomon and Taylor (1989) found that granitic rocks in a broad region in the northern Great Basin are characterized by relatively high $^{18}O/^{16}O$ ratios that can be subdivided further into two subzones on the basis of their $^{87}Sr/^{86}Sr$ initial ratios: <0.708 on the west, and >0.710 on the east. Solomon and Taylor (1989) suggest that these two subzones are bounded by a north-south suture and that the two subzones respectively reflect a Phanerozoic volcanogenic or eugeoclinal terrane and a Proterozoic miogeoclinal terrane. The suture inferred to bound these two terranes passes to the east of the Battle Mountain Mining District (Solomon and Taylor, 1989). Additional U–Pb isotopic studies are currently (1999) underway to establish fundamental basement variations across north-central Nevada (Wooden and others, 1998).

Sonoma orogeny

The region subsequent to the Antler orogeny has been overridden by several structurally higher late Paleozoic and early Mesozoic thrusts (fig. 3) associated with the Sonoma orogeny (Silberling and Roberts, 1962; Gabrielse and others, 1983). One of these thrusts, the post-late Early Permian to Early Triassic Golconda thrust, crops out prominently in the Battle Mountain Mining District (Roberts, 1964) and appears to be buried under Tertiary and Quaternary gravels along the eastern and northeastern flanks of the Havallah Hills (pl. 1). The Golconda thrust is exposed spectacularly in the south wall of the open pit at the Eight South deposit. The present trace of the sole of the Golconda thrust is then inferred to continue to the northwest from this general area and to pass between the northernmost exposures in the Havallah Hills and Lone Tree Hill (pl. 1). The easternmost trace of the Golconda allochthon is approximately 50 km to the east of the Battle Mountain Mining District (fig. 6), and includes the latest Devonian to Early Permian Schoonover sequence in the Independence Mountains in northeastern Nevada (Miller and others, 1984). The Havallah basin, prior to its dismemberment and closure during the Sonoma orogeny, is inferred to have been present in what is today the general area of the Great Valley and Sierra Nevada of California (Miller and others, 1992).

The Golconda allochthon recently has been subdivided into four informally designated lithotectonic assemblages of rock (Tomlinson, 1990). The four assemblages designated by Tomlinson (1990) are: (1) the Corral Canyon assemblage, (2) the Reese River assemblage, (3) the Mount Tobin assemblage, and (4) the New Pass assemblage. The term "assemblage", as used by Tomlinson (1990), denotes "nonstratiform tectonic rock assemblages [that] do not have a stratiform geometry, and deformation obscures stratigraphic relationships, [and they moreover] are bodies of rock that are distinguishable from adjoining rock bodies on the basis of....lithologic content." As shown on figure 6, the bulk of the rocks in the Golconda allochthon in the Valmy and North Peak quadrangles belong to the Mount Tobin assemblage according to the classification scheme of Tomlinson (1990) or to lithotectonic unit 2 (LU-2) of the Havallah sequence of Murchey (1990). The base of LU-2 is tectonic (fig. 7), and in the southern part of the Battle Mountain Mining District, the base was determined to be at the Willow Creek thrust, which over-rides lithotectonic unit 1 (LU-1) of Murchey (1990). The Willow Creek thrust is well exposed in the northern and central parts of the North Peak quadrangle where, in many places, it juxtaposes Mississippian, Pennsylvanian, and Permian clastics and basinal radiolarian chert deposits of LU-2 onto Pennsylvanian and Permian mostly argillite and spicule chert sequences of LU-1 (pl. 2; see below). Rocks of LU-1 are inferred to represent slope deposits (Murchey, 1990). In this report we follow the designations of rocks of the Havallah sequence as belonging to LU-1 and LU-2 (Murchey, 1990), which units correspond respectively with the Reese River and Mount Tobin assemblages of Tomlinson (1990) (fig. 8). The trace of the Willow Creek thrust in the Antler Peak 7-1/2 quadrangle, immediately to the south of the North Peak quadrangle, is shown in detail by Doebrich (1992, 1995), and the trace of the thrust through the Valmy, North Peak, and Antler Peak quadrangles is disrupted significantly by a number of post-Golconda structures (fig. 9). The Willow Creek thrust cuts out increasing thicknesses of slope deposits belonging to LU-1 progressively from south to north until it reaches an area just east of the Late Cretaceous pluton at Trenton Canyon. In this area, the Willow Creek thrust appears to

have tectonically been juxtaposed across the trace of the Golconda thrust. Thus, the Willow Creek thrust appears to be regionally extensive in this part of the Golconda allochthon, and it may also be present elsewhere in the Golconda allochthon generally to the west.

The Golconda thrust at the sole of the Havallah sequence marks a late Paleozoic (Pennsylvanian and Permian) boundary that juxtaposes rocks of two different terranes as noted by Speed (1977): an ocean basin terrane on the west and a continental borderland terrane on the east. Uranium-lead isotopic data on zircon extracted from a monolithologic, granodiorite conglomerate in the Mississippian-Permian Pablo Formation of Ferguson and Cathcart (1954) that crops in the southern Toiyabe Range, approximately 100 km southeast of Battle Mountain, and that has been assigned by Speed (1977) and Tomlinson (1990) to the allochthon of the Golconda allochthon, yield a crystallization age of 264 Ma, Early Permian (Tomlinson, 1990). As Tomlinson further points out, this age is compatible with the inferred Late Permian age for the volcaniclastic sandstone that hosts the granodiorite conglomerate. Andesitic and rhyolitic flows and breccias of the McConnell Canyon area near Yerington, Nev., are part of the volcanic arc terrane of the Havallah basin (Speed, 1977); these flows and breccias are intruded by somewhat younger quartz porphyry and metadiorite, dated at 232 and 233 Ma, respectively, that are possibly cogenetic with arc accumulation (Dilles and Wright, 1988). However, this latter magmatism, late Middle Triassic in age, apparently postdates the Sonoma orogeny and the Golconda thrust (Dilles and Wright, 1988).

Stewart and others (1977) and Stewart and others (1986) questioned validity of regional correlations involving the Pumpernickel and Havallah Formations because over relatively short distances abundant changes in lithology are present and widespread tectonic imbrication of units is known within the formations. They suggested that the use of these formational designations be abandoned and that the term Havallah sequence be substituted in place of the Pumpernickel and Havallah Formations. In this report and in Doebrich (1995), we refer to these rocks as the Havallah sequence, and we show below that LU-2 or the Mount Tobin assemblage of the Havallah sequence in the two quadrangles includes a number of mappable thrust plates. The structural sole of the lowermost plate of LU-2 is the regionally extensive Willow Creek thrust, whereas the much more narrowly exposed LU-1 or Reese River assemblage comprises a single structural plate close to the mapped trace of the Golconda thrust in the study area (pls. 1, 2; Doebrich, 1995). LU-1 is much more widely exposed in the southeast part of the Battle Mountain Mining District where it shows an area of outcrop as much as 3.5 km wide (fig. 9). The Corral Canyon assemblage of Tomlinson

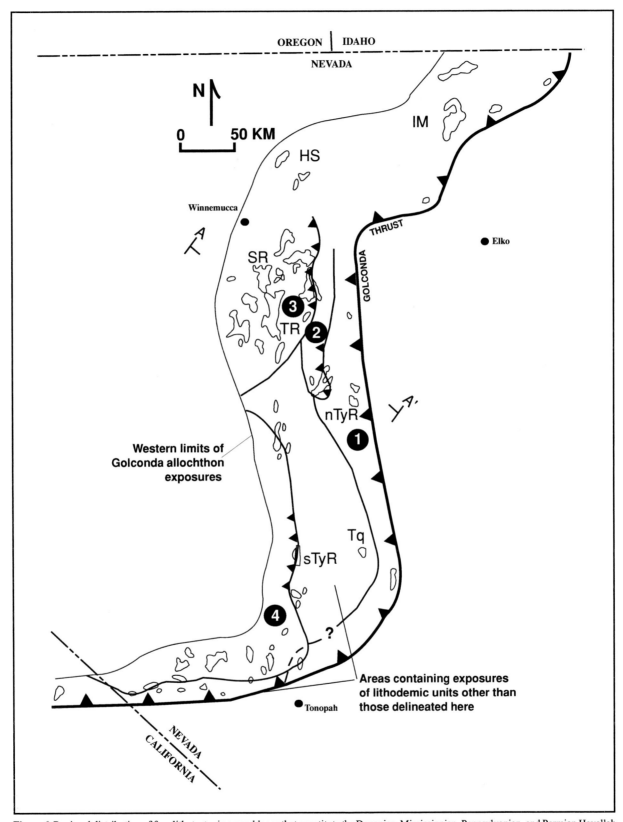

Figure 6 Regional distribution of four lithotectonic assemblages that constitute the Devonian, Mississippian, Pennsylvanian, and Permian Havallah sequence in Golconda allochthon, and the inferred easternmost limit of Golconda thrust. HS, Hot Springs Range; IM, Independence Mountains; SR, Sonoma Range; TR, Tobin Range; BM, Battle Mountain Range; Sh, Shoshone Range; nTyR, northern Toiyabe Range; sTyR, southern Toiyabe Range; Tq, Toquima Range. 1, Corral assemblage of Tomlinson (1990); 2, Reese River assemblage of Tomlinson (1990), and 3, Mount Tobin assemblage of Tomlinson (1990). AA', schematic line of section of figure 7.

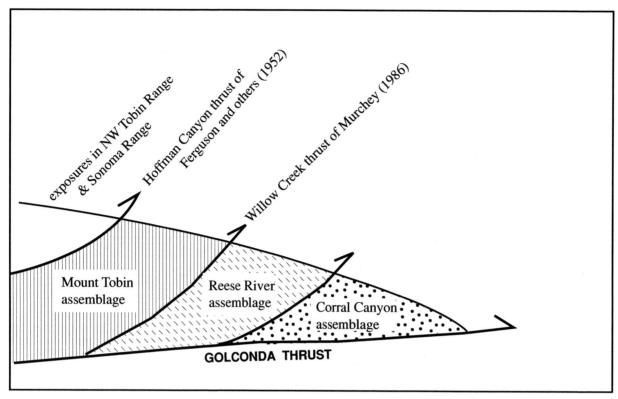

Figure 7 Schematic geologic cross section showing regional thrust fault relations among lithotectonic assemblages in Golconda allochthon. See figure 6 for line of section. Modified from Tomlinson (1990).

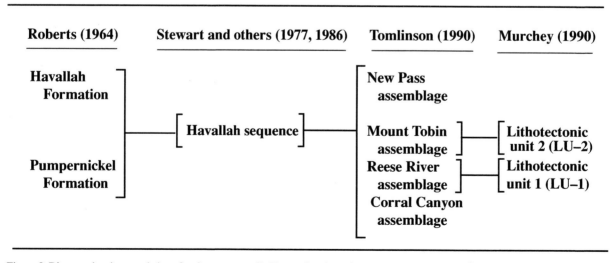

Figure 8 Diagram showing correlation of various terms applied by previous investigators to rocks of Golconda allochthon.

(1990) of the Havallah sequence (fig. 7) does not crop out in the Valmy and North Peak quadrangles.

Some prior local studies involving rocks of the Havallah sequence also demonstrated that the Golconda allochthon in the Battle Mountain Mining District consists of four to six tectonically interleaved, major rock assemblages that then were grouped (Miller and others, 1982; Brueckner and Snyder, 1985; Stewart and others, 1986) upward from the sole of the Golconda thrust in age as (1) late Early to early Late Permian; (2) Pennsylvanian and Early Permian; (3) Mississippian; and (4) undated rocks of the Trenton Canyon Member of the Havallah Formation of Roberts, (1964). Subsequent studies by Snyder and Brueckner (1983) and Brueckner and Snyder (1985) divided the Golconda allochthon in the Battle Mountain Mining District into several lithotectonic packets. The bulk of the internal deformation of these packets was ascribed by them as having occurred prior to final thrusting along the master sole thrust at the base, the Golconda thrust. Recognition of altered and mineralized basalts containing ridge-type hydrothermal systems suggested to them that ocean-basin spreading centers must have been active during much of late Paleozoic time at some sites of deposition in the Havallah basin-genetic implications of the chemistry of volcanic rocks in the Havallah sequence are discussed below. In-depth studies of the rocks of the Havallah sequence by Murchey (1990; unpub. data, 1999) allowed us to refine further the lithotectonic makeup of the allochthon.

The age of major eastward displacement(s) along the Golconda thrust must post-date age of the youngest deformed rocks (Wordian) known in the upper plate of the thrust. However, age of major movements along the Golconda thrust is still equivocal (Gabrielse and others, 1983)—if the thrust were emplaced as part of the Sonoma orogeny, then it must be of pre-late Early Triassic age; if undeformed rocks of the Koipato and Star Peak Groups that rest unconformably on deformed rocks of the allochthon were transported "piggyback" (Stewart and others, 1986), then it must be of Jurassic or even Cretaceous age (Ketner, 1984). Approximately 75 km to the east of the Battle Mountain Mining district in the area of Mount Ichabod and Major Gulch, Nev., Ketner (1991) and Ketner and others (1993b) have reported rocks of the Golconda and Roberts Mountains allochthons to overlie Triassic deep-water turbidites on low-angle faults. These may be faults younger than the Golconda and the Roberts Mountains thrust faults, and may be associated with north-south shortening associated with the Jurassic Elko orogeny (Thorman and others, 1990). The Havallah sequence is considered by Miller and others (1992) and Burchfiel and others (1992) to represent a marginal back-arc basin wherein continental slope deposits together with some basinal deposits are emplaced eastwards onto continental shelf deposits upon closing of the basin during latest Permian to Early Triassic time. Devonian- to Permian-age volcanic arc complexes are present in the Klamath Mountains and northern Sierra Nevada, Calif. Burchfiel and others (1992) further suggest that Triassic metavolcanic rocks of the Koipato Group were emplaced onto rocks of the Havallah sequence during final stages of the Sonoma orogeny while rocks of the Havallah sequence beneath the Koipato Group were already deformed during some of the earliest stages of emplacement of the Golconda allochthon. This could explain the metamorphic fabrics that are present in rocks of the Koipato Group in the East Range, Nev., whereas another more likely interpretation (Peters and others, 1996) is that the metamorphic fabrics in the East Range and in the Humboldt Range are Late Cretaceous and are associated with the continental-scale docking of the Jungo terrane of Silberling and others (1984, 1987).

Summary of Paleozoic regional tectonostratigraphy

Three major thrust plates of regional significance crop out in the Battle Mountain Mining District (fig. 3): the Golconda plate, the Roberts Mountains plate, and the Dewitt plate, which is part of the Roberts Mountains plate. The Roberts Mountains plate and the Dewitt plate are below an equally important autochthonous block (Roberts, 1964). The lowest plate exposed (the Roberts Mountains plate) is made up of chert, shale, argillite, and greenstone of the Devonian Scott Canyon Formation; it also includes quartzarenite and chert of the Ordovician Valmy Formation, which is present in a few isolated exposures at Treaty Hill and at Lone Tree Hill, but widely crops out farther to the south in the North Peak quadrangle (pls. 1, 2), and along the west border of the Snow Gulch quadrangle (pl. 3). Both of these units, the Scott Canyon and the Valmy Formations, make up the upper plate of the Roberts Mountains thrust (the Roberts Mountains plate). The Scott Canyon Formation crops out in the study area only in a small fault-bounded block near the southwest corner of the Snow Gulch quadrangle (pl. 3). These two formations are shown to be in fault contact at Galena Canyon, approximately 6 km south of the Snow Gulch quadrangle, along steeply dipping normal faults and along a shallow-dipping thrust fault, but additional tectonic blocks of Valmy Formation are probably present near Galena Canyon (fig. 2; see above). The Scott Canyon Formation in this area also includes at least two small, olistolithic bodies of limestone that have yielded archaeocyathan assemblages that are Cambrian in age (Roberts, 1964; Debrenne and others, 1990). The Scott Canyon and Valmy Formations were both, in turn, overthrust by sandstone and feldspathic sand-

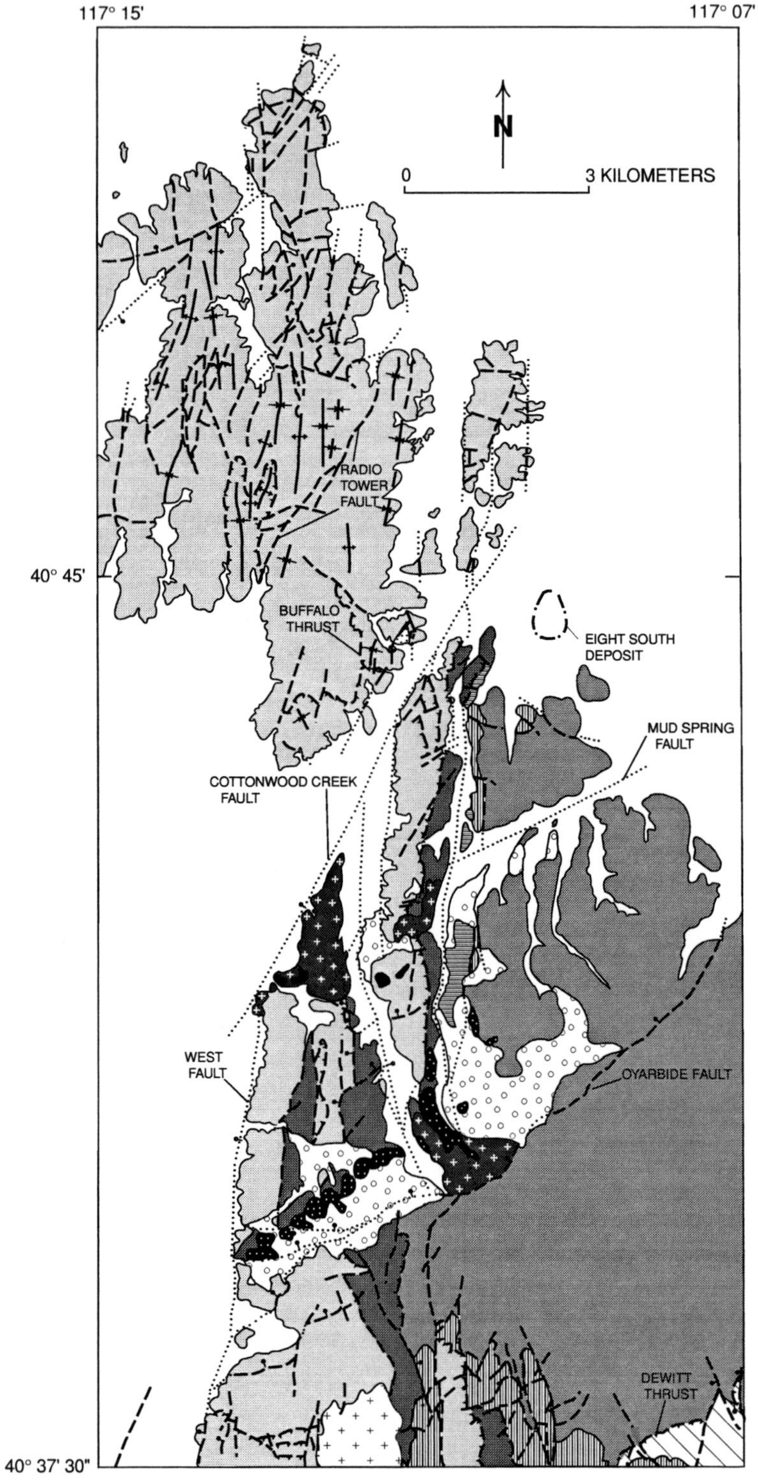

EXPLANATION

☐ Alluvium and fanglomerate deposits (Quaternary)

▨ Unconsolidated older alluvium and slope wash deposits (Quaternary)

■ Basaltic rocks (Pliocene and Oligocene)—Oliocene near Cottonwood Creek and Pliocene west-southwest of Willow Creek

▨ Calc-alkaline rhyolite tuff (Miocene or Oligocene)—Equivalent in age to Bates Mountain Tuff of Stewart and McKee (1977)

▨ Gravel deposits (Miocene or Oligocene and Oligocene)—Locally interbedded with basaltic rocks near Cottonwood Creek

▨ Pyroclastic rocks (Oligocene)— Equivalent in age to the Caetano Tuff of Gilluly and Masursky (1965)

▨ Granodiorite, altered granodiorite, monzogranite (Oligocene or Eocene)

▨ Grandiorite, grandiorite porphyry, monzogranite, and granite (Late Cretaceous)

Antler sequence of Roberts (1964) (Pennsylvanian and Permian): in this area consists of:

▨ Edna Mountain Formation (Permian) Mostly brown platy sandstone where shown separately south-southwest of the Eight South deposit

▨ Antler sequence undivided (Permian and Pennsylvanian)—Includes mostly chert-pebble conglomerate of Pennsylvanian Battle Formation and Permian and Pennsylvanian Antler Peak Limestone south-southwest of the Eight South deposit, and the Battle Formation, Antler Peak Limestone, and mostly well-sorted pebbly conglomerate and some limestone of the Edna Mountain Formation elsewhere

(Geology and explanation continued on the opposite page)

Figure 9 Sketch map showing diagrammatically traces of Golconda and Willow Creek thrusts, and distribution of lithotectonic units 1 (LU–1) and 2 (LU–2) of Devonian, Mississippian, Pennsylvanian, and Permian Havallah sequence in Valmy, North Peak, and Antler Peak 7–1/2 minute quadrangles. Geology modified from plates 1 and 2, and Doebrich (1992, 1995).

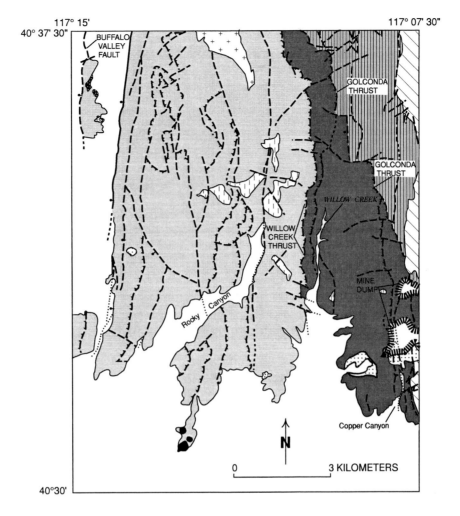

EXPLANATION *(continued)*

GOLCONDA THRUST PLATE
Havallah sequence of Silberling and Roberts (1962)
  (Permian, Pennsylvanian, Mississippian, and Devonian):
  In this area consists of:

Lithotectonic unit 1 of Murchey (1990)—Mostly
  argillite and sponge-spicule chert slope deposits

Lithotectonic unit 2 of Murchey (1990)—Mostly
  radiolarian chert and calcareous turbidite basinal
  deposits

ROBERTS MOUNTAINS THRUST PLATE

Valmy Formation (Middle and Late Ordovician)

Harmony Formation (Late Cambrian)

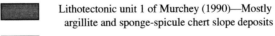

———— Contact

High-angle fault--Dashed where approximately located;
  short dashed where inferred, dotted where concealed.
  Bar and ball on downdropped block

Thrust fault--Dashed where approximately located; dotted
  where concealed or inferred. Sawteeth on upper plate

Figure 9 (second half)

stone of the Upper Cambrian Harmony Formation along the Dewitt thrust as described above (fig. 3; pls. 2, 3). The Harmony Formation, which makes up the middle (the Dewitt plate) of the three thrust plates, crops out near the southeast corner of the North Peak quadrangle (pl. 2), and is widely exposed in the Snow Gulch quadrangle (pl. 3). These two plates, the Roberts Mountains plate and the Dewitt plate, are overlain unconformably by the Antler sequence (fig. 9).

The upper Paleozoic Antler sequence makes up an autochthonous structural block that rests along a major unconformity on the Harmony Formation and the Valmy Formation (pls. 1–3). Rocks of the Antler sequence belong to the overlap assemblage that is associated with the Antler orogeny (Roberts, 1964). Three formations compose the Antler sequence in the area: (1) the Middle Pennsylvanian Battle Formation, (2) the Upper Pennsylvanian and Lower Permian Antler Peak Limestone, and (3) the Permian Edna Mountain Formation. However, Erickson and Marsh (1974) cautioned that the term "sequence" no longer be applied to this package of rocks because at Edna Mountain, about 35 km to the northwest, they found that the basal formations of the sequence are folded, then capped unconformably by the uppermost formation in the sequence, the Edna Mountain Formation. However, as will be described in the section below entitled "Geology of the Marigold Mine Area," widespread debris-flow deposits also are present in the Edna Mountain Formation as a probable result of their derivation, in part, from the advancing edge of the Golconda allochthon. These debris-flow deposits host some of the most economically important gold ore in the Eight South deposit. Approximately 250 km to the northeast in the general area of Mountain City, Nev., the overlap assemblage includes some Upper Mississippian strata that contain a 200- to 800-m-thick sequence of basaltic rock, the Mississippian Nelson Formation of Coats (1969), from which one can infer that submarine basaltic volcanism must have affected some parts of the continental margin during that time (Little, 1987). Altered basic volcanic rocks also make up a significant proportion of the Mississippian Goughs Canyon Formation in the Osgood Mountains approximately 30 km to the northwest of the Battle Mountain Mining District (fig. 1; Hotz and Willden, 1964). However, the Goughs Canyon Formation apparently is part of the Golconda plate (Sando, 1993). The Edna Mountain Formation crops out on Treaty Hill in the Valmy quadrangle where the formation, as mapped, rests unconformably on the Valmy Formation; the Edna Mountain Formation also crops out in the workings of the Lone Tree, Stonehouse, Eight South deposits, and at the Old Marigold Mine (pl.1). The formation also is present near the trace of the Golconda thrust in the north-central part of the North Peak quadrangle (pls. 2). At this latter locality, the Edna Mountain Formation apparently is in depositional contact with an unusually thin underlying sequence of rocks belonging to the Antler Peak Limestone. The Battle Formation crops out in a small area on Lone Tree Hill, and at various other places farther to the south in the general area of the Eight South deposit, in the central part of the North Peak quadrangle, and near the southeast corner of the North Peak quadrangle (pls. 1, 2). It is also present near the southeast corner of the Snow Gulch quadrangle (pl. 3). The Antler Peak Limestone crop outs mainly in three areas: to the immediate south of the Eight South deposit and near the southeast corner of the North Peak quadrangle (pl. 2), and near the southeast corner of the Snow Gulch quadrangle (pl. 3).

The uppermost of the three major thrust plates, the Golconda plate (fig. 3), originally was mapped in the quadrangles by Roberts (1964) as consisting of mostly chert and argillite belonging to his Pennsylvanian and Permian Pumpernickel Formation and sandstone, shale, quartzarenite, limestone, and chert belonging to his Middle Pennsylvanian and Lower Permian Havallah Formation. However, as described above, these rocks are now considered to be tectonic packages of rock that do not strictly represent valid "formations". The Havallah Formation was interpreted by Roberts (1964) as being in depositional contact with the underlying Pumpernickel Formation. The rocks mostly west of Willow Creek in the southern part of the Battle Mountain Mining District were assigned by Roberts to his Havallah Formation, as was the bulk of the rocks that crop in the Havallah Hills (see also, Doebrich, 1995). In the Valmy and North Peak quadrangles, rocks of the Golconda plate are in fault contact with generally sporadic exposures of the Antler sequence both on the Golconda thrust and also along steeply dipping Tertiary normal faults; the latter is especially true near Trenton Canyon (pl. 2).

Intrusive rocks

The four major structural blocks in the area of the Battle Mountain Mining District (fig. 3), made up of lower to upper Paleozoic rocks, have been intruded by a large number of stocks and dikes that range in composition from gabbro to granite (pls. 1–3). Most stocks and dikes are felsic in the quadrangles. This report uses the plutonic-rock-classification scheme of Streckeisen (1973), unless otherwise noted. Some stocks crop out in an area more than 2.5 km^2 as in the east-central part of the Battle Mountain Mining District (fig. 9). The northernmost outcrops of the largest intrusive body in the mining district crop out at Trenton Canyon and they extend approximately 1 km into the North Peak quadrangle at the south border of the quadrangle (pl. 2). Although felsic intrusive rocks throughout the mining district vary widely in composition, most are calc-alkaline probably emplaced as granodiorite and monzogranite (Theodore and others, 1973). Some granodiorite bodies have been altered by postmagmatic hydrothermal fluids to potassic, phyllic, and (or) intermediate argillic assemblages in the various porphyry systems that are present throughout the mining district.

Plutonism in this part of Nevada occurred during a number of periods of time from the Triassic to the middle Tertiary (McKee and Silberman, 1970; Silberman and McKee, 1971). Probable Triassic two-mica plutonic rocks are present in the Granite Mountain area of the East Humboldt Range, approximately 65 km southwest of the Battle Mountain Mining District. Jurassic plutonic rocks crop out at Buffalo Mountain (fig. 1). A plutonic body at Trenton Canyon was originally reported to be Late Cretaceous (87 Ma) in age, whereas all other plutonic rocks in the Battle Mountain Mining District dated by Theodore and others (1973) are late Eocene or early Oligocene (41 to 37 Ma). Recalculation of the Late Cretaceous age using currently (1999) accepted K–Ar constants yields an age of 89 Ma. Potassium-argon and $^{40}Ar/^{39}Ar$ studies of various alteration and intrusive phases of the Buckingham stockwork molybdenum system yield Late Cretaceous to early Tertiary ages in the range 88–61 Ma (McKee, 1992). The most likely age for the emplacement of this system is 86 Ma. Seven of the radiometrically dated Tertiary igneous bodies in the Buckingham area also are late Eocene or early Oligocene (35–39 Ma) (McKee, 1992). This chemically intermediate calc-alkaline plutonism is related to subduction of the Pacific (Farallon) plate beneath the North American plate (McKee, 1996).

Some plutonism of Cretaceous age has been documented from other significant metal deposits elsewhere along the Battle Mountain-Eureka mineral belt. Silberman and McKee (1971) dated biotite from a granitic rock at depth below the Gold Acres sedimentary-rock-hosted gold deposit at 99 Ma, and sericite at 93 Ma using the K–Ar method. Sericite from a quartz porphyry cropping out within the alteration zone associated with the Old Gold Acres deposit also was dated by them at 94 Ma and is inferred to be associated with a buried molybdenum-bearing porphyry system under the deposit (Wrucke and Armbrustmacher, 1975). Earlier studies by Rytuba (1985), suggested that gold-mineralized rocks at Gold Acres and the nearby Horse Canyon and Cortez deposits (fig. 1) is most likely related to rhyolite porphyry dikes that are comagmatic with the approximately 34.5–Ma Oligocene Caetano Tuff. However, subsequent mining operations at Gold Acres failed to document presence of rhyolite porphyry dikes previously assigned genetically to introduction of gold (Stan Foo, Bob Hayes, and Chris Gillette, oral communs., 1993). Gold-bearing fluids at Gold Acres apparently were channeled along high-angle faults into a

structurally prepared, carbon-rich imbricate west-dipping thrust zone containing mostly rocks of the Silurian and Devonian Roberts Mountains Formation. Furthermore, Hayes and others (1994) suggest that some mineralized structures in the general area of the Gold Acres and Cortez deposits are roughly parallel to the trend of the northern Nevada rift, and that these mineralized structures may actually pre-date development of the rift—these structures may be reactivated late Paleozoic faults that formed initially along a "proto" northern Nevada rift (Theodore and others, 1998). Finally, Arehart and others (1993) suggest that most sedimentary-rock hosted gold deposits along the Carlin trend may be Cretaceous in age, on the basis of sericite potassium-argon and $^{40}Ar/^{39}Ar$ ages in the 85–Ma to 125–Ma range from mineralized structures, which time span coincides with a period of regional compression (Seedorff, 1991). However, ages obtained from micas in Carlin-type deposits may not be indicative of the age of gold mineralization in the deposits (Folger and others, 1995), and Emsbo and others (1996) suggest that much of the Au may be late Eocene and (or) early Oligocene.

The middle Tertiary plutons at Battle Mountain are part of a regional alignment of middle Tertiary plutons that extends from Eureka to Battle Mountain (Silberman and McKee, 1971), but these plutons overall are relatively rare when one considers the entire area of Nevada. In addition, the middle Tertiary plutons in this part of Nevada appear to be preferentially clustered in a north-south elongate area that extends from Lone Tree Hill on the north to the McCoy Mining District (fig. 1) on the south. In the Battle Mountain Mining District, as well as in the McCoy Mining District, the middle Tertiary is an extremely important time from the standpoint of Au and Cu mineralization (Theodore and others, 1973; Brooks and others, 1991; see section below entitled "Potassium-argon Chronology of Cretaceous and Cenozoic Igneous Activity, Hydrothermal Alteration, and Mineralization"). Some Au–mineralized rocks along the Carlin trend appear to have formed sometime during the middle Tertiary (<38.5 Ma, Emsbo and others, 1996), and they appear to have formed between 38 and 34 Ma at Tonkin Springs, Nev., near the southeast terminus of the Battle Mountain-Eureka mineral belt (Maher and others, 1993).

Some evidence in the surrounding region, especially to the east of Battle Mountain, indicates a former fairly widespread occurrence of volcanic rock that is approximately the same age as mineralized middle Tertiary plutons in the mining district (Brooks and others, 1994, 1995; Thorman and Brooks, 1994). In addition, Wallace and McKee (1994) infer presence of a possible Eocene volcanic field north of the Battle Mountain Mining District and centered approximately 20 km north of the Osgood Mountains and 10 km west of the Snowstorm Mountains (fig. 1). However, evidence is quite fragmentary in the immediate region of the Battle Mountain Mining District for presence of middle Tertiary volcanic rocks. Tuffaceous sedimentary rocks near the base of the Tertiary sequence close to Beowawe Geysers, Nev., contain hornblende-biotite andesite flows that have yielded a 38.7±1.3–Ma potassium-argon age on primary biotite (Struhsacker, 1980). These tuffaceous sedimentary rocks also contain some cobbles of biotite dacite from which a separate of primary biotite yielded a 38.8±1.3–Ma age. Both of these ages have been obtained from exposed rock in N. 1/2 sec. 15, T. 31 N., R. 48 E., approximately 3 km east of Beowawe Geysers (fig. 1; Struhsacker, 1980). In addition, hornblende-biotite andesite of apparently the same middle Tertiary age is present at similar stratigraphic positions in tuffaceous sedimentary rocks in both of the deep geothermal wells, Ginn 1–13 and Rossi 21–19, drilled near Beowawe Geysers (Struhsacker, 1980). Farther to the east, re-examination by $^{40}Ar/^{39}Ar$ techniques of volcanic rocks in the Tertiary Elko Formation yield 36– to 38–Ma ages (C.H. Thorman, oral commun., 1994). Thus, some middle Tertiary plutons in the region, at least one of which has been documented to have had quite shallow-seated levels of emplacement (Nash and Theodore, 1971), must have vented onto middle Tertiary paleosurfaces.

Tertiary volcanic rocks

At least five suites of Tertiary volcanic rocks crop out in the general area of the Battle Mountain Mining District. Major outcrops of calc-alkaline welded tuffs are present to the south and southeast as erosional outflow-facies remnants of the previously much more extensive Oligocene Caetano Tuff (Roberts, 1964; Gilluly and Masursky, 1965; Stewart and McKee, 1977; Doebrich, 1995), whose age in the general area of the mining district is approximately 34 Ma (recalculated from McKee and Silberman, 1970). A small outcrop of the Caetano Tuff is present near the southeast corner of the Snow Gulch quadrangle (pl. 3). Augite basaltic andesite that is Oligocene in age (approximately 31 Ma), crops out in the North Peak quadrangle just to the west of Cottonwood Creek (pl. 2). Late Oligocene or early Miocene calc-alkaline rhyolite tuff, late Oligocene or early Miocene biotite crystal tuff, and probably age-equivalent lacustrine tuff—all about 23 to 25 Ma—also are present in the North Peak quadrangle. Approximately 12–Ma Miocene olivine basalt crops out in the Treaty Hill area of the Valmy quadrangle (pl. 1). Finally, Pliocene basalt, approximately 3 Ma, crops along the southern edge of the mining district (fig. 9). Oligocene augite basalt (approximately 31 Ma, see section below entitled "Potassium-argon Chronology of Cretaceous and Cenozoic Igneous Activity, Hydrothermal Alteration, and

Mineralization") is present as small bodies that are interbedded with gravel in the the area of the middle stretches of Cottonwood Creek (unit Tg, pl. 2). Furthermore, some basaltic rock that has an approximate 30– to 35–Ma age also has been encountered interbedded with lake beds at moderate depths under alluvium west-southwest of Lone Tree Hill (see section below entitled "Lone Tree Gold Deposit"). These Oligocene basaltic rocks are evidence of previously undocumented mafic magmatism that apparently must have been extent in the region approximately 10 m.y. after onset of crustal extension, which has been established as being roughly 40 Ma in the Copper Canyon area of the Battle Mountain Mining District (Theodore and Blake, 1975).

Late Oligocene or early Miocene calc-alkaline rhyolite tuff (unit Tt, pl. 2) is present in the central part of the North Peak quadrangle, where it crops out in a number of discontinuous bodies along a total strike length of approximately 8 km in the hanging wall of the Oyarbide fault. Some apparent lacustrine tuffs also have been encountered below valley-fill gravels in several drill holes put down in Buffalo Valley west-southwest of the quadrangles (J.L. Doebrich, oral commun., 1992). The age of these lacustrine tuffs is unknown at this writing (1999), but some may be as old as the 23– to 25–Ma calc alkaline rhyolite tuff interbedded with gravels at the Eight South deposit. A small body of late Oligocene or early Miocene biotite crystal tuff rests unconformably on mineralized debris-flows of the Permian Edna Mountain Formation and is also, in places, interbedded with Tertiary valley-fill gravels on the 4,620–ft bench of the Eight South deposit (D.H. McGibbon, oral commun., 1992; see section below entitled "Geology of the Marigold Mine Area"). Farther to the east, Fleck and others (1998) determined that lacustrine air-fall tuff units in the Miocene Carlin Formation are 15.1 to 14.4 Ma.

Miocene 12–Ma basalt at Treaty Hill is roughly correlative with some of the youngest basalt flows associated with the northern Nevada rift (Zoback and others, 1994), and is about 3 m.y. younger than rhyolite flows correlated with the informally named Craig rhyolite of Bartlett and others (1991) at the Hollister Mine (fig. 1). Elsewhere, in the general area of the Battle Mountain Mining District, many Tertiary basaltic rocks are younger than this age. A sample of basalt from the southern part of the Battle Mountain Mining District, near Copper Canyon, has an age of about 3.2 Ma using the K–Ar method (McKee, 1992), and another basalt body also dated by the K–Ar method from the southern part of the Antler Peak 7–1/2 minute quadrangle is approximately 2.8 Ma (see section below entitled "Potassium-argon Chronology of Cretaceous and Cenozoic Igneous Activity, Hydrothermal Alteration, and Mineralization"; see also, Doebrich, 1995). Basalt flows in the area of Rock Creek, approximately 30 km to the northwest, have an age of about 5 Ma (Erickson and Marsh, 1974).

Relations among trace-element ratios and isotopic ratios in Cenozoic mafic lavas in the Basin and Range of the southwestern United States recently have been utilized to make inferences about the nature of the lithosphere underlying this part of the Great Basin (McKee and mark, 1971; Mark and others, 1975; McKee and others, 1983; Ormerod and others, 1988; Fitton and others, 1988; Kempton and others, 1991; Fitton and others, 1991). Generally, these authors conclude that Cenozoic mafic lavas in the Basin and Range, mostly less than 5–Ma basalt, show marked geochemical differences when compared with similar lavas in other provinces in the southwest, and that all these less than 5–Ma mafic lavas in the Basin and Range are strikingly similar to ocean island-type basalt. These relations suggest thereby that their sources are within the asthenosphere. The older than 5–Ma mafic lavas, however, show no distinct intra-provincial geochemical differences and have relatively high La/Nb and Ba/Nb ratios, possibly reflecting involvement of subduction-enriched mantle as will be discussed below in the section entitled "Oligocene Augite Basalt." Although these studies rely mostly on data obtained from lava fields far to the south of the Battle Mountain Mining District, geochemistry and mineral chemistry of a small number of samples of Oligocene and (or) Miocene basaltic rocks from the district appear to be compatible with these conclusions.

GEOLOGY OF THE VALMY, NORTH PEAK, AND SNOW GULCH QUADRANGLES

The area of the Valmy, North Peak, and Snow Gulch quadrangles includes parts of the allochthonous Roberts Mountains and Dewitt tectonic blocks, the autochthonous Antler sequence, and the allochthonous Golconda block—about 20 percent of this 850–km² area contains exposures of rocks belonging to the Devonian, Mississippian, Pennsylvanian, and Permian Havallah sequence in the upper plate of the Golconda thrust (fig. 9). Although Havallah sequence-equivalent rocks of the Schoonover sequence in the Independence Mountains, Nev., contain rocks as old as Devonian (Miller and others, 1992), rocks of the Havallah sequence in the study area include rocks known only to be as old as Mississippian (B.L. Murchey, unpub.

data, 1999). Part of the Late Cretaceous granodiorite of Trenton Canyon crops out near the south border of the North Peak quadrangle, as described above, and much smaller outcrops of Cretaceous or Tertiary granitic stocks and dikes, as well as Tertiary dikes crop out locally. The monzogranite and granodiorite of Trenton Canyon was emplaced high into the upper plate of the Golconda allochthon (fig. 9), which shows a well-developed and extemely wide contact metamorphic aureole, mostly biotite hornfels. Indeed, the pluton at Trenton Canyon cuts the trace of the Golconda thrust (Roberts, 1964; Doebrich, 1995). Most small stocks and dikes have not been dated radiometrically, primarily because of the highly altered nature of their exposures. The late Eocene and (or) early Oligocene Elder Creek porphyry copper system was emplaced into rocks of the Upper Cambrian Harmony Formation, and this system is surrounded by a wide expanse of pyrite-bearing biotite hornfels (pl. 3). Although outcrops of late Oligocene or early Miocene rhyolite tuff and Oligocene and Miocene basaltic rocks are quite sparse (pls. 1, 2), the few available unequivocal geologic relations of these volcanic rocks to surrounding units and some major faults, nonetheless provide critical data that allow a concise reconstruction of the structural history during much of the Tertiary. Furthermore, geologic mapping has revealed complex structural configurations of packets of rock that make up the Golconda allochthon. The present geometry of the Paleozoic, Mesozoic, and Tertiary rocks must represent superposition of several major tectonic events.

Tectonic events documented include (1) deformation involving thrusting and crustal shortening during Devonian and (or) Mississippian time associated with emplacement of the Roberts Mountains allochthon, (2) additional crustal shortening associated with early Mesozoic thrusting of the Golconda plate, (3) arching and faulting of Paleozoic rocks probably either before or synchronous with Late Cretaceous magmatism, and (4) Tertiary extensional phenomena that may be quite protracted in age. The latter probably began at about 40 Ma, and culminated in major displacements along north-striking range-front faults at some time after 31 Ma. In addition, extension along the northeast-striking Oyarbide fault probably occurred during a time span no more precisely resolvable than sometime during the last 9 m.y.

Strata of the upper plate of the Roberts Mountains Thrust

HARMONY FORMATION (CAMBRIAN)

The Harmony Formation is allochthonous and widespread in the Battle Mountain Mining District (Roberts, 1964). The formation crops out in a small area near the southeast corner of the North Peak quadrangle (pl. 2), and across much of the Snow Gulch quadrangle (pl. 3). As reviewed previously in Theodore and others (1992), the Harmony Formation is considered middle and late Late Cambrian in age on the basis of contained trilobites found near the north end of the Hot Springs Range, 5 to 10 km west-northwest of the Osgood Mountains (Roberts and others, 1958; Hotz and Willden, 1964; C.A. Suczek, oral commun., 1992). Although Whitebread (1994) designated rocks of the Harmony Formation in the East Range, Nev., as being pre-Mississippian in age, ample evidence is available regionally to assign a more precise age than this to these enigmatic rocks, despite some apparent conflicts in the evidence. Some other rocks assigned previously to the Harmony Formation by Hotz and Willden (1964) in the Osgood Mountains, and also apparently containing trilobites at one locality, are probably fault-bounded olistostromal units containing large blocks of rock lithologically identical to undisturbed strata of the Harmony Formation (Suczek, 1977). The rocks in question are shown by Hotz and Willden (1964) as an approximately 3–km-long fault-bounded block in the upper stretches of Goughs Canyon, T. 38 N., R. 41 E., in the Osgood Mountains (fig. 1). They have been reassigned by McCollum and McCollum (1989) to an informally named Middle to Upper Pennsylvanian "Getchell" Formation on the basis of the presence of *Gondollella* aff. *elegantula* conodont fauna in the rocks. These rocks, also referred to as "Harmony mélange" by Jones (1991a), consist of many large obviously exotic blocks belonging to the Harmony Formation which are hosted in a sheared matrix of volcaniclastic rock, and some shale, granule sandstone, radiolarian chert, dolomite, and dolomitic sandstone (McCollum and McCollum, 1989). Willden (1979) also doubted the middle and late Late Cambrian age of the Harmony Formation because he felt that it lies unconformably above the Devonian Scott Canyon Formation in the Battle Mountain Mining District, as did McCollum and others (1987). McCollum and others (1987) suggested that rocks of the Harmony Formation may be equivalent to the Silurian Elder Formation, and McCollum and McCollum (1989) subsequently asserted that the Harmony Formation should be restricted generally in terms of its age only to the lower Paleozoic (see also, Debrenne and others, 1990). More-

over, geologic relations described above in the Galena Canyon area of the mining district demonstrate that rocks of the Harmony Formation must be older than rocks presently (1999) assigned to the Devonian Scott Canyon Formation. Madden-McGuire and others (1991) reported acritarchs and algal cysts in shale belonging to the Harmony Formation at its type locality in the northern Sonoma Range. These acritarchs and algal cysts apparently support a Late Cambrian or Early Ordovician age, although subsequent collection of approximately the same sites showed that many of the rocks there had been heated to such an extent that acritarchs could not be recovered (M.B. McCollum, written commun., 1992)—the original identification of the acritarchs has been questioned. Some limestone within the Harmony Formation in the Hot Springs Range recently has yielded Devonian-age conodonts (Jones, 1997). However, some caution must still be exercised because reworking of some Ordovician rocks during the Devonian and deposition of their detritus in minor basins is widespread in the Shoshone Range (C.T. Wrucke, oral commun., 1995). Three samples from the Harmony Formation, from outcrops near the southern border of the Snow Gulch quadrangle, were submitted for palynomorph age determination. As reported by Rosemary Askin-Jacobsen (written commun., 1992), all of these samples are barren of in-place palynomorphs, although two of the samples contain low concentrations of organic matter. It turns out that this organic matter is finely disseminated microdebris, and that one of the samples of shale also shows in-place organic matter that has a high thermal maturation.

Uranium-lead geochronologic studies based on single grains of detrital zircons from the Harmony Formation yield ages that range from 680 to 2,570 Ma (Smith and Gehrels, 1992, 1994). These ages cluster in the following narrow age ranges: 695–710 Ma, 1,015–1,225 Ma, approximately 1,330 Ma, approximately 1,745 Ma, 1,915 Ma, and 2,570 Ma. Therefore, the data are important in that they appear to rule out the 1,500–Ma Proterozoic Salmon River arch in Idaho as the main source for the largely feldspathic arenites of the Harmony Formation (J.H. Stewart, oral commun., 1992; Gehrels and others, 1993). Regardless of their source area, rocks of the Harmony Formation must be younger than the age of the youngest detrital zircon. Smith and Gehrels (1992) further suggest that middle and late Proterozoic igneous rocks were widespread along a Cordilleran margin that must have predated postulated rifting in the latest Proterozoic or Early Cambrian. Gehrels and others (1993) suggest that detritus comprising the rocks of the Harmony Formation was derived from off-shore regions rather than from the craton, and subsequently Smith and Garrels (1994) suggest that rocks of the Harmony Formation may have been derived largely from either 670 to 750– and 1,000 to 1,350–Ma basement terranes to the north (Kootenay, Yukon-Tanana, and others), or, less likely, the Proterozoic Mendenhall Gneiss of southern California. Recent regional interpretations of available sources for the Harmony Formation suggest that it may have been derived from the North American craton (J.H. Stewart, oral commun., 1999).

To date (1999), however, only trace fossils have been identified definitely from the Harmony Formation in the Battle Mountain Mining District (Suczek, 1977). These fossils are present on bedding planes of sandstone and are considered to be feeding tracks of *Zoophycos* and *Asterosoma*. Nonetheless, the distinctive lithologies of the Harmony Formation in the North Peak and Snow Gulch quadrangles are the same as those that crop out in the Hot Springs Range (fig. 1) and elsewhere in other nearby ranges. Furthermore, the regionally distinctive lithology of the Harmony Formation was recognized in modal analyses of clasts hosted by the Ordovician Valmy Formation, and possibly even in the Devonian Slaven Chert, in the Shoshone Range, 30 km southeast of the Valmy-North Peak area (Madrid, 1987). This relation suggests that the Harmony Formation may be older than Ordovician in age (see also, Miller and others, 1992). Nevertheless, the age of the Harmony Formation remains uncertain.

The Harmony Formation is exposed widely in many parts of the Battle Mountain Mining District, and it consists of first cycle interbedded quartzarenite, subarkose, arkose, granule and pebbly quartzarenite, shale, and limestone (Roberts, 1964). Most of the formation probably is a medium-grained subarkose, although silt and calcareous shale commonly crop out widely in some parts of the Buckingham area, in the east-central part of the mining district (Theodore and others, 1992). The thickness of the formation cannot be measured directly, because the rocks are, for the most part, poorly exposed. Roberts (1964) estimated total thickness of the formation in the mining district to be about 900 m, or approximately near the middle of the estimate by Stewart and Suczek (1977) for regional variation in thickness (600 to 1,300 m). The provenance of sandstone belonging to the Harmony Formation has been classified as transitional continental or belonging to a suite transitional between interior craton and uplifted basement suites (Dickinson and Suczek, 1979; Madrid, 1987). Rowell and others (1979) previously had suggested derivation of the Harmony Formation as a proximal turbidite in the middle part of a submarine fan derived from Proterozoic granitic rocks to the northeast in the central Idaho, Salmon River arch (Armstrong, 1975), but this seems now unlikely because of ages of detrital zircons in the Harmony Formation (see this section above). On the one hand, the overall compositional and textural immaturity of sand, grit, and conglomerate in the Harmony Formation suggested to Palmer (1971) that the Harmony Formation was derived from the northeast; on the other hand, this same composi-

tional and textural immaturity suggested to Ketner (1977) that these rocks were derived locally. Erickson and Marsh (1974), who also believed that the rocks were derived locally, thought that they came from an emergent highland somewhere between the Sonoma Range to the west and the Battle Mountain Mining District. The uplift of this highland may be also indicated by the presence of west-verging folds in the Cambrian Preble Formation and Cambrian(?) Osgood Mountain Quartzite according to their regional reconstructions. However, in the Hot Springs Range, the Harmony Formation is itself deformed in places into tightly appressed west-verging folds (Hotz and Willden, 1964; Jones, 1993). The clastic sedimentary rocks of the Harmony Formation are interpreted by Willden (1979) as having been derived from a part of the emergent continental crust in present-day northwestern Nevada. Finally, Madrid (1987) and Poole and others (1992), on the basis of their similar detrital modes, suggested that the first-cycle mostly feldspar-rich sandstones of the Harmony Formation were the sources of many younger Paleozoic second-cycle sandstones present in the Roberts Mountains allochthon.

The Harmony Formation crops out in a small area of approximately 1 km² in secs. 21 and 22, T. 32 N., R. 43 E. of the North Peak quadrangle where much of the formation is in fault contact with rocks of the tectonically underlying Valmy Formation along a fairly steep-dipping, northeast-striking segment of the Dewitt thrust (pl. 2). The Harmony Formation near the south fork of Trout Creek also is in contact with the Valmy Formation along the northwest-striking, Lucky Strike-Lauritzen high-angle normal fault that, in its western block, has faulted downwards rocks of the Harmony Formation relative to the Valmy Formation. Exposures of the Harmony Formation are fairly subdued throughout this area (fig. 10), but distribution of distinctive grayish olive-green feldspathic arenite fragments, which are highly diagnostic of the presence of the formation under colluvium, allows reasonably well-constrained placement of contacts. At most, a thickness of several hundred meters of the Harmony Formation is exposed in this area, generally in sequences that are typically not continuous for more than several tens of meters away from individual exposures; the stratigraphic base of the formation is not exposed in the area. The Harmony Formation is overlain unconformably by the Battle Formation which is exposed in several narrow outcrops along the south border of the North Peak quadrangle (pl. 2) and near the southeast corner of the Snow Gulch quadrangle (pl. 3).

The Harmony Formation, where it crops out near the southeast corner of the North Peak quadrangle, is remarkably similar to the rocks of the formation elsewhere in the Battle Mountain Mining District with some exceptions (Roberts, 1964). In large part, rocks of the formation

Figure 10 Photograph showing generally subdued nature of exposures of Upper Cambrian Harmony Formation (f h) near upper reaches of Cottonwood Creek. Rugged exposures on skyline are basal conglomerate of Middle Pennsylvanian Battle Formation (d b).

in this area consist of generally slope-forming, olive gray-green feldspathic arenite that is poorly sorted and immature. Many exposures are present along ridgelines where greasy-gray, detrital quartz fragments are quite angular— some of the quartz fragments are various shades of pale drab blue. Where these rocks are altered, as along one of the minor mineralized splays in the SW 1/4 sec. 22, T. 32 N., R. 43 E., that parallels the Lucky Strike-Lauritzen fault system (pl. 2), feldspathic arenite of the Harmony Formation is bleached and clay-altered for distances as much as 10 to 15 m away from the trace of the fault at the surface. In addition, although the actual trace of the fault zone is marked by quartz-veined breccia only about 1 m wide, distinctive ochre- to red-brown Fe oxides are common throughout the entire area of bleaching. Nearby bedding, where well exposed, shows approximately 0.6 m between parting surfaces that are distinctly plane parallel and show no evidence of basal fluting and grading, although graded beds are fairly common elsewhere in the rocks of the formation in the mining district (Roberts, 1964). Weathered surfaces are typically greenish brown, and red-brown Fe oxides are concentrated along fractures. In addition, some outcrops of the Harmony Formation include enhanced concentrations of detrital white mica in the olive-green feldspathic arenite. However, in marked contrast to relative abundance of shale in many thick sequences of the Harmony Formation exposed elsewhere in the mining district, particularly in its northeast part near Copper Basin (Theodore and others, 1992), the relatively limited exposures of the Harmony Formation in the North Peak quadrangle do not show a comparable proportion of shale.

The rocks of the Harmony Formation in the Snow Gulch quadrangle have been folded broadly into a large number of synclines and anticlines, whose hingelines show generally north-trending attitudes (pl. 3). These rocks also have been affected widely by epigenetic alteration associated with many of the large number of mineralized Tertiary plutons emplaced into the formation. Much of this alteration, mostly to biotite hornfels, surrounds either the Elder Creek porphyry system in the central part of the quadrangle, or the alteration is associated with the large number of Tertiary stocks that are clustered near the southeast part of the quadrangle. In addition, some conversion of rocks of the Harmony Formation to biotite hornfels, with or without accompanying dispersed pyrite and pyrrhotite, at the latter occurrence may also be the result of emplacement of the Late Cretaceous stockwork Mo system, which is centered just south of the south border of the quadrangle (Theodore and others, 1992).

Chemical analyses of seven samples of feldspathic arenite are available from the small area of outcrop of the Harmony Formation in the North Peak quadrangle (table 1). One of these analyses (analysis 2) is from a narrow sequence of rock bleached and clay altered near one of the fault strands associated with the base metal-mineralized, mostly sphalerite enriched Lucky Strike-Lauritzen fault system (pl. 2; Roberts and Arnold, 1965). As such, analysis 2 includes the highest K_2O/Na_2O ratio and the highest Zn and As contents among the seven analyzed rocks, respectively 360 and 28 ppm. Clay alteration in the sample may be masking an earlier stage of potassic alteration associated with base-metal mineralization along the fault system. In addition, the analyzed sample of altered rock, contains less Ce, Co, Cr, La, Nd, Pb, Sc, and Sr than the apparently unaltered samples. These metals may have been leached from the mineralized fault by supergene downward percolating acidic fluids. These seven analyses also show relatively wide-ranging contents of SiO_2, from 72.2 to 87.7 weight percent, which range moreover incudes the reported SiO_2 contents (approximately 78 and 83 weight percent) in two other samples of unaltered feldspathic arenite of the Harmony Formation from the Galena Canyon area of the Battle Mountain Mining District (Theodore and Blake, 1975, table 9, analyses 11, 14). However, other altered partly shale-bearing rocks of the Harmony Formation analyzed from recrystallized portions of the Buckingham stockwork molybdenum system show SiO_2 contents in the range 88.3 to 57.8 weight percent (Theodore and others, 1992, table 1). Normalized K_2O/Na_2O ratios, calculated volatile free for the six unaltered samples from the North Peak quadrangle, are in the range 0.51 to 0.89 and thereby plot (fig. 11) near the join between the passive margin and active continental margin empirically derived fields of the tectonic discrimination diagram of Roser and Korsch (1986). The altered sample contains less than the lower limit of detection of Na_2O (analysis 2, table 1), but shows a normalized K_2O/Na_2O ratio approximately equal to 22, if one assumes 50 percent of the lower detection level as the concentration of Na_2O in the rock. The altered sample thus plots well into the passive margin tec-

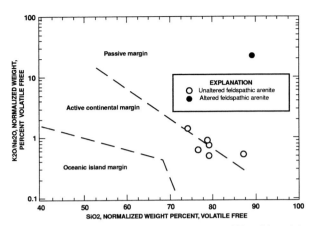

Figure 11 Diagram showing K₂O/Na₂O ratio versus SiO₂, all in weight percent normalized volatile free to 100 percent, for seven samples of feldspathic arenite from Upper Cambrian Harmony Formation. Dashed lines, approximate empirically derived boundaries of tectonic margins for sandstones and argillites of Roser and Korsch (1986). Data from table 1.

tonic field of Roser and Korsch (1986) and is shown on this diagram (fig. 11) only for comparative purposes. No implications of tectonic regime should be derived from the plotted position of the altered sample.

Gold was not detected in the six unaltered samples at a detection threshold of 4 ppb (parts per billion), whereas the altered sample along the Lucky Strike-Lauritzen fault system contains some detectable Au, but at a concentration less than 4 ppb (table 1). Background concentrations of Au in feldspathic arenite of the Harmony Formation thus are probably less than 4 ppb.

VALMY FORMATION (ORDOVICIAN)

The Valmy Formation is a regionally extensive formation in the upper plate of the Roberts Mountains thrust, and it crops out mostly in the North Peak quadrangle as well as in a narrow strip along the western border of the Snow Gulch quadrangle (pls. 2, 3; Roberts, 1964). Rocks of the Valmy Formation also are present elsewhere in a few isolated localities in the northern part of the area, namely at Treaty Hill, at Lone Tree Hill, and in a small fault-bounded sliver in NE 1/4 sec. 13, T. 33 N., R. 42 E., near the southeast corner of the Havallah Hills (pls. 1, 2). All rocks of the Valmy Formation are considered to be Early and Middle Ordovician in age on the basis of extensive collections of graptolites described by Roberts (1964) from the four members he delineated in the formation. In the Sonoma Range, approximately 20 km to the west, the Valmy Formation is speculated to be Early, Middle, and Late Ordovician (Gilluly, 1967), as it is in the Mount Lewis and Crescent Valley areas of the Shoshone Range, about

30 km to the southeast (Gilluly and Gates, 1965; see also, Madrid, 1987). Farther to the northeast in the Independence Mountains (fig. 1), the rocks of the Valmy Formation, or Group as the rocks are referred to there, also are assigned to the Early, Middle, and Late Ordovician by Watkins and Browne (1989; see also, Churkin and Kay, 1967). Nonetheless, it is possible that some rocks included with the Ordovician Valmy Formation in the Valmy and North Peak quadrangles are instead part of the Devonian Scott Canyon Formation. Many thick sequences of chert in the study area as of this writing (1999) have yielded no biostratigraphically diagnostic faunal assemblages. On the basis of U–Pb analyses of included zircons, the detritus that makes up the Valmy Formation is considered by Gehrels and others (1993) to have formed as a lateral facies of the Middle Ordovician Eureka quartzite, both formations apparently derived from the general area of the Peace River arch in northwestern Canada (see also, Smith and Garrels, 1994).

In this report, the Valmy Formation is divided into four units: unit 1 is essentially the same as the undifferentiated part of the Valmy Formation mapped by Roberts (1964) (compare pl. 2 with his pl. 1). However, the top of unit 1 is considered to be at the top of a sequence of well-exposed, thick-bedded quartzarenite (see McBride, 1963) in NE 1/4 sec. 33, T. 33 N., R. 43 E., near the east border of the North Peak quadrangle (pl. 2). At this locality, in the footwall of the Oyarbide fault, an apparently conformable, depositional contact between thick-bedded quartzarenite and an overlying sequence of thin-bedded, light gray-green ribbon chert is present (fig. 12)—the sequence of chert, herein assigned to unit 2 of the formation, extends about 0.5 km farther to the south. The thick-bedded quartzarenite is correlated on the basis of stratigraphic position and lithology with similar sequences of quartzarenite that make up most of the Valmy Formation in the hanging wall of the Oyarbide fault. Units 2, 3 and 4, as the Valmy Formation is herein delineated, thus approximately correspond respectively to members 1, 2 and 3 previously mapped by Roberts (1964). However, there are some additional minor differences between this report and that by Roberts (1964) in the packaging of the rocks of the Valmy Formation; for example, the contact between unit 3 and unit 2 is believed to be a low angle fault (pl. 2). The relatively large scale of geologic mapping in the current study also permitted several subunits within each of the four units of the Valmy Formation to be mapped separately and these subunits are described in some detail in the subsections that follow. The subunits show both depositional and tectonic contacts with one another.

Rocks of the Valmy Formation that crop out in the hanging wall of the Oyarbide fault, unit 1 according to the mapping scheme employed (pl. 2), are accordingly inferred to make up the lowest stratigraphic parts of the formation.

Table 1—Analytical data on feldspathic arenite of Upper Cambrian Harmony Formation, North Peak quadrangle, Humboldt County, Nevada.†

Analysis number	1	2	3	4	5	6	7
Field number							
90TT...	016	017	018	025	026	027	028
Chemical analyses (weight percent)							
SiO_2	86.2	87.7	77.1	74.5	77.2	77.0	72.2
Al_2O_3	5.94	6.46	10.7	10.7	10.1	9.26	12.5
Fe_2O_3	1.6	1.65	3.45	3.33	2.67	2.05	1.94
FeO	1.26	.02	.83	1.86	1.96	2.19	3.02
MgO	.76	.26	.93	1.4	1.16	1.12	1.56
CaO	.15	.14	.15	.9	.21	1.19	.33
Na_2O	1.15	<.15	1.99	2.17	2.37	2.07	1.93
K_2O	.8	1.65	1.78	1.36	1.2	1.56	2.81
H_2O^+	1.2	1.58	2.08	2.36	2.03	1.8	2.47
H_2O^-	.17	.25	.42	.36	.25	.2	.34
TiO_2	.31	.19	.49	.54	.55	.45	.63
P_2O_5	.08	.09	.08	.09	.11	.07	.1
MnO	.13	.08	.1	.2	.06	.13	.04
F	.02	.02	.03	.03	.03	.03	.03
Cl	.005	.004	.004	.005	.008	.005	.005
CO_2	.03	.03	.03	.57	.03	.82	.12
Total S	<.01	<.01	<.01	<.01	<.01	<.01	<.01
Subtotal	100.17	100.12	100.16	100.38	99.94	99.95	100.03
Less O = F, Cl, S	.01	.01	.01	.01	.01	.01	.01
Total	100.16	100.11	100.15	100.37	99.93	99.94	100.02
Inductively coupled plasma-atomic emission spectroscopy (total) (parts per million)							
As	—	26.0	—	—	—	—	—
Ba	310.0	310.	410.0	370.0	260.0	530.0	840.0
Be	—	—	1.	—	—	—	1.
Cd	—	4.	—	—	—	—	—
Ce	48.	27.	70.	59.	62.	60.	72.
Co	7.	3.	14.	11.	12.	11.	16.
Cr	29.	20.	64.	67.	60.	58.	82.
Cu	5.	2.	11.	11.	10.	8.	18.
Ga	6.	8.	14.	14.	11.	13.	17.
La	25.	14.	36.	29.	31.	30.	38.
Li	51.	59.	41.	52.	37.	34.	38.
Nb	—	—	4.	—	4.	—	5.
Nd	20.	8.	27.	23.	25.	23.	31.
Ni	14.	7.	27.	29.	25.	21.	30.
Pb	15.	9.	29.	22.	26.	26.	39.

Table 1 – (cont.)

Sc	4.	3.	7.	8.	7.	6.	9.
Sr	47.	29.	60.	63.	60.	76.	71.
Th	13.	5.	15.	9.	14.	15.	16.
V	23.	22.	49.	54.	49.	39.	62.

Inductively coupled plasma-atomic emission spectroscopy (partial) (parts per million)

Ag	—	0.08	—	—	—	—	—
As	0.69	28.	3.3	5.5	1.7	0.78	1.4
Cd	—	4.	—	.053	.068	.078	—
Cu	5.1	3.	12.	12.	11.	9.8	20.
Mo	.76	.36	.44	.29	—	2.3	.16
Pb	13.	4.1	27.	21.	28.	24.	36.
Sb	—	.96	—	2.	1.5	.86	1.2
Zn	28.	360.	55.	71.	69.	54.	29.

Chemical analyses (parts per million)

As	1.6	28.0	4.9	8.1	2.7	2.1	2.9
Au	—	<.004	—	—	—	—	—
Hg	—	.03	.02	.04	—	.03	.02
Te[1]	—	—	—	—	—	—	—
Tl	.2	.85	.45	.3	.25	.35	.5

[1]Detection level is 0.05 ppm.

1. Olive gray-green feldspathic arenite; abundant iron-oxide staining along detrital grain boundaries.
2. Bleached, clay altered feldspathic arenite; abundant textural evidence of recrystallization; some veining by quartz.
3. Drab olive-green feldspathic arenite; abundant detrital white mica.
4. Olive gray-green feldspathic arenite; relatively abundant framework grains of plagioclase, some detrital grains of calcite.
5. Olive gray-green feldspathic arenite, sparse detrital white mica; abundant framework grains of feldspar and monocrystalline quartz.
6. Do.
7. Drab olive gray-green feldspathic arenite, locality close to contract with uncomformably overlying Middle Pennsylvanian Battle Formation.

†[Chemical analyses of major oxides in weight percent by methods of Jackson and others (1987); analysts, J.S. Mee and D.F. Siems. FeO, H_2O^+, H_2O^- in weght percent by methods of Jackson and others (1987); analyst, N.H. Elsheimer. Total S in weight percent by combustion-infra-red method; analyst, N.H. Elsheimer. F, Cl in weight percent by specific ion electrode method; analyst, N.H. Elsheimer. Inductively coupled plasma-atomic emission spectroscopy (total and partial) in parts per million by methods of Crock and Lichte (1982), Lichte and others (1987), and Motooka (1988); analysts, D.L. Fey, and J.M. Motooka. Precision for concentration higher than 10 times the detection limit is better than +10-percent relative standard deviation; instrumental precision of the scanning instrument is +2-percent relative standard deviation. Looked for, but not found, at parts-per-million detection levels in parantheses: Au (8), Bi (10), Eu (2), Ho (4), Mo (2), Sn (5), Ta (40), Th (4), U (100). Chemical analysis for Au by combined chemical separation-atomic absorption method; analyst, E.P. Welsch, L.A. Bradley, A.H. Love, and B.H. Roushey. Chemical analysis for As, Te (0.05 ppm detection level), and Tl by graphite furnace atomic absorption spectrometry (Wilson and others, 1987; Aruscavage and Crock, 1987); chemical analysis for Hg by cold-vapor atomic absorption spectrometry; analysts, E.P. Welsch, L.A. Bradley, A.H. Love, and B.H. Roushey; —, not detected]

Figure 12 Conformable steeply dipping, depositional contact between quartzarenite (qtz) of unit 1 and chert (ch), at point of hammer, of unit 2 of Ordovician Valmy Formation. Locality in NE 1/4 sec. 33, T. 33 N., R. 43 E.

Units 2, 3, and 4 apparently do not crop out in the hanging wall of the Oyarbide fault, a relation which, on the one hand, may partly reflect that (1) the entire package of rocks of the Valmy Formation may have had a regional tilt to the east-southeast prior to Tertiary mostly normal faulting along the Oyarbide fault, and that (2) strata of units 2, 3, and 4 did not extend any farther to the northwest at the time of major offsets along the Oyarbide fault. These offsets along the Oyarbide fault are inferred to include a dextral component of slip. Or, on the other hand, colluvium-covered slopes along the northwest-facing scarp of the Oyarbide fault frankly may have obscured recognition of at least some parts of these units in the hanging wall block near the trace of the fault. Nonetheless, the stratigraphic base of the formation is not exposed anywhere in the quadrangles, although the Valmy Formation was at one time considered to be in contact along a low-angle thrust fault with underlying chert, argillite, and some basaltic and volcaniclastic rocks of the Devonian Scott Canyon Formation in the Galena Canyon area of the Battle Mountain Mining District (Roberts, 1964; Theodore and Roberts, 1971). Recent geologic mapping of these rocks suggests that the contacts between the formations are high-angle faults (Doebrich, 1994). The inferred Valmy thrust that comprises the basal contact of the Valmy Formation in the mining district is considered to have had regional-scale east-directed transport along it (fig. 3; see also, Madrid, 1987).

The four units of the Valmy Formation in the three quadrangles (pls. 1–3) consist generally of a diverse assortment of marine siliciclastic and minor volcanic rocks that apparently accumulated in a quiet environment that straddled the continental rise and abyssal ocean floor during the Ordovician (Rogers and others, 1974; Wrucke and others, 1978; Watkins and Browne, 1989). Unit 1 crops out at Treaty Hill, where it is mostly quartzarenite (subunit Ov1q, pl. 1); the unit is mostly quartzarenite and chert

at Lone Tree Hill (subunits Ov1q and Ov1c, pl. 1), and, mostly quartzarenite throughout an undivided part of the unit in the eastern one-half of the hanging wall block of the Oyarbide fault in the North Peak quadrangle (subunit Ov1u, pl. 2). Units 2–4 include significant proportions of chert and shaly argillite and they crop out only in the footwall of the Oyarbide fault, where these units show structurally complex relations with one another. Unit 3 is the most widely exposed unit of the Valmy Formation in the Snow gulch quadrangle (pl. 3).

UNIT 1

Unit 1 of the Valmy Formation crops out in various places, mostly in the hanging wall of the Oyarbide fault. An area less than 0.5 km^2 of generally glassy-appearing, medium-grained gray quartzarenite of the Valmy Formation is present in outcrop in the western half of Treaty Hill (subunit Ov1q, pl. 1). At Treaty Hill, quartzarenite is unconformably overlain by siliceous, mostly chert-pebble coarse sands of the Permian Edna Mountain Formation, and the quartzarenite also is faulted against rocks of the Edna Mountain Formation along a steeply dipping, northwest-striking fault. Rocks of the Valmy Formation are not that well exposed at Treaty Hill, but where bedding attitudes were obtained, these few attitudes suggest that the bulk of the Valmy Formation dips quite steeply. Quartzarenite of the Valmy Formation probably is much more extensive under the Quaternary silt and sand in this general area, because rocks belonging to the formation also crop out at three isolated small hills surrounded by Quaternary deposits several km to the northwest in the general area of the abandoned Ellison Ranch (T.G. Theodore, unpub. data, 1999).

Two subunits of unit 1, mostly chert and mostly quartzarenite (respectively Ov1c and Ov1q, pl. 1), are present at Lone Tree Hill (fig. 13*A*). The chert subunit crops out in the central part of the hill and masses of thick-bedded quartzarenite are present at the north and south ends of the hill. Quartzarenite is structureless between the widely spaced parting surfaces along bedding. Bedding generally dips quite steeply in each of the subunits. The two subunits also are faulted against each other along northeast- and northwest-striking, apparently high-angle normal faults, respectively near the north and south ends of the hill. Stratigraphic relations between the two subunits are not exposed, although the absence of conglomerate of the Battle Formation from the area of outcrop of the quartzarenite subunit suggests that the chert has been faulted downwards relative to quartzarenite and that the masses of quartzarenite at either end of the hill thus must belong to a part of the Valmy Formation that is lower stratigraphically than the chert subunit. It also is possible that the chert subunit instead is here part of unit 2 of the Valmy Formation to be described below.

The actual contact between the Valmy and the Battle Formations is not exposed on Lone Tree Hill, although in several places on the hill, the contact, which must dip shallowly to the west because of its map pattern, can be positioned confidently to less than 0.1 m between outcrops belonging unquestionably to each of the two formations. No apparent tectonic disruption of rocks is present on either side of the unconformity (fig. 13*B*). Nonetheless, in a small area just south of the area of outcrop of the Battle Formation on the hill, rocks of the chert subunit are brecciated and hydrothermally altered possibly during development of a breccia pipe; the extent of this brecciation and hydrothermal alteration is not shown on plate 1 primarily because of the small area of outcrop of the breccia, and the diffuse nature of the breccia with surrounding rocks. In addition, near the south end of the hill, the quartzarenite subunit has been intruded by a narrow, 39–Ma (see section below entitled "Potassium-argon Chronology of Cretaceous and Cenozoic Igneous Activity, Hydrothermal Alteration, and Mineralization") granodiorite porphyry dike that strikes approximately N. 70° W. Quartzarenite has been cut by numerous quartz-plus-minor-pyrite veins, particularly near the highest parts of the hill (fig. 13*C, D*).

Unit 1 of the Valmy Formation also crops out in an areally confined, narrow sliver, apparently bounded by two, northeast-striking high-angle faults near the south end of the Havallah Hills; this sliver is approximately 2.5 km west-southwest of the Eight South deposit (pl. 2). In this area, rocks of the Valmy Formation consist of bold outcrops of massive, thick-bedded gray quartzarenite probably dipping to the east. The quartzarenite is recrystallized due to emplacement of nearby Tertiary or Cretaceous monzogranite, but the quartzarenite nonetheless includes, hand-specimen scale, wispy zones of dark gray quartzarenite resulting from local enhanced concentrations of organic material or solid carbon that are common throughout the Valmy Formation. In addition, this relatively isolated occurrence of the Valmy Formation in the Havallah Hills, may owe its immediate tectonic relations to drag in part during emplacement of the nearby Tertiary or Cretaceous monzogranite, and in part due to offsets along the faults bounding the unit. These faults must have been active penecontemporaneous with major offsets along the nearby Cottonwood Creek fault (pl. 2). In addition, presence of these rocks of the Valmy Formation suggests that the Golconda thrust and its footwall comprised of rocks belonging to the Antler sequence may also be relatively close to the surface in this general area.

Unit 1, mostly quartzarenite (Ov1u), moreover crops out widely in the northeastern part of the North Peak quadrangle (pl. 2) and extends about 0.2 km into the adjoining

Snow Gulch quadrangle where it is finally terminated by a steeply dipping, north-striking fault (pl. 3). Some of the best exposures of this unit are present in a small canyon in SW 1/4 sec. 27, T. 33 N., R. 42 E. near the eastern border of the North Peak quadrangle. However, throughout much of the remainder of this area, rocks of unit 1 weather to generally smooth-rounded slopes that are not marked by prominent outcrops (fig. 14*A*). Quartzarenite, nonetheless, is the most resistant lithology in the unit and it is generally thick-bedded, showing in many places 1 to 3 m between parting surfaces of bedding. Distribution of these resistant ledges of quartzarenite suggests that a large part of the unit is made up of lensoid discontinuous masses of quartzarenite that show no systematic variation in bedding thickness either laterally or vertically. The generally massive or structureless character of many of the bold outcrops of quartzarenite is such that in many places it is extremely difficult to measure a strike and dip directly on the outcrops themselves. Miller and Larue (1983) suggested that similar quartzarenite beds in their facies 3 of the Valmy Formation in the Independence Mountains (fig. 1) result from deposition in channels, a conclusion based on lack of thickening-upwards megasequences of bedding there and the abrupt, sharp contacts between quartzarenite and chert-argillite beds. Both of these relations also are present in unit 1 of the Valmy Formation in the North Peak quadrangle. Some quartzarenite appears to be "floating" structurally in the surrounding sequences of mostly chert and argillite.

Exposures along roads and open-cuts of mines indicate much more complexly interbedded sequences of rock in the colluvial-covered areas of unit 1 here than one would infer from slope-wash fragments (fig. 14*B*). These sequences, in places, include interbedded sequences of thin beds of quartzarenite and shale; interbedded yellowish-white sandstone, light gray shale, and black carbonaceous shale and argillite; and gray-to-olive green shale and minor amounts of gray-black chert. In the uppermost bench of the East Hill/UNR Au deposit, sheared interbedded shale and quartzarenite are contorted complexly into tightly appressed, ptygmatic folds showing highly variable axial-plane attitudes and numerous faults suggestive of intense tectonic disruption (fig. 14*C*). In some other places, such as in the NE 1/4 sec. 20, T. 33 N., R. 43 E. approximately 0.5 km north of Mud Spring (pl. 2), thin sequences of black carbonaceous shale apparently localized zones of increased strain and provided sites along which subsequent faulting occurred. At this particular locality, shale is deformed into rod-like structures whose long axes are parallel with the surface trace of the northeast-striking fault there. Nonetheless, overall the unit, quartzarenite is the predominant lithology even in those well-exposed sequences that show a number of rock types present, and it typically makes up about 60 to 70 volume percent of the rocks. Generally,

Figure 13 Relations at Lone Tree Hill. *A*, Lone Tree Hill, at head of arrow, viewed to S. 60° W. from Treaty Hill. Main channel of Humboldt River floodplain in foreground and Buffalo Mountain on skyline; *B*, Undisturbed contact between Middle Pennsylvanian Battle Formation (d b) and Ordovician Valmy Formation (Ov), at base of hammer; *C*, Typical density of quartz veins developed in quartzarenite of Valmy Formation near south end of Treaty Hill; *D*, Closeup of *C*.

bedding attitudes in this part of the Valmy Formation strike northwesterly, and the rocks dip somewhat shallowly to the southwest. Some relatively thin sequences of quartzarenite have been folded into broad open folds showing roughly north-south trends of their hinge lines (fig. 14*D*), although most individual beds of quartzarenite in the unit cannot be traced for great distances beyond their limited outcrops on ridge crests.

A poorly exposed sequence of rock, at one time thought to include some bedded barite, is present in unit 1 near the upper reaches of a northwest-draining gulley in the SW 1/4 sec. 17, T. 33 N., R. 42 E. of the North Peak quadrangle. This barite apparently has been recrystallized to a certain extent, possibly by fluids associated with emplacement of ores at the Top Zone deposit, and the barite also is interlayered with quartzarenite. However, isotopic values of sulfur for the barite are so highly reduced relative to those for other Paleozoic bedded barites elsewhere in the region that the more likely interpretation is that the

barite is epigenetic and not part of bedded strata of the formation (S. Howe, oral commun., 1995).

In all, 16 selected samples of unit 1 were examined petrographically. Most samples of unit 1 are from that part of the Valmy Formation that crops out south-southeast of Mud Spring (pl. 2). The fine- to coarse-grained sands examined are all submature to mature in the classification scheme of Folk (1974) in that they include less than 5 volume percent clay cement and more than 95 volume percent, mostly monocrystalline quartz at those framework-mineral sites that show minimal amounts of recrystallization; most sands are medium-grained. As such, the sands are quartzarenite, using McBride's (1963) sandstone classification, and they are subrounded to rounded with a high sphericity (Powers, 1953). The bimodal character of size distribution of the framework grains in some samples results in a reduction of the depositional fabric from a mature to an immature level. Additional detrital fragments in trace amounts include tourmaline, zircon, white mica, biotite, and chert; some of the tourmaline and

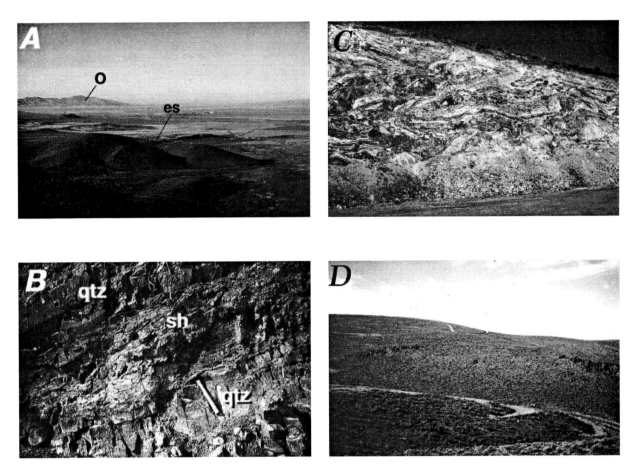

Figure 14 General character of Ordovician Valmy Formation in northeastern part of North Peak quadrangle. *A*, Generally smooth rounded slopes of unit 1 near Mud Spring. O, Osgood Mountains; es, Eight South deposit. Viewed north-northwest; *B*, Interbedded quartzarenite (qtz) and shale (sh) exposed in road cut in SE 1/4 sec. 29, T. 33 N., R. 43 E.; *C*, Complex folds in slump deposits of shale and quartzarenite at uppermost bench East Hill/UNR deposit. Bench face is approximately 8 m high. Viewed west; *D*, View to S. 10° W. along hingeline of broad open anticline in unit 1 present in north-central sec. 32, T, 33 N., R. 43 E. Note roads in central part of photograph for scale.

biotite crystals are included within monocrystalline quartz. Even in those samples affected sparsely by recrystallization, only rare textural indications of partial authigenic overgrowths are present. Various degrees of post-authigenic recrystallization are extremely widespread in quartzarenite in unit 1 and this recrystallization results in a dramatic reduction of overall porosity of the rocks (fig. 15). Much of this recrystallization is reflected also by an increase of complexity of suturing at quartz-quartz grain boundaries and by an accompanying increase in amount of neocrystallization. Eventually, some domains of rocks that show advanced stages of recrystallization contain an abundance of 120° dihedral quartz-quartz boundaries in their fabrics. In addition, deformation lamellae are visible throughout many individual grains, especially in domains of quartzarenite obviously affected by other types of brittle deformation, including undulose extinction and annealed microcracks. Annealed microcracks typically are intergranular; they show increased abundances of fluid inclusions along their traces, of which many fluid inclusions

also exhibit increased abundances of opaque minerals therein; and these microcracks, in places, parallel relatively wide quartz-sulfide mineral-bearing microveins in the rocks. Finally, planar narrow zones of recrystallization course irregularly through some quartzarenite and these zones are commonly various shades of gray in hand sample. All of these secondary effects are especially well developed in areas that show some nearby indications of structurally controlled epigenetic Au mineralization, such as in the general area of the Valmy Au deposits (pl. 2). Although modal analyses of quartzarenite from unit 1, and from other units of the Valmy Formation as well, were not performed for the present investigation, if such analyses had been completed they undoubtedly would provide a strong clustering of data in the interior craton provenance field of Dickinson and others (1983) because of exceptionally high proportions of detrital monocrystalline quartz in the rocks.

Petrographic examination of a limited number of samples of chert and shale from strata of unit 1 reveals that radiolarian chert in the unit is highly variable in its content of detrital components; some chert contains a significant proportion of silt-sized fragments of quartz whereas other chert contains no quartz fragments and only minor amounts of clay. Some essentially unrecrystallized gray-green chert that is interbedded with extensive sequences of quartzarenite shows, at high magnification, abundant filling of fractures by Fe oxide and variable mottling to various shades of reddish-brown due to differing concentrations of Fe oxide.

More in-depth and wide-ranging studies of quartzarenite of the Valmy Formation than mine concluded previously that these rocks are deep-sea fan deposits that formed in a predominantly siliceous graptolitic chert and shale belt during the Ordovician (Churkin, 1974; Wrucke and others, 1978; Miller and Larue, 1983). Most previous workers agree that quartzarenite in the Valmy Formation reflects second-cycle deposits derived from an uplifted sedimentary terrane, and that these deposits accumulated in a pelagic environment based in part on the presence there of depth-diagnostic trace fossils (Roberts, 1964; Gilluly and Gates, 1965; Ketner, 1966; 1977; Miller and Larue, 1983; Madrid, 1987). Careful observation at high magnifications by the author reveal that many quartzarenite samples show presence of chert cement as a major component of the matrix in their otherwise framework-supported fabrics. This relation suggests that some quartzarenite must have been deposited in short pulses in a basinal environment. In addition, exposures of quartzarenite belonging to unit 1 in the area are such that it was not possible to determine a consistent cyclic arrangement of bed thicknesses anywhere and thus discriminate between a middle (thinning upwards cycle) or an outer (thickening upwards cycle) fan environment (see also, Mutti, 1974; Mutti and Ricci Lucchi, 1978). Generally throughout the formation, however, in a few of the better exposed sequences of quartzarenite, upward thinning cycles appear to be more common.

In fact, some controversy still exists as to the ultimate source terrane of the quartzarenite of the Valmy Formation. Kay (1960) suggested that quartz-rich sands streamed westwards from the miogeocline or shelf into the deep-water environment of the eugeocline during the middle Ordovician. Ketner (1966, 1977) and Gilluly and Gates (1965) pointed out that quartzarenite in the Valmy Formation could not represent a spillage from eastern miogeoclinal quartz sands of the same age, such as the Ordovician Eureka Quartzite, because at the time of those studies it was considered that some of the lowest Ordovician strata in the miogeocline were not quartz rich and that quartzarenite in the miogeoclinal facies was apparently better sorted and finer grained than that in the Valmy Formation. Since those studies, however, some early Ordovician quartzarenite has been recognized in the miogeoclinal environments of central Nevada and western Utah (Churkin 1974). Madrid (1987), in a reconstruction of Ordovician paleogeography along the continental margin, suggested a dual source for quartzarenite of the Valmy Formation that involves: (1) an elevated platform somewhere along the northern California-Nevada border consisting of detrital feldspar-rich rocks of the Harmony Formation, and (2) continentally derived sands from areas to the east of the region. Nonetheless, it is not likely that rocks of the Harmony Formation provided a source terrane for quartzarenite of unit 1, nor for other units of the formation, because careful petrographic examination of every quartzarenite sample that was sectioned from this unit for the present study failed to reveal presence of even a single framework grain of K–feldspar or plagioclase. Evidently, quartzarenite sands of the Valmy Formation in the Battle Mountain Mining District were derived largely from continental regions to the east.

Minor-element concentrations were determined for 23 select rock samples, mostly quartzarenite, from unit 1 of the Valmy Formation (table 2). One of the samples (analysis 1, table 2) is from the Treaty Hill area of the formation, three (analyses 2–4) are from Lone Tree Hill, and the remaining 19 samples are from an approximately 2–km long traverse line (A–A', fig. 17). The high silica and low alkali content apparently common to most quartzarenite from the Valmy Formation is indicated by concentrations of 98.3 weight percent SiO_2 and 0.26 weight percent total alkalis in an analysis reported by Roberts (1964). Interpretations of geochemistry of sedimentary rocks, however, inherently involve major uncertainties. Generally, ratios of major elements of sandstone containing more than 5 weight percent Al_2O_3 have been used in the past by some (see also, Middleton, 1962; Pettijohn and others, 1972; Blatt and others, 1980) to categorize sandstones from known provenances into broad clans that, nonetheless, show a significant amount of overlap in their ratios of major oxides. However, the limitations of such data are also discussed by Middleton (1960), Blatt and others (1980), and by many others, who all point out that some ratios important for determination of tectonic provenance of source regions, such as Na/Ca ratio, can be so strongly affected during diagenetic introduction of calcite cement and its associated minor elements that chemistry of sandstones by itself does not provide reliable indications of provenance. These indications are readily available through standard modal studies of framework grains. Furthermore, minor-element chemistry of sandstones has been used by an even smaller number of investigators (see also, Bhatia, 1983; 1985) in attempts to discriminate between active and passive continental margins, but conclusions of these studies have been questioned (McCann, 1991), primarily because of the small number of samples in the data base and because of probability of adsorption of minor elements by phyllosilicate minerals directly pro-

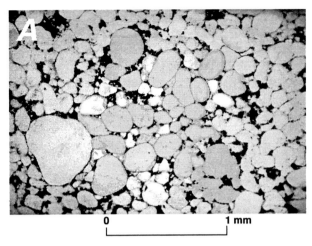

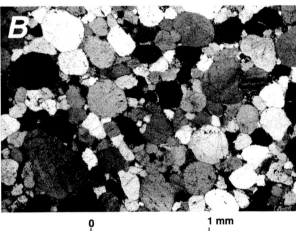

Figure 15 Photomicrographs showing progressive increase in recrystallization in quartzarenite from unit 1 of the Ordovician Valmy Formation. *A*, Essentially unrecrystallized fabric in quartzarenite east of Red Rock gold deposits. Partially crossed nicols. Sample 90TT364. *B*, Moderately recrystallized quartzarenite showing overgrowths and moderate concentration of secondary fluid inclusions along annealed microcracks. Partially crossed nicols. Sample 90TT366. *C*, Intensely recrystallized quartzarenite showing abundant 120° dihedral angles among quartz grains and abundant annealed microcracks. Crossed nicols. Sample 90TT370.

portional to modal amounts of phyllosilicate minerals in the rocks. Nonetheless, minor-element chemistry of select rocks (table 2) of the Valmy Formation, and other rocks described in various subsections below, was determined primarily to provide local geochemical backgrounds for comparisons with epigenetic mineralization here and elsewhere. Although few of the samples (analyses 2–4, and 12, table 2) show abundances of some elements that probably reflect nearby effects of mineralization, the remaining rocks provide a base to which we can compare minor-element signatures of mineralized rocks.

Relations among concentrations of elements in rocks of the Valmy Formation were examined using standard "spider" graph-type plots (fig. 17*A–C*) for nine elements after first grouping the nine elements by geochemical association and normalizing to a selected sample (analysis 23, table 2) that shows relatively low-level concentrations for many elements. In addition, a small number of substitutions, shown paranthetically, using values at 50 percent of the lower limit of sensitivity (Sanford and others, 1993) were made initially for some of the samples that are reported to contain elemental concentrations less than lower limits of sensitivity: Sr (1), V (2), As (2), Pb (3), Zn (1), and Mo (3). Spider-graph plots of four mineralized samples

from Lone Tree Hill and from the immediate area of the Valmy gold deposit (analyses 2–4, and 12, table 2) show some of the highest normalized enrichment patterns for As, Cu, Pb, Zn, and Ba. In addition, the plots reveal somewhat similar enrichment patterns for Ba, Sr, and V for these four samples that differ substantially from patterns of the remaining unmineralized samples of the Valmy Formation (fig. 17*A*, *B*). Such plots provide a graphical representation of elemental enrichments relative to local geochemical thresholds, which here are defined quite well by those samples of the Valmy Formation that delimit a broad, low-level, roughly one-to-ten enrichment pattern across the plots for all nine elements. Furthermore, the four samples with high Ba, Sr, V, As, and Cu do not show enrichments of other base-metal ratios normalized to local thresholds comparable to enrichments present in jasperoids in the Elephant Head area of the Battle Mountain Mining District (Theodore and Jones, 1992). Nonetheless, Ba, As, and Cu, and, of course, Au and Ag are diagnostic parts of the geochemical signatures of both of the Lone Tree and Valmy mineralized systems. Some north-south striking mineralized faults close to the geochemical sampling traverse (fig. 16) have elevated concentrations of Pb and Ba, together with some other elements.

Examination by SEM of a few samples of mineralized fault breccia in the Valmy Formation from SW 1/4 sec. 28, T. 33 N., R. 43 E. approximately 0.5 km northeast of the traverse of figure 16, reveals that fault breccia contains euhedral hydrothermal quartz, barite, jarosite, Fe oxide, galena, and an unknown phosphate- and sulfate-bearing hydrated Fe–Pb oxide mineral. The latter mineral possibly may be a phosphate-bearing alunite, some of which has been reported from the Lepanto Au–Ag deposits in the Philippine Islands (J.W. Hedenquist, oral commun., 1994), and which recently has been documented to include the mineral kintoreite, $PbFe_3(PO_4)_2(OH,H_2O)_6$ as part of the jarosite-alunite family (Pring and others, 1995). Galena is present in the fault breccia as minute, less than 1–μm-wide crystals concentrated within Fe oxide rims that surround euhedral quartz crystals. Barite is the hydrothermal variety, and is generally somewhat smaller, roughly 15– to 30–μm wide, than hydrothermal quartz in the fault breccia. The unknown phosphate-bearing sulfate mineral, presumably alunite, is present in irregularly shaped masses as much as 40 μm in long dimension, and the mineral shows a moderately strong elemental zoning with Fe concentrated in the rim regions of the mineral. Rye and others (1992) have noted that one of the unique characteristics of the acid-sulfate hyrothermal environment is presence of phosphate-bearing alunite in altered rocks (see also, Albino, 1994). These rocks were collected on the fringes of alteration surrounding the Valmy Au deposit, and the reported alunite is compatible with the deposit being part of a high-level acid-sulfate system similar to that at the Eight South deposit (see section below entitled "Geology of the Marigold Mine Area").

UNIT 2

Strata of unit 2 of the Valmy Formation are present only in the footwall of the Oyarbide fault and these strata can be subdivided into three subunits that crop out in an area of about 3 to 4 km² in the east-central part of the North Peak quadrangle: the subunits are mostly chert (Ov2c, pl. 2), mostly quartzarenite (Ov2q), and an undivided subunit that includes a wide variety of rock types (Ov2u). The original stratigraphic relations among the subunits are unknown inasmuch as contacts among them apparently are structural. In addition, the basal contact of unit 2 with underlying quartzarenite of unit 1 is quite sharp (fig. 12), although some minor shearing along the contact may have occurred in response to deformation that primarily includes folding of the rocks of unit 2 and faulting of the rocks of unit 1. As mapped, the basal contact of unit 2 appears to define a broad open syncline because the contact dips at progressively more shallow angles as it is traced to the south (pl. 2). The chert subunit is best exposed along a ridge near the east border of sec. 33, T.33 N., R.43 E. In this area, well-bedded strata of chert show outcrop-scale structures, folds and minor thrust faults, that probably formed at time of emplacement of the Roberts Mountains allochthon during the Antler orogeny (fig. 18A, B). Typically, the minor thrust faults in this general area show a west-to-east sense of tectonic transport; additional details of the geometry and possible kinematics of these deformed

rocks are described below in the subsection entitled "Antler Orogeny Deformation." Attitudes of bedding in unit 2 are generally quite variable, although near the mouth of the North Fork of Trout Creek, strata of the unit show fairly uniform northerly strikes, and they dip to the west-southwest. Farther to the northeast, in an area that extends from the mouth of the North Fork of Trout Creek to the steep slopes leading up to North Peak, strata of unit 2 strike northeasterly and dip to the southeast and they are overlain structurally by rocks, mostly quartzarenite, of unit 3 (fig. 18C).

The chert subunit of unit 2 crops out in an area of approximately 1.5 km² near the east border of the North Peak quadrangle (pl. 2). In this general area, light gray-green chert crops out near the base of the subunit in a conformable depositional relationship with underlying quartzarenite of unit 1 as described above. The light gray-green chert is succeeded upwards by a thick sequence of thin-bedded and highly deformed black cherty argillite that includes thin interbeds of quartzarenite; stratigraphic thickness of this part of the subunit cannot be determined with any degree of confidence primarily because of the folded nature of the rocks and absence of readily mappable marker beds. Nonetheless, black cherty argillite may be as much as 100 to 200 m thick. In turn, the sequence of black cherty argillite is overlain by olive gray-green shale that is well laminated; the olive gray-green shale probably is about 50 to 60 m in stratigraphic thickness. Finally, at the top of the subunit, where it is well exposed approximately 1 km northwest of North Peak, the subunit includes a 30–m-thick sequence of brownish-gray siltstone and black siltstone. The contact of the brownish-gray siltstone and black siltstone with quartzarenite of unit 2 is covered by colluvium, but it can be placed roughly within 3 m of well-exposed rocks on either side of the contact. The map pattern of quartzarenite assigned to unit 3 is highly discordant to attitudes in the chert subunit of unit 2, suggesting that the contact between these two units is structural. Quartzarenite is intensely fractured near the contact, whereas the underlying brownish-gray siltstone and black siltstone is not deformed intensely.

Quartzarenite subunit of unit 2 crops out in three high-angle-fault-bounded blocks; two of these blocks are on the north flank of the North Fork of Trout Creek and the other one also is a high-angle-fault-bounded block, approximately 1.5 km northwest of North Peak (pl. 2). In total area, these blocks probably make up no more than 0.3 km². The petrographic character of the quartzarenite, all in all, is quite similar to quartzarenite of unit 1 described above and will not be repeated here.

The undivided subunit of unit 2 crops out best along a major northeast-trending ridge leading up to North Peak in S1/2, sec. 5, T. 32 N., R. 43 E. (pl. 2). Along this ridge, the undivided subunit includes abundant exposures of dark gray-black argillite plus interbedded gray chert and some

117°07'30"

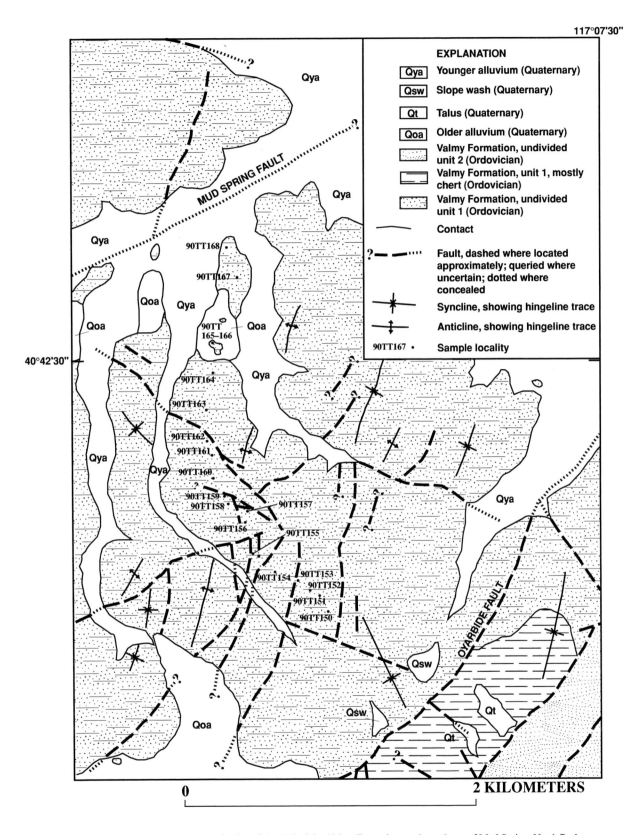

EXPLANATION

Qya	Younger alluvium (Quaternary)
Qsw	Slope wash (Quaternary)
Qt	Talus (Quaternary)
Qoa	Older alluvium (Quaternary)
	Valmy Formation, undivided unit 2 (Ordovician)
	Valmy Formation, unit 1, mostly chert (Ordovician)
	Valmy Formation, undivided unit 1 (Ordovician)

Contact

? Fault, dashed where located approximately; queried where uncertain; dotted where concealed

Syncline, showing hingeline trace

Anticline, showing hingeline trace

90TT167 • Sample locality

MUD SPRING FAULT

OYARBIDE FAULT

40°42'30"

0   2 KILOMETERS

Figure 16 Sample locality map for 19 rocks of unit 1 of the Ordovician Valmy Formation south-southeast of Mud Spring, North Peak quadrangle, analyzed in table 2 (analyses 5–23). Geology modified from plate 2. Explanation for geology same as plate 2.

Table 2—Analytical data on select rock samples of quartzarenite from unit 1 of Ordovician Valmy Formation, Valmy and North Peak quadrangles, Nevada.†

Analysis No.Field no.		Ag	As	Ba	Be	Bi	Ce	Cd	Co	Cr	Cu	Ga	La	Li
		Inductively coupled plasma-atomic emission spectroscopy (total) (parts per million)												
1.	85TT246	N.d.	<10.0	140.0	<1.0	<10.0	5.0	<2.0	<1.0	6.0	3.0	<4.0	3.0	2.0
2.	88TT013	2.0	<200.	500.	<1.	15.	N.d.	<20.	<10.	<10.	30.	<5.	<50.	N.d.
3.	88TT017	5.	200.	2000.	1.	<10.	N.d.	<20.	<10.	50.	70.	10.	<50.	N.d.
4.	85TT254	N.d.	20.	700.	<1.	<10.	<4.	<2.	1.	48.	83.	5.	4.	7.
5.	90TT150	<2.	<10.	84.	<1.	<10.	9.	<2.	<1.	1.	2.	<4.	5.	3.
6.	90TT151	<2.	<10.	140.	<1.	<10.	9.	<2.	<1.	<1.	2.	<4.	4.	4.
7.	90TT152	<2.	<10.	38.	<1.	<10.	9.	<2.	<1.	<1.	4.	<4.	4.	3.
8.	90TT153	<2.	20.	360.	<1.	<10.	9.	<2.	1.	<1.	9.	<4.	5.	5.
9.	90TT154	<2.	<10.	47.	<1.	<10.	8.	<2.	<1.	<1.	1.	<4.	4.	3.
10.	90TT155	<2.	<10.	100.	<1.	<10.	6.	<2.	1.	<1.	2.	<4.	3.	3.
11.	90TT156	<2.	<10.	53.	<1.	<10.	8.	<2.	1.	<1.	2.	<4.	4.	4.
12.	90TT157	<2.	82.	1100.	1.	<10.	35.	<2.	3.	23.	78.	15.	21.	18.
13.	90TT158	<2.	10.	58.	<1.	<10.	14.	<2.	<1.	<1.	5.	<4.	8.	3.
14.	90TT159	<2.	<10.	71.	<1.	<10.	9.	<2.	<1.	<1.	4.	<4.	4.	2.
15.	90TT160	<2.	10.	79.	<1.	<10.	9.	<2.	<1.	1.	3.	<4.	5.	3.
16.	90TT161	<2.	10.	70.	<1.	<10.	9.	<2.	<1.	<1.	5.	<4.	5.	3.
17.	90TT162	<2.	10.	82.	<1.	<10.	11.	<2.	<1.	1.	3.	<4.	6.	6.
18.	90TT163	<2.	<10.	75.	<1.	<10.	5.	<2.	1.	<1.	4.	<4.	3.	4.
19.	90TT164	3.	72.	690.	<1.	<10.	7.	<2.	2.	<1.	14.	<4.	3.	6.
20.	90TT165	<2.	<10.	230.	<1.	<10.	<4.	<2.	<1.	1.	2.	<4.	<2.	120.
21.	90TT166	<2.	32.	400.	<1.	<10.	8.	<2.	2.	1.	13.	<4.	5.	14.
22.	90TT167	<2.	<10.	790.	<1.	<10.	7.	<2.	1.	2.	3.	<4.	5.	2.
23.	90TT168	<2.	<10.	42.	<1.	<10.	9.	<2.	<1.	4.	3.	<4.	5.	3.

Analysis No.Field no.		Mn	Mo	Nb	Nd	Ni	Pb	Sc	Sr	V	Y	Yb	Zn
		Inductively coupled plasma-atomic emission spectroscopy (total) (parts per million)											
1.	85TT246	130.0	<2.0	N.d.	<4.0	<2.0	<4.0	<2.0	6.0	6.0	<2.0	<1.0	<4.0
2.	88TT013	10.	<5.	<20.0	N.d.	<5.	20.	<5.	<10.	15.	<10.	N.d.	<200.
3.	88TT017	70.	<5.	<20.	N.d.	7.	30.	7.	300.	150.	15.	N.d.	<200.
4.	85TT254	85.	<2.	N.d.	6.	4.	7.	<2.	63.	270.	10.	<1.	5.
5.	90TT150	46.	<2.	<4.	<4.	<2.	<4.	<2.	7.	4.	4.	<1.	7.

Table 2—Analytical data on select rock samples of quartzarenite from unit 1 of Ordovician Valmy Formation, Valmy and North Peak quadrangles, Nevada.†

Inductively coupled plasma-atomic emission spectroscopy (total) (parts per million)

Analysis No. Field no.	Mn	Mo	Nb	Nd	Ni	Pb	Sc	Sr	V	Y	Yb	Zn
6. 90TT151	33.	<2.	<4.	<4.	2.	<4.	<2.	16.	5.	5.	<1.	11.
7. 90TT152	9.	<2.	<4.	<4.	<2.	<4.	<2.	4.	3.	3.	<1.	6.
8. 90TT153	4.	<2.	<4.	<4.	6.	<4.	<2.	34.	12.	12.	<1.	18.
9. 90TT154	7.	<2.	<4.	<4.	<2.	<4.	<2.	7.	3.	3.	<1.	6.
10. 90TT155	26.	<2.	<4.	4.	3.	<4.	<2.	8.	7.	7.	<1.	9.
11. 90TT156	62.	<2.	<4.	<4.	10.	<4.	<2.	4.	8.	8.	<1.	36.
12. 90TT157	20.	2.	9.	13.	16.	10.	9.	96.	150.	150.	<1.	35.
13. 90TT158	11.	<2.	<4.	5.	<2.	<4.	2.	6.	7.	7.	<1.	8.
14. 90TT159	15.	<2.	<4.	<4.	<2.	<4.	<2.	17.	5.	5.	<1.	6.
15. 90TT160	20.	<2.	<4.	<4.	3.	<4.	<2.	14.	7.	7.	<1.	7.
16. 90TT161	35.	<2.	<4.	<4.	3.	<4.	<2.	10.	7.	7.	<1.	8.
17. 90TT162	45.	<2.	<4.	5.	4.	<4.	<2.	9.	18.	18.	<1.	11.
18. 90TT163	75.	<2.	<4.	<4.	4.	<4.	<2.	9.	8.	8.	<1.	8.
19. 90TT164	100.	<2.	<4.	<4.	9.	<4.	<2.	20.	19.	19.	<1.	29.
20. 90TT165	55.	<2.	<4.	<4.	<2.	6.	<2.	25.	<2.	<2.	<1.	2.
21. 90TT166	97.	<2.	<4.	4.	4.	9.	<2.	24.	15.	<2.	<1.	23.
22. 90TT167	77.	<2.	<4.	<4.	<4.	<2.	<2.	23.	5.	2.	<1.	16.
23. 90TT168	9.	<2.	<4.	<4.	<2.	<4.	<2.	6.	4.	<2.	<1.	<2.

Inductively coupled plasma-atomic emission spectroscopy (partial) (parts per million)

Analysis No. Field no.	Ag	As	Bi	Cd	Cu	Mo	Pb	Sb	Zn
1. 85TT246	N.d.	N.d.	N.d.	N.d.	N.d.	N.d.	N.d.	N.d.	N.d.
2. 88TT013	0.039	47.	5.2	0.035	9.4	0.2	36.	1.5	15.
3. 88TT017	1.8	210.	1.	.37	160.	1.9	31.	140.	8.2
4. 85TT254	N.d.	N.d.	N.d.	N.d.	N.d.	N.d.	N.d.	N.d.	N.d.
5. 90TT150	<0.067	3.	<0.67	<0.05	2.	.26	1.1	.76	3.8
6. 90TT151	<.067	3.6	<.67	<.05	1.5	.35	.72	<.67	5.9
7. 90TT152	<.067	1.5	<.67	<.05	1.3	.26	<.67	<.67	1.4
8. 90TT153	<.067	22.	<.67	.11	8.8	1.3	1.9	.93	12.
9. 90TT154	<.067	2.4	<.67	<.05	1.8	.24	.71	<.67	1.1
10. 90TT155	<.067	4.3	<.67	<.05	1.5	.4	1.4	<.67	4.1

Table 2—Analytical data on select rock samples of quartzarenite from unit 1 of Ordovician Valmy Formation, Valmy and North Peak quadrangles, Nevada.†

Inductively coupled plasma-atomic emission spectroscopy (partial) (parts per million)

Analysis No.	Field no.	Ag	As	Bi	Cd	Cu	Mo	Pb	Sb	Zn
11.	90TT156	<.067	7.	<.67	<.05	1.6	.73	1.1	.83	26.
12.	90TT157	.32	83.	<.67	.19	70.	3.7	5.8	2.3	22.
13.	90TT158	<.067	7.7	<.67	<.05	3.	.34	1.2	<.67	2.7
14.	90TT159	<.067	6.1	<.67	<.05	2.4	.18	.95	<.67	2.6
15.	90TT160	<.067	14.	<.67	.057	4.9	.38	1.8	<.67	5.1
16.	90TT161	<.067	12.	<.67	.061	2.6	.29	1.2	<.67	5.4
17.	90TT162	<.067	15.	<.67	<.05	3.	.67	3.2	.72	8.2
18.	90TT163	<.067	9.	<.67	.066	4.	.4	.89	<.67	5.9
19.	90TT164	<.067	<.67	<.67	<.05	<.02	<.06	<.67	<.67	<.02
20.	90TT165	.28	.83	<.67	<.05	1.5	.17	5.2	1.	1.7
21.	90TT166	.65	32.	<.67	.29	11.	.41	7.2	7.	19.
22.	90TT167	<.067	6.1	<.67	<.05	1.8	.51	1.4	<.67	14.
23.	90TT168	<.067	1.5	<.67	<.05	1.3	.41	1.1	<.67	.82

Chemical analyses (parts per million)

Analysis No.	Field no.	S	As	Au	Hg	W
1.	85TT246	N.d.	N.d.	N.d.	N.d.	N.d.
2.	88TT013	N.d.	N.d.	N.d.	N.d.	N.d.
3.	88TT017	N.d.	N.d.	N.d.	N.d.	N.d.
4.	85TT254	N.d.	N.d.	N.d.	N.d.	N.d.
5.	90TT150	100.0	3.3	—	0.21	—
6.	90TT151	500.	4.6	—	.04	—
7.	90TT152	<100.	2.6	—	.07	—
8.	90TT153	<100.	22.	—	.17	—
9.	90TT154	<100.	2.1	—	.02	—
10.	90TT155	200.	4.8	—	.04	—
11.	90TT156	<100.	7.8	—	.03	—
12.	90TT157	<100.	80.	<.05	.24	1.0
13.	90TT158	<100.	8.1	—	—	—
14.	90TT159	<100.	5.9	—	.04	—
15.	90TT160	<100.	14.	—	.04	—

Table 2—Analytical data on select rock samples of quartzarenite from unit 1 of Ordovician Valmy Formation, Valmy and North Peak quadrangles, Nevada.†

Chemical analyses (parts per million)

Analysis No.	Field no.	S	As	Au	Hg	W
16.	90TT161	100.	12.	—	.03	—
17.	90TT162	<100.	15.	—	.05	—
18.	90TT163	400.	9.6	—	.02	—
19.	90TT164	200.	68.	—	.05	—
20.	90TT165	400.	2.4	—	.02	—
21.	90TT166	400.	33.	.05	.08	—
22.	90TT167	700.	6.5	—	—	—
23.	90TT168	<100.	2.5	—	.02	—

1. From area of Treaty Hill.

2.–4. From area of Lone Tree Hill.

5.–23. From area south of Mud Springs. Location of field nos. shown on figure 16.

†[Field numbers same as figure 17. Inductively coupled plasma-atomic emission spectroscopy (total and partial) in parts per million by methods of Crock and Lichte (1982), Lichte and others (1987), and Motooka (1988); analysts, P.H. Briggs, D.L. Fey, and J.M. Motooka. Precision for concentration higher than 10 times the detection limit is better than +2-percent relative standard deviation; instrumental precision of the scanning instrument is +2-percent relative standard deviation. Looked for, but not found, at parts-per-million detection levels in parantheses: Au (8), Bi (10), Eu (2), Ho (4), Sn (5), Ta (40), Th (4), U (100). Total S in weight percent by combustion-infra-red method; analyst, N.H. Elsheimer. Chemical analysis for Au by flame atomic absorption spectrophotometry method; chemical analysis for As by graphite furnace atomic absorption spectrometry (Wilson and others, 1987; Aruscavage and Crock, 1987); chemical analysis for Hg by cold-vapor atomic absorption spectrometry; chemical analysis for W by spectrophotometric method of Wilson and others (1987); analysts, E.P. Welsch, B.H. Roushey, N.d., not determined; —, not detected]

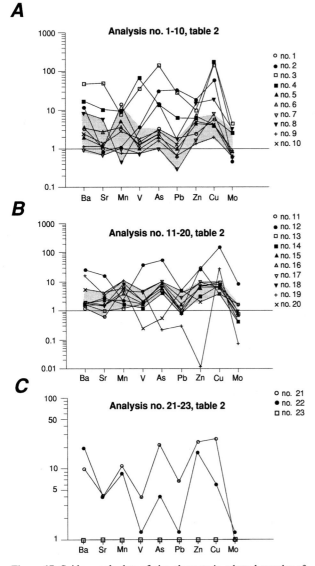

A

Analysis no. 1-10, table 2

○ no. 1
● no. 2
□ no. 3
■ no. 4
▲ no. 5
△ no. 6
▽ no. 7
▼ no. 8
+ no. 9
× no. 10

B

Analysis no. 11-20, table 2

○ no. 11
● no. 12
□ no. 13
■ no. 14
▲ no. 15
△ no. 16
▽ no. 17
▼ no. 18
+ no. 19
× no. 20

C

Analysis no. 21-23, table 2

○ no. 21
● no. 22
□ no. 23

Figure 17 Spider-graph plots of nine elements in selected samples of Ordovician Valmy Formation. Normalized to analysis 23 (table 2; see text). *A*, analyses 1–10 (table 2); *B*, analyses 11–20 (table 2); *C*, analyses 21–22 (table 2).

butt sandstone and quartzarenite. The gray chert weathers to light gray, and it is thinly bedded, showing 2 to 5 cm between parting surfaces along bedding. In places through this sequence of rocks, additional concentrations of light gray to dark gray chert, including well-developed compaction structures, comprise the bulk of the rocks. However, an apparent increase in overall abundance of buff sandstone and quartzarenite is present in the stratigraphically low parts of the subunit; the quartzarenite commonly forms 2– to 4–m–wide resistant ledges along the ridge. Petrographic examination of four samples of quartzarenite revealed that feldspar is not present in any of them as a framework grain.

Unit 3

Unit 3 of the Valmy Formation is the most widely exposed of the four units in the Valmy Formation, and it crops out across a broad, approximately 10– to 12–km² area of the North Peak quadrangle, including North Peak itself (pl. 2), and a roughly similar-sized area in the Snow Gulch quadrangle (pl. 3) . The unit, as described above, is generally equivalent to member 2 of the Valmy Formation of Roberts (1964), and it herein includes four subunits: mostly quartzarenite (Ov3q), mostly greenstone or altered basalt (Ov3g), mostly shale and chert (Ov3sc), and an un-divided part (Ov3u). The undivided part is the subunit that shows the largest area of outcrop. Although age relations are essentially not resolvable among the subunits, the shale and chert subunit makes up one of the highest plates in the vertical tectonic stacking pattern of unit 3 (pl. 2). The base of unit 3, as previously described, is tectonic as is the top, inasmuch as the upper contact is marked by a shallow east-dipping thrust fault that, in places, is generally coplanar with attitude of bedding in unconformably overlying con-glomerate of the Battle Formation that locally caps the tops of ridges nearby. The thrust at the top of unit 3 has been traced northeast approximately 4 to 5 km from the general area of the South Fork of Trout Creek to as far as the east edge of the North Peak quadrangle (pl. 2). To the east in the adjacent Snow Gulch quadrangle, strata of unit 3 make up some of the most widely exposed parts of the Valmy Formation, and, in this area, the thrust bounding the top of the unit is overturned along a strike-length of approximately 3 km such that rocks of unit 3 now rest tectonically on rocks of unit 4 (pl. 3). In addition, strata of unit 3 near Trout Creek, mostly the shale and chert sub-unit, are highly disrupted structurally within about 20 to 30 m of the thrust at the base of unit 4, the next higher structural unit in the stacking order of the Valmy Forma-tion. Rocks assigned to the undivided subunit and some other large masses of quartzarenite of unit 3 are both over-lain unconformably by conglomerate of the Battle Forma-tion near the south-central part of the North Peak quadrangle (pl. 2). In addition, the various subunits of unit 3 are intruded by a small number of narrow granodiorite dikes that are either Tertiary or Cretaceous in age. The rocks of unit 3 are also intruded by Oligocene granodior-ite porphyry at a few places in the Snow Gulch quadrangle (pl. 3).

The shale and chert subunit of unit 3 (Ov3sc) shows some fairly clear structural relations with other subunits of the Valmy Formation. The base of the shale and chert subunit is marked by a shallow-dipping thrust fault that is quite conspicuous on hillsides in NW 1/4 sec. 17, T. 32 N., R. 42 E., primarily because of the smooth, slope-form-ing character of the subunit in the upper plate of the thrust

Figure 18 Various geologic relations of unit 2 of the Ordovician Valmy Formation. *A*, general character of outcrops of chert subunit along approximately north-trending ridge near east-central border of sec. 33, T. 33 N., R. 43 E. View due north from 7,480–ft elevation; *B*, minor thrust, at hammer point, showing west-to-east sense of tectonic transport in chert subunit across a small, upright anticline in footwall of thrust. View to N. 45 W. near area of figure 12; *C*, chert subunit (ch) viewed to southeast from North Peak mineralized area (see text) showing quartzarenite (Q) of unit 3 on skyline.

as exemplified on the northeast-facing slopes of the South Fork of Trout Creek (fig. 19). The actual thrust surface everywhere is covered by colluvium. In this general area, Madrid (1987) found that a sequence of thin-bedded black cherty argillite and black argillite, roughly 30 to 40 m thick, containing abundant boudinage structures crops out at the top of unit 2, just below the thrust at the base of the shale and chert subunit (pl. 2). This distinctive argillite map unit of his, which is not shown on plate 2 because of its small size, follows closely the trace of the thrust at the base of the shale and chert subunit. Black cherty argillite and black argillite probably acted as a zone of preferential slip along which much of the strain attendant with thrusting could then be concentrated.

Figure 19 Intraformational thrust fault in Ordovician Valmy Formation showing shale and chert subunit of subunit Ov3sc of Valmy Formation in thrust contact with underlying undivided subunit Ov3u in NW 1/4 sec. 17, T. 32 N., R. 43 E. Bold outcrops of undivided subunit in right-central part of photograph are mostly quartzarenite. Viewed to S 70° W.

The shale and chert subunit of unit 3, where examined in detail, is lithologically heterogeneous and contains a wide variety of minor rock types, some of which result from post-diagenetic deformational events. In addition, the shale and chert subunit includes a small lenticular body of greenstone in the SW 1/4 sec. 16, T. 32 N., R. 43 E. that also is not shown on plate 2 because of its limited areal extent, but is shown on the large-scale map of the southeast part of the North Peak quadrangle by Madrid (1987). Paleozoic diabase (unit c d) also is present along the trace of the thrust at the base of the shale and chert subunit on the north-facing slope of the South Fork of Trout Creek. In the SW 1/4 sec. 9, T. 32 N., R. 43 E., some excellent exposures of the shale and chert subunit on ridges leading up to the 7,895–ft hill show interbedded chert and quartzarenite with well-exposed, plane-parallel, depositional contacts between these two rock types (fig. 20*A*). Petrographic examination of these quartzarenite interbeds, particularly in the chert strata and others elsewhere (fig. 20*B*), reveals that the quartzarenite is mostly well-sorted, mature chert-cemented sands. More than 95 to 97 volume percent of the quartzarenite contains essentially unstrained, monocrystalline quartz at the framework sites; the remaining 3 to 7 volume percent of the rock is chert cement and minor amounts of detrital chert. No textural evidence is present for overgrowths on quartz framework grains, some of the larger grains show size-specific increases in rounding. Sprinkled throughout the chert matrix are various Fe oxide minerals that have probably replaced diagenetic Fe sulfide minerals, and there also are some chert-filled rhombic outlines preserved in the matrix suggesting replacement of authigenic dolomite by silica. Recrystallization is quite minor overall, however, and is confined to narrow zones that are co-planar with a small number of fractures that cut quartzarenite. Overall sedimentary fabric of these rocks suggests that chert-cemented quartzarenite resulted from a sporadic influx or spillover of extrabasinal quartz sands into a basinal environment in which chert was being deposited. This is the same relation between framework quartz and matrix chert found in unit 2 of the Valmy For-

Figure 20 Interbedded chert and quartzarenite in Ordovician Valmy Formation. Hammer point on quartzarenite. *A*, Shale and chert subunit at 7,654-ft elevation in SW 1/4 sec. 9, T. 32 N., R. 43 E.; *B*, Undivided subunit on ridge at 7,010-ft elevation near SE corner sec. 7, T. 32 N., R. 43 E.

mation and described briefly above. Furthermore, the non-erosive character of basal contacts of quartzarenite with

underlying chert suggests an outer fan or proximal basinal environment according to turbidite facies criteria of Mutti and Ricci Lucchi (1978).

In NW 1/4 sec. 16, T. 32 N., R. 43 E., abundant black-to-gray argillaceous shale is the predominant rock in the chert and shale subunit of unit 3 and this argillaceous shale shows highly contorted bedding surfaces that even become phyllonitic and well lineated near the basal thrust bounding the subunit (pl. 2). Near the basal thrust in this area, clay minerals increase in abundance toward the trace of the thrust, resulting in conversion of black-to-gray argillaceous shale to highly variable shades of light gray. The shale and chert subunit in this general area also includes some thin beds of quartzarenite, mostly less than 1 m thick, that weather as prominent boulders on otherwise smooth slopes of the subunit. These boulders, in places, are quite abundant. One large mass of quartzarenite, nevertheless, is present within the shale and chert subunit in the central part of sec. 16, T. 32 N., R. 43 E., and it is assigned to a quartzarenite subunit (Ov3q, pl. 2).

Elsewhere near the east border of the North Peak quadrangle along the North Fork of Trout Creek, some thin, folded chert-rich strata of unit 3 also include exceptionally—or perhaps more appropriately termed "spectacularly"—well-developed and preserved compaction structures that at first appear similar to systematically arranged cannonballs (fig. 21*A–C*). Many of these compaction structures show a symmetrically developed bulbous part of the structure on the undersides of individual beds when viewed down dip. The overall arrangement of these compaction structures in a more-or-less uniform distribution areally across several square meters of individual chert beds suggests that there may have been some type of prediagenetic differential flow of siliceous ooze toward the now preserved thick parts of the structures. In addition, no apparent development of a tectonic fabric is present in these compaction structures, from which one cannot infer that they reflect poorly developed, post-diagenetic boudinage structures or boudinage structures in their initial stages of formation. Madrid (1987) presents evidence for development of similar structures prior to lithification; the structures he studied in the Shoshone Range are cut by soft-sediment dikes.

Greenstone or altered basalt crops out in four, relatively large masses approximately 1 to 1.5 km east-south-east of the confluence of the North and South Forks of Trout Creek (subunit Ov3g, pl. 2). From description of these rocks in Roberts (1964), Watkins and Browne (1989) suggested that these masses of basalt may be some type of seamount deposits. However, the basalt is relatively thin compared to piles of basalt that normally comprise seamounts on the ocean floor today. Basalt and some highly sheared volcaniclastic sandstones are the most widespread types of rock in the greenstone subunit, which overall is

Figure 21 Folded, chert-rich sequence of shale and chert subunit of unit 3 of Ordovician Valmy Formation that includes exceptionally well-developed compaction structures. Locality in North Fork of Trout Creek near east border of North Peak quadrangle. *A*, Small N. 25° W.-trending, horizontal anticline in thin-bedded chert assigned to shale and chert subunit. Viewed to southeast; *B*, Well-developed compaction structures, apparently confined preferentially to some chert beds of shale and chert subunit; *C*, Closeup of *B*.

commonly weathered deeply. Best exposures of basalt are present near the easternmost parts of the northernmost of the four masses of greenstone mapped north of the North Fork of Trout Creek (pl. 2). Generally, contacts of these rocks with surrounding ones are obscured by colluvium. However, olive-brown to greenish-brown soils on green-

stone are quite distinctive and allow fairly precise placement of contacts even in slope-wash-covered areas. Where not disrupted by some of the minor high- or low-angle faults that cut it, rocks of the greenstone subunit appear to show generally conformable relations with other lithologies in unit 3, including some rocks assigned to the shale and chert subunit and various other rocks included with the undivided subunit. The greenstone or altered basalt subunit also includes a small, roughly 5-m-wide plug of gabbro that was first described by Roberts (1964); this plug is present in the greenstone subunit on the north side of the North Fork of Trout Creek. Gabbro intrudes some of the well-exposed pillow-bearing flows of the subunit (fig. 22). As shown, the bulbous downward tapering and cusped lower contact versus the relatively flat upper contact of many of the individual pillows (Macdonald, 1967), as they are presently oriented, suggests that the sequence dip of the pillow-bearing flows is not inverted. Other minor rocks included with the greenstone subunit are light-gray silty shale, gray-brown quartzarenite, as well as some poorly exposed sequences of thin-bedded gray to olive-green chert and shale. Some gray limestone, also in relatively minor amounts, is present in the body of greenstone that crops out in the bottom of the South Fork of Trout Creek. Greenstone has been intruded in a few places by narrow dikes of Tertiary or Cretaceous granodiorite (pl. 2).

The undivided subunit of unit 3 includes a mixture of most of the rock types described above, but it includes only a minor amount of mappable greenstone (pl. 3). In this subunit, there is no preponderance of a single lithology that can be used as a rock designation for the subunit. Nonetheless, locally some rocks are more abundant than others. For example, in SE 1/4 sec. 8, T. 32 N., R. 43 E., gray-to-black ribbon chert is interbedded with similarly colored cherty argillite over a fairly expansive area of the ridges. Some exposures of these rocks include well-developed compaction structures that impart an overall wavey appearance to bedding surfaces. The compaction structures typically show variable diameters, and they are present in concentrations of as many as 20 to 40 mounds and depressions on approximately 1 m^2 of bedding surface. Thus, these compaction structures are not as large as those shown above in figure 21. In addition, some outcrops of the undivided subunit show thin-bedded chert to be interbedded with somewhat thicker beds of more resistant, massive quartzarenite showing plane-parallel bedding surfaces with surrounding chert (fig. 20*B*).

The undivided subunit in the Snow Gulch quadrangle is a heterogeneous mixture of chert, argillite, shale, and quartzarenite (pl. 3). General presence of quartzarenite in the subunit can be used as diagnostic mapping criteria to separate these rocks from those of unit 4. The overall amount of quartzarenite present in subunit 3 appears to diminish from west to east, particularly in the area west-

Figure 22 Pillow basalts in greenstone subunit belonging to unit 3 of Ordovician Valmy Formation. *A*, General character of pillow basalt flows in outcrop near northernmost contact of the subunit in the SE 1/4 sec. 5, T. 32 N., R. 43 E.; *B*, Closeup of one of pillows in *A*.

to-southwest of Elder Creek. Where black chert strata are well exposed, they define homoclinal sequences showing minimal amounts of tectonic disruption. Individual bedding surfaces of chert commonly show development of compaction structures and parting surfaces, the latter marked by thin seams of gray-to-gray-black argillite. Some argillite seams may be as much as 1 cm wide. The subunit also includes some greenish-gray chert, some of the best exposures of which contain apparently well-preserved radiolarians and are present near the upper bounding thrust in SE 1/4 sec. 3, T. 32 N., R. 43 E. (pl. 3). Some gray shale strata in the subunit are as much as 1 m thick and contain less than 5 volume percent silt-sized detrital fragments. Petrographic examination of 10 thin-sectioned samples of quartzarenite reveals that these rocks are largely mature medium sands whose framework sites are occupied by more than 95 volume percent monocrystalline quartz. Framework feldspar was not noted in any of the thin-sectioned rocks. Chert cement is usually 5 volume percent or less, and, where altered or recrystallized, is replaced by variably complex quartz-quartz sutured grain boundaries. Where such altered rocks also have a "sanded" appear-

ance in hand sample, the rocks in thin section show presence of secondary clay(s) ± white mica along the grain boundaries of detrital quartz grains.

UNIT 4

Unit 4 of the Valmy Formation crops out in an irregularly-shaped area of approximately 3 to 5 km² near the southeast corner of the North Peak quadrangle (pl. 2), and it extends from this area as an elongate tectonically bounded body along the west border of the Snow Gulch quadrangle (pl. 3). As described above, neither the stratigraphic base nor the stratigraphic top of the unit is exposed because these contacts are structural, the top being limited by the Dewitt thrust which juxtaposed rocks of the Harmony and Valmy Formations during the Antler orogeny. Unit 4 also is overlain unconformably by conglomerate and pebbly sandstone of the Battle Formation near the south edge of the North Peak quadrangle, where the Battle Formation is present in a small number of mappable bodies, mostly capping some of the high ridges there. Unit 4 is equivalent to member 3 of the Valmy Formation of Roberts (1964), and unit 4 in this area is mapped as an undivided subunit (Ov4u). However, the most common strata in the subunit are interbedded argillite and chert. The Middle Ordovician age of the unit was established by Roberts (1964) who lists two localities in the unit from which age-specific graptolite collections were obtained. These two localities are: (1) SW 1/4 sec. 21, T. 32 N., R. 43 E., at a 7,050–ft elevation on a small knoll, and (2) NE 1/4 sec. 20, T. 32 N., R. 43 E., at a 6,850–ft elevation on the northeast side of Cottonwood Creek, approximately 70 m above the road. As pointed out by Roberts (1964), age of unit 4 apparently is approximately the same as that of unit 3, although at one time he considered rocks of unit 4 to be part of the Ordovician Comus Formation (Ferguson and others, 1952) rather than the Valmy Formation. Indeed, overall character of strata that compose unit 4 differs from that of the other three units, and many sequences of rock in unit 4 are surprisingly similar in overall lithologic appearance to many sequences of the Devonian Scott Canyon Formation as it crops out in the southeastern part of the Battle Mountain Mining District (Roberts, 1964; Doebrich, 1994). This is especially true with regards to the common presence of thick sequences of black chert, in places intensely deformed, in each of these formations. Additional attempts were made during early stages of the present investigation to obtain biostratigraphically suitable radiolarians from six sites for dating selected black chert sequences of unit 4 in the upper parts of Trout Creek, but these attempts yielded radiolarians far too recrystallized for purposes of age designation (B.L. Murchey, written commun., 1987). Subsequent collections of chert for radiolarians were also made from along almost the entire north-south extent of unit 4 late in 1993, and are described by Murchey (1994). She reports that none of the eight samples examined from unit 4 contain radiolarians that can be used biostratigraphically to establish age of the rocks.

Unit 4 makes up the highest tectonic package of the Valmy Formation, although unit 4 is over-ridden by rocks of unit 3 along an overturned portion of the basal thrust near the west border of the Snow Gulch quadrangle (pl. 3). Typically, colluvium-covered slopes near outcrops are characterized by a silvery-gray sheen mostly resulting from bleached argillaceous chips weathering from topographically subdued outcrops of the unit. Many outcrops near the basal thrust are dark gray cherty argillite, in places highly deformed into tight small-scale folds whose fold axes are parallel to the mapped trace of the basal thrust. Near this thrust, rocks also are intensely Fe–oxide stained to various shades of ochre-brown to orange-brown, and many associated fracture surfaces are filled by milky-white vein quartz that commonly dips steeply, but that also strikes parallel to the trace of the thrust. The strata of unit 4 seem to be more carbonaceous than underlying strata making up subjacent unit 3. Furthermore, many exposed sequences of rock in undivided unit 4 are characterized by wavey, discontinuous parting surfaces along bedding.

The undivided rocks of unit 4 immediately below the Dewitt thrust in approximately N 1/2 sec. 21, T, 32 N., R. 43 E., near the southeast corner of the North Peak quadrangle (pl. 2), show a preponderance of highly deformed dark gray cherty argillite with a minor amount of recrystallized, carbonaceous gray to dark gray quartzarenite. Quartzarenite makes up less than 10 volume percent of rock strata exposed there near the trace of the thrust. Many exposures of unit 4 near the thrust also exhibit numerous outcrop-scale, minor slip surfaces generally parallel to bedding but also including some minor faults that transect bedding at high angles. In addition, narrow milky white quartz veins, generally 2 to 4 mm wide, are relatively abundant and they typically also cut bedding at high angles. Most ridge-forming resistant knobs of unit 4 result from presence of fairly thick sequences of contorted masses of gray-black chert. This chert is commonly thin-bedded and Fe–oxide stained and shows steeply plunging, tightly appressed, outcrop-scale antiforms that trend almost at right angles to the trace of the nearby Dewitt thrust; some chert also is bleached to various shades of light gray. In addition, some sequences of chert show the presence of fairly abundant, low-amplitude compaction structures on their bedding surfaces.

The first exposures of the undivided subunit of unit 4 that show presence of notable concentrations of gray to gray-brown quartzarenite as interbeds in largely thin-bedded, black carbonaceous chert and cherty argillite are present in NW 1/4 sec. 21, T. 32 N., R. 43 E., approximately 600 m northwest of the trace of the Dewitt thrust (pl. 2). These quartzarenite interbeds are internally structureless and commonly show thicknesses in the 1.5– to 3.0–m range; quartzarenite is interbedded conformably with surrounding greenish-gray chert that has 2– to 4–cm intervals between parting surfaces. Some sequences of quartzarenite and chert also are folded into tight chevron folds that plunge shallowly to the south in this area.

Rocks of unit 4 in the Snow Gulch quadrangle are generally slope-forming, and strata of the unit are in many places poorly exposed (pl. 3). Typically, these rocks include relatively thick sequences of bluish-gray-to-silvery argillite, locally as much as hundreds of meters thick and highly contorted, and thin-bedded black chert that in many exposed sequences weathers to light gray or was bleached light gray during subsequent alteration. Rocks of the undivided subunit as a whole show much more outcrop-scale deformation and internal disruption than those of unit 3. However, some exposures of relatively undeformed thick-bedded black chert are present in unit 4, particularly near the trace of the fault at the base of the overthrust rocks belonging to unit 3, which chert shows prominent 8– to 10–cm-wide intervals between parting surfaces along bedding. Near this fault, unit 4 includes some minor green chert interbeds with bioturbated black argillite. Overall amount of quartzarenite in the undivided subunit of unit 4 is quite sparse, although some moderately sorted, medium sands are uncharacteristically quite immature because of high contents of organic-rich cement that includes abundant angular fragments of fine-grained quartz. Some exposures in the undivided unit also include well-developed, penetrative rod-like structures formed by intersection of two sets of cleavage, and other 5– to 8–cm-wide beds of black chert show poorly developed, 1– to 2–cm-wide compaction structures on bedding surfaces, which compaction structures are common in chert of the formation. Close to the trace of the Dewitt thrust, some strata of the subunit apparently have been rotated into parallelism with the general attitude of the plane of the thrust.

Greenstone crops out only in one area as a small body of olive gray-to-grayish green, deeply weathered basaltic rock in NE 1/4 sec. 10, T. 32 N., R. 43 E., where it is cut off by rocks of the Upper Cambrian Harmony Formation in the upper plate of the Dewitt thrust (subunit Ov4g, pl. 3). These rocks are poorly exposed, but they appear, in part, to be fragmental. Contact with adjoining gray-black chert of undivided subunit Ov4u shows presence of shearing. The greenstone subunit also includes some minor beds of gray micrite.

Petrographic examination of six samples including three samples of quartzarenite from the undivided subunit of unit 4 in NW 1/4 sec. 21, T. 32 N., R. 43 E. near the southeast corner of the North Peak quadrangle, failed to reveal presence of any detrital feldspar at framework sites. Much black cherty argillite and dolomitic chert in this area appears to be microfolded at the scale of a thin section and includes locally elevated concentrations of Fe–oxide and quartz as infillings of probable tension gashes that opened during displacements along nearby minor faults and fractures. Carbonaceous chert at the first exposure in the undivided subunit closest to the trace of the Dewitt thrust shows abundant veining by quartz. The samples of quartzarenite examined petrographically are generally framework supported, fine-to-coarse sands that are commonly well sorted, including probably more than 99 volume percent detrital quartz. Most unstrained monocrystalline quartz at framework sites is well rounded, although some domains of the sands show subangular fabrics. Size-specific rounding is fairly common. Some wispy silt-sized domains in the quartzarenites are poorly sorted and not as clean as their fine- to coarse-grained counterparts.

BASALT AND GABBRO

Petrogaphic examination of a limited number of thin sections of disparate volcanogenic rocks, including basalt and gabbro, that make up much of the greenstone subunit of unit 3 of the Ordovician Valmy Formation provides additional insight into petrogenetic development of the seafloor environment these rocks represent. The overall geologic relations of the rocks of this subunit to the surrounding ones have been described in the subsections immediately above. The types of volcanogenic rock examined petrographically include representative samples of fairly fresh basaltic submarine flows, the previously mentioned 5–m-wide plug of gabbro that intrudes flows in the northernmost of the four mapped masses of greenstone in the subunit as it crops out in the North Peak quadrangle (pl. 2), and intensely altered fragmental flow rocks. Fine-grained vesicular basaltic flows mostly include a fabric of 0.15–mm-long microlites of plagioclase, either sodic oligoclase or albite, tightly intergrown with similarly-sized crystals of augite, in places showing spherulite texture (fig. 23*A*). Opaque minerals are ubiquitous. Chlorite is present as a close spatial associate of calcite in narrow veins or as an alteration mineral in groundmass near calcite veins. The size of most of the vesicles, many now filled with chalcedonic quartz or chlorite, is in the 0.25– to 0.40–mm range, which respectively corresponds to ocean depths during Ordovician time of 1 or 2 km according to the ocean depth-versus-vesicle size curves in Wrucke and others (1978). These ocean depths are much less than the 5–km depths inferred by Wrucke and others (1978) from their exami-

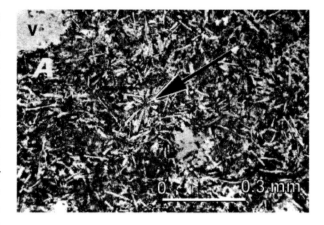

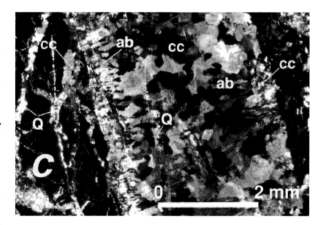

Figure 23 Photomicrographs of rocks from greenstone subunit of unit 3 of Ordovician Valmy Formation. *A*, Basaltic flow showing vesicles (V) and spherulite texture (at head of arrow). Sample 90TT417; *B*, Gabbro showing large phenocrysts of augite (A) partly enclosing tabular crystals of oligoclase-andesine (oa); some chlorite (chl) also in field of view. Partly-crossed nicols. Sample 90TT392. *C*, Fragmental flow rock veined by albite (ab), quartz (Q), and calcite (cc). Partly-crossed nicols. Sample 90TT419.

nation of other samples of basaltic rock of the Valmy Formation collected at approximately the same sites.

Gabbro in the greenstone subunit of unit 3 is quite a bit coarser grained than its associated flows, and it commonly shows optically continuous phenocrysts of augite, as much as 5 to 6 mm wide, enclosing or partly enclosing tabular, euhedral laths of plagioclase (fig. 23*B*). Many plagioclase crystals, oligoclase or sodic andesine, are approximately 1.0– to 1.5–mm long, and extend beyond the outermost borders of the large phenocrysts of augite. The generally enhanced sodic composition of plagioclase in the gabbro, relative to gabbros elsewhere in general, would be reflected in overall elevated abundances of Na in these rocks if chemical analyses of these rocks were available. Interstices among plagioclase crystals, which are themselves generally clouded by dispersed fine-grained white mica and clay, are filled typically by pale-green chlorite.

The intensely altered fragmental flow rocks in the greenstone subunit include hypocrystalline brown fragments essentially surrounded by complexly interwoven veins that include highly variable amounts of albite, quartz, calcite, and chlorite (fig. 23*C*). Most quartz in the veins is relatively coarse-crystalline, shows complexly sutured intravein grain boundaries, and, in places, appears to be growing from earlier stage, finely-crystalline, cross-fiber-textured chalcedonic quartz. This latter type of quartz, actually in quite sparse concentrations in the fragmental flow rocks, may be relict exhalative silica associated with submarine basaltic volcanism. If these interpretations are more or less correct for the vein minerals, then the likewise coarsely crystalline albite, showing tabular outlines in many domains, may have formed from a previously crystallized Na–rich zeolite such as mordenite, which has been reported as being present in finely-crystalline forms in some modern sea-floor spreading centers (R.A. Zierenberg, oral commun., 1992). A number of the walls of the veins are marked by coronas defined by prismatic crystals, possibly amphibole, growing from the brown hypocrystalline fragments that have been engulfed by the veins. These relations suggest that the coronas result from a chemical reaction of vein-related, silica-saturated fluids with essentially silica-depleted fragments. Vein calcite seems to be of at least two paragenetic ages in the veins. Some early-stage calcite is tightly intergrown with albite, quartz, and chlorite, whereas late-stage calcite is present in nearly monomineralic veins that cut diagonally across the fabric of many of the paragenetically earlier veins.

Chemical analyses of select samples of greenstone in unit 3 of the Valmy Formation were not attempted because all samples initially collected specifically for such chemical studies contain widespread concentrations of many secondary minerals, including abundant secondary calcite.

SCOTT CANYON FORMATION (DEVONIAN)

The Scott Canyon Formation crops out only in an approximately 0.6–km^2 fault-bounded block near the southwest corner of the Snow Gulch quadrangle, and was separated from the surrounding rocks of the Valmy Formation primarily on the basis of contained Devonian-age radiolarians described by Murchey (1994) at a locality in NW 1/4 sec. 22, T. 32 N., R. 43 E. (pl. 3). The Scott Canyon Formation is much more widely exposed in the southern part on the Battle Mountain Mining District (Roberts, 1964; Doebrich, 1994). Rocks assigned to the Scott Canyon Formation in the Snow Gulch quadrangle are mostly black chert and argillite that are poorly exposed throughout much of the map area of the unit. In addition, they are intensely recrystallized where black chert and argillite of the unit comprise a window through the tectonically overlying Upper Cambrian Harmony Formation in NE 1/4 sec. 22, T. 32, R. 43 E. The rocks exposed in this window are deformed highly and cut by abundant discontinuous veins of creamy white quartz.

Strata of the overlap assemblage

BATTLE FORMATION (PENNSYLVANIAN)

Rocks of the Middle Pennsylvanian Battle Formation, lowermost strata of the Antler sequence (Roberts, 1964) of the overlap assemblage, crop out mostly as small erosional remnants in four places: (1) in a single body at Lone Tree Hill; (2) across a broad area near the trace of the Golconda thrust to the south-southwest of the Eight South deposit; (3) near the south edge of the area, east-northeast of Trenton Canyon, where rocks of the formation are more widely exposed than in either of the preceding two areas (pls. 1, 2); and, finally, (4) near the southeast corner of the Snow Gulch quadrangle (pl. 3). In the first two areas, rocks of the Battle Formation rest unconformably only on the Valmy Formation and, in the third area, the rocks of the formation are unconformably resting on either the Valmy or the Harmony Formations. In the fourth area, rocks of the Battle Formation are in depositional contact with underlying rocks of the Harmony Formation only along a short strike length of contact (pl. 3). The Battle Formation also is present at depth below Tertiary and Quaternary gravels in some down-dropped fault blocks in the Lone Tree deposit (B.L. Braginton, oral commun., 1992; Bloomstein and others, 1993), where conglomeratic rocks of the formation apparently also have been deposited unconformably on quartzarenite of the Valmy Formation. The Battle Formation also is probably present at the Eight South deposit where it makes up a basal conglomerate and coarse hematitic sandstone of the Antler sequence (unit Pab of Graney and McGibbon, 1991; see also, section below entitled "Geology of the Marigold Mine Area").

The rocks of the Battle Formation near the south edge of the study area, that is south of the Oyarbide fault, are parts of a highly disjointed, west-dipping limb used to define a major post-Permian fold in the Battle Mountain Mining District, the Antler anticline (Roberts, 1964; Madrid, 1987; Madrid and Roberts, 1992). The northwest-trending hingeline of this anticline is inferred to pass through the quadrangles somewhere in the general area of the South Fork of Trout Creek (pl. 2). The rocks of the Battle Formation near the southeast corner of the Snow Gulch quadrangle are east dipping and make up part of the eastern limb of the Antler anticline.

The Middle Pennsylvanian age (Atokan Epoch) of the Battle Formation in the Battle Mountain Mining District initially was established by Roberts (1964) on the basis of an age designation by L.G. Henbest for a fossiliferous lens of limestone in conglomerate belonging to the formation at its type section, approximately 0.8 km east of Antler Peak. Discovery of additional *Profusulinella* sp. foraminifera from this same lens of limestone by George Verville suggested to him that the rocks belong to the upper Lower Atokan Epoch (Saller and Dickinson, 1982). Supplementary age-diagnostic fossils were obtained from rocks of the Battle Formation in the North Peak quadrangle during the present investigation—a sandy fossil packstone thin-sectioned a number of times (sample no. 90TT382; lat 40°41'43.29"N and long 117°10'32.65"W) where it crops out in N 1/2 sec. 31, T.33 N., R. 43 E. contains abundant macrofauna bioclasts of mollusca, ostracodes, fusulinids, and echinoderms (A.K. Armstrong, written commun., 1992). This particular sample was subsequently examined by Bernard L. Mamet (written commun., 1992) who reported that the sample shows "extensive leaching and includes abundant reworked, chertified fragments [with] the following microfossil list:

Apterinellids
Biseriella sp.
Calcivertella sp.
Cuneiphycus sp.
Endothyra sp.
Eoschubertella sp.
Eostaffella sp.
Globivalvulina sp.
Millerella sp.
cf *Profusulinella ?* sp. (primitive?)
Priscella sp.
Pseudoendothyra sp.
Pseudostaffella sp.
Tetrataxis sp. "

The age of above examined sample of the Battle Formation belongs also to the Early to Middle Atokan Epoch, or Baschkirian in the international scale, and such an age corresponds to foraminiferal Mamet Zone 21 or possibly Mamet Zone 22 (Mamet and others, 1965; Mamet and others, 1971), if the presence of *Profusulinella ?* sp. at the locality is confirmed by subsequent investigations (B.L. Mamet, written commun., 1992). Thus, at least two widely separate localities in strata of the Battle Formation in the mining district are now (1999) known for which a Middle Pennsylvanian age has been established.

Rocks of the Battle Formation also are present in some other nearby mountain ranges. In the general area of Edna Mountain, approximately 20 km to the northwest of the study area, rocks of the Battle Formation have been assigned by J.W. Huddle to the Lower Pennsylvanian on the basis of a collection of conodonts there (Erickson and Marsh, 1974). In that same area, however, some sequences of the Middle and Early Pennsylvanian Highway Limestone are approximately the same age as the Battle Formation in the Battle Mountain Mining District. Thus, sequences of the Highway Limestone probably represent a marine offshore facies equivalent to similar-aged rocks of the Battle Formation in these two areas. To southeast of the study area in the Shoshone Range, Gilluly and Gates (1965) refer to rocks lithologically similar and stratigraphically akin to those of the Battle Formation in the Battle Mountain Mining District as the Battle Conglomerate in order to emphasize the highly conglomeratic makeup of the limited exposures of the formation there. In a regional study of the Battle Formation, Saller and Dickinson (1982) concluded from paleocurrent data and stratigraphic relations that the largely alluvial facies of the Battle Formation represents paleotributaries of a broad, single paleovalley that drained to the southwest from the Antler highlands. An intriguing postulate may be that regional distribution of some of these conglomeratic rocks in the lowermost part of the Antler sequence may have been controlled partly by the angle of the initial decollement and development of "piggyback" basins within the allochthonous terrane prior to eventual accumulation of the conglomeratic rocks at the foreland terminus, if sedimentation were similar to that suggested for the late Mesozoic fold and thrust belt in Utah, Wyoming, and Idaho (Lawton and others, 1994).

Lithologically varied conglomeratic rocks of the Battle Formation and especially its basal contact with underlying lower Paleozoic rocks are well exposed in several places in the Valmy and North Peak quadrangles as described above. Although Roberts (1964) described three units with a combined thickness of about 220 m as comprising the Battle Formation in the Battle Mountain Mining District, in the present study rocks of this formation are undivided; however, in the three quadrangles, rocks of the formation probably are all equivalent to the lower unit as defined by Roberts (1964). However, some sedimentary facies of the formation are quite similar to rocks of the middle unit in the Copper Canyon area of the mining district. In the east-central part of the mining district, Theodore and others (1992) mapped and described the lower and middle units of the Battle Formation where they crop out north-northwest of the Surprise Mine; the upper unit of the formation apparently is not present in that area. At Lone Tree Hill, quartzarenite cobbles and pebbles of the Valmy Formation make up the bulk of a tightly compact conglomerate, at most approximately 50 m thick, belonging to the Battle Formation (fig. 24A). The map pattern of the conglomerate at Lone Tree Hill suggests that the basal contact of the Battle Formation with underlying chert of the Valmy Formation dips at a shallow angle to the west. Moreover, angularity of quartzarenite fragments in these rocks of the Battle Formation suggests derivation from a relatively nearby source of Valmy Formation. In addition, rocks of the Battle Formation at Lone Tree Hill are notably depleted in detrital Fe oxide compared to generally red-bed-like character of the formation, particularly its lowermost strata, at its type section and elsewhere in the mining district (Roberts, 1964). However, basal rocks of the Battle Formation near the south edge of the study area contain a much higher proportion of detrital hematitic material in pebbly coarse sands than stratigraphically equivalent rocks at Lone Tree Hill (fig. 24B). Thus, rocks of the Battle Formation in the southern part of the study area are somewhat alike those rocks that crop out at its type section near Antler Peak and in the Copper Canyon and Galena Canyon areas of the mining district (Roberts, 1964; Theodore and Blake, 1978). A marked angular discordance of pebbly conglomerate of the Battle Formation with underlying chert of the Valmy Formation is especially well preserved in a small sliver of the Battle Formation that crops out in SE 1/4 sec. 12, T. 32 N., R. 42 E. (fig. 24C). At this locality, in the footwall of the Oyarbide fault (pl. 2), thin-bedded chert of the undivided subunit of unit 3 of the Valmy Formation dips steeply to the west under a thin erosional remnant of the Battle Formation exposed along the ridgeline. The contact between the two formations is knife-edge and it traces on the outcrop a more-or-less convex-upwards pattern across the underlying beds. Moreover, the unconformity at the base of the Battle Formation is extremely important from a regional perspective because it contributes an important component to overall evidence used to establish that the bulk of the intense internal deformation present in various lower Paleozoic formations making up the upper plate of the Roberts Mountains thrust occurred during the pre-Pennsylvanian Antler orogeny (Roberts and others, 1958; Roberts, 1964).

Additional well-documented evidence for a pre-Battle Formation shortening event is present elsewhere in the region, and one of the places where this evidence is especially well-preserved is in the Osgood Mountains where massive pebble conglomerate of the Battle Formation laps across a number of relatively tight chevron folds formed in upper Proterozoic or Lower Cambrian Osgood Mountains Quartzite (Saller and Dickinson, 1982).

To the immediate south of the locality figured above showing rocks of the Battle Formation resting unconformably on upturned beds of quartzarenite of the Valmy Formation, rocks of the Battle Formation are cut out structurally by overlying rocks of the Havallah sequence present in the upper plate of the Golconda thrust, or possibly even along a high-angle fault that follows essentially the trace of the Golconda thrust (fig. 24D). Ultimately, approximately several hundred meters farther to the south of the site shown in this figure, rocks of the Havallah sequence have been juxtaposed along the fault directly against rocks of the Valmy Formation.

Where roughly 1.5–m-thick beds of the Battle Formation rest directly on rocks of the Harmony Formation near the south edge of the study area in SW 1/4 sec. 21, T. 32 N., R. 43 E. (pl. 2), lithology of the detrital clasts in the Battle Formation is dominated by clasts derived from underlying rocks of the Harmony Formation; cobbles of quartzarenite from the Valmy Formation are extremely rare although angular fragments of chert are relatively more common. Much poorly sorted cobble and boulder conglomerate of the Battle Formation in this general area is drab greenish gray reflecting thereby the overwhelming presence locally of mica- and feldspathic arenite-rich debris derived from underlying rocks of the Harmony Formation. Nonetheless, some sequences of the Battle Formation here include approximately 3–m-thick interbeds of brick-red calcareous sandstone of indeterminate parentage.

Rocks of the Battle Formation in NW 1/4 sec. 20, T. 32 N., R. 43 E. include some strata that are remarkably similar to sequences of the formation assigned by Roberts (1964) and Theodore and Blake (1975) to the middle unit of the formation in the Copper Canyon area of the Battle Mountain Mining District. In this part of sec. 20, brown to red-brown shaly hornfels crops out widely and some thin interbeds of coarsely crystalline white marble are interbedded with well-indurated cobble- to boulder-sized conglomerate including mostly cobbles of green chert and some gray quartzarenite. Matrix of the conglomerate is largely siliceous medium- and coarse-grained sands. Marble appears to be confined to a single thin bed, approximately 1 to 2 m thick, that crops out over a strike length of about 50 m. In places, however, the shaly hornfels shows possibly soft-sediment compaction features that contain a bedding-plane-parallel pinching and swelling of creamy-white calcareous interbeds within brown to red-brown shaly hornfels. In addition, some trenches dug into

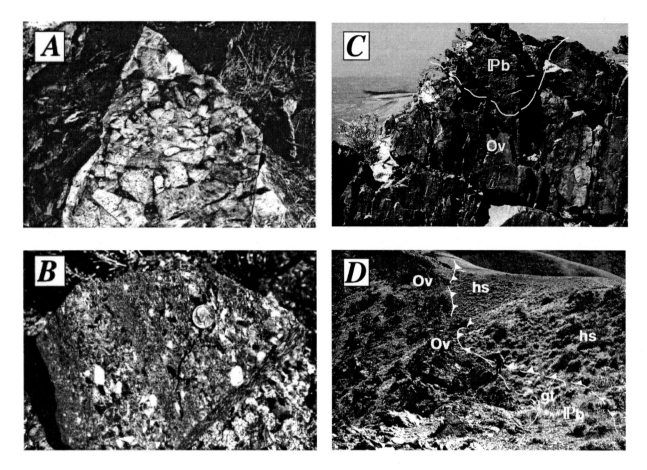

Figure 24 Photographs of the Middle Pennsylvanian Battle Formation. ɑ b, Battle Formation. *A*, Slightly rounded angular, framework-supported cobbles of quartzarenite derived from Ordovician Valmy Formation. Note 10–cm-long pocket knife for scale at lower left. On top of Lone Tree Hill, above locality 90TT177; *B*, Boulder consisting of pebbly, coarse hematitic sands near basal contact of Battle Formation. Locality in SE 1/4 sec. 20, T. 32 N., R. 43 E. close to 91TT059; *C*, Pebbly conglomerate of Battle Formation viewed to north showing a marked angular dis cordance with vertically-dipping chert beds of undivided subunit of unit 3 of Valmy Formation (Ov). Note 10–cm-long pocket knife for scale in central part of photograph. Locality 91TT019 in SE 1/4 sec. 12, T. 32 N., R. 42 E.; *D*, View to south from same locality as *C* showing thin sliver of the Battle Formation tectonically cut out by the Golconda thrust (gt) which juxtaposes Devonian, Mississippian, Pennsylvanian, and Permian rocks of Havallah sequence (hs) and Valmy Formation (Ov). Note person in central part of photograph for scale.

the hillside show presence of thin, but areally persistent, sequences of gray-brown to maroon silty and sandy shale and mudstone. These colors are highly suggestive of sequences of the Battle Formation that have been assigned to the middle unit in the Copper Canyon area of the Battle Mountain Mining District. Some mudstone also is various shades of purple and includes clasts of white marble that are locally mantled by thin reaction rims of green chlorite, resulting from the widespread contact metamorphism these rocks have undergone. Allochems in some of the carbonate beds are barely recognizable because their outlines are blurred by the effects of recrystallization. Although these shaly and silty facies of the Battle Formation in this area could be interpreted as belonging to the middle unit of the formation, it is more likely that they belong to the lower member as previously defined by Roberts (1964) inasmuch as the basal contact of the formation is present at a very shallow depth below the above-described exposures. However, thicknesses of the formations that make up the Ant-

ler sequence are highly variable, which is consistent with their depositional environment. Extremely narrow and intensely altered dikes of apparent biotite quartz monzogranite of unknown age are present in some road cuts in SW 1/4 sec. 20, T. 32 N., R.43 E. (T.J. Thomas, oral commun., 1993). These dikes may be associated temporally with some of the widespread contact metamorphic effects in this general area.

Petrographic examination of a number of rocks from the Battle Formation shows that some are relatively unaltered even though they may be in the general area of weak porphyry-style mineralization, whereas others within the contact aureole of the granodiorite of Trenton Canyon contain epigenetic mineral assemblages indicative of widespread and even locally intense contact metamorphism. Some pebbly calcareous lithic arenite in an area prospected in the past for the presence of a nearby porphyry-copper system and its associated Au–bearing rocks, specifically in the N. 1/2 sec. 31, T. 33 N., R. 43 E., nonetheless contains chert, quartzarenite, monocrystalline quartz, and calcite detrital grains all set in a micrite matrix showing no development of calc-silicate contact metamorphic minerals. The rock sample contains approximately 30 to 35 volume percent detrital fragments of several origins: subangular quartz sands derived from a weathered land mass of predominantly igneous rock, large angular fragments 2 to 6 mm wide of spicule chert derived from basinal bedded chert, and chlorite hornfels derived from a metamorphic terrane (A.K. Armstrong, written commun., 1992). As further pointed out by Armstrong, the carbonate part of the rock is of an open marine, shallow shelf, subtidal origin containing bioclasts of packstone. The bioclasts are in a matrix of pelloid lime mud or micrite composed of 2– to 6–μm-wide calcite crystals that do not show any evidence of neomorphism. The fabric of the rock is typical of limestone that has not been substantially affected by geologic events since diagenetic cementation during the Middle Pennsylvanian. Voids or pores in the pelloidal matrix part of the rock were filled by blocky calcite cement typical of late Paleozoic shallow marine shelf environments (A.K. Armstrong, written commun., 1992). In this general area, the most widespread rock of the Battle Formation is hematitic, unmetamorphosed well-bedded conglomerate that shows 5 to 7 cm between parting surfaces along bedding. In contrast, almost all rocks of the Battle Formation near the granodiorite of Trenton Canyon contain abundant concentrations of diopsidic pyroxene and (or) tremolite-actinolite. Early prograde diopside-quartz-K–feldspar-sphene contact-metamorphic assemblages, in places, show partial replacement by secondary biotite, possibly reflective of porphyry-style K–silicate alteration. The secondary biotite, in turn, has been partially converted to chlorite during a retrograde stage of alteration (fig. 25). Other rocks show crystallization of minor amounts of epi-

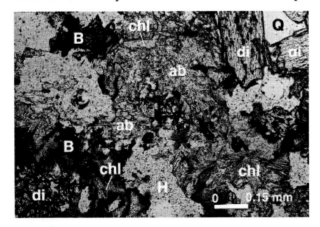

Figure 25 Photomicrograph of rocks of Middle Pennsylvanian Battle Formation showing intensely recrystallized pyroxene hornfels including a prograde diopside (di)-quartz (Q) assemblage partly replaced by secondary biotite (B) that is, in turn, replaced partly by chlorite (chl)-albite (ab) assemblage. H, hole in polished thin section. Sample 90TT425, from SW 1/4 sec. 19, T. 32 N., R. 43 E.

dote to be compatible with chlorite. Thus, the incomplete reactions relict in some of these individual samples provide a record of the entire petrogenetic evolution of the large nearby, magmatic-hydrothermal system at Trenton Canyon that will be described in much greater detail below.

ANTLER PEAK LIMESTONE (PENNSYLVANIAN AND PERMIAN)

The Pennsylvanian and Permian Antler Peak Limestone crops out in two general areas of the North Peak quadrangle: in an area immediate to the Red Rock and Top Zone gold deposits, and in a narrow approximately 1–km-long strip near the south edge of the study area (pl. 2; fig. 26A, B). The Antler Peak Limestone also crops out near the southeast corner of the Snow Gulch quadrangle (pl. 3). Rocks of the Antler Peak Limestone likewise are present in the open pit at the Eight South deposit (see section below entitled "Geology of the Marigold Mine Area").

The Upper Pennsylvanian and Lower Permian age of the Antler Peak Limestone was established by Roberts (1964) from extensive collections of fossils obtained by him and others from several localities in the Battle Mountain Mining District. Recently, approximately 5–cm-wide unusual spiraled tooth crowns of a fish were found by Russell T. Ashinhurst of Battle Mountain Gold Co. in Antler Peak Limestone, near the top of hill 5301 in SE 1/4 sec. 21, T. 32 N., R. 44 E. on the east side of the Battle Mountain Mining District (pl. 3). The locality is close to the stratigraphic top of the Antler Peak Limestone. This specimen subsequently was identified by R.A. Hanger (written commun., 1993, to Russell T. Ashinhurst; see also, Hanger and others, 1994) as tooth crowns of the Edestid shark *Helicoprion* sp., of which one genus was described by Wheeler (1939) from the Triassic Koipato Group. As reported by Wheeler (1939), the holotype of the *Helicoprion nevadensis* he described apparently was collected from a tuffaceous shale approximately 1,000 m below the top of the Triassic Rochester Trachyte of the Triassic Koipato Group in the Humboldt Range, Nev., or roughly from about the middle of the Koipato Group. However, this specimen of *Helicoprion nevadensis* is considered by Silberling (1973) not to have been originally obtained from the Koipato Group in the Humboldt Range because of discrepancies between lithology of rock fragments attached to the fossil and the diagnostic lithologies found in the Koipato Group. *Helicoprion* also has been reported from the Permian Phosphoria Formation (Williams and Dunkle, 1948; Bendix-Almgreen, 1966) and from the Permian Skinner Ranch Formation in west Texas (Kelly and

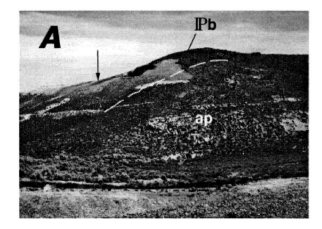

Figure 26 General appearance in outcrop of Upper Pennsylvanian and Lower Permian Antler Peak Limestone (ap) near Top Zone and Red Rock gold deposits. *A*, Block of Antler Peak Limestone resting on Middle Pennsylvanian Battle Formation (ᑯ b) near open cut of workings at Top Zone gold deposit at head of arrow. Viewed to S. 45° E.; *B*, Discontinuous outcrops of Antler Peak Limestone at Red Rock gold deposit. Viewed to southeast.

Zangerl, 1976). Kelly and Zangerl (1976) reported that *Helicoprion* has been used as an index fossil that ranges from the Uralian Stage (Pennsylvanian) through the Artinskian Stage (Permian); its age apparently is thus compatible with the range in age of the Antler Peak Limestone originally reported by Roberts (1964). The strata from which the *Helicoprion* sp. was obtained in the Snow Gulch quadrangle have been established as Wolfcampian in age from a collection of fusulinids examined from near the top of hill 5301 by Verville and others (1986).

The Antler Peak Limestone shows a variety of lithologies in the quadrangles. In the general area of the Red Rock Au deposit, the otherwise more commonly gray-colored Antler Peak Limestone is altered moderately to various shades of buff and orange because of the introduction and oxidation of Fe–sulfide minerals associated with emplacement of Au in the deposit; calcite veins also are quite common near the deposit. Essentially

unmetamorphosed gray micrite of the Antler Peak Limestone near the south edge of the study area is somewhat more thick-bedded than that near the Top Zone and Red Rock deposits, typically showing approximately 1–m thicknesses between parting surfaces of bedding. Approximately 1.5 km southwest of Mud Spring in the north-central part of the North Peak quadrangle, a few relatively good exposures of thin sequences of the Antler Peak Limestone also are present near the nearby projected trace of a covered normal fault that probably now follows the prior trace of the Golconda thrust (pl. 2). In this area, exposures of gray micrite of the Antler Peak Limestone include on their fracture or joint surfaces some sporadic 2– to 4–cm-wide macrofossils that are probably corals. In addition, gray micrite includes some sparse local ovoid clots of chert and sporadic concentrations of vein calcite. Some micrite close to throughgoing fractures has been converted to dolomite. Generally, strata of the Antler Peak Limestone, specifically those strata in basal parts of the formation near the underlying Battle Formation, appear to contain more dolomite and more siliciclastic detrital components than nearby exposed sequences that are stratigraphically high in the Antler Peak Limestone. Finally, overall thickness of Antler Peak Limestone in the western part study area is significantly less than that in the east-central part of the Battle Mountain Mining District where Roberts (1964) reports that as much as 600 m of the Antler Peak Limestone may be present north of the promontory at Elephant Head (see also, Theodore and Jones, 1992). This latter area is in the general area of the Antler Peak Limestone that crops out near the southeast corner of the Snow Gulch quadrangle (pl. 3). As shown on the map of the Snow Gulch quadrangle, the Antler Peak Limestone is divided into a micrite-dominant facies and a siliclastite-dominant facies, together with an undivided subunit which constitutes those parts of the formation that are poorly exposed.

Petrographic examination of six samples of Antler Peak Limestone, two from an area approximately 1.5 km south of the Red Rock deposit and four samples from near the south edge of the North Peak quadrangle, reveals that the rocks are generally quite similar. Rocks from the area south of the Red Rock deposit typically are sparry biomicrites that include numerous cross-cutting veins of calcite and some ovoid allochems. Fragments of macrofossils are abundant throughout the rocks. Rocks of the Antler Peak Limestone from near the southern border of the area commonly include sparry biomicrite and micrite, in places veined by calcite. Some biomicrite also contains as much as three volume percent angular, coarse silt-sized fragments of terrigenous monocrystalline quartz. Many calcite veins show evidence of post-vein strain in the form of a weakly developed strain-slip cleavage that cuts the veins. Other than the apparently metal-barren calcite veins themselves, no other evidence is present in the rocks suggesting circulation of hydrothermal fluids near the trace of the Golconda thrust where it crops out in the southern part of the area.

EDNA MOUNTAIN FORMATION (PERMIAN)

The Permian Edna Mountain Formation, the uppermost formation of the overlap assemblage of Roberts and others (1958) and Roberts (1964), crops out in three areas relatively far removed from one another, and that, in aggregate, span about two-thirds of the entire north-south length of the Valmy and North Peak quadrangles (pls. 1, 2). At Treaty Hill, to the north of the Humboldt River, rocks of the Edna Mountain Formation rest unconformably on quartzarenite of the Valmy Formation (pl. 1). Near Lone Tree Hill, rocks of the Edna Mountain Formation are exposed in the open pits at the Lone Tree and Stonehouse Au deposits; they also are present below the Golconda thrust at the Eight South deposit and at the Old Marigold Mine. In all but the latter of these four occurrences, rocks of the Edna Mountain Formation previously were covered by Tertiary and Quaternary gravels prior to the startup of mining operations (B.L. Braginton and D. H. McGibbon, oral communs., 1992; see sections below titled "Lone Tree Gold Deposit" and "Geology of the Marigold Mine Area"). Rocks of the Edna Mountain Formation in these two Au deposits are described in detail in the sections below, which enumerate various aspects of metallization and the origin of many sequences of rock near the base of the formation as debris flows. Debris flows belonging to the Edna Mountain Formation apparently are regionally extensive because they also apparently are present near the south end of the Shoshone Range, roughly 80 km south of the Battle Mountain Mining District (Moore and Murchey, 1998).

An approximately 1.6–km-elongate area in the north-central part of the North Peak quadrangle along Trout Creek contains the most widely exposed strata belonging to the Edna Mountain Formation (pl. 2). In this area, west-dipping brown platy siltstone and fine sandstone of the Edna Mountain Formation show discontinuous exposures that aggregate approximately 200 to 250 m of stratigraphic section; this area is fairly close to a prior trace of the Golconda thrust which apparently was faulted downwards along the Trout Creek fault (pl. 2). Although strata near the base of the Edna Mountain Formation include prominent debris-flow and well-sorted pebble conglomerate units respectively in the Eight South deposit (see section below entitled "Geology of the Marigold Mine Area") and in the Fortitude deposit (J. Lamana and R.T. Ashinhurst, oral communs., 1993; Doebrich and others, 1995), exposed strata of the formation along Trout Creek are made up of siltstone and fine-grained sandstone. The siltstone and sandstone strata of the Edna Mountain Formation com-

monly include ripple marks, worm trails, and widespread indications of bioturbation indicative of deposition in shallow water conditions. The debris-flow units and pebble conglomerate-bearing strata of the Edna Mountain Formation probably were never deposited in the area along Trout Creek. Along Trout Creek, rocks herein assigned to the Edna Mountain Formation previously were mapped as part of the Pennsylvanian Pumpernickel Formation by Roberts (1964). In addition, rocks previously assigned by Roberts (1964) to the Battle Formation at the Old Marigold Mine are now considered to belong instead to the Edna Mountain Formation based on discovery by D.H. McGibbon of fragments of colonial corals in cobbles undoubtedly derived from the Antler Peak Limestone (see also, section below entitled "Geology of the Marigold Mine Area").

An Upper Permian age, Ochoan, of the Edna Mountain Formation in the Battle Mountain Mining District was assigned originally by Roberts (1964) on the basis of a collection of brachiopods obtained from a limey shale unit that at one time cropped out about 70 m north of the Nevada Mine in the upper reaches of Copper Canyon (see Theodore and Blake, 1975, pl. 1, for the location of the mine). However, this age may be somewhat young in light of recent evaluations of the regional distribution of Permian strata in the Great Basin (B.L. Murchey, oral commun., 1993). As pointed out by Roberts (1964), the Edna Mountain Formation at one time was considered to show only a small number of patchy exposures overall the Battle Mountain Mining District. The brachiopod-bearing locality discovered by Roberts in the southern part of the mining district no longer exists because of encroachment of mining activities associated with the Fortitude Au deposit near the former bottom of Copper Canyon (Doebrich, 1995). Nonetheless, the faunal collection obtained from the general area of the Nevada Mine apparently contained many of the same species of brachiopods found in the type section of the formation at Edna Mountain and described initially by Ferguson and others (1952). However, brachiopods from the locality near the Fortitude Au deposit were not examined and described by a paleontologist specializing in the upper Paleozoic. As mapped by Theodore and Blake (1975, pl. 1), the Edna Mountain Formation in the Copper Canyon area of the mining district included brown calcareous sandstone and fairly well-sorted chert-pebble conglomerate. Some conglomerate belonging to the Edna Mountain Formation is still (1999) preserved at the surface in narrow fault-bounded blocks near the trace of the Virgin fault at Copper Canyon (Theodore and Blake, 1975).

Elsewhere in the region, rocks of the Edna Mountain Formation show geologically significant relations to other Paleozoic formations. Near the type section of the formation at Edna Mountain, calcareous grits of the formation lap across some of the older Paleozoic rocks, including the Antler Peak Limestone, the Middle and Lower Pennsylvanian Highway Limestone, the Battle Formation, and the Upper and Middle Cambrian Preble Formation (Erickson and Marsh, 1974). As further pointed out by Erickson and Marsh (1974), the Edna Mountain Formation in the general area of Edna Mountain is chiefly a brown, coarse, calcareous grit including abundant detrital grains of black chert and locally abundant brachiopod faunas. In the Bruneau River region, Nev., north of Elko and near the border between Nevada and Idaho, Permian heterogeneous conglomerate described by Ketner and others (1993b) may be equivalent to debris-flow units in the Battle Mountain Mining District assigned herein to the Edna Mountain Formation. In addition, the Permian heterogeneous conglomerate in the Bruneau River region contains boulders that include abundant phosphatic material.

During the present investigation, Lower and Middle Permian ages, as indicated by faunal assemblages 8 (Leonardian) and 9 (early Guadalupian, probably Wordian) of Murchey (1990), were obtained by B.L. Murchey (unpub. data, 1994) from five localities examined along the 1.6–km-long exposure of Edna Mountain Formation east of Trout Creek (pl. 2). Demosponge-spicule faunas with distinctive rhax spicules were obtained at these five localities, which are at approximately the same stratigraphic position along the entire exposed strike length of the Edna Mountain Formation (Murchey and others, 1995). As Murchey and others (1995) further point out, although these faunas have a possible age range from Leonardian to Wordian, only Wordian megafossils are now considered to be present in strata assigned to the Edna Mountain Formation. The brown, platy siltstone and fine sandstone of the Edna Mountain Formation are in depositional contact with underlying Antler Peak Limestone in SE 1/4 sec. 19, T. 33 N., R. 42 E. Therefore, the unconformity between these two formations must span all or part of the Leonardian.

Rocks of the Edna Mountain Formation at Treaty Hill are mostly well-sorted coarse grits containing framework-supported grains of drab olive-green and black chert. The grits mostly weather to greenish golden gray-brown to gray-brown, but they also show shades of greenish gray with tints of drab red on fresh surfaces. The rocks of the Edna Mountain Formation at Treaty Hill rest unconformably on quartzarenite of the Valmy Formation, and they are in turn overlain by almost flat-lying Miocene basalt flows that have an approximate 12–Ma age as described in the section below entitled "Potassium-argon Chronology of Cretaceous and Cenozoic Igneous Activity, Hydrothermal Alteration, and Mineralization." It is on the basis of the lithologic similarity of these grits resting on the Valmy Formation with those of the Edna Mountain Formation in SW 1/4 sec. 14, T. 35 N., R. 41 E. (Erickson and Marsh, 1974) near its type locality 20 km to

the northwest that the grits on Treaty Hill are herein assigned to the Edna Mountain Formation. Fossils have not been found in the Edna Mountain Formation on Treaty Hill at the time of this writing (1999). Generally, the shallow-dipping rocks of the Edna Mountain Formation on Treaty Hill show smooth, rounded, blocky-weathering slopes because exposures are fairly sparse except near the cliff faces incised by the nearby Humboldt River (fig. 27A). Where exposed, some beds in the formation are as much as 1.5 m thick, but generally the beds are somewhat thinner than this (fig. 27B).

The rocks of the Edna Mountain Formation on Treaty Hill locally show some signs of epigenetic alteration, especially near the trace of the north-south striking fault in SE 1/4 sec. 31, T. 35 N., R. 43 E. (pl. 1). Near the trace of this fault, sets of quartz veins parallel the strike of the fault, the widths of the veins commonly are approximately 0.5 to 1.0 cm, and they are present in concentrations of about ten veins per 0.5–m-stretch of outcrop. Elsewhere, also near the trace of the fault, the generally siliciclastic rocks of the Edna Mountain Formation show abundant evidence

Figure 27 Character of Permian Edna Mountain Formation at Treaty Hill. *A*, Gentle smooth rounded hillsides of formation along west-facing slopes of Treaty Hill. Outcrops of quartzarenite of Ordovician Valmy Formation at head of arrow. View to north; *B*, Closeup view of typical bed of chert-pebble grit of Edna Mountain Formation.

of recrystallization and introduction of some sulfide minerals, together with minor amounts of epigenetic barite. The effects of this recrystallizaton are such that detrital grains of chert in the grits in essence blend into recrystallized matrix of the rocks. In thin section, the consequence of recrystallization is quite pronounced. Many detrital siliceous fragments show partial replacement by newly grown cross-fiber-textured, presumably chalcedonic quartz, and much of the silica in the groundmass is recrystallized into complexly sutured, feathery-appearing aggregates. These rocks assigned to the Edna Mountain Formation at Treaty Hill contained much less carbonate material prior to introduction of epigenetic silica than the rocks of the Edna Mountain Formation at either its type section to the northwest or that to the southeast in the Copper Canyon area of the Battle Mountain Mining District. Discontinuous stringers of 2– to 5–mm-wide vein quartz, in places with a somewhat more northwesterly strike than the mapped structure, are quite common in many local domains that also show visible signs of recrystallization in outcrop at Treaty Hill.

Although only a small number of samples were collected for geochemical orientation purposes from the Edna Mountain Formation at Treaty Hill, the relatively high-detection levels for some elements, for example 50 ppb Au detection limit, precluded determination of any meaningful background concentrations for those elements. Chemical analyses of six samples of the Edna Mountain Formation, nonetheless, seem to show a low level of introduction of base metals in two of the samples (247 and 252 ppm Zn; analyses 1, 6, table 3). These two samples also show contents of 590 and 940 ppm Ba, respectively. The interesting relation here is that the two Zn–enriched samples are located closer to the trace of the northwest-striking faults than the north-south striking one on Treaty Hill that seems to show the preferred spatial association with widespread effects of recrystallization. In addition, these six samples of the Edna Mountain formation at Treaty Hill contain 0.29 to 0.43 weight percent P, concentrations of P apparently somewhat less than that of the spiculitic brown siltstone of the Edna Mountain Formation that crops out along Trout Creek.

Rocks of the Edna Mountain Formation at the Old Marigold Mine include some thin sequences of dark gray silty shale that directly rest unconformably on quartzarenite of the Valmy Formation near the south end of the open cut. In addition, at this locality some reddish brown conglomerate now (1999) is present mostly in piles of waste rock, that contains abundant cobbles of limey siltstone and limestone. Some of these cobbles include well-preserved corals derived from the Antler Peak Limestone. On the exposed wall of the open cut at the mine just below the trace of the Golconda thrust, also preserved are some 2– to 3–m-thick debris-flow units correlative with much thicker debris flows present in the Eight South deposit (see

section below entitled "Geology of the Marigold Mine Area"). The debris-flow units at the Old Marigold Mine also include round cobble- and boulder-size blocks of quartzarenite derived from the Valmy Formation.

The rocks of the Edna Mountain Formation in the north-central part of the North Peak quadrangle, as well, comprise a narrow, north-south trending, approximately 1.6–km-long block of west-dipping strata described above that is bordered by Trout Creek on its west side, and apparently by another normal fault on the east (pl. 2). Near the north end of this area of outcrop of the Edna Mountain Formation, rocks of the formation are in depositional contact with underlying gray micrite of the Antler Peak Limestone. Farther to the south along east-bounding fault, strata of the Edna Mountain Formation are in structural contact with pebbly conglomerate of the Battle Formation near the south border of sec. 30, T. 33 N., R. 43 E. (pl. 2). The Antler Peak Limestone is missing from this general area either owing to its being structurally cut out or to its lack of having been deposited; the latter may be more likely because where the Antler Peak Limestone is present on the west side of the range, it is much thinner than that on the east side.

The rocks of the Edna Mountain Formation along Trout Creek are quite varied lithologically, and, as described above, their outcrops overall are quite poor, with a few notable exceptions in NW 1/4 sec. 31, T. 33 N., R. 43 E. The formation in this area, approximately 200 to 250 m thick as inferred from discontinuous outcrops, homoclinally dips to the west and includes black to brownish-black and gray to brownish-gray laminated platy siltstone and fine sandstone, mudstone including some silt-sized quartz fragments and some dolomite rhombs, chert-pebble sedimentary breccia, calcareous fine sands that in places are also quite platy, and minor amounts of green shale. Some brownish-black siltite weathers to a drab orange-brown color. The sedimentary breccia shows 8 to 10 cm between parting surfaces along bedding and includes well-sorted, gray-black chert fragments that are on average 2 to 4 mm wide. This sedimentary breccia is quite similar to some rocks of the Edna Mountain Formation in the Copper Canyon area of the Battle Mountain Mining District (Theodore and Blake, 1975). Some siltstone has been contact metamorphosed to a brown biotite metasiltite by emplacement of nearby narrow Tertiary or Cretaceous dikes that show signs of intense phyllic alteration in NW 1/4 sec. 31, T. 33 N., R. 43 E.

Petrographic and electron microscopic examination of brown siltstone from the Edna Mountain Formation where it crops out along Trout Creek reveals that these rocks contain abundant sponge spicules, many of which are detected at high magnifications to be filled by brownish-appearing collophane when viewed under plane-polarized light (fig. 28*A–D*). However, moderate etching by

Table 3—Analytical data on six select samples of Permian Edna Mountain Formation at Treaty Hill, Humboldt County, Nevada.†

Analysis number	1	2	3	4	5	6
Field number 85TT...	247	248	249	250	251	252
Inductively coupled plasma-atomic emission spectroscopy (total) (parts per million)						
As	90.0	50.0	50.0	50.0	40.0	30.0
Ba	590.	260.	990.	970.	490.	940.
Ce	—	6.	9.	11.	14.	5.
Co	1.	—	2.	2.	2.	1.
Cr	62.	57.	52.	71.	48.	50.
Cu	23.	25.	12.	10.	9.	11.
La	5.	6.	12.	14.	16.	6.
Li	8.	6.	12.	11.	21.	11.
Mo	2.	—	—	—	—	—
Nd	7.	7.	9.	13.	12.	9.
Ni	12.	11.	14.	9.	8.	11.
Pb	18.	4.	6.	5.	—	5.
Sr	72.	70.	72.	78.	45.	81.
V	61.	54.	53.	55.	41.	54.
Y	20.	13.	16.	23.	14.	15.
Zn	190.	61.	32.	36.	21.	140.
Chemical analyses (parts per million)						
Ag	<3.0	<3.0	<3.0	<3.0	<3.0	<3.0
As	110.	75.	70.	63.	38.	41.
Au	<.05	<.05	<.05	<.05	<.05	<.05
Sb	5.2	3.4	15.	7.	4.1	2.4
Tl	.27	—	.13	.13	.16	.13

1.-6. Pebbly grit of Permian Edna Mountain Formation containing detrital chert fragments and highly variable amounts of iron oxide replacing previously introduced pyrite.

†[Inductively coupled plasma-atomic emission spectroscopy (total) in parts per million by methods of Crock and Lichte (1982), Lichte and others (1987), and Motooka (1988); analysts, P.H. Briggs. Precision for concentration higher than 10 times the detection limit is better than +10-percent relative standard deviation; instrumental precision of the scanning instrument is +2-percent relative standard deviation. Looked for, but not found, at parts-per-million detection levels in parentheses: Au (8), Be (1), Bi (10), Cd (2), Eu (2), Ga (4), Ho (4), Sc (2), Sn (20), Ta (40), Th (4), U (100), Yb (1). Chemical analysis for Au by combined chemical separation-atomic absorption method involving heating 10-g sample to 700° C to remove sulfides and then decomposing sample in HBr and Br_2. Au is extracted into methyl isobutyl Ketone, and organic layer analyzed for Au by graphite furnace-atomic absorption spectrometry (Meier, 1980); analysts, unknown, K. Kennedy, J.G. Crock. Chemical analysis for Ag by HNO_3 digestion; analysts, unknown, K. Kennedy, J.G. Crock. Chemical analysis for As, Sb, and Tl by graphite furnace atomic absorption spectrometry (Wilson and others, 1987; Aruscavage and Crock, 1987); —, not detected]

hydrofloric acid removes much of the fine-grained siliceous material in these rocks, and assists observation by SEM of the abundant sponge spicules that are present in many of them (fig. 28*E–F*). Much of the matrix among silt-sized fragments in the siltstone is quite siliceous, and, locally, the matrix is even cherty. Extensive chemical analyses of rocks of the Edna Mountain Formation in the general area of the Eight South deposit show that these rocks are quite phosphatic, in places containing as much as 0.9 weight percent phosphorous (D.H. McGibbon, oral commun., 1993; Murchey and others, 1995). The regional implications of these phosphorous-bearing spiculitic rocks

and how they provide a tie between the overlap assemblage of the Roberts Mountains allochthon, the edge of the continent in upper Paleozoic time, and rocks of the partly coeval Havallah sequence are discussed by Murchey and others (1994).

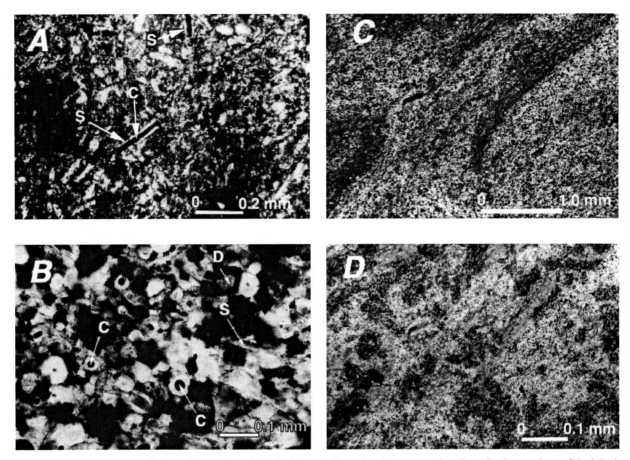

Figure 28 Spiculitic character of brown siltstone of Permian Edna Mountain Formation that crops out along Trout Creek, central part of North Peak quadrangle. S, spiculite. *A*, Spiculites showing collophane-filled cores (C) in weakly calcareous siltstone. Locality 91TT046, plane-polarized light; *B*, Spiculites in local cherty matrix of siltstone also showing presence of relict outlines of rhombic dolomite (D) together with collophane-filled cores (C) of spiculite (S). Locality 91TT064, plane-polarized light; *C*, General appearance of wavy, discontinous character of some bedding traces in siltstone at high magnification. Dark areas include high proportion of organic material and dolomite. Locality 91TT066, plane-polarized light; *D*, Close-up view of *C*.

Strata of the upper plate of the Golconda thrust

HAVALLAH SEQUENCE (DEVONIAN TO PERMIAN)

The rocks of the Devonian, Mississippian, Pennsylvanian, and Permian Havallah sequence probably make up about 50 areal percent of the bedrock of the Valmy and North Peak quadrangles in the upper plate of the Golconda thrust (pls. 1, 2). Although rocks of the Havallah sequence are known to contain no rocks older than Mississippian in the Battle Mountain Mining District, the Havallah sequence is designated throughout this report as being as old as Devonian because equivalent rocks elsewhere in Nevada contain Devonian-age sequences (Miller and others, 1984; Tomlinson, 1990; Miller and others, 1992). As described above, the lowest lithotectonic packet of rocks in the Havallah sequence in the Willow Creek area of the Battle

Mountain Mining District belongs to lithotectonic unit 1 (LU–1) (Murchey, 1990) and this unit crops out discontinuously from Willow Creek on the south to just south of the Stonehouse deposit where a small isolated patch of LU–1 is present at the north end of the Havallah Hills (fig. 9). Whereas rocks of LU–1 are quite limited areally in the Valmy and North Peak quadrangles, rocks assigned to lithotectonic unit 2 (LU–2) of Murchey (1990) are widespread throughout the Havallah Hills part of the quadrangles, and elsewhere farther to the south (Doebrich, 1992, 1995). Finally, there is a large block of interbedded chert and calcareous siltstone belonging to LU–2 south of the Oyarbide fault, near the south edge of the study area, that has been intensely contact metamorphosed by emplacement of granodiorite and monzogranite of Trenton Canyon and other related minor dikes and sills (subunit hcs, pl. 2). The unmetamorphosed equivalent of this large block of rocks crops out widely up to approximately 5 km south of the south edge of the study area (Doebrich, 1992, 1995).

Figure 28 (cont.) *E-F,* Scanning electron micrographs at variable magnifications of surface of brown siltstone etched by hydrofloric acid. Locality 91TT046.

LITHOTECTONIC UNIT 1

The lowermost rocks in the Havallah sequence in LU–1 are mostly grayish-brown, Pennsylvanian and Permian argillite and interbedded argillite and chert, mapped separately as two subunits (ha1, ha2), and black cherty shale (hbc) together with their contact metamorphosed equivalents in the general area of Trenton Canyon (pls. 1, 2). Subunit ha1 crops out almost along the entire length of Trout Creek, where it comprises the footwall of the Willow Creek thrust, and subunit ha2 crops out west of Cottonwood Creek, where it is present in an approximately 3–km-long north-trending, fault-bounded block. Subunit ha2 partly makes up the footwall of several north-striking normal faults that contribute to development of north-trending horsts and grabens in this general area west of Cottonwood Creek (pl. 2). Subunits ha1, ha2, and hbc yield radiolarians from a small number of localities that respectively belong to the following of Murchey's (1990) radiolarian biostratigraphic assemblages: (1) biostratigraphic assemblages 8 and 9 (Leonardian to early Guadalupian);

(2) biostratigraphic assemblage 6 and 7 (late Atokan to Wolfcampian); and (3) biostratigraphic assemblage 8 and 9 (Leonardian to early Guadalupian) (B.L. Murchey, written commun., 1993). Thus, some sequences of argillite in ha1 may be younger than those established in argillite and chert of subunit ha2 (B.L. Murchey, unpub. data, 1994). However, subsequent collections of biostratigraphically useful fauna from the currently two separately mapped subunits of argillite and argillite and chert in LU–1 may require additional refinements of the structural positions of these two subunits relative to one another. Some of the youngest sequences of argillite in LU–1 may in part be correlative with some of the Lower Permian reddish brown argillite map unit of the Reese River assemblage of Tomlinson (1990), which crops out in the northern Shoshone Mountains and southern Shoshone Range, Nev. Subunit ha1 is probably correlative with a sequence of mostly argillite that also includes some Chester and early Atokan sequences near the upper reaches of Willow Creek (Doebrich, 1992, 1995).

Leonardian and early Guadalupian black cherty shale of LU–1, subunit hbc, is stratigraphically below subunit ha1 based on relations at the Old Marigold Mine. As described above, subunit ha1 contains radiolarians of approximately the same age as subunit hbc. Argillite subunit ha1 also includes a small number of interlayered masses of basalt that are mapped separately (subunit hb, pl. 2). Black cherty shale of subunit hbc likewise hosts a minor amount of basalt in masses too small to be shown on the geologic map (B. Matlack, oral commun., 1992), and it includes a few 1–m-thick beds of chert-micrite pebble-conglomerate near the lowermost sequences exposed near the Golconda thrust (fig. 29). However, the black cherty shale subunit is not as continuously exposed as subunit ha1 along the entire mapped length of LU–1. Best exposures of black cherty shale are in the footwall of the Oyarbide fault, in secs. 12 and 13, T. 32 N., R. 42 E., generally along cuts in numerous drill roads (pl. 2). In this general area, black cherty shale shows an exposure width of as much as approximately 0.8 km that then narrows dramatically as the subunit is progressively cut out structurally by contact metamorphosed chert and siltstone, subunit hcs, that has been thrust over black cherty shale along the Willow Creek thrust. In fact, black cherty shale is cut out entirely by the Willow Creek thrust close to the southern border of the North Peak quadrangle. Approximately 1.2 km farther to the east, a north-narrowing sliver of presumably subunit ha1 crops out between the trace of the Willow Creek thrust and the trace of the Golconda thrust that, in turn, apparently cuts out both the rocks of LU–1 and the Willow Creek thrust itself (pl. 2). The Willow Creek thrust has been traced by Doebrich (1992, 1995) continuously to the south to the area where the thrust was originally recognized by Miller and others (1982) and Murchey (1990) (fig. 9).

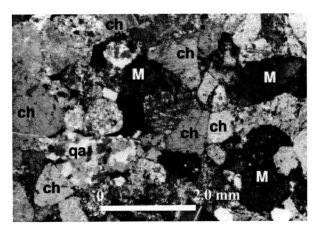

Figure 29 Chert-micrite, pebble-conglomerate in black cherty shale, subunit hbc, belonging to lithotectonic unit 1 of Devonian, Mississippian, Pennsylvanian, and Permian Havallah sequence. M, micrite; ch, chert; qa, quartzarenite; plane-polarized light. Locality 91TT019 in SE 1/4 sec. 12, T. 32 N., R. 42 E.

The northernmost exposure of LU–1 in the quadrangles is present near the north end of the Havallah Hills in a small, approximately 0.15–km-wide area in NE 1/4 sec. 14, T. 34 N., R. 42 E., just south of the Stonehouse deposit (subunit ha1, pl. 1). This area of subdued topography and poor exposures is underlain by argillite that in one outcrop at least strikes northeasterly and dips moderately to the northwest, the attitude of bedding thus striking almost at right angles into the projected contact with adjoining black, ribbon-type chert to the south (pl. 1). By 1998, however, the exposure of argillite was removed by stripping operations associated with mining of the Antler and Sequoia Au deposits (pl. 1). In addition, excellent exposures resulting from these stripping operations suggest that pebbly limestone exposed near the north end of the Havallah Hills (subunit hpls, pl. 1) may instead belong stratigraphically with subunit ha1 below the Willow Creek thrust (K. Kunkle, oral commun., 1998). These pebbly units in LU–1 may be some of the channels that provided entry sources for turbiditic sands deposited in the LU–2 basinal environment. The fairly thick sequence of ribbon-type radiolarian chert south of the argillite, however, is inferred to be Mississippian in age and belonging to LU–2. Argillite of LU–1 is next encountered in the quadrangles about 8 km farther to the southeast just to the north-northwest of the Old Marigold Mine (pl. 2).

In the general area of the Old Marigold Mine, LU–1 includes argillite as well as subunit hbc consisting of recrystallized black cherty shale that appears to be stratigraphically below argillite subunit ha1 (pl. 2). Dark-gray to black recrystallized cherty shale (subunit hbc, pl. 2), including some sequences that appear to be largely chert, also is tectonically the lowest unit mapped in the upper plate of the Golconda thrust in the quadrangles, and its age is inferred to be Leonardian and early Guadalupian on the basis of a microfaunal collection obtained approxi-

mately 0.1 km north of the workings at the Old Marigold Mine (B.L. Murchey, written commun., 1993). Many outcrops of recrystallized black cherty shale in this area are dark brown possibly reflecting oxidation of sulfide minerals associated with introduction of precious metals in the nearby Old Marigold Mine, and the somewhat more distant Eight South and Top Zone deposits.

Leonardian to early Guadalupian siliceous argillite (subunit ha1, pl. 2) is stratigraphically above black cherty shale and is well-exposed in a drainage ditch approximately 0.1 km southwest of the main open cut at the Old Marigold Mine. Here, argillite is drab olive gray-green and shows a preponderance of ochre-colored Fe oxides and yellow jarosite throughout its highly fractured exposures that, in places, are also highly contorted possibly as a result of their position close to the sole of the Golconda thrust.

As one continues farther to the south, subunit ha1, as well as an alluvial-covered short segment of the Willow Creek thrust, are encountered next, where they have both been downfaulted along the Trout Creek fault (pl. 2). From exposures near the Old Marigold Mine, argillite of ha1 can be traced almost continuously for approximately 7 km to the south where it apparently has been folded into a broad, open anticline whose hinge-line trends approximately N. 35° W. This hingeline is located roughly 1.5 km northwest of the site of the abandoned Oyarbide ranch (pl. 2). This is presumably a Mesozoic-age anticline (Theodore, 1991c), and its regional structural associations are discussed below in the subsection entitled "Mesozoic Structures." Subunit ha1 also is well-exposed in most of the minor gulleys near the central part of sec. 19, T. 33 N., R. 43 E., generally along the low hills west of the main, north-flowing drainage through Trout Creek. However, away from gulleys, exposures are extremely difficult to find because the subunit is slope-forming. Near the central part of sec. 19, argillite is various shades of olive gray-green and it includes less than 10 volume percent olive-green interbeds of chert, which typically show 2– to 3–cm intervals between parting surfaces. In many outcrops west of Trout Creek, much of subunit ha1 is well layered, steeply dipping, and includes a strongly-developed lamination with 4 to 5 mm partings that are further enhanced by cleavage that seemingly parallels lithologic layering (fig. 30*A*, *B*). Some outcrops show highly-deformed and crumpled fabrics that are almost phyllonitic in character (fig. 30*C*). Subunit ha1, where it crops out west-northwest of the abandoned Oyarbide ranch, specifically in the W. 1/2 sec. 1, T. 32 N., R. 42 E., also includes a minor apparently interbedded basalt flow and some cross-cutting basalt dikes that presumably must also be Pennsylvanian or Permian in age (subunit hb, pl. 2). These rocks are described below in greater detail in the subsection entitled "Basalt of Lithotectonic Unit 1 of the Havallah Sequence."

Figure 30 Photographs showing general character of Pennsylvanian and Permian argillite (subunit ha1) of lithotectonic unit 1, as it crops out west of Trout Creek in northern part of North Peak quadrangle. *A, B* Steeply dipping chert-bearing argillite at locality 91TT067; *C*, Intensely disrupted layering in argillite at locality 91TT068.

Farther to the west of the abandoned Oyarbide ranch, presumably late Atokan and Wolfcampian argillite and chert (subunit ha2, pl. 2) is present in an elongate horst, elevated tectonically by two approximately north-striking faults, the western of which is mineralized along its south-

ernmost trace (see also, subsection below entitled "Altered Rocks in Area of Section 11 Gold Occurrence"). Subunit ha2 appears to contain a somewhat higher proportion of minor interbedded chert than subunit ha1. The stratigraphic base of subunit ha2 is not exposed, although its base may be tectonic based on relations mapped by Doebrich (1992, 1995) in the northern part of the major drainage through Willow Creek.

A limited number of various lithologies from argillite subunits ha1 and ha2, including volcaniclastic sedimentary rock, sandy limestone, gray-black flinty chert, cherty shale, and olive-green shaly argillite—listed roughly in order of increasing abundance—were examined petrographically. Fragments in volcaniclastic sedimentary rocks include exhalative silica, multicrystalline and monocrystalline plagioclase generally imbedded in Fe– and Mn–oxides, other rounded grains of Fe– and Mn–oxides, and microcrystalline basalt all set in a complexly flow-banded, calc-lithite fabric. Sandy limestone, in places, includes 20 to 30 volume percent mostly monocrystalline quartz that is subangular and is the size of fine sand. Other siliciclastic fragments include K–feldspar and plagioclase. Allochems make up the remaining 70 to 80 volume percent of the sandy limestone and they consist of packed biomicrite now partially recrystallized to an unsorted biosparite. Almost all previously present micrite cement in these rocks has been converted to sparry overgrowths on allochems. Some chert interbedded in argillite subunit ha1 in NE 1/4 sec. 36, T. 33 N., R. 42 E. is extremely free of coarser-than-silt-sized fragments compared to many other sequences of chert in the Havallah sequence (see this subsection below), and it shows sparse multiple generations of vein quartz including early-stage quartz-albite-barite-pyrite veins followed by late-stage, apparently low-temperature, cross-fiber-textured quartz. Some cherty shale contains abundant monaxon sponge spicules and much lesser quantities of other varieties of spicules (B.L. Murchey, unpub. data, 1999) that are as much as 0.09 mm long. The relatively high proportion of sponge spicules in these rocks is consistent with previous observations by Murchey (1990) that rocks of LU–1 accumulated at sea-floor depths generally less that 1,000 m; as pointed out by Murchey, shallow depths favor high proportions of sponge spicules. Some shaly argillite contains a uniform distribution of 1 to 2 volume percent Fe oxide as small equant grains. Each grain of Fe oxide averages 0.09 mm wide, and each grain still retains a framboidal-appearing outline probably relict from diagenetic pyrite. Almost all these rocks in subunits ha1 and ha2 of LU–1 in the quadrangles contain the same suite of heavy minerals as the rocks of LU–2 to be described below. This suite of heavy minerals includes tourmaline, zircon, apatite, and opaque Fe–oxide minerals.

LITHOTECTONIC UNIT 2 PLATES MOSTLY NORTH OF COTTONWOOD CREEK FAULT

Lithotectonic unit 2 (LU–2) can be separated into a small number of tectonostratigraphic plates, a total of five plates in all, that are designated plate A at the structural base through plate E at the structural top of the stacking order (fig. 31). Based on abrupt lithologic intraplate changes (see correlation diagram for plate 1), the plates appear to have only local consequence and should be not considered as having any regional significance. Furthermore, assignments of some subunits to the proper plate is still subject to considerable interpretation, and such assignments should be considered provisional, at best, until additional collections of biostratigraphically diagnostic faunal assemblages become available.

The overall stacking pattern of rocks in LU–2 generally is similar to the stacking pattern of rocks equivalent to LU–2 described by Tomlinson (1990) elsewhere in the region, except that rocks of LU–2 in the Valmy and North Peak quadrangles appear to be imbricated tectonically a number of times causing recognizable sequences in the sedimentary cycles to be repeated several times. Near the base of several described sequences of LU–2 in the Tobin Range, Tomlinson (1990) reports the presence of a Devonian (?) to Permian chert and argillite unit (his DPca unit), which is succeeded upwards by a Carboniferous (upper Upper Mississippian to lower Middle (?) Pennsylvanian) lithic-rich sandy calcarenite unit (his Cls unit), and which, finally, is followed near the uppermost parts of the stacking order by a Permian calcareous quartzarenite (his Pq unit) and a Permian argillite and subordinate chert unit (his Pac unit). The Carboniferous lithic-rich sandy calcarenite of Tomlinson (1990) in the Tobin Range also contains his lithofacies C, which is described by him as brown and black, silty calcareous chert and interbedded argillite, that includes chert derived from a sponge-spicule ooze and some silt beds derived from a silty calcarenite. Parts of these two latter lithologies probably are correlative with separately mapped subunits of interbedded gray micrite and black spicular chert of LU–2 in the Valmy and North Peak quadrangles (subunit hlc, pls. 1, 2). The Havallah sequence in the Reese River area, Nev., as described by Tomlinson (1990), also includes a Pennsylvanian and Permian gray limestone unit—designated DPls by him—that is lithologically equivalent to subunit hlc of plates 1 and 2.

In the area of the Valmy quadrangle (pl. 1), the stratigraphic succession of the rocks of LU–2 has been structurally shortened by folding in an east-west direction primarily during presumably latest Late Permian to Early Triassic emplacement of the Golconda allochthon. The rocks in the Havallah Hills also show their most expansive distributions roughly parallel to the covered trace of the Golconda thrust that is projected to pass just to the east-northeast of the Havallah Hills (pl. 1). This relation partly may be a reflection of a number of broad, open anticlines and synclines that show north-trending hingelines in this area (fig. 31). The base of LU–2 is the Willow Creek thrust, as described above, and, although only a short segment of this thrust is present near the north end of the Havallah Hills (pl. 1), much better exposures of the Willow Creek thrust and its lower plate are present in the North Peak quadrangle (pl. 2). However, in contrast to Mississippian rocks in the hanging wall of the Willow Creek thrust at Willow Creek (Murchey, 1990; Doebrich, 1992, 1995), the upper plate of the Willow Creek thrust in the general area of Trout Creek has Leonardian and Guadalupian basaltic trachyandesite and trachyandesite, together with an overlying sequence of radiolarian chert of the same age in its upper plate (pl. 2). Mississippian rocks, however, are present somewhat higher in the upper plate of the Willow Creek thrust, principally in the upper plate of the successively higher Buffalo thrust, which crops out near the south end of the Havallah Hills (fig. 31). Finally, geologic relations suggest that LU–2 becomes progressively thinner from north to south in the Havallah Hills, possibly as a result of the Willow Creek thrust being emplaced on a surface that had some topographic relief. The entire package of rocks that makes up LU–2 is especially thin south of the Cottonwood Creek fault because plate E crops out within approximately 0.3 km of the trace of the Golconda thrust in the footwall block of the Cottonwood Creek fault.

Tectonostratigraphic Plate A

Tectonostratigraphic plate A includes a structurally and stratigraphically complex assortment of radiolarian ribbon chert, volcaniclastic sands, and basalt—all part of subunit hc (pl. 1)—as well as relatively large volumes of mixed carbonate units and siliciclastic sandstone, siltstone, green shale, and pebbly limestone (subunits hu, hls, hys, hsh, and hpls, pl. 1) . The lithofacies of these rocks provides sedimentary evidence for a fairly typical basin-filling succession during deposition of LU–2, wherein apparently somewhat oxygen-starved radiolarian chert and minor basalt and basaltic andesite flows and sills near the base, in places, are followed upwards by green shale, including some silt and sand, and then finally by clean-washed, largely extra-basinal turbidites made up of fine-grained calcareous sandstone and siltstone. Some pebbly conglomerate is present near the top of the plate. Recognition of locally and laterally chaotic nature of these basin-filling cycles is critical to unraveling stratigraphic and structural complexities in the lithotectonic unit, although absense of well-exposed, extensive vertical se-

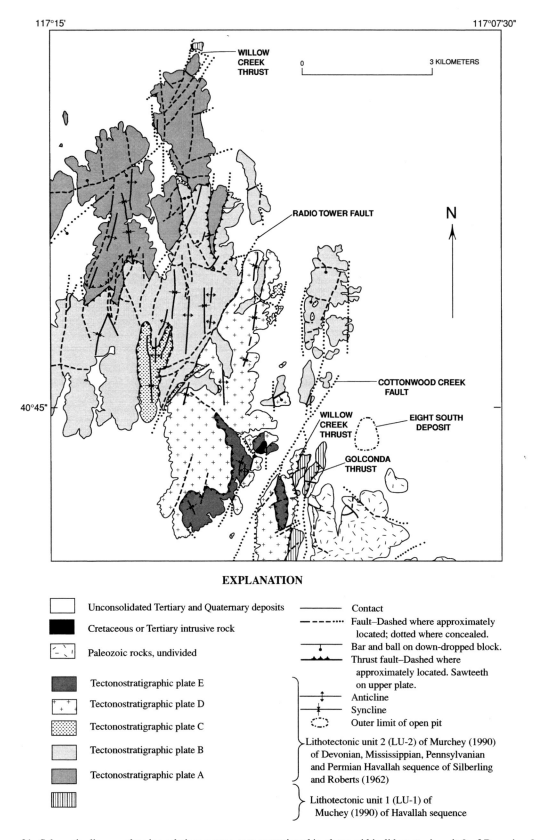

Figure 31 Schematic diagram showing relations among tectonostratigraphic plates within lithotectonic unit 2 of Devonian, Mississippian, Pennsylvanian, and Permian Havallah sequence in Havallah Hills and in major tectonic blocks south of Cottonwood Creek fault. Geology modified from plates 1 and 2.

quences of rock hampers application of modern facies analysis (Rupke, 1978; Lowe, 1982; Anderton, 1985; Stow, 1985; Arribas and Arribas, 1991) to reconstruct confidently details of much of the pre-tectonic paleogeography of the basins that comprise the Havallah sequence.

The lowest subunit of plate A is pebbly limestone that crops out in two small areas near the northern terminus of the Havallah Hills (subunit hpls, pl. 1). It is structurally below chert and limestone, chert, and some of the undivided subunit (respectively subunits hlc, hc, and hu, pl. 1). Stratigraphic relations of the pebbly limestone subunit to other subunits in LU–2 is unknown because pebbly limestone is covered by Quaternary deposits along its northern and western flanks. Where geologic relations of subunit hpls to other subunits can be established, pebbly limestone is present everywhere in the Havallah sequence as part of LU–2 (pls. 1, 2; see also, Doebrich, 1995).

Plate A includes near its base a poorly exposed sequence of Mississippian chert, volcaniclastic sands, minor basalt and basaltic andesite (using the classification scheme of Le Bas and others, 1986), as well as argillite and some minor terrigenous or continentally-derived quartz sands (subunit hc, pl. 1). Subunit hc crops out in an area of approximately 3 km² in the northernmost part of the Havallah Hills (pl. 1). The area of outcrop of subunit hc defines a northeast-trending, 4– to 5–km-long belt that extends from the western border of the Valmy quadrangle to an area close to the northernmost part of the Havallah Hills, just south of the Stonehouse Au deposit. Such a distribution of rocks could be defined quite well on the basis of the presence within the subunit of distinctive black to greenish-black, well-layered, radiolarian chert that includes locally prominent compaction structures. The basal contact of the radiolarian chert subunit, as described above, appears to be a low-angle fault, which places radiolarian chert on top of pebbly limestone (subunit hpls, pl. 1). This low-angle fault is related undoubtedly to imbrication associated with tectonic emplacement of the Golconda allochthon. Overall thickness of the chert subunit cannot be measured directly because of extremely poor exposures and the topographically subdued nature of those few areas where exposures are present.

In places, the chert subunit is followed upwards in an apparent conformable stratigraphic succession by either a subunit of mostly green shale (subunit hsh, pl. 1), also of relatively limited exposure, by yellow-to-orange calcareous sandstone (subunit hys), by some undivided parts of LU–2 (subunit hu), or quite locally even by interlayered gray micrite and black spicular chert (subunit hlc). These geologic relations attest to the complex lithological makeup of LU–2 and apparently rapid changes that affected the environment(s) of deposition in the Havallah sequence during the Devonian (see also,

Tomlinson, 1990; Miller and others, 1992), the Mississippian, the Pennsylvanian, and the Permian periods.

Petrographic examination of nine rock samples from the Mississippian chert subunit in tectonostratigraphic plate A, included four samples from a singular, well-exposed locality along a bulldozer cut near the north end of the Havallah Hills. This examination revealed that although these rocks are, for the most part, heterogeneous lithologically, they nonetheless are dominated throughout their areas of outcrop by well-layered primary radiolarian or spiculitic radiolarian chert that has possibly starved, deep water affinities. In places, some blue-green to black radiolarian ribbon chert includes well-developed argillite-bearing partings along bedding every 2 to 4 cm, and chert is bleached to various shades of light gray where cut by sporadic veins of quartz that show either diffuse or sharp-walled boundaries with enclosing chert. Individual radiolarians commonly measure approximately 0.16 mm in diameter and radiolarians are typically recrystallized to complexly intergrown fabrics with the enclosing fine-grained silica matrix. Elsewhere in the subunit, some spiculitic chert is gray-brown to yellowish-brown in hand sample and shows average grain sizes of about 0.04 mm including complexly reticulated microfabrics. Recrystallization is quite evident in thin section as reduction of grain size of chert is widespread in areas of concentrated secondary calcite and in those areas where masses of cross-fiber textured quartz are present either in barite-bearing veins or in diffuse clots. Furthermore, some spiculitic chert includes ovoid, pale tan in thin section, approximately 0.3–mm-wide fragments of basalt or basaltic andesite undoubtedly derived from nearby flows which also are present locally in the subunit. In addition, significant variability is present in the amounts of terrigenous components in silty chert. Some silty chert includes as much as 20 to 25 volume percent mostly angular fragments of terrigenous monocrystalline quartz showing an average grain size of approximately 0.1 mm with long dimensions of quartz grains typically aligned parallel with traces, in thin section, of low-amplitide, wavey foliation surfaces. Quartz veins in some silty chert also may include sodic oligoclase. Typically, spicules in some spiculitic chert show mud-filled cores; most, however, show no cores to be present.

Some basalt and basaltic andesite, not large enough to show on the geologic map, likewise are present in subunit hc, and are described in detail below in the subsection entitled "Basalt of Lithotectonic Unit 2 of the Havallah Sequence."

The Pennsylvanian and Permian (late Atokan and Wolfcampian) green shale subunit is commonly quite poorly exposed throughout its area of outcrop but, locally, the subunit is extremely well exposed (fig. 32*A*, *B*). The subunit includes some interbeds of calcareous sandstone near the uppermost parts of the stratigraphic sequence of green shale. Where best exposed in SE 1/4 NE 1/4 sec. 23, T. 34 N., R. 42 E. green shale is probably on the order of at most 100 to 150 m in stratigraphic thickness (pl. 1). Three samples (91TT041, 91TT042, and 91TT043) from a locality in this quarter-quarter section near the stratigraphic base of the subunit all contained radiolarians diagnostic of biostratigraphic radiolarian assemblages 6 and 7 of Murchey (1990) (B.L. Murchey, written commun., 1993). Some green shale has also been mapped, mostly using distribution of chips of green shale in colluvium, as discontinuous lenses near the base of one of the next stratigraphically higher subunits in tectonostratigraphic plate A, a Pennsylvanian and Permian(?) predominantly yellow-to-orange calcareous sandstone (subunit hys, pl. 1). This intercalation of yellow-to-orange calcareous sandstone and green shale near the base of the sandstone suggests a locally sporadic initiation of major influx of extrabasinal terrigenous sands that eventually become quite thick higher in the overall stratigraphic sequence of plate A.

Figure 32 Pennsylvanian and (or) Permian (late Atokan to Wolfcamp) green shale, subunit hsh, in northern Havallah Hills. *A*, General overview to northeast of gulley in SE 1/4 NE 1/4 sec. 28, T. 34 N., R. 42 E., that has well exposed sequences of subunit hsh; *B*, typical exposures of subunit hsh at site 91TT041 described in *A*.

Mostly green to blue-green shale (subunit hsh, pl. 1) in places overlies the basal chert subunit in tectonostratigraphic plate A, as described above, and this subunit showed a basal contact with underlying chert that is nowhere exposed sufficiently well to provide a conclusive determination as to the type of relation between the two subunits. Nonetheless, the contact appears to be conformable with attitudes of bedding in the underlying chert near the northernmost part of the green shale in the NE 1/4 sec. 23, T. 34 N., R. 42 E., and, as such, the relation is provisionally considered to be a normal, albeit gradational depositional contact between the two subunits. The gradation between the two subunits takes place across a stratigraphic interval of approximately 20 to 30 m. Moreover, a significant amount of variation in the strike of bedding is present in the central parts of the green shale subunit where it is best exposed in some of the more sharply incised gulleys through the subunit. Such variation is possibly due to generally low overall angles of dip in the green shale. The green shale subunit is important from a sedimentologic standpoint in that it marks a transition from predominantly ribbon chert-rich sequences below to fine-grained sands and silts above that are dominated by large volumes of silt- and sand-sized terrigenous quartz, together with significant components of intrabasinal and extrabasinal carbonate minerals. In addition, green shale also includes some interbedded brown calcareous sandstone and some sparse amounts of drab brick-red to maroon shale. These shaly units in LU–2 generally lack the strain-slip cleavage that is present in lithologically similar rocks in LU–1. Petrographic examination of four samples of green shale from LU–2 reveals that these spicule- and conodont-bearing rocks show minimal signs of alteration adjacent to some mm–sized, anastomosing quartz-barite-sulfide mineral-chlorite veins that, in places, cut the well-laminated green shale. Some green shale also includes small fragments of basalt or basaltic andesite undoubtedly derived from the underlying chert subunit.

Yellow-to-orange calcareous sandstone crops out in an area of about 3 km^2 in the east-central part of the Havallah Hills, mostly in E 1/2 sec.27 and W 1/2 sec. 26, T. 34 N., R. 42 E., where the subunit is warped into some broad open folds showing generally north-trending hingelines (subunit hys, pl. 1). Typically, rocks of this subunit weather to orange-brown subangular blocks that are strewn profusely on colluvium-covered slopes from exposures that commonly show partings between bedding surfaces of about 10 to 20 cm. Rocks of subunit hys rest depositionally on radiolarian ribbon chert belonging to subunit hc in SW 1/4 sec. 27, and the subunit includes some unmapped thin sequences of green shale near its base. Subunit hys is in turn overlain by subunits hu and hlc to the south and east, respectively, where in places, such as in SE 1/4 sec. 27, rocks of subunit hlc appear to lap unconformably with a slight angular discordance across underlying strata of subunit hys (pl. 1). In addition, subunit hys includes some discontinuous interbeds of green shale, subunit hsh, that show their maximum area of outcrop, about 0.2 km², along the border between secs. 26 and 27. These rocks are essentially the same as subunit hsh described above, and rocks of subunit hys are intruded in SE 1/4 sec. 22, T. 34 N., R. 42 E. by a small number of presumably Tertiary felsite dikes that are altered intensely to phyllic mineral assemblages. The predominantly medium-grained sandstones of subunit hys also include variable proportions of other rocks, including fine-grained sandstone and silt in increasing abundances near the top of the sequence, and platy thin beds of sandy micrite that appear to increase in abundance to the south. Micrite contained in subunit hys displays abundant frosted granules of detrital quartz on its weathered surfaces. Size-specific rounding of approximately equal intensity is a characteristic of both monocrystalline quartz grains and extrabasinal carbonate micrite grains suggesting shorter distances to micrite source(s).

In the modal studies described in this section below, optically sharp, somewhat oxidized crystal edges of extrabasinal carbonate grains, including both monocrystalline and polycrystalline varieties, are one of the criteria used to distinguish extrabasinal grains from intrabasinal carbonate grains which are typically much less in abundance in these rocks (see also, Zuffa, 1980; 1985). In addition, extrabasinal carbonate grains commonly show grain sizes similar to the obviously terrigenous quartz grains in the same thin sections. Other detrital minerals and fragments in the rocks include polycrystalline quartz of various types, K–feldspar, plagioclase, white mica, tourmaline, zircon, opaque minerals, dolomite (in places showing multiple rhombic overgrowths on rounded dolomite grains and on rounded micrite grains), and polycrystalline carbonate grains, including some micrite and fragments of various macrofossils. In one of the samples, sparse millimeter-sized open cavities are filled by secondary hydrothermal barite and euhedral crystals of calcite. No other signs of hydrothermal alteration are present in the rock containing the trace amounts of secondary barite.

Monocrystalline quartz is the dominant framework mineral in five samples of variably calcareous, fine-grained sandstone analyzed modally from subunit hys using some modifications of standard counting procedures (fig. 33A). The quantitative detrital modes of these rocks were determined generally using the counting technique of Dickinson (1970, 1985). As suggested by Dickinson (1985), counting procedures concerning lithic fragments should be restricted to those fragments whose aphanitic matrix is less than 0.0625 mm in mean diameter, that is, less than the size of fine sand in the classification scheme of Folk (1968), and, furthermore, all intrabasinal grains generally should

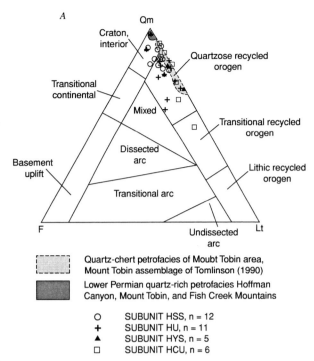

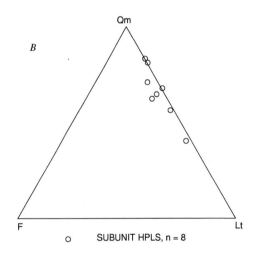

Figure 33 Modes of detrital framework grains from various calcareous, mostly sand units mapped in the Devonian, Mississippian, Pennsylvanian, and Permian Havallah sequence. Classified according to provenance scheme of Dickinson and others (1983); Qm, monocrystalline quartz, F, total feldspar, and Lt, total lithics, including extrabasinal carbonate grains, polycrystalline quartz, and silicate rock fragments. *A*, Modal analyses of samples from subunits hss, hu, hys, and hcu. *B*, Modal analyses of samples of sand- and silt-sized facies in subunit hpls, plate B. Lt, also includes phosphatic fragments.

be ignored. Contrary to the suggestion of Dickinson (1985), detrital extrabasinal carbonate grains were counted in this report and included here as part of the total population of lithic grains (Lt, fig. 33A) in the compositional modes (see also, Mack, 1984; Arribas and Arribas, 1991). Two samples of subunit hys were stained for K–feldspar, calcite, and porosity. As shown by data plots on figure 33A, these five samples plot in either the quartzarenite or sublitharenite field of McBride (1963), and they show modal compositions suggestive of either a cratonic interior provenance or a quartzose recycled orogen provenance. The compositional field defined by the samples from subunit hys also shows a substantial overlap with that of subunit hu to be described below (fig. 33A). However, a much wider range in lithology of lithic fragments is present in subunit hys than in subunit hu. Extrabasinal carbonate grains make up approximately 25 to 55 volume percent of lithic fragments of samples from subunit hys that plot in the quartzarenite field, in contrast to their making up roughly 91 to 95 volume percent of the remaining three samples that plot in the sublitharenite field. In addition, the two samples of quartzarenite contain a relatively small overall proportion of lithic fragments at their framework sites, namely 2 to 4 volume percent. One of the samples analyzed modally also contains a higher proportion of detrital chert fragments than extrabasinal carbonate fragments. Similar to subunit hu, however, heavy minerals present in subunit hys include tourmaline, Fe oxide, zircon, biotite, and white mica as described below; some micas also are present as small euhedral inclusions in monocrystalline quartz grains. All of these relations, nevertheless, attest to broad differences in source terranes from which rocks of subunit hys were derived.

Chemical analyses of two samples of fine-grained quartzarenite and litharenite from subunit hys are shown in table 4. The analyzed sample of quartzarenite (analysis 1, table 4) contains approximately 88 volume percent monocrystalline quartz at its framework sites, which value compares remarkably well with the 91.1 weight percent SiO_2 in the chemical analysis. Analysis 1 (0.29 weight percent CO_2) contains roughly 0.01 volume percent extrabasinal carbonate grains, whereas analysis 2 (7.41 weight percent CO_2) contains approximately 16 volume percent extrabasinal carbonate grains. Both of these analyses plot in the passive margin field in the K_2O/Na_2O ratio versus SiO_2 tectonic discrimination diagram of Roser and Korsch (1986) (fig. 34). Minor elements in analyzed samples of subunit hys generally are the same as those found in analyzed samples of calcareous sands from subunit hu with exception of vanadium contents. Based on available analyses, vanadium appears to be higher slightly in sands of subunit hys, 26 to 27 ppm versus 7 to 17 ppm (compare tables 4 and 5).

Yellow-to-orange calcareous sandstone in places is succeeded upwards by a sequence of limestone and chert (subunit hlc, pl. 1), shown as Pennsylvanian and Permian in age, that somewhat more specifically is made up of an irregularly alternating, but distinctive interbedded black sponge-spicule-bearing chert and gray micrite. Gray micrite in places is quite sandy from an included detrital component of mostly monocrystalline quartz. Moreover,

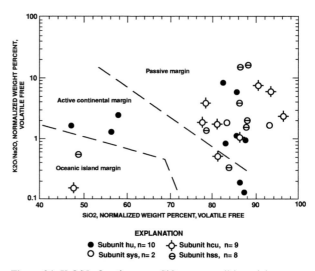

Figure 34 K_2O/Na_2O ratio versus SiO_2 content, all in weight percent normalized volatile free to 100 percent, for 29 calcareous sands and sandy limestones from four subunits of Devonian, Mississippian, Pennsylvanian, and Permian Havallah sequence. Dashed lines, approximate empirically derived boundaries of tectonic margins for sandstones and argillites of Roser and Korsch (1986). Data from tables 5–8.

as shown on plate 1, yellow-to-orange calcareous sandstone of plate A must be the lateral equivalent of some of the stratigraphically lowermost parts of the lithologically more variegated Pennsylvanian and Permian undivided subunit of tectonostratigraphic plate A that has been mapped in S 1/2 sec. 23, T. 34 N., R. 42 E. (subunit hu, pl. 1). In addition, near tops of some of the better exposed sequences of yellow-to-orange calcareous sandstone in plate A, yellow-to-orange calcareous sandstone shows gradational contacts with a sequence of rocks that are included with the undivided map subunit.

The undivided part of tectonostratigraphic plate A is quite varied lithologically and it includes, in some of the better exposed sequences, the following lithologies: (1) thin-bedded to laminated, fine-grained calcareous sandstone, (2) gray platy micrite and black massive micrite, (3) brown-to-black shale, (4) gray calcareous siltite, and (5) minor dolomitic chert, possibly of a replacement-type origin. Locally, near the base of the undivided subunit in SE 1/4 sec. 23, T. 34 N., R. 42 E., some calcareous sandstone also includes relatively abundant concentrations of the Nereites facies ichnofossils *Scalarituba* and *Lophoctenium* (fig. 35), the latter of which is generally accepted as being indicative of deep-water environments (Stewart and others, 1977). Thus, tectonostratigraphic plate A can be characterized as showing a fairly standard basin-filling cycle, including (1) a generally thin sequence of ribbon-type radiolarian chert overlain, in places near its base, by green shale, and (2) interbedded gray micrite and black spicular chert near its top. Between these two parts of the cycle there is commonly a highly variable, both lithologically and in overall thickness, intervening sequence

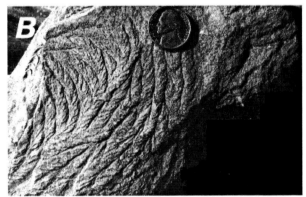

Figure 35 Nereites-facies ichnofossils collected from Pennsylvanian and Permian undivided subunit hu (pl. 1) of lithotectonic unit 2 (Murchey, 1990) of the Devonian, Mississippian, Pennsylvanian, and Permian Havallah sequence as it crops out at locality 90TT186 in the SE 1/4 sec. 23, T. 34 N., R. 42 E. *A, Scalarituba; B, Lophoctenium.*

of clean-washed calcareous sands and silts that, in detail, show widespread gradational relations among the aggregate sands and silts. Although sequences of interbedded gray micrite and black spicular chert most commonly are present near tops of these apparent stratigraphic successions, some interbedded gray micrite and black spicular chert seem to rest depositionally directly on the basal chert subunit in SE 1/4 sec. 14, T. 34 N., R. 42 E., further attesting thereby to a probable complex intergradational relation of interbedded gray micrite and black spicular chert with all of the lithologies that make up plate A above the basal chert subunit (pl. 1).

Rocks assigned to undivided subunit hu in plate A also crop out below an east-dipping apparently minor thrust in SW 1/4 sec. 25 and NW 1/4 sec. 35, T. 34 N., R. 42 E. (pl. 1). The rocks in the upper plate of this thrust are a sequence of radiolarian ribbon chert overlain depositionally by calcareous siliciclastic rocks that are also assigned to undivided subunit hu.

Sequences of lithologic subunits showing similar sedimentologic cycles or lithofacies to those listed above are found elsewhere throughout the entire area of outcrop of LU–2 in the Golconda allochthon in the Valmy and North Peak quadrangles. In places, these subunits are fur-

ther repeated tectonically either along plate-bounding low-angle faults, or they are repeated by high-angle faults that are present interior to the successively higher tectonostratigraphic plates B through E to be described below in the sections that follow (pls. 1, 2).

Petrographic study of rocks from the undivided subunit in the Havallah Hills involved examination by standard thin section of 21 rocks from various localities (pl. 1). This subunit is, in places, depositionally above green shale in plate A. From these 21 rocks, an additional 12 thin sections were stained to expedite petrographic studies of modal mineralogy. Splits from six of the stained samples also were analyzed chemically, and results of these studies are described in this section below. Most thin-sectioned rocks of the undivided subunit are from the northern part of the Havallah Hills, mostly in secs. 14 and 23, T. 34 N., R. 42 E.

Generally, rocks that make up undivided subunit hu are an assortment of fine- to medium-grained, submature to mature calcareous sands, according to the classification scheme of Folk (1968, 1974). These sands usually show parallel lamination of bedding surfaces. The rocks contain an abundance of monocrystalline, terrigeneous quartz that most commonly has a framework-supported fabric, and they also contain extrabasinal and intrabasinal detrital carbonate grains in variable proportions. Many samples show size-specific rounding wherein large grains of terrigenous quartz, approximately 0.4 mm in size, have higher sphericities than small grains of quartz. These large grains of quartz are not matched in the rocks by similarly large-sized fragments of detrital carbonate fragments, resulting locally in a textural inversion suggestive of multiple sources. In addition, some rocks examined contain more than 50 volume percent crystalline carbonate minerals whose sparry textures do not allow easy recognition of original depositional fabrics; these latter rocks would be termed limestone in the classification scheme of Dunham (1962). All calcareous sands have porosities that are essentially zero, largely either because of advanced stages of silica overgrowth at many of the quartz-quartz grain boundaries or because of presence of sparry calcite cement that does not seem to have affected texturally most of the many monocrystalline extrabasinal carbonate grains (EC, in the classification of Zuffa, 1980, or Lc, in the classification of Dickinson, 1985). The outer edges of extrabasinal carbonate grains typically are well preserved and optically sharp although many geologic events subsequent to diagenesis have affected them. Commonly, these grain boundaries are preferentially discolored by increased concentrations of Fe oxide probably reflecting preferred oxidation of Fe contents of extrabasinal carbonate grains relative to calcite cement. Calcite cement in the rocks seems to be relatively free of cations other than Ca. This relation suggests that a chemical distinction is present between the

Table 4—Analytical data on calcareous litharenite of subunit hys of lithotectonic unit 2 of the Devonian, Mississippian, Pennsylvanian, and Permian Havallah sequence, Humboldt County, Nevada.†

Analysis number	1	2
Field number 90TT..	215	216
Chemical analyses (weight percent)		
SiO_2	91.1	76.7
Al_2O_3	2.83	3.2
Fe_2O_3	.56	.44
FeO	.08	.17
MgO	.28	1.36
CaO	1.02	7.85
Na_2O	.55	.61
K_2O	.9	1.16
H_2O^+	.48	.49
TiO_2	.21	.2
P_2O_5	.1	.09
MnO	.04	.03
F	.04	.04
Cl	.003	.004
Organic C	.22	.12
Carbonate C	.08	2.02
CO_2	.29	7.41
Total S	.08	<.01
Subtotal	99.13	102.03
Less O=F, Cl, S	.02	.02
Less carbonate C	.08	2.02
Total	99.03	99.99
Inductively coupled plasma-atomic emission spectroscopy (total) (parts per million)		
As	10.0	—
Ba	2500.	1200.0
Ce	13.	14.
Co	4.	4.
Cr	13.	13.
Cu	9.	6.
La	9.	10.
Li	6.	6.
Nd	7.	9.
Ni	10.	7.
Pb	12.	5.
Sr	69.	180.
V	27.	26.
Y	6.	8.
Zn	26.	21.

Table 4 – (cont.)

Inductively coupled plasma-atomic emission spectroscopy (partial) (parts per million)		
Ag	.16	.31
As	14.	7.5
Cd	.16	.53
Cu	5.9	4.8
Mo	.2	.32
Pb	6.1	3.
Sb	.85	.95
Zn	22.	23.

Energy-dispersive X-ray fluorescence spectroscopy (parts per million)		
Rb	18.0	24.0
Sr	72.	172.
Zr	495.	215.
Y	17.	17.
Ba	2400.	1100.

Chemical analyses (parts per million)		
As	14.0	8.7
Au	.002	<.002
Te	—	—
Tl	.1	.1
Hg	.02	.07

†[Chemical analyses of major oxides in weight percent by methods of Jackson and others (1987); analyst, unknown. FeO, H_2O^+, H_2O^- in weght percent by methods of Jackson and others (1987); analyst, N.H. Elsheimer. Total C in weight percent by combustion-thermal conductivity method; analyst, N.H. Elsheimer. Carbonate C in weight percent by extraction-coulometric titration method; analyst, C.S. Papp. Organic C by difference. Total S in weight percent by combustion-infra-red method; analyst, N.H. Elsheimer. F, Cl in weight percent by specific ion electrode method; analyst, N.H. Elsheimer. Inductively coupled plasma-atomic emission spectroscopy (total and partial) in parts per million by methods of Crock and Lichte (1982), Lichte and others (1987), and Motooka (1988); analysts, D.L. Fey, and J.M. Motooka. Precision for concentration higher than 10 times the detection limit is better than +10-percent relative standard deviation; instrumental precision of the scanning instrument is +2-percent relative standard deviation. Looked for, but not found, at parts-per-million detection levels in parantheses: Au (8), Be (1), Bi (10), Cd (2), Eu (2), Ga (4), Ho (4), Mo (2), Nb (4), Sc (2), Sn (5), Ta (40), Th (4), U (100). Energy-dispersive X-ray fluorescence spectrometry in parts per million by method of Johnson and King (1987); analyst, J. Kent. Looked for, but not found, at parts-per-million detection levels in parantheses: Nb (10), Ce (30), La (30). Chemical analysis for Au by combined chemical separation-atomic absorption method involving heating 10-g sample to 700° C to remove sulfides and then decomposing sample in HBr and Br_2. Au is extracted into methyl isobutyl Ketone, and organic layer analyzed for Au by graphite furnace-atomic]

two types of carbonate minerals in the rocks. In addition, some detrital carbonate grains show relatively coarse-crystalline sparry fabrics that obviously predate rounding of grains, and brown microcrystalline, granular fabrics in these grains differ from clear platelets of calcite commonly found in surrounding cement.

Monocrystalline quartz is the predominant framework mineral in 11 samples studied modally from the undivided subunit of plate A (fig. 33*A*), and all of these samples would be termed sublitharenite in the sandstone classification scheme of McBride (1963). Lithic fragments are subordinate to monocrystalline quartz, but lithic fragments are present in abundances greater than total feldspar. Extrabasinal carbonate grains make up 62 to 90 volume percent of total lithic fragments in the rocks and are far more abundant than detrital fragments of chert or phyllosilicate-bearing fragments. Detrital K–feldspar-to-plagioclase ratios are extremely divergent in these rocks as well. As a result, all samples from the undivided subunit analyzed (fig. 33*A*) plot in the compositional field characteristic of sands derived from quartzose recycled orogens of Dickinson and others (1983). These data suggest that source terrane(s) for generally fine-grained calcareous sands of the undivided subunit may have been generally constant over time, and that the terrane(s) probably included both carbonate sequences and some probably minor granitic sources.

In comparison, sandstone of the Havallah sequence in the Mount Tobin area (fig. 1) described by Tomlinson (1990) and assigned by him to his Mount Tobin assemblage, equivalent to LU–2 of this report, is at first appearance remarkably similar modally to sandstone of the undivided subunit in the Havallah Hills. Sandstones in both areas contain approximately the same proportion of monocrystalline detrital quartz fragments. However, in the Mount Tobin area, quartz-chert petrofacies sands (Tomlinson, 1990) apparently are derived predominantly from basinal sedimentary rocks that include quartzose clastic rocks, chert, argillite, and minor compositionally basic volcanic rocks as lithic fragments in contrast to the apparently carbonate-rich sedimentary protoliths from which sublitharenites in subunit hu in the Havallah Hills were mostly derived. Some sublitharenites of the undivided subunit, moreover, appear to contain an overall slightly higher proportion of total feldspar than quartz-chert petrofacies sands in the Mount Tobin area.

Chemical analyses of 10 calcareous sublitharenites of subunit hu, nine of which are in plate A close to the northern end of the Havallah Hills and the remaining sample is from a locality near the east-central part of the Havallah Hills, are shown in table 5. The extreme variation in SiO_2, CaO, and CO_2 contents in these rocks results primarily from a wide range in proportions of siliciclastic and calcareous minerals, the latter of which includes both

framework minerals and cement as described above. The amounts of SiO_2, CaO, and CO_2 respectively show 33.8 to 81.4, 4.78 to 33, and 5.43 to 27.3 weight percent ranges in composition. Although modal analysis of one of the samples with 22.7 weight percent CO_2 (analysis 3, table 5) revealed that more than 90 volume percent of the framework sites are occupied by extrabasinal carbonate grains (total lithics are approximately 18 volume percent), actual modal carbonate content of the rock is much higher because of abundant crystalline carbonate grains present primarily as sparry calcite cement that was not counted modally. In addition, chemical analyses of these unaltered rocks provide a local geochemical background to which effects of decarbonatization and potassic alteration in similar rocks metallized at the Lone Tree deposit can be compared (see section below entitled "Lone Tree Gold Deposit"). Contents of K_2O of the 10 unaltered samples from subunit hu are in the range 0.13 to 1.37 weight percent (table 5). The sample with the highest content of K_2O (analysis 6, table 5) modally contains approximately 10 volume percent detrital K–feldspar. Major-element compositions of litharenites of subunit hu, and sands from other subunits to be discussed in the subsections that follow below, cannot be readily compared with previously published summary analyses of sandstones by Pettijohn (1963) because the latter analyses exclude carbonate sands.

Major-element geochemical signatures of 10 samples analyzed from subunit hu widely span the passive margin and active continental margin settings of the tectonic discrimination diagram of Roser and Korsch (1986) (fig. 34). For these plots, a value of approximately 50 percent of the lower limit of determination was used for seven Na_2O analyses, and for one P_2O_5 analysis (see Sanford and others (1993) for a discussion of the suitability of such substitutions). Although five analyzed samples from subunit hu plot in the active continental margin part of the tectonic discrimination diagram and are apparently compatible with the known active margin setting of the Havallah basin, the remaining five samples plot in the passive margin part of the diagram as do most other sands analyzed chemically from LU–2 to be described below. The SiO_2–rich character of many of these calcareous sands would be further enhanced if the analyses were recalculated volatile free to 100 weight percent because of their generally elevated carbon dioxide contents (table 5). Indeed, if the analyses are calculated volatile free and CaO free, then all samples show greater than 86 weight percent SiO_2 (normalized to 100 percent), and would plot in the passive margin part of the tectonic discrimination diagram of Roser and Korsch (1986). However, overall geochemical character of these sands primarily reflects a recycling of detritus from the Antler orogenic belt, the lower Paleozoic platform, and possibly also including a significant component of material from the interior craton, as well as some detritus from the advancing Golconda allochthon. McCann (1991) also found major discrepancies in tectonic provenance of sands in the southern Welch basin, Wales, when the provenance of sands, established by other methods, was examined using standard chemical plots purported to discriminate among tectonic regimes. McCann (1991) further pointed out that geochemical signatures of sands he examined instead may be recording primarily the chemistry of the land mass from which the sands were derived and not the type of tectonic regime of the basin into which the sands were being shed.

Minor-element chemistry of the sands of subunit hu were also studied using standard "spider"-graph-type plots after grouping elements selected for study by geochemical association. One sample (analysis 6, table 5) was used as the base to which the remaining nine samples were normalized to examine thereby variations in contents of Ba, Sr, Mn, Pb, Zn, Cu, As, Ag, and Au among the samples (fig. 36A, B). For these plots, a value of 50 percent of the lower limit of determination again was used for those elemental analyses below detection. As is apparent from the plots, Sr is present in elevated normalized ratios as high as 3 in three samples (analyses 3, 5, and 10, table 5) that contain the most abundant modal carbonate minerals, and two metals (As and Ag) are present at elevated normalized ratios in most samples relative to analysis 6. The overall pattern of low-level but inconsistent normalized enrichment ratios for Mn, As, and Ag in these samples is one that is repeated for many of other analyzed sands in the Havallah sequence to be described below. The increased normalized ratios of Mn are present in samples that show elevated abundances of modal carbonate minerals, both as framework grains and as cement. Although one of the samples (analysis 1, fig. 36A) shows a normalized ratio for As approximately 6 times the assigned geochemical threshold, concentration of As in this sample (9.7 ppm, table 5) is roughly 10 times its average concentration in sandstones as estimated by Turekian and Wedepohl (1961). Many samples that show elevated normalized ratios of As also show elevated normalized ratios of Ag. Reported "background" geochemical abundances in four types of rock in the Havallah sequence near the Lone Tree Mine are in the range 51 to 184 ppm As (see section below entitled "Lone Tree Gold Deposit"). Inasmuch as these latter concentrations of As near the mine are approximately one to two orders of magnitude higher than the concentrations of As in unmineralized analyzed samples of subunit hu well away from the area of the mine (table 5), the reported concentrations near the mine must reflect effects of alteration and mineralization near fringes of that large mineralized system. In addition, from concentration data for Au in table 5, it is apparent that background contents in these rocks are probably in the range 1 to 2 ppb (parts per billion) Au, although contents as high as 10 ppb Au are present

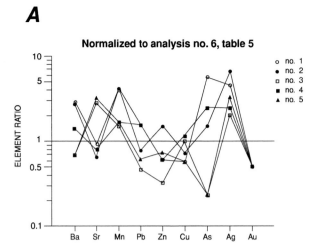

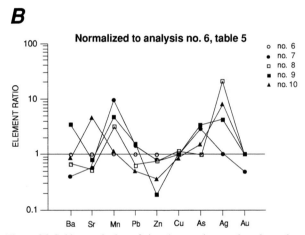

Figure 36 Spider-graph plots of nine elements in ten selected samples of subunit hu in tectonostratigraphic plate A of the Devonian, Mississippian, Pennsylvanian, and Permian Havallah sequence. Analysis numbers same as table 5. Normalized to analysis 6 using 50 percent of the lower limit of determination for those elements that are below detection (table 5; see text). A, Analyses 1–5; B, Analyses 6–10.

elsewhere in some other unmineralized sands in rocks of the Havallah sequence in the mining district.

The undivided subunit hu of tectonostratigraphic plate A, in places, apparently is also laterally equivalent stratigraphically to thin mappable sequences of rock consisting of interbedded black spicular chert and gray micrite or sandy micrite (subunit hlc, pl. 1). Some rocks lithologically equivalent to these, but cropping out in the Antler Peak 7–1/2 minute quadrangle, have been shown to contain early Permian (Wolfcampian and Leonardian) fusulinids (Doebrich, 1995). Subunit hlc of plate A crops out in two general areas, in total making up about 1.5 km² of the plate: the first area straddles the border between secs. 14 and 23, and the second area is mostly in the central and east parts of sec. 26—both areas in T. 34 N., R. 42 E. In the latter area, chert and micrite of subunit hlc crop out stratigraphically above subunit hys, mostly yellow-to-

Table 5—Analytical data on calcareous litharenite of subunit hu of lithotectonic unit 2 of Devonian, Mississippian, Pennsylvanian, and Permian Havallah sequence, Humboldt County, Nevada.†

Analysis number	1	2	3	4	5	6	7	8	9	10
Field number 90TT..	179	181	186	188B	190A	190B	191	193	197	254
					Chemical analyses (weight percent)					
SiO$_2$	78.1	81.7	44.4	76.3	33.8	74.6	79.7	79.7	81.4	42.4
Al$_2$O$_3$	2.15	2.69	1.86	4.11	2.	3.63	2.48	3.37	2.33	1.49
Fe$_2$O$_3$	.63	.38	.21	.49	.26	.38	.37	.71	.5	.37
FeO	.14	.11	.13	.31	.16	.19	.48	.35	.04	.16
MgO	.58	1.13	.89	2.48	1.57	1.86	2.2	2.12	.38	.72
CaO	8.79	5.7	28.	5.72	33.	7.9	5.64	4.78	7.13	29.4
Na$_2$O	<.15	.63	.31	1.05	.35	.17	.89	.76	1.01	.29
K$_2$O	.41	.59	.74	.88	.56	1.37	.17	.85	.13	.37
H$_2$O$^+$	.73	.48	.31	.65	.42	.77	.41	.54	.28	.41
H$_2$O$^-$	.2	.17	.08	.26	.1	.22	.12	.16	.11	.12
TiO$_2$	.11	.18	.09	.31	.13	.22	.14	.34	.15	.17
P$_2$O$_5$	.08	.06	.07	.08	.1	.16	.08	.08	.07	.21
MnO	.05	.06	.02	.03	.02	.02	.12	.04	.06	.02
F	.02	.02	.02	.04	.02	.04	.02	.03	.02	.04
Cl	.004	.004	.003	.004	.004	.005	.003	.004	.003	.006
Organic C	.27	.2	.35	.19	.6	.18	.16	.17	.15	.51
Carbonate C	2.01	1.48	6.2	1.84	7.44	2.19	1.88	1.59	1.61	6.33
CO$_2$	7.37	5.43	22.7	6.75	27.3	8.03	6.89	5.83	5.9	23.2
Total S	<.01	<.01	<.01	<.01	<.01	<.01	<.01	<.01	<.01	<.01
Subtotal	101.64	101.01	106.4	101.49	107.83	101.49	101.75	101.42	101.27	106.22
Less O=F, Cl, S	.01	.01	.01	.02	.01	.02	.01	.02	.01	.02
Less carbonate C	2.01	1.48	6.2	1.84	7.44	2.19	1.88	1.59	1.61	6.33
Total	99.62	99.52	100.19	99.63	100.38	99.73	99.86	99.81	99.65	99.87
			Inductively coupled plasma-atomic emission spectroscopy (total) (parts per million)							
Ba	1900.0	1800.0	450.0	920.0	450.0	650.0	260.0	440.0	2300.0	560.0
Ce	7.	7.	5.	17.	9.	13.	8.	24.	5.	18.
Co	5.	4.	3.	4.	3.	3.	3.	2.	4.	4.
Cr	13.	17.	15.	18.	18.	28.	9.	18.	7.	27.
Cu	4.	5.	7.	8.	4.	7.	7.	8.	7.	6.
La	5.	5.	6.	10.	8.	10.	5.	14.	5.	15.
Li	8.	5.	4.	8.	6.	13.	6.	7.	4.	6.
Nd	5.	5.	—	9.	8.	7.	4.	11.	—	13.
Ni	7.	6.	4.	5.	6.	6.	5.	5.	5.	6.
Pb	9.	7.	—	5.	—	6.	7.	5.	4.	—

Table 5 – (cont.)

	1	2	3	4	5	6	7	8	9	10
Analysis number	1	2	3	4	5	6	7	8	9	10
Field Number 90TT..	179	181	186	188B	190A	190B	191	193	197	254
Sr	140.	100.	430.	120.	490.	150.	87.	80.	120.	670.
V	7.	8.	8.	17.	9.	12	9.	12.	10.	11.
Y	6.	6.	4.	8.	6.	10.	6.	10.	5.	12.
Yb	—	—	—	—	—	—	—	1.	—	1.
Zn	19.	47.	10	19.	23.	31.	24.	23.	6.	11.

Inductively coupled plasma-atomic emission spectroscopy (partial) (parts per million)

	1	2	3	4	5	6	7	8	9	10
Ag	0.15	0.22	0.067	0.08	0.11	—	—	0.67	0.14	0.26
As	8.	2.1	—	3.4	—	0.83	2.2	—	3.9	1.2
Cd	.15	.2	.2	.25	.76	—	.16	.062	.073	.5
Cu	3.2	1.8	2.9	5.7	3.8	4.5	6.9	8.2	7.7	3.8
Mo	.99	.4	.12	.36	.24	.29	.35	.3	.38	.42
Pb	4.4	2.2	1.3	4.3	1.7	2.8	3.9	1.8	4.3	1.4
Sb	.84	—	—	.92	—	—	—	—	1.1	—
Zn	21.	27.	10.	17.	20.	24.	18.	19.	7.8	16.

Energy-dispersive X-ray fluorescence spectrometry (parts per million)

	1	2	3	4	5	6	7	8	9	10
Rb	11.0	11.0	19.0	22.0	17.0	29.0	—	18.0	—	13.0
Sr	130.	108.	390.	122.	445.	140.	91.0	83.	120.0	600.
Zr	188.	335.	104.	425.	225.	380.	270.	650.	355.	235.
Y	15.	15.	14.	19.	17.	19.	15.	24.	14.	20.
Ba	1700.	1600.	405.	840.	415.	620.	255.	420.	2100.	510.

Chemical analyses (parts per million)

	1	2	3	4	5	6	7	8	9	10
As	9.7	5.8	1.1	4.7	1.1	1.4	4.1	1.4	4.7	2.1
Au	<.002	<.002	<.002	<.002	<.002	.002	<.002	.002	.002	.002
Te	—	—	.05	.1	—	—	.05	—	—	—
Tl	.05	.05	—	—	.1	.1	.05	.1	—	—
Hg	.16	.06	—	.09	.02	.02	.05	.05	.03	—

†[Chemical analyses of major oxides in weight percent by methods of Jackson and others (1987); analyst, N.H. Elsheimer. Total C in weight percent by combustion-thermal conductivity method; analyst, N.H. Elsheimer. Carbonate C in weight percent by extraction-coulometric titration method; analyst, C.S. Papp. Organic C by difference. Total S in weight percent by combustion-infra-red method; analyst, N.H. Elsheimer. F, Cl in weight percent by specific ion electrode method; analyst, N.H. Elsheimer. Inductively coupled plasma-atomic emission spectroscopy (total and partial) in parts per million by methods of Crock and Lichte (1982), Lichte and others (1987), and Motooka (1988); analysts, D.L. Fey, and J.M. Motooka. Precision for concentration higher than 10 times the detection limit is better than +-10-percent relative standard deviation; instrumental precision of the scanning instrument is +-2-percent relative standard deviation. Looked for, but not found, at parts-per-million detection levels in parentheses: As (10), Au (8), Be (1), Bi (10), Cd (2), Eu (2), Ga (4), Ho (4), Mo (2), Nb (4), Sc (2), Sn (5), Ta (40), Th (4), U (100). Energy-dispersive X-ray fluorescence spectrometry in parts per million by method of Johnson and King (1987); analyst, J. Kent. Looked for, but not found, at parts-per-million detection levels in parentheses: Nb (10), Ce (30), La (30). Chemical analysis for Au by combined chemical separation-atomic absorption method involving heating 10-g sample to 700° C to remove sulfides and then decomposing sample in HBr and Br$_2$. Au is extracted into methyl isobutyl Ketone, and organic layer analyzed for Au by graphite furnace-atomic absorption spectrometry (Meier, 1980); analyst, H. Smith. Chemical analysis for As, Te (0.05 ppm detection level), and Tl by graphite furnace atomic absorption spectrometry (Wilson and others, 1987; Aruscavage and Crock, 1987); analysts, A.H. Love, L.A. Bradley, and P.L. Hageman; chemical analysis for Hg by cold-vapor atomic absorption spectrometry; analysts, A.H. Love, L.A. Bradley, and P.L. Hageman; —, not detected]

orange limey sands, whereas subunit hys approximately 1 km to the southwest of this area has been shown to have above it a highly varied assortment of calcareous silt, micrite, and calcareous sandstone assigned to subunit hu (pl. 1). Thus, some parts of subunit hu are stratigraphically in the same position as some sequences of black spicular chert and gray micrite. The top of plate A is marked by a low-angle thrust fault that is east dipping in the southern part of sec. 26 and predominantly south-dipping somewhat farther to the southwest where the fault between plates A and B apparently has been subsequently overridden by rocks of subunit hu along a minor northwest-dipping thrust, the Section 34 thrust fault, that has a northeast strike in sec. 34, T. 34 N., R. 42 E. (pl. 1). In this latter area, the fault initially bounding plates A and B also is inferred to be overridden by the Section 34 thrust fault which juxtaposes subunit hu of plate A and black spicular chert and gray limestone presumably of plate B. Exposures are relatively poor and it is extremely difficult in this area to clarify fully some of the ambiguities remaining in the geologic relations.

Exposures of subunit hlc are commonly quite poor but locally some geologic relations within the subunit are discernible. Some of the better exposures of subunit hlc are in NE 1/4 sec. 26, T. 34 N., R. 42 E., where the subunit includes discontinuous pod-like masses of thin-bedded, flinty chert that are 3–5 m thick and can be traced approximately 50 m almost continuously in outcrop; the subunit also includes minor amounts of green shale and some gray arenite. In thin section, flinty chert contains abundant rhombs of authigenic dolomite, some phosphatic fragments, and approximately 10 to 15 volume percent, subrounded to subangular grains of monocrystalline quartz. Grains of monocrystalline quartz average approximately 0.1 mm wide. The suite of heavy minerals includes sparse grains of tourmaline as do sandstone and silt of subunit hu described above. Some laminated gray micrite from subunit hlc includes a microcrystalline sparry fabric containing small, 0.05–mm-wide, generally angular grains of detrital quartz, narrow blades of detrital white mica, sparse grains of tourmaline and phosphatic material, and rare calcareous fossil fragments. Rocks of subunit hlc also crop out in some of the plates to the south successively higher than plate A, and they are described in more detail below.

Tectonostratigraphic Plate B

Rocks assigned to tectonostratigraphic plate B show the most widespread area of outcrop of the four plates defined in the central part of the Havallah Hills, probably measuring on the order of about 8 to 10 km^2 (fig. 31), and rocks of plate B show vertical and lateral relations among

mappable subunits comparable in complexity to those of plate A (pl. 1). As mapped, relations among subunits in plate B, many of which also are admittedly quite poorly exposed, appear to include intraplate depositional contacts, some of which are gradational, minor unconformities, and some low-angle imbricate thrust faults. Near the border between secs. 25 and 26, T. 34 N., R. 42 E., the base of plate B is placed at a shallow west-dipping thrust fault that juxtaposes undivided calcareous clastic rocks and interbedded limestone and black spicular chert (subunits hcu and hlc, respectively, pl. 1) with underlying mostly radiolarian ribbon chert in subunit hc or calcareous sublitharenite in subunit hu. Farther to the southwest, the base of plate B is placed at the lowermost exposures of a thin sequence of radiolarian ribbon chert and interbedded limestone and black spicular chert that crop out near the border between secs. 34 and 35, T. 34 N., R. 42 E. These two subunits are missing in the area of secs. 25 and 26 as they apparently have been structurally cut out. In secs. 34 and 35, the base of plate B also has been disturbed structurally as it has been covered by subsequent minor overthrusting, presumably to the south, by the Section 34 thrust fault as discussed above (pl. 1). Lateral stratigraphic relations among three widely separate areas that make up plate B (west, central, and east, as shown on plate 1), are complex, and stratigraphic stacking patterns of sequences of rock in each of these areas are schematic. The easternmost stacking pattern is derived from relations in the low hills south-southeast of the radio tower in sec. 31, T. 34 N., R. 43 E. Furthermore, similarity of the easternmost sequence with upper portions of the sequence in the central part of plate B suggests that the easternmost sequence has been downdropped relative to the central one. The top of plate B is placed at lowermost exposures of a sequence of interbedded radiolarian ribbon chert and shale that crop out mostly in secs. 2 and 11, T. 33 N., R. 42 E. and have been here designated as plate C (fig. 31). These latter rocks appear to have been folded into a number of broad, open synclines and anticlines together with underlying rocks of plate B.

All subunits that make up plate B are not discussed individually in comparable detail below inasmuch as many subunits are lithologically the same as those that make up plate A. However, there are some differences. A fairly thick sequence of pebbly limestone or bioclastic limestone, subunit hpls, crops out in the southwest part of plate B and is much more widely exposed there than near the north end of the Havallah Hills (pl. 1). Additional exposures of pebbly limestone also make up much of plate E near the southeast corner of the Havallah Hills (pl. 2), and these rocks will be discussed in greater detail below. In approximately the central part of the mapped area of plate B, specifically along a ridgeline in SE 1/4 sec. 35, T. 34 N., R. 42 E., excellent exposures are present showing a conformable stratigraphic succession of a fairly complete sedimentary cycle that extends from radiolarian ribbon chert at the base, through a thin sequence of calcareous fine sandstone and silt, and, finally, interbedded gray sandy micrite and black spicular chert at the top that constitutes subunit hlc. This succession of rocks was used to define the central diagrammatic stratigraphic stacking pattern of plate B (respectively, subunits hc, hcu, and hlc, pl. 1). Although map relations of the basal ribbon chert subunit suggest that the base of the chert must be a shallow-dipping fault, the actual trace of the fault at the base of the plate apparently has been offset by a number of high angle faults that have resulted in exposure of the basal thrust only in a small number of places (pl. 1).

Pebbly limestone in the southwestern part of plate B shows a variety of lithologies to be present and also shows complex relations of these lithologies to surrounding ones. In this area, pebbly limestone includes brownish-gray bioclastic limestone, gray calcareous lithic arenite and lithic arenite, grayish-brown sandy micrite, brown siltstone, as well as the pebbly limestone and conglomerate facies that are so diagnostic throughout the subunit (subunit hpls, pl. 1). Many outcrops of the subunit include highly disorganized interdigitations of calcareous poorly sorted submature sands and siliceous sands, commonly highlighted by exposures showing frosted medium-to-coarse quartz granules and granules of black chert in dull brownish-gray rocks. In many places, there are flame-like intergrowths between these two types of rock. The calcareous sands commonly are matrix supported by calcite cement, and they are more porous than the siliceous sands which show mostly framework-supported monocrystalline quartz and lithic fragments in a tight siliceous, in places, recrystallized matrix. Where sandy micrite is best exposed along some ridges, plane parallel bedding cycles, approximately 1– to 2–m thick, are present and they are defined by normally bedded grain-size distributions of detritus such that coarse-grained basal parts of cycles make up roughly 80 percent of individual cycles. Planar tops and bottoms of cycles show no evidence of scour and fill structures, and many fractures in the rocks show presence of shallow plunging slickensides indicating post-diagenetic tectonic disruption or jostling of the subunit. Where brown siltstone is present in the subunit, it constitutes generally the upper 20 percent of 0.5–m-thick cycles, the basal 80 percent typically being gray coarse-grained to medium-grained calcareous lithic arenite. Parallel planar bedding surfaces are present throughout the cycles.

Such thin siltstone-bearing cycles generally further diminish in thickness towards the top of the subunit and the proportion of lithologies making up the individual cycles also changes. Near the exposed top of the subunit, coarse basal parts of individual cycles decrease in average proportion such that they make up usually 50 to 60 percent of the cycle. In addition, near the top of the subunit some evidence of scouring appears to be present at the base of the siltstone segments of the cycles and a corresponding increase in undulosity of individual bedding surfaces in the coarse-grained basal parts of individual cycles. Generally, rock sequences of subunit hpls exposed in the central part of plate B appear overall to be not as coarse-grained as those exposed in plate E in the southeast part of the Havallah Hills (see subsection below entitled "Tectonostratigraphic Plate E").

Modal analyses of eight samples of litharenite representative of sand-sized facies of the pebbly limestone subunit emphasize by their widespread compositions along the monocrystalline quartz-lithic fragment join the increased amounts of lithic fragments in many of these rocks compared to other sands in LU–2 (fig. 33B). The center of gravity of plotted compositions of these eight samples is much closer to the total lithic corner (Lt) of the ternary diagram. In addition, six of the eight samples plot in the quartzose recycled orogen field of Dickinson and others (1983) whereas the two remaining samples plot in their transitional recycled orogen field. Many sands in subunit hpls contain abundant concentrations of subangular-to-subrounded phosphatic fragments that are dark brown when viewed in transmitted light by microscope (fig. 37). Many of these fragments contain nucleii of skeletal material, quartz, small granules of apatite, and even some calcite. Many subangular fragments show evidence of reworking or abrasion of otherwise concentric layers that suggest transport rather than in-situ crystallization of the phosphatic material. In five of the eight sand samples analyzed modally from subunit hpls, phosphatic fragments make up 13 to 42 volume percent of total population of lithic fragments; the remaining three samples contain no modal phosphatic fragments at all. It is difficult to ascertain overall proportion of these rocks containing phosphatic

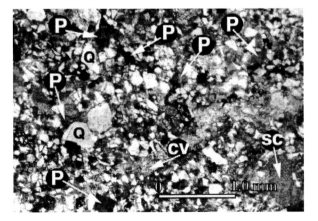

Figure 37 Photomicrograph of phosphatic fragments (P) in subunit hpls of Devonian, Mississippian, Pennsylvanian, and Permian Havallah sequence in west-central part of Havallah Hills. Detrital quartz grains (Q), calcite vein (cv), and clot of sparry calcite (sc) also in field of view. Plane-polarized light, locality 92TT075.

fragments because sparry calcite cement in the matrix of these rocks was not counted modally. However, phosphatic fragments probably make up at least several volume percent of the rocks.

Phosphatic fragments, the associated fusulinids with North American affinities, and the age of their host sands are important in that they appear to establish another late Paleozoic tie between rocks of the Havallah basin and cratonal North America faunal provinces (Stevens, 1987; Murchey, 1990). Late Permian phosphatic sedimentary rocks, exemplified by the Late Permian Phosphoria Formation of southeastern Idaho and adjoining states (Mansfield, 1927; Cressman and Swanson, 1964; McKelvey, 1967; Piper and Medrano, 1994), are part of the passive-Paleozoic-margin sequences present in subsiding shelfal basins along the the western margin of the Cordillera (Stevens, 1977; Oldow and others, 1989; Miller and others, 1992). However, rocks containing phosphatic fragments in the Havallah sequence appear to be no younger than early Permian, from which relation it is not possible to presume their derivation from the late Permian Phosphoria Formation or its age-equivalent rocks in Colorado and Utah. The Edna Mountain Formation, in the autochthonous block of Antler sequence rocks tectonically below the Havallah sequence, in places near Lone Tree Hill contains as much as 3 weight percent P_2O_5 (see section below titled "Lone Tree Gold Deposit"). Leonardian and early Guadalupian microfauna are now known to be present in rocks herein assigned to the Edna Mountain Formation (see section above titled "Edna Mountain Formation"). Furthermore, as further described above, sponge spicules make up a significant proportion of brown siltstone belonging to the Edna Mountain Formation where it crops out along Trout Creek in the central part of the North Peak quadrangle (pl. 2).

Other rocks are present in the region that could have provided phosphatic material for eventual deposition in the Havallah basin. The Silurian Elder Sandstone, which is part of the Roberts Mountains allochton (Gilluly and Gates, 1965; Madrid, 1987), also contains some relatively coarse, brown phosphatic material quite similar in overall appearance under the microscope to phosphatic fragments in subunit hpls in the Havallah Hills (C.T. Wrucke, oral commun., 1993; unpub. data, 1999). Presence of these phosphatic fragments, however, is not discussed in a provenance study of the Elder Sandstone by Girty and others (1985). The Elder Sandstone presumably must have been part of the Antler highlands and orogenic belt, and the Elder Sandstone must have been available as a source from which material could have been shed westerly into the depositional basin of the Havallah sequence. The important point here is not so much that the Elder Sandstone itself was the source for all or even a significant proportion of the phosphatic fragments, but rather that there are

sequences of phosphatic-bearing rocks in the allochthon of the Roberts Mountains thrust from which phosphatic detritus in the Havallah sequence could have been derived. Upwelling of phosphatic spicules probably occurred along the margins of the Havallah basin prior to accretion and closure of the basin during the Sonoma orogeny, and resulted from a westward stepping of the upper Paleozoic shelf margin and its coincident phosphorite-spiculite belt (B.L. Murchey, written commun., 1995; see also, Piper and Medrano, 1994).

As with the other calcareous silt and sand subunits in the Havallah sequence in plate A previously discussed above, monocrystalline quartz also is the predominant detrital grain at framework sites in lithologically equivalent silts and sands of plate B (fig. 33A). Modal analyses of six stained thin sections of the clastic rocks undivided subunit (hcu, pl. 1), mostly calcareous submature coarse silt and fine sandstone, show a relatively broad spread of almost 50 volume percent in content of total lithic fragments. Extrabasinal carbonate grains make up anywhere from 5 to 98 volume percent of these lithic fragments. In contrast, total detrital feldspar content is relatively low, reaching at most 5 to 6 volume percent in two of the samples analyzed modally. Because of the breadth of the variation in content of total lithic fragments in these rocks, plots of the six analyzed samples range from the field of cratonic interior provenance to the field of transitional recycled orogen provenance (fig. 33A). Nonetheless, the samples do not show a presence of volcanically-derived lithic fragments comparable to that reported by Tomlinson (1990) in quartz-chert petrofacies sands in the Mount Tobin area, and by Whiteford (1990) in the Havallah-equivalent Schoonover sequence to the east. In fact, modally analyzed samples from the clastic rocks undivided subunit of plate B contain only sparse-to-trace concentrations of detrital grains of chert at framework sites, and they also contain a suite of heavy minerals similar to all the calcareous sands and silts from the study area previously described above. Some samples also include fairly abundant concentrations of broken phosphatic fossil fragments that are quite common in litharenite of subunit hpls as described above.

Chemical analyses of nine samples, four of which were analyzed modally, from the clastic rocks undivided of plate B (subunit hcu, pl. 1) also show wide ranges in major element composition, mostly including broad ranges in contents of SiO_2, CaO, and CO_2 (table 6). Some differences in minor-element chemistry are present when compared to analyses of samples of subunit hu. In the nine analyzed samples of subunit hcu, SiO_2 ranges from 34.6 to 92.3 weight percent, the highest silica-containing sample represents a silty facies that has a significant proportion of probable jasperoidal material as replacement of its pre-alteration detrital minerals. In addition, this particular

sample (analysis 5, table 6) shows narrow cross-fiber textured chalcedonic quartz veins in its matrix. In the $K_2O/$
Na_2O ratio versus SiO_2 tectonic discrimination diagram of
Roser and Korsch (1986) for sandstone and argillite, eight
of the samples analyzed from subunit hcu cluster in their
passive margin field. Possible contributing factors to the
apparent discrepancy between the known active margin
tectonic regime of the Havallah basin and the chemistry
of its sands is discussed above. Contents of the three major-element oxides in eight of the nine analyzed samples
of subunit hcu show significant overlap with analyzed
samples from subunits hu and hys, described above, and
with subunit hss of tectonostratigraphic plate C to be described below.

Relations among Ba, Sr, Mn, Pb, Zn, Cu, As, Ag,
and Au were examined in standard "spider"-graph-type
plots, normalized for comparative purposes to the same
sample (analysis 6, table 5) as were the preceding similar
plots for analyses of samples of subunit hu (fig. 38). The
"spider"-graph-type plots reveal overall similar patterns
of normalized elemental enrichments and depletions in
subunits hcu and hu (compare figs. 36 and 38). One of the
samples analyzed from subunit hcu, however, is strongly
depleted in its normalized ratio of Ba relative to the other
samples (81 ppm Ba, analysis 7, table 6). This particular
sample also contains a high concentration of carbonate
minerals and also the highest concentration of Au (10 ppb)
in the entire suite of samples of sandstone analyzed from
the Havallah sequence. All normalized ratios for other elements, with the exception of the normalized ratio for Ag,
also are relatively low in analysis 7.

Tectonostratigraphic Plate C

Rocks of tectonostratigraphic plate C, mostly radiolarian ribbon chert and shale, are in thrust contact with
underlying rocks of plate B, and they are the least widely
exposed rocks of the variously designated plates in the
Havallah Hills (pls. 1, 2; fig. 31). These rocks crop out in
four places: (1) in a fairly restricted, north-south elongate
topographically high area that straddles the border between
the Valmy and North Peak quadrangles, (2) in two small,
hill-capping, poorly exposed masses along the Radio Tower
fault in NW 1/4 sec. 11, T. 33 N., R. 42 E., (3) in two other
small occurrences in low hills east of a narrow Quaternary
gravel-filled valley just to the east of the main mass of the
Havallah Hills, and, finally, (4) in a small occurrence near
the east-central part of the Havallah Hills (fig. 31). The
chert and shale of plate C in the first area shows an increasing abundance of shale and non-calcareous silt from
north to south (pl. 1). In places, strata of plate C include
some fine-grained sandstone. An another isolated locality

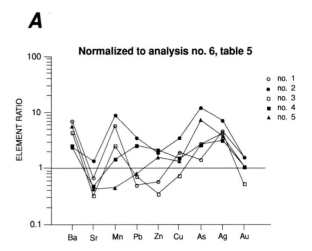

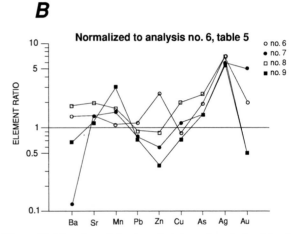

Figure 38 Nine elements in nine selected samples of subunit hcu in
tectonostratigraphic plate B of the Devonian, Mississippian,
Pennsylvanian, and Permian Havallah sequence. Analysis numbers same
as table 6. Normalized to analysis 6 of table 5 using 50 percent of the
lower limit of determination for those elements that are below detection.
A, Analyses 1–5; *B*, Analyses 6–9.

of chert and shale is present in NW 1/4 sec. 12, T. 33 N.,
R. 42 E., and may be part of plate C (pl. 1). At this locality,
chert and shale are overlain unconformably by rocks belonging to plate D.

Six thin sections of chert from plate C were examined petrographically and these samples all show black,
bluish-green, and brownish-black laminated radiolarian
ribbon chert to be quite dolomitic and to contain variable
proportions of extrabasinal-derived detrital material, including quartz, plagioclase, and white mica. Presence of
this extrabasinal material in these rocks indicates that some
terrigenous debris reached the deepest basinal parts of the
Havallah sequence and foreshadowed an eventual filling
by widespread, stratigraphically high-silt and sandstone
accumulating subsequently in the basin.

Table 6—Analytical data on calcareous litharenite of subunit hcu of lithotectonic unit 2 of Devonian, Mississippian, Pennsylvanian, and Permian Havallah sequence, Humboldt County, Nevada.†

Analysis number	1	2	3	4	5	6	7	8	9
Field number 90TT..	230	233	234	235	247	272	273	276	277
Chemical analyses (weight percent)									
SiO_2	81.6	69.9	90.2	86.7	92.3	73.5	34.6	68.1	71.5
Al_2O_3	4.52	3.42	1.69	3.03	1.07	1.98	1.22	1.36	.79
Fe_2O_3	1.03	2.64	.68	1.1	.62	.54	.25	.53	.24
FeO	.66	.24	.12	.14	.16	.32	.16	.15	.14
MgO	1.23	1.05	.54	.17	.11	.38	.25	.65	.55
CaO	2.94	10.3	2.38	3.55	.65	11.9	34.8	14.8	14.1
Na_2O	.92	.55	—	—	—	.48	.45	—	—
K_2O	.95	.99	.41	.51	.16	.25	.07	.27	.12
H_2O^+	.94	1.06	.41	.99	.37	.53	.25	.56	.44
H_2O^-	.14	.31	.15	.18	.21	.09	.06	.21	.12
TiO_2	.46	.46	.1	.21	.1	.21	.14	.12	.08
SiO_2	81.6	69.9	90.2	86.7	92.3	73.5	34.6	68.1	71.5
P_2O_5	.2	.46	.12	.08	.41	.88	.49	.81	.53
MnO	.07	.11	.03	.02	.01	.02	.02	.03	.04
F	.04	.08	.03	.03	.06	.12	.07	.12	.07
Cl	.005	.004	.003	.009	.005	.004	.004	.006	.011
Organic C	.1	.2	.08	.15	.16	.35	.95	.42	.41
Carbonate C	.8	2.16	.59	.71	.02	2.37	7.35	3.04	2.91
CO_2	2.93	7.92	2.16	2.6	.07	8.69	27.	11.1	10.7
Total S	<.07	<.01	<.01	<.01	.34	.09	<.01	.11	<.01
Subtotal	99.61	101.85	99.69	100.19	96.83	102.7	108.1	102.39	102.75
Less O=F, Cl, S	.02	.03	.01	.01	.29	.07	.03	.08	.03
Less carbonate C	.8	2.16	.59	.71	.02	2.37	7.35	3.04	2.91
Total	98.79	99.66	99.09	99.47	96.52	100.26	100.72	99.27	99.81
Inductively coupled plasma-atomic emission spectroscopy (total) (parts per million)									
As	—	20.0	—	—	—	—	—	—	—
Ba	4500.0	1500.	2800.0	1600.0	3600.0	880.0	81.0	1200.0	440.0
Ce	42.	71.	6.	17.	10.	22.	10.	11.	7.
Co	8.	8.	4.	6.	4.	3.	2.	4.	3.
Cr	36.	32.	7.	15.	13.	33.	39.	44.	26.
Cu	13.	24.	5.	10.	9.	6.	8.	14.	5.
Ga	6.	—	—	4.	—	—	—	—	—
La	23.	45.	4.	11.	15.	22.	11.	18.	10.
Li	23.	20.	37.	14.	8.	6.	4.	10.	6.
Mo	—	2.	—	—	—	—	—	—	—
Nd	21.	39.	5.	9.	13.	17.	11.	11.	8.
Ni	17.	33.	8.	11.	6.	9.	4.	10.	4.

Table 6 — (cont.)

Analysis number	1	2	3	4	5	6	7	8	9
Field number 90TT..	230	233	234	235	247	272	273	276	277
Inductively coupled plasma-atomic emission spectroscopy (total) (parts per million) (cont.)									
Pb	—	13.	—	9.	—	—	—	—	—
Sc	3.	3.	—	—	—	—	—	—	—
Sr	95.	200.	39.	61.	49.	210.	230.	310.	180.
Th	5.	7.	—	—	—	—	—	—	—
V	25.	32.	9.	27.	7.	17.	13.	20.	8.
Y	17.	38.	5.	12.	15.	36.	15.	25.	16.
Yb	2.	3.	—	1.	—	2.	—	1.	1.
Zn	17.	57.	10.	63.	46.	79.	18.	27.	11.
Inductively coupled plasma-atomic emission spectroscopy (partial) (parts per million)									
Ag	0.15	0.23	0.14	0.1	0.12	0.23	0.19	0.23	0.18
As	—	15.	2.	2.3	8.4	1.2	—	2.2	.77
Cd	—	.19	.20	.31	.22	1.6	.38	.48	.36
Cu	11.	18.	3.7	8.5	7.1	6.4	5.5	7.4	4.9
Mo	.47	1.6	.22	.6	.59	.48	.28	.59	.29
Pb	1.3	9.4	1.9	6.9	2.1	3.2	2.2	2.5	2.
Sb	.74	1.4	1.1	1.2	3.1	—	—	—	—
Zn	21.	56.	15.	61.	43.	72.	21.	29.	16.
Energy-dispersive X-ray fluorescence spectrometry (parts per million)									
Nb	11.0	10.0	—	—	—	—	—	—	—
Rb	21.	26.	—	16.0	—	—	—	—	—
Sr	95.	198.	45.0	68.	58.0	205.0	210.0	295.0	170.0
Zr	570.	1000.	164.	154.	136.	180.	104.	96.	62.
Y	31.	50.	14.	19.	20.	36.	17.	26.	20.
Ba	4300.	1300.	2400.	1400.	>5000.	790.	79.	1000.	420.
Ce	—	63.	—	38.	—	—	—	—	—
Chemical analyses (parts per million)									
As	1.9	17.0	3.6	3.5	10.0	2.7	2.0	3.5	2.0
Au	.003	.003	<.002	.002	.002	.004	.01	<.002	<.002
Te	—	—	—	—	—	—	—	—	—
Tl	.1	.1	.05	.05	.05	—	—	.05	.05
Hg	—	.03	.02	.08	.35	.07	.02	.05	.03

†[Chemical analyses of major oxides in weight percent by methods of Jackson and others (1987); analyst, unknown. FeO, H_2O^+, H_2O^- in weight percent by methods of Jackson and others (1987); analyst, N.H. Elsheimer. Total C in weight percent by combustion-thermal conductivity method; analyst, N.H. Elsheimer. Carbonate C in weight percent by extraction-coulometric titration method; analyst, C.S. Papp. Organic C by difference. Total S in weight percent by combustion-infra-red method; analyst, N.H. Elsheimer. F, Cl in weight percent by specific ion electrode method; analyst, N.H. Elsheimer. Inductively coupled plasma-atomic emission spectroscopy (total and partial) in parts per million by methods of Crock and Lichte (1982), Lichte and others (1987), and Motooka (1988); analysts, D.L. Fey, and J.M. Motooka. Precision for concentration higher than 10 times the detection limit is better than +10–percent relative standard deviation; instrumental precision of the scanning instrument is +2–percent relative standard deviation. Looked for, but not found, at parts-per-million detection levels in parantheses: Au (8), Be (1), Bi (10), Cd (2), Eu (2), Ho (2), Nb (4), Sn (5), Ta (40), U (100). Energy-dispersive X–ray fluorescence spectrometry in parts per million by method of Johnson and King (1987); analyst, J. Kent. Looked for, but not found, at parts-per-million detection levels in parantheses: Ce (30). Chemical analysis for Au by combined chemical separation-atomic absorption method involving heating 10–g sample to 700° C to remove sulfides and then decomposing sample in HBr and Br_2. Au is extracted into methyl isobutyl Ketone, and organic layer analyzed for Au by graphite furnace-atomic absorption spectrometry (Meier, 1980); analyst, H. Smith. Chemical analysis for As, Te (0.05 ppm detection level), and Tl by graphite furnace atomic absorption spectrometry (Wilson and others, 1987; Aruscavage and Crock, 1987); analysts, A.H. Love, L.A. Bradley, and P.L. Hageman; chemical analysis for Hg by cold-vapor atomic absorption spectrometry; analysts, A.H. Love, L.A. Bradley, and P.L. Hageman; —, not detected]

Tectonostratigraphic Plate D

In contrast to well-defined fault-bounded bases of the three successively lower plates in the Havallah sequence described above, the base of plate D is generally not well exposed. It is placed, nonetheless, at the unconformity at the base of a large expanse of Pennsylvanian and (or) Permian calcareous sandstone, which rests on strata of interbedded limestone and black chert, and which crops out at the locality in NW 1/4 sec. 12, T. 33 N., R. 42 E. described above, north of the Cottonwood Creek fault. The base of this tectonostratigraphic plate apparently is structural south of the Cottonwood Creek fault (see relations of calcareous sandstone subunit hss, pls. 1, 2). The geologic age of the calcareous sandstone is probably Wolfcampian on the basis of two fusulinid localities in NE 1/4 sec. 2, T. 32 N., R. 42 E. (B.L. Murchey, written commun., 1993); or no more specific than early Permian on the basis of a number of fusulinids in the same area as described in this section below. Unconformities apparently also have been recognized in some of the highest packages of the Havallah-equivalent Schoonover sequence in the Independence Mountains (fig. 1; Miller and others, 1984; Whiteford, 1990). As will be described below, offsets along a mineralized normal fault, the Section 11 fault (pl. 2), marking the eastern boundary of subunit hss to the west-northwest of the site of the abandoned Oyarbide ranch, are probably quite substantial. This relation in turn suggests that a significant proportion of LU–2, including much of plates A through C, has been cut out structurally along the Section 11 fault where this fault crops out in the hanging wall block of the Oyarbide fault.

Overall the Havallah Hills part of its area of outcrop, subunit hss extends approximately 5 km in a northeasterly direction, and it has a maximum width of outcrop of about 2 km in a northwesterly direction. Subunit hss in the Havallah Hills also is bounded partly on the northwest by the northeast-striking Radio Tower fault, a normal fault that has faulted downwards its southeastern block, which includes mostly rocks of subunit hss (pl. 1). In places, subunit hss is interbedded with sequences of black spicular chert and gray sandy micrite (subunit hlc, pl. 2) near the south end of the Havallah Hills. Moreover, subunit hss is intruded by a small number of Tertiary or Cretaceous altered felsite dikes or monzogranite along the southeastern flank of the Havallah Hills. None of these intrusions appears to have altered significantly the surrounding country rock.

Monocrystalline quartz is the predominant detrital framework grain in three samples of immature calcareous sublitharenite from subunit hss analysed modally from the Havallah Hills part of the subunit (fig. 33A). Modal data from all three of these samples plot in a fairly tight cluster

in the quartzose recycled orogen provenance field, and extrabasinal carbonate grains at framework sites make up 72 to 86 volume percent of the total lithic fragments is chert. As such, these rocks are not notablydifferent from other calcerous clastic subunits of the Golconda allochthon. However, the calcareous component of some samples from this subunit shows somewhat increased abundances of intrabasinal carbonate material that in thin section typically appears to be surrounded by rounded to well-rounded grains of monocrystalline quartz. The monocrystalline quartz grains appear to have been imbedded in calcerous mud.

Chemical analyses of eight samples of calcerous litharenites from subunit hss are generally similar to those discussed above for subunits hu, hys, and hcu (compare tables 4–7; see also, fig. 35). However, mean content of Al_2O_3 in eight analyzed samples is approximately 0.7 weight percent higher than mean content of Al_2O_3 from the three other subunits, possibly reflecting an increased concentration of feldspar or clay. The mean content of K_2O also is higher in the analyzed samples of subunit hss: 0.84 compared with 0.56 wight percent in the 21 samples from other subunits analyzed. This relation seems to support the judgment that rocks of subunit hss probably contain overall higher concentrations of feldspar and (or) clay(s) than the other sandstone and siltstone subunits described above. An enhanced abundance of feldspar rather than clay(s) appears to be supported by a limited number of modally analyzed samples, three in all, from subunit hss north of Cottonwood Creek fault, which, in the ternary Qm–F–Lt diagram, all fall on the feldspar side of the center of gravity for other samples analyzed modally (fig. 33A). As will be discussed below, rocks from subunit hss south of the Cottonwood Creek fault also show somewhat increased contents of framework feldspar relative to other subunits of LU–2.

Spider-graph plots for Ba, Sr, Mn, Pb, Zn, Cu, As, Ag, and Au of subunit hss are shown in figure 39, normalized for comparative purposes to the same sample (analysis 6, table 5) as were the preceding spider-graph plots for analyses of samples of subunits hu and hcu. A pattern showing normalized ratios enriched for As and Ag, and nominally flat for the base-metal part of the diagram, is present for most of the analyzed samples from subunit hss—a pattern essentially equivalent to those determined for the other subunits discussed above. The analogous patterns of minor-element normalized ratios throughout the carbonate-bearing siliciclastic facies of the Havallah sequence, regardless of stratigraphic position of the samples analyzed, suggests that the same cratonal and continental margin sources(s) continued to shed debris without significant interruption into the basin of the Havallah sequence throughout much of its depositional history.

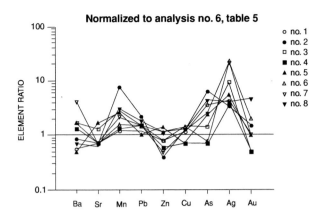

Figure 39 Nine elements in eight selected samples of subunit hss in tectonostratigraphic plate D of the Devonian, Mississippian, Pennsylvanian, and Permian Havallah sequence. Analysis numbers same as table 7. Normalized to analysis 6 of table 5 using 50 percent of the lower limit of determination for those elements that are below detection.

Tectonostratigraphic Plate E

The rocks of tectonostratigraphic plate E, structurally the highest of the thrust plates in LU–2 of the Havallah Hills and including mostly pebbly limestone and conglomerate, crop out in the three following areas of the North Peak quadrangle: (1) the largest area of outcrop, approximately 1.5 km², in the southeast part of the Havallah Hills; (2) the second area is an approximately 0.2–km²-area southeast of the trace of the Cottonwood Creek fault and 0.8 km southwest of the Old Marigold Mine; and (3) the third area comprises a narrow, north-south elongate mass of pebbly limestone and conglomerate that crops out in the east-central part of the North Peak quadrangle (fig. 31; pl. 2). The structural base of plate E is herein termed the Buffalo thrust fault, a low-angle, southeast dipping fault, that cuts across attitudes of mostly east-west striking, north-dipping strata in plate D near the southeast corner of the Havallah Hills.

The pebbly limestone unit in the southeast corner of the Havallah Hills previously was mapped by Roberts (1964) as part of the Jory Member of his Havallah Formation, at that time presumed to be Pennsylvanian and Permian (?) in age and at that time assumed to be the lowest member of the Havallah Formation. Absence of cross beds, ripple marks, and other shallow water features suggested to Roberts (1964) that deposition of rocks of subunit hpls occurred below wave base. These rocks were reported by Roberts (1964) to contain an assemblage of fusulinid fauna in the southeast part of the Havallah Hills, specifically in SE 1/4 sec. 12, T. 33 N., R. 42 E., that were considered to be Atokan or early Middle Pennsylvanian in age. However, these fusulinid collections also were described as having been reworked and actual age of the rocks could

not be established more precisely than Pennsylvanian and Permian (?). Subsequently, re-examination by Calvin H. Stevens of San Jose State University (cited in Murchey, 1990) of additional fusulinid collections from two other localities of subunit hpls in the North Peak quadrangle resulted in discovery of abundant fusulinids of probable Wolfcampian age, Early Permian. These latter two localities are described as being hosted by coarse-grained sandy bioclastic limestone. However, subunit hpls in plate E also contains some thin sequences of green radiolarian chert near its base that are unequivocally part of the stratigraphic succession of subunit hpls.

Additional evidence that the base of plate E is structural is provided by presence of Mississippian (Osage-Merramec) sequences of green radiolarian ribbon chert in the lowermost strata of plate E, which green radiolarian chert in turn rests on rocks of subunit hss. The latter rocks are as young as Leonardian and early Guadalupian in age as described above. The age of the pebbly limestone subunit, which constituted much of the Jory member of Roberts (1964), thus is now known to range widely in age, from as old as Mississippian to as young as Early Permian. Minor green radiolarian chert beds near the base of the subunit contain microfauna from which Mississippian (assemblages 1 through 4, Osage to Meramec; B.L. Murchey, written commun., 1993) have been obtained at four localities. In addition, three other localities of radiolarian chert beds in strata assigned to plate E include Mississippian and Pennsylvanian microfauna (assemblage 5, Chester to early Atokan; B.L. Murchey, written commun., 1993). In SE 1/4 sec.14, T. 33 N., R. 42 E., *Archaediscus* ? sp., a small possibly reworked fusulinid, is present in some well exposed pebbly limestone, and has been assigned a Mississippian and (or) Permian range in age by C. H. Stevens (written commun., 1994).

Rocks of subunit hpls in plate E display many post-diagenetic, outcrop-scale structures associated with a brittle-style of deformation. Many outcrops of pebbly limestone and conglomerate show abundant concentrations of slickensides on joint surfaces near the trace of the Buffalo thrust fault. Strata in plate E in this area furthermore exhibit a small number of broad open folds and variable attitudes of bedding; a high proportion of attitudes in plate E show north-south strikes and moderate dips to the east. Rocks of plate E along the common border between secs. 1 and 2, T. 32 N., R. 42 E., apparently form a thin klippe that lies astride relatively steep-dipping chert and interbedded chert and micrite probably belonging to plate A (pl. 2). If these subunit assignments of the rocks are generally correct, then rocks belonging to plates B through plate C probably are missing here primarily as a result of their being cut out structurally by displacements and ramping along the Buffalo thrust (pl. 2). By a similar line of reasoning, considered together with absence of strata

Table 7—Analytical data on calcareous litharenite of subunit hss of lithotectonic unit 2 of Devonian, Mississippian, Pennsylvanian, and Permian Havallah sequence, Humboldt County, Nevada.†

Analysis number	1	2	3	4	5	6	7	8
Field number 90TT..	248	259	264	280	290	303	305	249
				Chemical analyses (weight percent)				
SiO_2	83.2	76.1	70.2	80.7	35.1	84.1	82.1	80.6
Al_2O_3	3.12	1.97	4.34	3.32	.92	4.17	2.51	3.28
Fe_2O_3	.41	.78	.79	.33	.48	.9	.41	.46
FeO	.08	.05	.2	.14	.37	.18	.11	.09
MgO	.64	.16	1.5	1.03	.4	.52	.77	.52
CaO	5.4	10.5	9.91	6.04	34.3	2.69	6.07	6.68
Na_2O	—	.54	.78	.32	.22	.69	.44	—
K_2O	1.11	.18	1.06	1.23	.12	1.36	.68	1.03
H_2O+	.78	.5	.81	.55	.4	.67	.42	.88
H_2O-	.15	.21	.42	.23	.05	.25	.18	.2
TiO_2	.17	.23	.33	.24	.07	.67	.18	.23
P_2O_5	.07	.12	.16	.13	—	.13	.07	.14
MnO	.02	.09	.04	.02	.03	.03	.03	.04
F	.03	.04	.03	.04	.03	.04	.03	.04
Cl	.004	.006	.004	.005	.006	.004	.004	.006
Organic C	.09	.27	.23	.15	.35	.17	.14	.25
Carbonate C	1.26	2.2	2.45	1.52	7.41	.59	1.48	1.43
CO_2	4.62	8.07	8.98	5.57	27.2	2.16	5.43	5.24
Total S	<.01	<.01	<.01	<.01	<.01	<.01	<.01	<.01
Subtotal	101.15	102.02	102.23	101.57	107.46	99.32	101.05	101.12
Less O=F, Cl, S	.01	.02	.01	.02	.01	.02	.01	.02
Less carbonate C	1.26	2.2	2.45	1.52	7.41	.59	1.48	1.43
Total	99.88	99.8	99.77	100.03	100.04	98.71	99.56	99.67
			Inductively coupled plasma-atomic emission spectroscopy (total) (parts per million)					
Ba	370.0	560.0	1100.0	840.0	320.0	1100.0	2700.0	450.0
Ce	9.	11.	24.	15.	10.	52.	9.	16.
Co	3.	6.	3.	3.	3.	4.	4.	3.
Cr	11.	12.	23.	24.	16.	35.	9.	21.
Cu	8.	9.	10.	5.	5.	10.	8.	9.
Ga	—	—	—	—	—	4.	—	—
La	6.	8.	16.	10.	9.	28.	6.	9.
Li	12.	7.	9.	5.	4.	7.	4.	9.
Nd	4.	6.	13.	11.	11.	23.	5.	9.
Ni	6.	14.	10.	3.	4.	9.	6.	3.

87

Analysis number	1	2	3	4	5	6	7	8
Field number 90TT..	248	259	264	280	290	303	305	249
Pb	6.	7.	6.	6.	—	10.	7.	8.
Sc	—	—	—	—	—	2.	—	—
Sr	110.	99.	200.	110.	280.	89.	110.	87.
Th	—	—	—	—	—	6.	—	—
V	16.	14.	17.	9.	14.	30.	10.	15.
Y	6.	9.	15.	10.	11.	17	6.	9.
Yb	—	1.	1.	—	—	2.	—	1.
Zn	19.	6.	19.	11.	38.	32.	8.	29.

Inductively coupled plasma-atomic emission spectroscopy (partial) (parts per million)

	1	2	3	4	5	6	7	8
Ag	0.14	0.11	0.31	0.12	0.18	0.73	0.69	0.13
As	3.8	7.5	.69	—	1.4	4.6	2.2	4.3
Cd	.28	.21	.26	.4	.71	.3	.18	.62
Cu	6.9	7.	6.7	5.3	5.2	7.2	7.8	5.8
Mo	.23	.64	.39	.21	.26	.57	.3	.32
Pb	3.3	6.2	4.	4.3	2.8	4.3	4.4	5.1
Sb	.8	—	—	—	—	1.5	1.3	1.2
Zn	24.	12.	25.	18.	43.	34.	15.	33.

Energy-dispersive X-ray fluorescence spectrometry (parts per million)

	1	2	3	4	5	6	7	8
Nb	—	—	11.0	—	—	13.0	—	—
Rb	23.0	—	28.	21.0	—	28.	10.0	19.0
Sr	106.	102.0	194.	106.	255.0	92.	106.	91.
Zr	215.	1100.	325.	475.	89.	1600.	520.	350.
Y	14.	26.	23.	19.	17.	41.	17.	18.
Ba	350.	520.	930.	750.	300.	960.	2400.	430.
Ce	—	—	36.	—	—	35.	—	—

Chemical analyses (parts per million)

	1	2	3	4	5	6	7	8
As	5.0	8.7	2.0	1.0	3.3	1.1	3.5	5.9
Au	.002	.003	<.002	<.002	.002	.004	.002	.009
Te	—	—	—	—	—	—	—	—
Tl	.1	.05	.05	.05	—	.15	.05	.1
Hg	.08	.05	.05	.06	.02	.08	.03	.09

†[Chemical analyses of major oxides in weight percent by methods of Jackson and others (1987); analyst, unknown. FeO, H_2O^+, H_2O^- in weight percent by methods of Jackson and others (1987); analyst, N.H. Elsheimer. Total C in weight percent by combustion-thermal conductivity method; analyst, N.H. Elsheimer. Carbonate C in weight percent by extraction-coulometric titration method; analyst, C.S. Papp. Organic C by difference. Total S in weight percent by combustion-infra-red method; analyst, N.H. Elsheimer. F, Cl in weight percent by specific ion electrode method; analyst, N.H. Elsheimer. Inductively coupled plasma-atomic emission spectroscopy (total and partial) in parts per million by methods of Crock and Lichte (1982), Lichte and others (1987), and Motooka (1988); analysts, D.L. Fey, and J.M. Motooka. Precision for concentration higher than 10 times the detection limit is better than ±10-percent relative standard deviation; instrumental precision of the scanning instrument is ±2-percent relative standard deviation. Looked for, but not found, at parts-per-million detection levels in parantheses: As (10), Au (8), Be (1), Bi (10), Cd (2), Eu (2), Ho (4), Mo (2), Nb (4), Sn (5), Ta (40), U (100). Energy-dispersive X-ray fluorescence spectrometry in parts per million by method of Johnson and King (1987); analyst, J. Kent. Looked for, but not found, at parts-per-million detection levels in parantheses: La (30). Chemical analysis for Au by combined chemical separation-atomic absorption method involving heating 10-g sample to 700° C to remove sulfides and then decomposing sample in HBr and Br_2. Au is extracted into methyl isobutyl Ketone, and organic layer analyzed for Au by graphite furnace-atomic absorption spectrometry (Meier, 1980); analyst, H. Smith. Chemical analysis for As, Te (0.05 ppm detection level), and Tl by graphite furnace atomic absorption spectrometry (Wilson and others, 1987; Aruscavage and Crock, 1987); analysts, A.H. Love, L.A. Bradley, and P.L. Hageman. Chemical analysis for Hg by cold-vapor atomic absorption spectrometry; analysts, A.H. Love, L.A. Bradley, and P.L. Hageman;—, not detected]

immediately below pebbly limestone and conglomerate belonging to the Pennsylvanian period, pebbly limestone and conglomerate making up plate E in the latter area must also have been juxtaposed directly onto Mississippian radiolarian chert belonging to plate A during shortening associated with emplacement of the Golconda allochthon. In addition, such shortening must also have been accomodated, in part, by major displacements along the Willow Creek thrust that must be present at shallow depths below a down-dropped klippe, only about 0.5 km to the east of the trace of the Section 11 fault (pl. 2).

Poorly sorted pebbly limestone and conglomerate of subunit hpls in the southeast part of the Havallah Hills crops out mostly as drab gray, resistant rubble-rich exposures that cap ridges (fig. 40*A, B*). In many places, 1– to 2–m-thick beds show abundant concentrations of fairly equant dark, angular chert fragments and subrounded quartzarenite fragments probably derived mostly from lower Paleozoic sequences in the Roberts Mountains allochthon. The quartzarenite fragments commonly are oriented dimensionally with their long axes parallel to

bedding (fig. 40*C*). Although actual exposures are poor even in some of the best preserved sequences here, it appears that tops and bottoms of beds are relatively planar and sharp, and that individual beds show some local inversions in their grading. These strata are further interpreted to reflect incomplete, high-density turbidites, possibly including some R and S_2 intervals of the idealized Lowe (1982) high-density turbidite. Ratios between proportions of quartzarenite and chert fragments are highly variable throughout the subunit, as is amount of calcareous material, including framework micrite grains, intrabasinal carbonate grains, calcite veins, and carbonate cement. An increased abundance of calcite veins appears to be present near the trace of the north-east striking fault in the central part of sec. 13, T. 33 N., R. 42 E., which fault has been intruded by a small Tertiary or Cretaceous dike approximately 0.6 km to the northeast (pl. 2). In addition, the subunit includes some thin sequences of siliciclastic-to-somewhat-calcareous pebbly sandstone and quartzarenite, and a few minor apparent interbeds of radiolarian chert beds.

A small number of other rocks of varying lithologies from the pebbly limestone subunit were examined petrographically. Under the microscope, calcareous pebbly sandstone shows poorly sorted submature fabrics including highly undulose monocrystalline quartz, polycrystalline quartz (some reflecting metamorphosed quartzarenite protoliths), chert, micrite, siltite (some calcareous), packed biosparite, various opaque and heavy minerals, and fossil fragments (some phosphatic) at framework sites. Although only nine thin sections from the subunit were studied petrogaphically, it appears that feldspar grains at framework sites in these rocks are rare, in fact, probably much rarer than in any of the calcareous-siliciclastic rocks described in any of the plates above.

Figure 40   Chert-micrite, pebble limestone and conglomerate, subunit hpls, plate E, of Devonian, Mississippian, Pennsylvanian, and Permian Havallah sequence in southeast part of the Havallah Hills. *A*, General character of hill-forming, rubbly exposures at locality 91TT029 viewed to northeast; *B*, closeup of typical 1– to 2–m-thick discontinuous beds, locality same as *A*, note hammer in center of photograph for scale; *C*, weathered joint surface of pebble conglomerate showing dark brownish-black chert fragments, at head of arrow, and light-colored quartzarenite fragments dimensionally oriented parallel to bedding, locality 91TT033. Knife is approximately 10 cm long.

LITHOTECTONIC UNIT 2 PLATES BETWEEN COTTONWOOD CREEK AND OYARBIDE FAULTS

Several of the tectonostratigraphic plates delineated above as belonging to LU–2 and making up a significant areal part of the Golconda allochthon south of the Cottonwood Creek fault and north of the Oyarbide fault (pl. 2) will be described together in this part of the report. Detailed descriptions of pebbly limestone and conglomerate of plate E in two small areas between these two faults are included in the subsection immediately above and are not repeated here. Aside from aforementioned rocks of plate E between the Cottonwood Creek and Oyarbide faults, rocks of LU–2 apparently belong to plates A and D. Plates B and C are missing evidently because they have been cut out structurally. Plate A here includes fairly continuous exposures of presumably late Pennsylvanian and early Permian (late Atokan and Wolfcamp Epochs, assemblages 6 and 7 of Murchey, 1990) basaltic trachyandesite and trachyandesite and its interbedded volcaniclastic sedimentary rocks along an almost 8–km-long strike length, of which the base of the basaltic trachyandesite and trachyandesite is the Willow Creek thrust (subunit hba, pl. 2). Although stratigraphic base of subunit hba is not exposed, estimates of total thickness of volcanic rock in the subunit also are limited by absence of well-exposed complete sequences of rock. Nonetheless, there appear, at most, to be no more than approximately several 100 meters of volcanic strata present in subunit hba. Furthermore, the structural relation between subunit hba and underlying rocks belonging to LU–1 is remarkably similar to structural relations reported by Miller and others (1984) in the Schoonover sequence in the Independence Mountains, Nev. (fig. 6). In the Independence Mountains, the base of greenstone (basalt), which is overlain depositionally by latest Devonian to Early Mississippian chert, is everywhere mapped as a thrust fault (Miller and others, 1984). Some late Atokan and Wolfcampian chert similarly is present depositionally above basaltic trachyandesite and trachyandesite in the area south of the Cottonwood Creek fault (pl. 2; B.L. Murchey, unpub. data, 1994).

Basaltic trachyandesite and trachyandesite between the Cottonwood Creek and Oyarbide faults are overlain sequentially from base to top by ribbon chert (subunit hc, pl. 2), calcareous sandstone and siltstone (subunit hss, pl. 2), and interbedded black chert and gray micrite (subunit hlc, pl. 2). Subunits hc and hlc are the most continous laterally of these three relatively thin subunits, and variably-elongate masses of subunits hc and hlc extend from the general area of the Old Marigold Mine on the north almost to the expansive Quaternary deposits present near the Oyarbide fault on the south.

Structurally above these rocks are some prominent Early Permian calcareous sandstones, including some se-quences of calcareous siltstone, and interbedded black chert and gray micrite (respectively subunits hss and hlc, pl. 2). Although there are some additional apparently minor thrust faults at the base of some of the mapped subunits above the large masses of calcareous sandstone, rocks in the upper plates of these minor thrusts (subunits hlc and hc, pl. 2) are not lithologically the same as pebbly limestone and conglomerate found farther to the north in the same tectonostratigraphic position. Therefore, it is difficult to assign lateral continuity to these thrust faults beyond their mapped traces. Moreover, offsets along these thrust faults are probably quite minor.

Early Permian calcareous sandstones, including some calcareous siltstone and mapped together as subunit hss (pl. 2), are present in two fairly continuous masses mostly in secs. 24 and 25, T. 33 N., R. 43 E., and in an approximately 5–km-long mass that extends north from the Oyarbide fault in the general area of the Section 11 fault (pl. 2). Modal analyses of nine fine-grained, mostly submature sandstone samples from these masses show a fairly tight clustering of data points in the sublitharenite field of McBride (1963), and five of the modes are within the quartzose recycled orogen field of Dickinson and others (1983), while the remaining four are close to the join between that field and the interior craton provenance field (fig. 33*A*). As such, data plots of the sandstones in this tectonic block are close to the center of gravity of all of the calcareous sandstones analyzed modally from the Havallah sequence in the study area, although the nine data points appear to cluster slightly towards the feldspar corner of the ternary diagram. Modal analyses of six of nine samples also are within the compositional field determined for quartz-chert petrofacies sandstones of the Mount Tobin assemblage of Tomlinson (1990) as described above. However, only one of these six samples analyzed from subunit hss south of the Cottonwood Creek fault shows a compositional facies containing abundant chert fragments and minimal extrabasinal carbonate grains at the framework sites, and is thus remarkably similar lithologically to the quartz-chert petrofacies of Tomlinson (1990). That one sample is from a locality in SW 1/4 sec. 11, T. 32 N., R. 42 E., near the North Peak Au deposit, and is the southernmost sample analyzed modally. In fact, no extrabasinal carbonate grains were found in the sample. Although a linkage appears to be present in provenance between the general area of that one sample and and similar sandstones from the Mount Tobin area, primarily because of absence of extrabasinal carbonate grains, nonetheless a significant proportion of carbonate rock is present elsewhere in the immediately surrounding sequence of rocks making up subunit hss near the North Peak Au deposit. The quartz-chert petrofacies rock under discussion is interbedded with much more widespread carbonate strata in the immediate area of its sampling site. These carbonate strata, where well exposed along drill roads in sec. 11, include well-bedded brownish-gray calcareous siltstone and fine-grained calcareous sandstone showing partings in the range 5 cm to 0.5 m. These strata also have plane-parallel or flat bedding contacts, suggesting that the calcareous siltstone and fine-grained sandstone sequence in this area may be a Tb interval of the classic Bouma (1962) turbidite and thus indicative of low-density traction sedimentation (Lowe, 1982). Across approximately 30–m exposures along roadcuts, three-to-five cycles of alternating thin-bedded and thick-bedded siltstone typically are present that overall appear to show weak thickening upwards of individual beds across the entire exposure.

Three samples analyzed modally are from the immediate area of the North Peak Au deposit, which, in all, appears to underlie approximately 0.3 km² of visibly altered rocks at the surface. Two samples, 92TT089 and 92TT091, are from the northernmost parts of the area drilled during exploration by several companies, and these two samples were collected from rock strata that are altered minimally. However, one of the samples, 92TT056, is from some of the most altered rock surrounding this deposit. This sample includes apparently elevated abundances of detrital grains of monocrystalline quartz, which elevated abundances result from decalcification reactions that depleted the rock of both detrital grains of extrabasinal sparry micrite and calcite cement. The rock also shows an approximately 7 volume percent porosity, an extraordinary porosity when one considers the generally nonporous nature of most unaltered rocks of the Havallah sequence. Pore spaces now are filled partially by secondary K–feldspar and clay minerals probably related to Au mineralization at the deposit.

Another fine-grained calcareous quartz sandstone of subunit hss, sample 90TT356, collected from the southernmost of two large masses of the subunit that crop out between the Cottonwood Creek and the Oyarbide faults, specifically in NE 1/4 sec. 2, T. 32 N., R. 42 E., is important both from a provenance standpoint and biostratigraphically. This particular sample contains subrounded to rounded grains of quartz cemented by calcite. Fossil fragments are fairly abundant in the sample and they are randomly distributed throughout the rock (fig. 41). The fragments include rugosa corals, bryozoans, brachiopods, echinoderms, and ostracodes (A.K. Armstrong, written commun., 1992). In addition, according to Armstrong, several smaller fragments of foraminifera and fusulinids are present, one of which appears to have the wall structure of the genus *Triticites*, which suggests a Late Pennsylvanian or even Early Permian age. A.K. Armstrong (written commun., 1992) further describes this sample as follows:

Many of the fossil fragments have oncolitic coatings, which are formed by algae and a foraminifera. These grow in a carbonate-sediment setting with "minor amounts" of terrigenous clastic input, typically in shallow water less than 20 m deep and in an environment that includes strong agitation, such as tidal currents or waves. Some of the oncoliths do contain terrigenous quartz grains in their centers and some banding so that the environment of carbonate biologic activity was not free [totally] of clastic quartz. The fossil fragments appear to have not been part of a lithified rock previously but [were instead] loose fossil fragments in a carbonate sand with pellets of uncemented lime mud and some quartz sand. ***the fossil fragments were deposited on a carbonate platform first, rattled around as carbonate sands, and were later carried by storm waves off the shelf and deposited in deeper water in a quartz sand environment. The finer, soft lime mud pellets [then] formed the cement for the quartz sand

Further examination of this particular sample by B.L. Mamet (written commun., 1992) revealed that the following fossils are in the rock:

Attached Apterrinellidae
Climacammina sp.
Ellesmerella oncolites
Epimastoporid fragments
Hemigordius sp.
Osagia oncolites *Pseudoglomospira* sp.
Schwagerinidae

As reported by Mamet, the age of the sample is approximately early Permian, and "There are certainly many sources of carbonates involved [in the provenance of this sample]. The abundance of *Ellesmerella-Osagia* is characteristic of Late Pennsylvanian-early Permian carbonate platforms in the [western United States]".

The apparent shallow depositional environment in the provenance history of this particular sample seems to corroborate the conclusions of many others that the depositional environment of the carbonate platform region from which the fossil fragments were derived was quite shallow. This particular sample was collected near the west edge of the subunit, and probably represents the uppermost stratigraphic part of the subunit in the stacking pattern of cycles in the Havallah sequence. However, a significant part of the Havallah sequence appears to be missing stratigraphically below the actual sampling site for 90TT356 and the projected trace at depth of the sole of

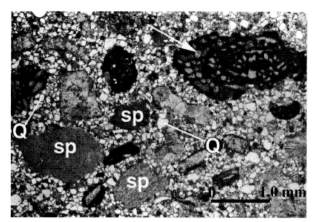

Figure 41 Calcite-cemented, fine-grained, Late Pennsylvanian or Early Permian quartz sandstone from subunit hss between Cottonwood Creek and Oyarbide faults. Shows fragments of sparry calcite (sp) and late Paleozoic fusulinid, at head of arrow, set in a calcite-cemented matrix that includes abundant grains of fine-grained quartz (Q). Locality 90TT356, plane polarized light; NE 1/4 sec. 2, T. 32 N., R. 42 E.

the Willow Creek thrust, apparently including a thick sequence of missing Mississippian radiolarian ribbon chert (pl. 2). The amount of the Havallah sequence missing is difficult to calculate, although paleobathymetric depositional depths during sedimentation of radiolarian chert in LU–2 is estimated to be approximately greater than or equal to 2,000 m (Murchey 1990). That same carbonate platform alluded to above by Armstrong and Mamet upon which the fossil fragments were first deposited also is the upper Paleozoic platform which probably provided carbonate lithic fragments that make up a significant proportion of this subunit and many of the other calcareous siliciclastic subunits in the Havallah sequence described previously.

No chemical analyses are available for rocks from subunit hss that crop out between the Cottonwood Creek and the Oyarbide faults.

LITHOTECTONIC UNIT 2 PLATES SOUTH OF OYARBIDE FAULT

Plates belonging to LU–2 south of the Oyarbide fault include near their base a fairly thick succession of generally west-dipping, mostly interbedded chert and calcareous or siliceous siltstone (subunit hcs, pl. 2) that have been contact metamorphosed intensely in a broad area largely surrounding the Late Cretaceous granodiorite and monzogranite of Trenton Canyon. The age of subunit hcs, radiolarian assemblage 5 (Chester to early Atokan) of Murchey (1990), was established by B.L. Murchey (written commun., 1993) on the basis of radiolarians obtained from a single locality south of the study area, and well outside the metamorphic aureole surrounding the pluton at Trenton Canyon (J.L. Doebrich, oral commun., 1993; Doebrich, 1995). Some rocks of the subunit, moreover,

have been intensely contact metamorphosed locally also by late Paleozoic or early Mesozoic gabbro that crops out in two small masses in NE 1/4 sec. 14, T. 32 N., R. 42 E. (pl. 2). In addition, rocks of subunit hcs are intruded by a narrow, northwest-striking, presumably Tertiary, phyllic-altered felsite dike in SW 1/4 sec. 19, T. 32 N., R. 43 E. (pl. 2). Rocks of subunit hcs apparently also were overthrust along the Willow Creek thrust onto black cherty shale belonging to LU–1, and rocks of subunit hcs, in turn, are overlain tectonically by thin sequences of calcareous sandstone (subunit hss) and interbedded gray micrite and black chert (subunit hlc) between Trenton Canyon and the south edge of the North Peak quadrangle. Offsets along these thrusts apparently are quite minor because J.L. Doebrich (oral commun., 1993; 1995) finds some rocks of subunit hss in depositional contact with rocks of subunit hcs a few km south of Trenton Canyon. Because of structural position of rock strata that make up subunit hcs, subunit hcs must be equivalent structurally to plate A described above (fig. 31). Rocks belonging to subunit hcs have not been recognized in the hanging wall of the Oyarbide fault in as much as they may have been cut out tectonically by displacements along the Willow Creek thrust.

Some of the best exposures of subunit hcs, albeit metamorphosed, are along the bottom of Trenton Canyon where thin-bedded sequences, approximately 0.5 km west of the Late Cretaceous pluton of Trenton Canyon, are almost entirely metachert (fig. 42A). These sequences of mostly metachert grade toward the north into sequences that contain greater abundances of metasiltstone. Other exposures of the subunit in SW 1/4 sec. 19, T. 32 N., R. 43 E., also within the contact metamorphic aureole and approximately 0.8 km east of the nearest exposures of the pluton, contain much higher proportions of metasiltstone than the ones in Trenton Canyon. These exposures have alternating layers of pale grayish-tan metachert and brown metasiltstone, now a biotite hornfels, that display variably thick beds and fairly continuous bedding surfaces (fig. 42B, C). These exposures are close to the trace of the Willow Creek thrust, and some exposures within approximately 0.5 km of the trace of the thrust include well-developed outcrop-scale folds that have north-northwest trending fold axes generally parallel to the local strike of the thrust (pl. 2). Proportion of metachert overall the subunit appears to be greater in its western parts than in its eastern parts.

Close to the small outcrops of late Paleozoic or early Mesozoic gabbro near the northwesternmost exposures of subunit hcs, strata of the subunit are characterized by undulating bedding surfaces that separate sucrosic, light-gray-to-white metachert beds from brown metasiltite or biotite hornfels. Metachert is present typically in 3- to 15-cm-thick beds, some of which are deformed into boudins. Fairly thick veins, as much as 1.5 m wide and consisting

Figure 42 General character of bedding in interbedded chert and calcareous siltstone, subunit hcs, of lithotectonic unit 2 belonging to Devonian, Mississippian, Pennsylvanian, and Permian Havallah sequence. Near southeast corner of North Peak quadrangle. *A*, Thin-bedded contact metamorphosed sequence of subunit hcs containing high proportion of original chert. Locality 92TT103; *B*, interbedded metachert (light bands) and biotite hornfels, in places containing calc-silicate minerals (dark bands), derived from variably calcareous siltstone. Locality 92TT100; *C*, closeup view of metachert (light) and biotite hornfels (dark) showing planar and continuous nature of contacts between lithologies, together with apparent grading across bedding relict from variabilities in original clay mineral content. Locality 92TT100.

of mats of coarse-crystalline tremolite-actinolite, locally cut metamorphosed rocks of the subunit along the east-west trending ridge immediately north of Trenton Canyon. Unfortunately, unmetamorphosed equivalents of rocks that make up this subunit do not appear to crop out anywhere in the North Peak quadrangle.

Petrographic examination of a limited number of samples from subunit hcs reveals that metachert in the subunit apparently is derived largely from sponge spicule-bearing protoliths and that the various metasiltstone facies present in the unit show metamorphic assemblages indicative either of siliceous or calcareous provenances. Metachert in the subunit also is in places quite heterogeneous lithologically as it contains abundant silt-sized fragments of monocrystalline quartz and abundant phosphatic material. Siliceous metasiltstone shows well-developed hornfelsic fabrics throughout the subunit, which fabrics typically show quartz±biotite±white mica±K-feldspar±tourmaline (trace)±rutile (trace)±chlorite±pyrite composite contact-metamorphic assemblages. Calcareous metasiltstone shows quartz-tremolite-sphene±biotite (trace)±chlorite±pyrite composite assemblages. Additional petrographic details are provided below in conjunction with discussion of the pluton at Trenton Canyon.

Some additional minor thrust plates present in the narrow strip of ground south of Trenton Canyon and north of the southern border of the North Peak quadrangle appear to be structurally emplaced on top of subunit hcs, and they include sequences of rock assigned to subunits hss and hlc (pl. 2). Rocks making up the upper plates of these minor thrust plates probably are part of plate E (fig. 31). Calcareous sandstone and siltstone of subunit hss in this area include some weakly contact metamorphosed, massive, generally light-gray micrite in exposures approximately 3– to 5– m -thick. These sequences of micrite crop out high on the ridge tops, and blocks of micrite, which are weathered-out in colluvium derived from these sequences, typically show prominent pinkish-red to brick-red colors on their weathered surfaces. Some sequences of subunit hss also include black siltstone and gray sandy micrite, particularly in its upper parts of the along the ridgeline in SW 1/4 sec. 23, T. 32 N., R. 42 E.

DISCUSSION

Rocks in the Havallah sequence in the Valmy and North Peak quadrangles show some broad similarities and some sharp differences with rocks of the sequence that crop out elsewhere. Only one of the most monocrystalline-quartz-rich samples analyzed from the calcareous clastic subunits in the study area plots within the field defined for Lower Permian monocrystalline quartz-rich petrofacies rocks in the Golconda allochthon from the Mount Tobin area, from the Hoffman Canyon area, and from the northern Fish Creek Mountains area by Tomlinson (1990). Major differences in the 16 samples analyzed by Tomlinson (1990) from these three areas when compared with samples from the Battle Mountain area primarily involve a uniformly much higher concentration of monocrystalline quartz, approximately 95 volume percent, in the samples from Battle Mountain. The total range in monocrystalline quartz at the framework sites at Battle Mountain is about 50 volume percent (fig. 33*A*). When compared to the quartz-chert petrofacies of LU–2 in the Mount Tobin area (Tomlinson, 1990), similarities there in overall proportions of framework monocrystalline quartz to the study area suggest similar maturities, although samples from subunit hss in the Valmy and North Peak quadrangles show elevated abundances of feldspar, particularly monocrystalline K–feldspar. However, samples from the quartz-chert petrofacies in the Mount Tobin area also differ significantly from samples of LU–2 at Battle Mountain in overall proportions of other framework grains primarily because of elevated abundances of chert and depleted abundances of extrabasinal carbonate grains in the former area (fig. 33*A*). Compared with sandstones in the Schoonover sequence (Whiteford, 1990), samples from subunit hss in the Battle Mountain area contain generally much less total feldspar and significantly less volcanic fragments, and similar ratios of monocrystalline quartz versus total lithics. In the Schoonover sequence, volcanogenic sandstones, ranging in composition from arkose to feldspathic litharenite, are interbedded with massive quartzose sandstones in Lower Mississippian-age parts of the section (Whiteford, 1990). In addition, proportion of extrabasinal carbonate fragments in the lithic population of subunit hss sandstones at Battle Mountain is much higher and more widespread than in the Schoonover sequence.

All modally analyzed samples, presumably mostly Pennsylvanian and (or) Permian in age from all four areas (Mount Tobin, Hoffman Canyon, northern Fish Creek Mountains, as well as Battle Mountain), essentially show no volcanic fragments present thus indicating that none of these four areas, during those periods of time, were within the submarine depositional aprons shed from the Sonoma-associated magmatic arcs much farther to the west. In marked contrast, sandstones in the Schoonover sequence, the northernmost exposures of the Golconda allochthon in Nevada, contain abundant concentrations of detrital valcanic fragments (Whiteford, 1990). However, B.L. Murchey (oral commun., 1993) has obtained some Devonian radiolarians from some detrital fragments of chert in pebbly limestone suggesting that some debris may have been shed into the Havallah basin from uplifted highlands of rock belonging to the Antler orogenic belt.

Nonetheless, it appears that all these provenance data closely tie depositional environments of the rocks of the

Golconda allochthon to the continental margin of upper Paleozoic times as proposed by many others before (Tomlinson, 1990; Miller and others, 1984, 1992; Whiteford, 1990; Jones, 1991b; Piper and Medrano, 1994; Murchey and others, 1995). As suggested previously by Whiteford (1990), much of the debris shed into the Havallah basin in the Battle Mountain area probably was derived from the Antler highlands. This conclusion appears to be verified by presence of relatively widespread sponge-spicule-bearing sequences in the Edna Mountain Formation of the overlap assemblage from which a source can be inferred for the spicular-chert facies that is so prominent in the Havallah sequence (B.L. Murchey, unpub. data, 1999; Murchey and others, 1995). However, Jones (1991b) noted that the Permian Poverty Peak terrane of the Havallah sequence in the Osgood Mountains is extremely mature. As described above, sands that constitute the Havallah sequence at Battle Mountain also are mature and require a significant contribution from a cratonal source.

Most extrabasinal carbonate grains in the Havallah sequence possibly could have been derived from lower Paleozoic carbonate shelf rocks in the lower plate of the Roberts Mountains thrust east of the Antler highlands as suggested by the paleogeographic reconstructions of Tomlinson (1990). However, the westernmost exposures of lower Paleozoic shelf rocks known in the region crop out in the East Range, west of the study area, and west of the inferred position of the Antler highlands. Moreover, it seems just as likely that some, and I emphasize some, extrabasinal carbonate material, particularly in the youngest rocks in the Golconda allochthon, were derived, initially in part subaerially but largely as submarine turbidites, from partly coeval upper Paleozoic block-faulted carbonate sequences in the overlap assemblage. But this creates a dilemma in that some extrabasinal carbonate fragments are in rocks of LU–2 as old as Middle Pennsylvanian (Murchey, 1990), but they do range in age to as young as early Permian. The oldest rocks in the overlap assemblage in the mining district also are Middle Pennsylvanian, Atokan Epoch, in age (Roberts, 1964), and Lower Pennsylvanian in the area of Edna Mountain. Furthermore, it is difficult to envision a paleogeomorphologic regime wherein drainage(s) supplying terrigenous debris from widespread lower Paleozoic shelf terranes distant to the east preferentially would widely breach the north-south Antler highlands at a scale apparently required by makeup of calcareous clastic rocks cropping out for over 32 km in a north-south direction in the Golconda allochthon. These rocks are present in the Golconda allochthon both in the two quadrangles of the study area and in the Antler Peak quadrangle on the south (J.L. Doebrich, oral commun., 1992; Doebrich, 1995). It seems likely that many of these essentially east-west transport pathways would have been diverted to north-south ones along the east margin of the

Antler highlands and thus would not have breached the highlands to a significant degree. These relations lead to the suggestion that some extrabasinal carbonate grains at framework sites in rocks of LU–2 may have been derived from elevated fault blocks or local platforms of the continental shelf west of the Antler highlands–fault blocks that were uplifted prior to late Paleozoic time. If this suggestion is generally correct, then some of these uplifted fault blocks may be present under Tertiary and Quaternary gravels in the valleys between the Battle Mountain Mining District and the East Range. Whatever the source(s) of the terrigenous and carbonate debris in the Havallah basin is, the source(s) appear to have been fairly constant for a protracted period of time based on their fairly constant patterns of normalized ratios for minor elements, regardless of location of the samples in the tectonostratigraphic succession of LU–2. Nonetheless, the major conclusions of Tomlinson (1990) and Miller and others (1984, 1992) apparently remain valid in that many rocks in the Golconda allochthon show ample evidence in their provenance for having attachments to a nearby continental margin.

BASALT AND BASALTIC ANDESITE OF LITHOTECTONIC UNIT 2 (MISSISSIPPIAN)

Basaltic rocks of presumably Mississippian age are present in one area of LU–2 in the quadrangles. That occurrence is near the north end of the Havallah Hills where a small isolated mass of basalt and basaltic andesite (BBA), too small to show on the geologic map, is present in one of the lowermost subunits of the upper plate of the Willow Creek thrust in NE 1/4 sec. 23, T. 34 N., R. 42 E. (pl. 1). As will be described in the following sections below, basaltic rocks from this occurrence show chemical affinities with similar rocks in the Havallah sequence elsewhere; in particular, in an area a number of km farther to the west, generally south of Golconda summit, and in the Mill Canyon area, just to the south of the southern border of the North Peak quadrangle. The chemistry of these rocks is markedly different from more extensive Pennsylvanian and Permian basaltic trachyandesite and trachyandesite (BTT) present between the Cottonwood Creek and Oyarbide faults (see subsection below entitled "Basaltic Trachyandesite and Trachyandesite of Lithotectonic Unit 2"), although the latter occurrences are approximately at the same tectonostratigraphic position as the BBA near the north end of the Havallah Hills. As will be documented fully below, submarine volcanic rocks in the Havallah sequence comprise two chemically distinct suites. Furthermore, these rocks provide local petrologic and geochemical background to which age-equivalent, mineralized basalts in

LU–2 within the mineralized aureole of the Lone Tree deposit can be compared (see subsection below titled "Lone Tree Gold Deposit").

No known occurrences of volcanogenic massive sulfide are present in Mississippian BBA or in Pennsylvanian and Permian BTT suites in either the Valmy quadrangle or in the North Peak quadrangle. The closest occurrence to the study area of such massive sulfide in Mississippian-age rocks is approximately 0.5 km south of the study area where pyrrhotite-rich massive sulfide in pillowed basalt is present in a small outcrop in the bottom of Mill Creek in SW 1/4 sec. 26, T. 32 N., R. 42 E., almost at the border with adjoining sec. 27 (Doebrich, 1992, 1995).

GEOLOGIC SETTING AND PETROGRAPHY

The occurrence of BBA is an isolated one present in Mississippian radiolarian ribbon chert, subunit hc, near the north end of the Havallah Hills. As such, these rocks are present only as an approximately 1– to 2–m-wide augite BBA flow or sill, in a sequence of intensely folded thin-bedded ribbon chert exposed during 1990 in a narrow bulldozer cut at the locality. The bulldozer cut has since been reclaimed. This flow or sill is presumably Mississippian in age, the same age as the chert subunit that encloses it (pl. 1). Moreover, strata exposed in the cut include some exhalative silica and a few beds of medium-grained sandstone dominated by terrigenous quartz.

Stubby laths of fresh plagioclase (labradorite, An$_{50-65}$, normally zoned), in-places showing slightly to highly curved crystal outlines, measure approximately 0.8 mm in long dimension, and, together with somewhat smaller crystals of subcalcic augite (see subsection below entitled "Mineral Chemistry of Augite"), they define a well-developed subophitic texture in the BBA (fig. 43A). Curved crystals of plagioclase are relatively common elsewhere in many quenched domains of basaltic rock (Cox and others, 1979). In addition, many interstices among plagioclase and subcalcic augite crystals are filled by what probably constitutes devitrified glass that is now replaced by orange-brown—as seen in transmitted light—patchy concentrations of a fine-grained chlorite-group mineral(s), titaniferous magnetite, and ilmenite, of which the latter two minerals probably make up as much as about 2 to 3 volume percent of the thin BBA outcrop (fig. 43B).

The BBA near the north end of the Havallah Hills appears to show mineral assemblages characteristic of an early magmatic stage that then was followed either by a weakly developed greenschist regional metamorphic event or, alternatively, a local hydrothermal event that affected significantly primary magmatic textures in the basalt. A greenschist event is probably more likely of the two, and this event shows no presence of an associated penetrative fabric recognizable in outcrop. However, textural relations among magmatic phases and those mineral phases associated temporally with the greenschist event are extremely complex, even at the thin-section scale, and they can be resolved only at extremely high magnifications by SEM methods. Some augite in the rock, in fact, shows chemical zoning in back-scattered imagery by SEM most probably indicative of a complex crystallization history during its magmatic stages of growth (fig. 44A). Cores of many augite crystals show slightly elevated abundances of Ti and Al relative to 20–μm-wide, planar-bounded zones near the margins of the crystals, which zones are interior to similar-sized areas that are enriched in Fe near the rim of the crystals. These zonal relations are probably indicative of an intermediate-stage residency of the augite crystals in a magma chamber prior to their being expelled onto the floor of the Havallah basin during the Mississippian, at which time rims of the augite crystals finally equilibrated (see

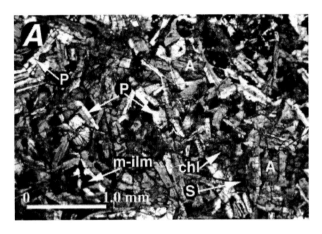

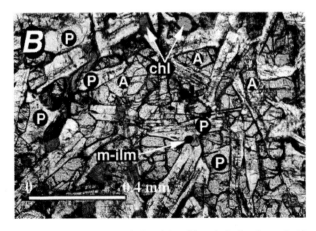

Figure 43 Mississippian augite basalt-basaltic andesite interlayered with chert and some quartz sand in subunit hc of tectonostratigraphic plate A of LU–2, belonging to Devonian, Mississippian, Pennsylvanian, and Permian Havallah sequence. A, augite; P, plagioclase; chl, chlorite; S, sphene; m-ilm, titaniferous magnetite or ilmenite. Locality 90TT286, near north end of Havallah Hills. *A*, Texture of basalt-basaltic andesite, partly crossed nicols; *B*, closeup view showing textural relations between augite and plagioclase.

the subsection below entitled "Mineral Chemistry of Augite"). Early magmatic-stage titaniferous magnetite or ilmenite in BBA apparently crystallized initially subsequent to augite inasmuch as titaniferous magnetite or ilmenite has not been found to be included within any augite crystals even after exhaustive protracted searches by SEM; these minerals are instead intergrown only with magmatically crystallized, calcic plagioclase (fig. 44A–C).

Some question still remains as to whether opaque minerals contemporaneous with magmatic stages of BBA are ilmenite, titaniferous magnetite, or some mixture of the two. These early-stage opaque grains display Fe contents consistently well in excess of Ti, a reversal of typical reported compositions of ilmenite (Deer and others, 1962a), and upon which the judgment is made that most early-stage opaque grains are most likely titaniferous magnetite. Chromite has not been found to be included in augite. However, some opaque grains in BBA show a patchy appearance where intergrown with paragenetically later lamellae of probable greenschist-stage albite and associated nearby chlorite (fig. 44D). Alternatively, some opaque grains also may show smooth surfaces, but with Ti-higher-than-Fe contents indicative of ilmenite. Thus, ilmenite appears to be compatible with both chlorite and albite, as is sphene, which is in places intergrown with chlorite (fig. 44E). Small, 1– to 2–μm-wide grains of zircon are rare in BBA (fig. 44F). Some apatite in BBA also is probably part of the superposed greenschist event because, in places, prismatic crystals of apatite cut albite. Thus, the weak greenschist event is reflected in BBA of the Havallah sequence by an albite-chlorite-ilmenite±sphene±rutile±apatite composite assemblage. Petrogenetic implications of the chemistry of three samples from BBA near the north end of the Havallah Hills are described below in the subsection entitled "Whole-rock Chemistry."

Sandstone interbedded with BBA in the bulldozer cut is lithologically heterogeneous, and it has an average grain size of approximately 0.3 mm; it includes a poorly sorted, submature fabric of mostly framework quartz showing all degrees of rounding from well-rounded to angular broken fragments. Chert fragments also make up approximately 5 to 10 volume percent of the rock. Many quartz grains are replaced partially by wormy microveins of calcite that course randomly through the rock. In addition, some parts of the sandstone are matrix supported by Fe oxide, and some other parts seem to show some sparse concentrations of green, devitrified (chlorite?) glass from which one might infer some association with the thin BBA layer nearby in the bulldozer cut.

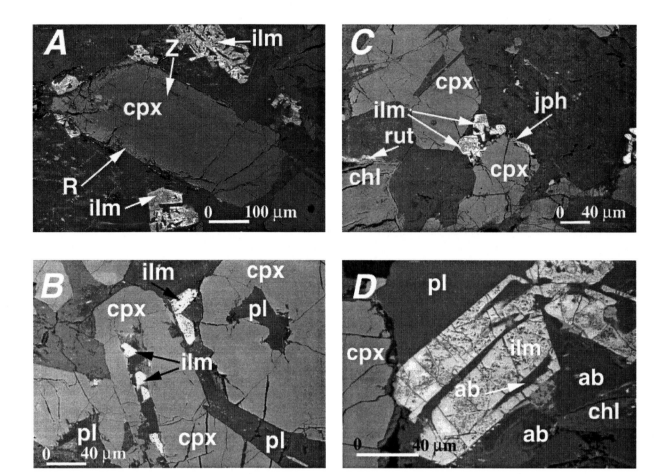

Figure 44 Back-scattered scanning electron micrographs of Mississippian basalt and basaltic andesite in lithotectonic unit 2 of Devonian, Mississippian, Pennsylvanian, and Permian Havallah sequence. Sample locality same as figure 43 near north end of Havallah Hills; cpx, augite; chl, chlorite; ilm, ilmenite; sph, sphene; rut, rutile; pl, plagioclase (labradorite); ab, albite. *A*, Subhedral crystal of augite showing zone (Z) near edge of crystal depleted in aluminum and titanium relative to core. Rim (R) of crystal enriched in iron relative to core; *B*, ilmenite of variable compositions associated paragenetically with plagioclase, and absent from interior parts of augite crystals; *C*, same as *B*, but also showing some rutile, chlorite, and sphene in field of view; *D*, relatively large crystal of ilmenite showing euhedral crystal boundary with plagioclase and including narrow laminae of albite. Some chlorite and augite also in field of view.

Mineral chemistry of augite

by Robert L. Oscarson and Ted G. Theodore

An electron microprobe was used to examine mineral chemistry and cation zoning of augite in one sample of the BBA suite from the northern Havallah Hills. As described above, the site, a bulldozer cut, from which all samples of BBA were obtained in the Valmy quadrangle has since been reclaimed. In addition, major- and minor-element data for this particular sample (90TT286) are described in the immediately succeeding subsection below titled "Whole-rock Chemistry." Preliminary petrographic and SEM examination of three samples of BBA rocks indicated that only one of the samples from the site contained pyroxene of a size sufficient to be studied by electron

microprobe. Presence of pyroxene could not be confirmed by SEM even after protracted search at extremely high magnifications in the two remaining samples of BBA. Mineral chemistry of augite in one polished thin section (90TT286), however, was studied using a Jeol electron microprobe, model JXA 8800, equipped with five wavelength dispersive spectrometers. Analysis conditions for standards and unknowns were 15kV accelerating voltage and 30nA beam current. Wet chemically analyzed silicate minerals or pure oxides were used as standards; diopside, PX–1 for Si and Ca, San Carlos olivine for Mg, synthetic fayalite for Fe, jadeite for Na and Al, chromite for Cr, Mn_2O_3 for Mn, and TiO_2 for Ti. Background measurements were made off peak for each analysis. Matrix corrections were made using Jeol-supplied ZAF software. Detection limits in augite for the major oxides are typically 0.1 weight percent. The nomenclature followed for classification of pyroxene is that of Morimoto and others

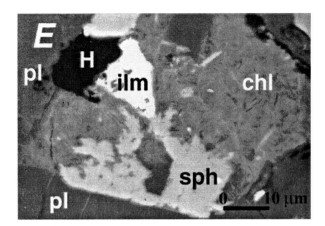

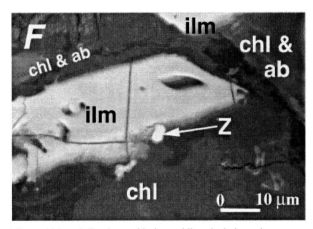

Figure 44 (cont.) *E*, sphene, chlorite, and ilmenite in interstices among crystals of plagioclase. Dark area in upper left quadrant of field of view is hole (H) in polished thin section; *F*, manganiferous ilmenite associated texturally with albite and chlorite. Small grain of zircon (Z) also in field of view.

(1988), and tests of acceptability of microprobe data generally complied with those suggested by Robinson (1980). To preserve compositional differences recorded by low-level differences in chemical compositions of core-, intermediate-, and rim-domains of individual augite crystals, all analyses reported represent single points rather than averages. In all, three points were analyzed on each of 17 representative well-formed augite crystals; these points were chosen to characterize three domains of the augite crystals, which overall typically are in the 200- to 400- μm range in long dimension. Representative microprobe analyses are presented in table 8, and all 51 microprobe analyses are plotted as individual points in diagrams to follow.

Some assumptions are required for conversion of microprobe analyses reported as oxide weight percentages, with total iron reported as FeO and no analyses for H_2O available by microprobe, to formulas based on a fixed number of oxygens per formula unit. For pyroxene, it is common practice to calculate formulas on the basis of four cations and six oxygens (Vieten and Hamm, 1978). How-

ever, some complete wet-chemical analyses of augite in igneous rocks commonly show trace amounts of H_2O and as much as 4 to 5 weight percent Fe_2O_3 (Deer and others, 1978). The convention of Morimoto and others (1988) used in this report with regards to adjusting the Fe^{3+}/Fe^{2+} ratio in the standard (M2) (M1) T_2O_6 pyroxene formula is as follows: (1) initially to sum the tetrahedral (T) site to 2.00 cations; (2) then to sum the M1 site to 1.00 using the required amount of Fe^{3+} cations; and (3), finally, to assign all remaining iron to Fe^{2+} in the M2 site. As shown in table 8, this results in the M2–group cations totaling slightly more, however roughly on average by only 2 percent more, than the 1.00 available positions in the M2 site, and a slight amount of Al being assigned to the M1 site. Approximately 75 to 84 percent of the M2–group site apparently is filled by Ca, and approximately 2 to 4 percent of the site is also occupied by Na (table 8), of which the latter element does not calculate to an amount larger than the number of cations (0.1 cations of Na) requiring the "sodian" adjectival modifier to the mineral name (Morimoto and others, 1988). However, the amount of Cr_2O_3 in the cores of some of the augite crystals is in concentrations such that Cr^{3+} cations make up more than 0.01 cations, actually as many as 0.03 cations, and thus meeting concentration levels in the augite that could be designated as "chromian".

Fifty of the 51 analyses of pyroxene in the analyzed sample of BBA plot in the augite compositional field of Morimoto and others (1988) (fig. 45). The other pyroxene analysis contains 45 molecular percent Wo and thus plots directly on the join between the augite and diopside fields of their classification diagram. In addition, augite crystals in the examined sample show some fairly systematic differences in chemical composition between rim- and core-domains of individual crystals analyzed by microprobe (table 8). These differences include a general low-level enrichment in Cr_2O_3 and Al_2O_3 contents apparently

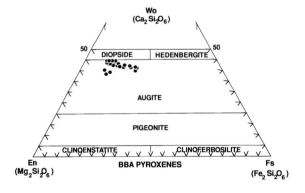

Figure 45 Fifty one microprobe analyses of augite from a select sample of BBA–suite rock from Devonian, Mississippian, Pennsylvanian, and Permian Havallah sequence plotted using ternary molecular proportions of end-member pyroxenes Wo ($Ca_2Si_2O_6$), En ($Mg_2Si_2O_6$), and Fs ($Fe_2Si_2O_6$). Classification scheme of pyroxenes from Morimoto and others (1988).

Table 8—Representative microprobe analyses of augite from Mississippian BBA–suite volcanic rocks in the Devonian, Mississippian, Pennsylvanian, and Permian Havallah sequence, northern Havallah Hills, Humboldt County, Nevada.†

Grain		A			K			L	
¹Domain	rim	inter.	core	rim	inter.	core	rim	inter.	core
Microprobe analyses (weight percent)									
SiO_2	49.68	52.12	49.86	48.99	49.36	49.03	48.28	48.85	50.53
Al_2O_3	3.72	2.22	4.43	3.81	4.71	5.07	2.72	4.1	4.09
FeO	9.76	7.57	6.56	10.8	6.85	6.17	13.52	7.8	6.52
MgO	14.51	17.47	15.5	13.69	15.27	15.09	11.87	14.81	15.64
CaO	20.1	18.8	20.57	19.52	20.77	20.94	19.32	20.51	21.17
Na_2O	.4	.24	.31	.45	.34	.33	.46	.37	.29
TiO_2	1.4	.71	1.11	1.49	1.23	1.2	1.66	1.02	.99
MnO	.21	.23	.21	.22	.15	.16	.32	.19	.18
Cr_2O_3	.04	.19	.49	.03	.22	.88	.02	.04	.2
Total	99.84	99.55	99.04	98.98	98.9	98.86	98.17	97.68	99.6
Cations (normalized to 4)									
Si T	1.86	1.93	1.86	1.86	1.85	1.83	1.88	1.86	1.87
AlIV	.14	.07	.14	.14	.15	.17	.12	.14	.13
AlVI	.03	.03	.06	.03	.06	.05	.01	.04	.05
Fe^{3+}	.12	—	.04	.15	.05	.05	.25	.09	.05
Ti M1	.04	.02	.03	.04	.03	.03	.05	.03	.03
Cr	—	.01	.01	—	.01	.03	—	—	.01
Mg	.81	.94	.86	.78	.85	.84	.69	.84	.86
Mg	—	.02	—	—	—	—	—	—	—
Fe^{2+} M2	.19	.23	.17	.19	.16	.14	.19	.17	.15
Ca	.81	.74	.82	.79	.83	.84	.81	.84	.84
Na	.03	.02	.02	.03	.03	.02	.04	.03	.02
Molecular proportions of end-member pyroxene									
En	42.13	49.58	45.62	40.52	44.85	44.9	35.59	43.65	45.31
Wo	41.96	38.36	43.54	41.54	43.83	44.8	41.65	43.45	44.09
Fs	15.91	12.06	10.84	17.94	11.29	10.31	22.75	12.9	10.6

¹Domain category of analyzed crystal grain; inter., intermediate.

†[Field number 90TT286; analysis no. 3, tables 10–11 (see text); —, not detected]

in many cores of the augite crystals, together with a corresponding enrichment of MgO contents in some core- and intermediate domains. In contrast, many rim domains of the augite crystals show enrichments in contents of Fe (mostly Fe^{3+} presumably), TiO_2, and possibly Na_2O, the latter being present in concentrations that yield extremely low-levels of elemental contrasts between cores and rims (table 8). Many of these differences in oxide components among rim-, intermediate-, and core domains are depicted graphically in figure 46A–D. As shown in these diagrams, cation concentrations at sites categorized as belonging to rim domains span the entire range of reported TiO_2, MgO, and FeO (total iron reported as FeO) compositions. However, most rims show elevated TiO_2 and FeO concentrations and depleted MgO concentrations relative to intermediate and core domains of the augite crystals. In addition, significant overlap apparently is present in concentrations of these oxides in the intermediate and core domains of the augite crystals. Finally, a relatively strong inverse relation exists between FeO and TiO_2 contents versus Cr_2O_3 contents overall the augite grains; many analyses at the cores of the augite crystals are elevated in Cr_2O_3 contents (fig. 46E, F). However, the highest Cr_2O_3 concentrations detected, approximately 1 weight percent, are present in two of the rim domains analyzed by microprobe, and Cr_2O_3 contents in all of the plotted analyses are unquestionably in crystal lattices of the augite grains. Because Cr and Ni are two elements that covary positively in many rocks, concentrations of these elements in this sample (330 and 70 ppm respectively) suggest that some Ni–bearing mineral phases might also be present in the rock. A few apparent grains of petlandite ([Fe, Ni]S), roughly 1 μm wide, were detected by SEM in the groundmass of sample 90TT286 together with sparse concentrations of equivalent-sized chalcopyrite grains. Copper content of sample 90TT286 is 91 ppm (analysis 3, table 9).

Qualitative distribution of a number of cations in several augite crystals of the sample of BBA–suite studied quantitatively above by microprobe (90TT286) was determined by preparation of elemental X–ray intensity maps also using the microprobe. Figure 47 shows distribution of Mg, Cr, Al, Fe, and Ca in a representative augite crystal (grain A, table 8). The X–ray intensity maps reveal striking differences between present outermost borders of many augite crystals and their former early-growth stages. These differences are reflected either by concentration of some cations and (or) by depletion of other cations in euhedrally-bounded intracrystalline domains that must indicate major differences of availability of cations at various times in the magmatic crystallization history of the BBA–suite rocks. These euhedrally bounded domains probably are indicative of residency of the typical augite crystal in the BBA–suite rocks effectively in a number of

magma reservoirs prior to final crystallization. In addition, although contents of Cr are concentrated in a wedge-shaped domain near the geometric center of the figured grain, this domain is not the earliest crystallization stage of grain A (fig. 47). As shown, the earliest stage of crystallization of this grain is instead one that is enriched in Mg, depleted in Al, and only contains a low-level concentration of Cr. Finally, contents of Fe are especially concentrated along only two of the borders of grain A, indicating perhaps some type of preferred concentration by direction of crystallographic growth.

WHOLE-ROCK CHEMISTRY

Chemical analyses of three samples of basaltic rock interlayered with chert in map subunit hc of LU–2 in the northern Havallah Hills indicate that these rocks are sodic ($Na_2O-2 > K_2O$) according to criteria of Le Bas and others (1986) (analyses 1–3, tables 9–10). In addition, according to their classification scheme, the three analyzed rocks would be termed basalt or basaltic andesite (BBA) based on data plots that show two of the samples plotting in their basalt field and the remaining sample plotting in their basaltic andesite field (fig. 48). Therefore, BBA from this lone locality appears to be the easternmost exposed occurrence in the Havallah sequence of a chemically coherent volcanic suite that widely crops out in at least two other areas: (1) at Buffalo Mountain, south of Golconda summit, and (2) in the lower Mill Canyon area (see section below entitled "Basaltic Rocks at Buffalo Mountain and in Mill Canyon Area"). They also are present in the upper parts of Timber Canyon and Mill Canyon (Doebrich, 1995). Furthermore, as shown on figure 49, the three samples from the northern Havallah Hills plot in the subalkaline fields as defined by both McDonald and Katsura (1964) and Irvine and Baragar (1971). However, in the K_2O content versus SiO_2 content classification of Peccerillo and Taylor (1976), the samples of the BBA suite plot in their low–K basalt field and in their low–K basaltic andesite field (fig. 50). Contents of K_2O in these three analyses are in the 0.32 to 0.33 weight percent range, contents of TiO_2 in the range 1.5 to 1.62 weight percent, and contents of Ba in the 2,000 to 4,500 ppm range (table 9). In an AFM diagram (where A is [$Na_2O + K_2O$], F is total iron as FeO, and M is MgO, all in normalized, volatile-free weight percent), the three samples of the BBA suite from the northern Havallah Hills plot close to the join between the tholeiitic and the calc-alkaline series as defined by Irvine and Baragar (1971) and many others, but somewhat towards the MgO corner of the diagram (fig. 51). In a Jensen (1976) ternary diagram (showing proportions of [$FeO+Fe_2O_3+TiO_2$], Al_2O_3, and MgO also in weight per-

Table 9—Analytical data of Mississippian basalt and basaltic andesite (BBA suite), Pennsylvanian and(or) Permian basaltic trachyandesite and trachyandesite (BTT suite), and Mississippian volcaniclastic sedimentary rocks from Devonian, Mississippian, Pennsylvanian, and Permian Havallah sequence, Humboldt County, Nevada.†

Analysis number	1	2	3	4	5	6	7	8	9	10	11	12
Field number 90TT...	284	285	286	289	308	310	312	314	319	328	306	307
	BBA Suite-Northern Havallah Hills				BTT Suite-South of Cottonwood Creek fault						Volcaniclastic Seds	
	Chemical analyses (weight percent)											
SiO_2	47.1	47.0	50.0	50.3	52.0	55.3	53.3	50.4	49.5	55.6	64.8	69.8
Al_2O_3	14.7	14.3	15.5	16.3	15.5	17.6	15.7	16.2	15.2	15.9	15.8	13.7
Fe_2O_3	5.02	5.45	4.77	6.72	11.68	3.86	10.5	7.6	9.38	11.7	4.16	2.81
FeO	4.15	4.5	3.11	3.44	1.47	3.32	.87	2.45	3.38	.06	.81	1.05
MgO	8.23	8.01	6.53	4.39	2.36	1.46	2.	2.58	3.14	.44	1.56	1.43
CaO	7.85	7.67	10.1	3.85	2.8	2.6	2.93	3.99	4.68	1.38	1.23	1.14
Na_2O	2.97	3.52	3.22	6.4	7.4	6.2	7.92	7.29	6.65	8.53	7.93	5.9
K_2O	.33	.32	.33	.77	.16	3.26	.3	.39	.13	.33	.42	.67
H_2O^+	4.04	3.52	1.96	2.82	1.86	2.13	1.28	1.97	2.35	1.19	1.	1.35
H_2O^-	1.39	2.18	1.34	.54	.42	.38	.34	.4	.33	.31	.21	.27
TiO_2	1.5	1.53	1.62	2.2	1.97	1.25	2.46	3.02	2.87	.95	.36	.31
P_2O_5	.16	.15	.17	1.44	.69	.38	.51	.74	1.02	.97	.11	.1
MnO	.17	.2	.23	.19	.16	.13	.14	.14	.17	.43	.05	.07
F	.05	.03	.16	.07	.11	—	—	—	—	—	.05	.05
Cl	.004	.004	.004	.012	.009	.008	.006	.007	.009	.012	.006	.006
CO_2	1.28	1.3	.86	.1	1.34	1.37	1.18	2.13	.73	.25	.82	.62
Total S	<.01	.02	<.01	<.01	<.01	<.01	.01	<.01	<.01	.01	<.01	.02
Subtotal	98.92	99.7	99.9	99.54	99.93	99.31	99.57	99.61	99.7	98.14	99.32	99.3
Less O = F, Cl, S	.03	.02	.07	.03	.05	.03	.05	.05	.07	.03	.02	.03
Total	98.89	99.68	99.83	99.51	99.88	99.28	99.52	99.56	99.63	98.11	99.3	99.27
	Inductively coupled plasma-atomic emission spectroscopy (total) (parts per million)											
Ba	4500.0	4100.0	2000.0	790.0	260.0	1200.0	690.0	280.0	940.0	1000.0	730.0	1200.0
Be	—	—	—	.3	1.	2.	1.	1.	1.	1.	1.	1.
Ce	17.	15.	17.	120.	41.	110.	52.	73.	87.	140.	41.	42.
Co	52.	48.	43.	40.	52.	11.	38.	31.	47.	13.	10.	11.
Cr	330.	350.	330.	54.	480.	3.	190.	38.	120.	5.	4.	2.
Cu	110.	91.	91.	88.	36.	53.	100.	18.	49.	39.	5.	5.
Ga	—	—	—	3.	—	—	—	—	2.	2.	—	—
La	6.	5.	6.	60.	25.	67.	25.	39.	44.	87.	26.	25.
Li	110.	82.	53.	31.	32.	12.	18.	18.	23.	13.	19.	20.
Mo	—	—	—	—	—	—	—	—	—	3.	—	—
Nb	9.	7.	9.	48.	14.	100.	39.	22.	33.	59.	10.	8.

Table 9 — (cont.)

Nd	12.0	11.0	13.0	64.0	22.0	45.0	23.0	36.0	44.0	64.0	21.0	21.0
Ni	72.	70.	70.	29.	180.	2.	120.	34.	97.	57.	5.	3.
Pb	—	—	—	—	—	—	—	—	—	5.	7.	4.
Sc	45.	46.	47.	13.	20.	4.	19.	10.	19.	4.	15.	12.
Sr	320.	290.	310.	650.	310.	390.	400.	330.	380.	270.	290.	240.
Th	—	—	—	5.	—	5.	4.	—	4.	7.	5.	5.
V	320.	340.	330.	68.	310.	13.	160.	140.	190.	59.	84.	61.
Y	23.	24.	25.	25.	30.	13.	9.	19.	22.	25.	26.	18.
Yb	2.	2.	2.	2.	2.	1.	—	1.	2.	2.	3.	2.
Zn	63.	74.	76.	150.	270.	53.	140.	110.	200.	320.	25.	36.

Inductively coupled plasma-atomic emission spectroscopy (partial) (parts per million)

Ag	2.17	—	—	—	—	—	—	—	0.14	—	—	—
As	3.45	—	1.0	—	—	—	—	0.85	3.2	—	—	—
Cd	.061	.058	—	.06	—	—	—	—	.17	—	.11	.21
Cu	110.	92.	92.	100.	4.3	55.0	100.0	17.	58.	41.	4.3	5.1
Mo	.14	.27	.16	.92	.27	2.1	.45	.91	1.7	3.2	.27	.72
Pb	1.2	1.7	1.	1.6	2.7	1.3	2.1	2.3	2.7	4.5	2.7	1.6
Sb	—	—	—	—	—	—	—	—	—	—	—	—
Zn	59.	73.	59.	190.	30.	64.	100.	140.	260.	370.	30.	41.

Energy-dispersive X-ray fluorescence spectrometry (parts per million)

Nb	—	10.0	—	100.0	12.0	154.0	74.0	82.0	72.0	144.0	12.0	—
Rb	—	14.	14.	15.	14.	46.	11.	12.	10.	17.	14.	17.0
Sr	285.0	270.	300.0	620.	290.	375.	395.	325.	355.	265.	290.	230.
Zr	96.	95.	102.	415.	154.	610.	235.	310.	295.	670.	154.	120.
Y	24.	28.	39.	37.	36.	34.	20.	38.	35.	56.	36.	30.

Chemical analyses (parts per million)

As	4.7	1.3	3.0	1.8	1.9	2.2	4.1	3.3	5.8	9.8	1.9	.07
Au	—	—	—	—	—	—	—	—	—	—	—	—
Hg	—	—	—	—	—	—	—	.04	.04	.43	—	—
W	—	—	1.	—	—	—	—	—	2.	—	—	—

†[Chemical analyses of major oxides in weight percent by methods of Jackson and others (1987); analysts, J.S. Mee and D.F. Siems. FeO, H_2O^+, H_2O^-, CO_2 in weight percent by methods of Jackson and others (1987); analyst, N.H. Elsheimer. Total S in weight percent by combustion-infra-red method; analyst, N.H. Elsheimer. F, Cl in weight percent by specific ion electrode method; analyst, N.H. Elsheimer. Inductively coupled plasma-atomic emission spectroscopy (total and partial) in parts per million by methods of Crock and Lichte (1982), Lichte and others (1987), and Motooka (1988); analysts, P.H. Briggs, and J.M. Motooka. Precision for concentration higher than 10 times the detection limit is better than +10–percent relative standard deviation; instrumental precision of the scanning instrument is +2–percent relative standard deviation. Looked for, but not found, at parts-per-million detection levels in parantheses: As (10), Au (8), Bi (10), Cd (2), Eu (2), Ho (4), Nb (4), Sn (5), Ta (40), U (100). Energy-dispersive X–ray fluorescence spectrometry in parts per million by method of Johnson and King (1987); analyst, J. Kent. Chemical analysis for Au by graphite furnace-atomic absorption spectrometry (Meier, 1980); analysts, E.P. Welsch, B.H. Roushey, and P.L. Hageman. Chemical analysis for As by graphite furnace atomic absorption spectrometry (Wilson and others, 1987; Aruscavage and Crock, 1987); chemical analysis for Hg by cold-vapor atomic absorption spectrometry; chemical analysis for W by spectrophotometric method of Wilson and others (1987); analysts, E.P. Welsch, B.H. Roushey, and P.L.Hageman; —, not detected]

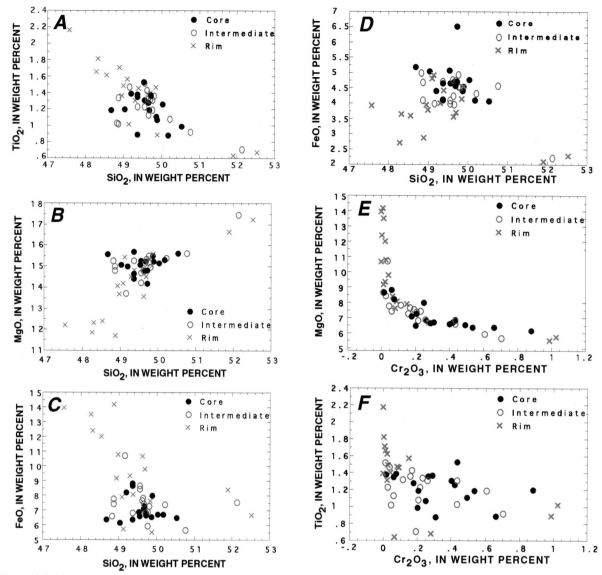

Figure 46 Oxide contents for 17 microprobe analyses each of rim-, intermediate-, and core-domains of augite crystals from a select sample of BBA–suite rock of the Devonian, Mississippian, Pennsylvanian, and Permian Havallah sequence. *A*, TiO_2 versus SiO_2 content; *B*, MgO versus SiO_2 content; *C*, FeO versus SiO_2 content; *D*, Al_2O_3 versus SiO_2 content; *E*, FeO versus Cr_2O_3 content; *F*, TiO_2 versus Cr_2O_3 content.

cent, normalized volatile free), which diagram is used commonly to discriminate volcanic rock series, two samples plot in the tholeiitic field, whereas the third plots in the calc-alkaline field (fig. 52). Nonetheless, all analyzed rocks from the Mississippian BBA suite in the Havallah sequence cluster in a fairly restricted domain near termini of trends of both calc-alkaline and tholeiitic series volcanic rocks elsewhere (fig. 52).

None of the CIPW norms of the three rocks analyzed from the BBA suite near the north end of the Havallah Hills, calculated volatile free and normalized to 100 weight percent with Fe_2O_3 and FeO recalculated from the original data such that Fe^{3+}/Fe (total) = 0.15 (Sawlan, 1991), show critical undersaturation with respect to silica resulting in presence of nepheline in the norms (table 10). Normative

plagioclase composition is approximately in the range 58 to 66 weght percent anorthite. In addition, the three rocks analyzed are strongly hypersthene (hy) and diopside (di) normative, including as much as roughly 35 weight percent of these two normative minerals combined. MgO normalized to 100 weight percent from the BBA suite near the north end of the Havallah Hills is in the range 6.8 to 8.9 weight percent (table 10). Only two of the samples of the BBA suite show presence of olivine (ol) in the CIPW norm (table 10). These contents of MgO suggest that basaltic rocks of the BBA suite in the Havallah Hills cannot be regarded as showing close affinities to primitive normal MORBs (mid-ocean ridge basalts) whose glasses elsewhere have been shown typically to have MgO contents greater than 9 weight percent (Elthon, 1990). As

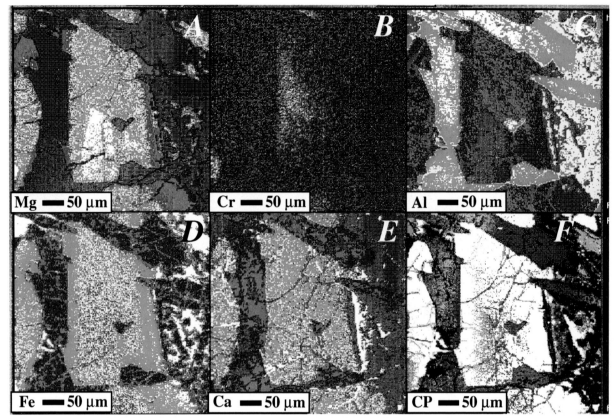

Figure 47 X–ray intensity maps showing distribution of elements in a select augite crystal of BBA–suite rock from Devonian, Mississippian, Pennsylvanian, and Permian Havallah sequence. Same as grain A (table 8). *A*, magnesium; *B*, chromium; *C*, aluminum; *D*, iron; *E* calcium; *F*, back-scattered electron image of grain A.

further pointed out by Elthon (1990), SiO_2 content (non-normalized) of most primitive MORB glasses are in the class interval 48.75 to 50 weight percent, somewhat higher than the mode of the rocks analyzed from the BBA suite in the Havallah Hills (range 47 to 50 weight percent SiO_2, table 10).

Silica variation diagrams for major and minor elements of the BBA suite in the northern Havallah Hills are shown in figure 53*A–J*, together with analyses of more widespread basaltic rocks south of Golconda summit and at Mill Canyon that are also part of the BBA volcanic suite in the Havallah sequence. The latter two occurrences are described below in the section entitled "Basaltic Rocks at Buffalo Mountain and in the Mill Canyon Area." Many silica variation plots show scatter in the data, undoubtedly reflecting, in part, presence of some strongly altered samples among those analyzed (fig. 53). Intensity of alteration affecting two of the BBA–suite rocks from Mill Canyon is readily apparent in the K_2O contents versus SiO_2 contents diagram wherein the two altered samples show normalized K_2O contents higher than 3 normalized weight percent (fig. 53*G*).

Some additional chemical relations can be used to distinguish further the BBA suite in the northern Havallah Hills, and, as will be discussed shortly below in a section entitled "Basaltic Rocks of Havallah Sequence at Buffalo Mountain and in the Mill Canyon Area," those chemically similar basaltic rocks in the Havallah sequence that are present west and south of the study area. The BBA suite in the northern Havallah Hills shows high values of *mg* ($100*$atomic $Mg/Mg+Fe^{2+}$, also setting Fe^{2+}/Fe total $=0.85$ as above) that are in the range 60.3 to 62.9, which suggests that the rocks may be referred to as being magnesian (table 10). In addition, this BBA suite of rocks is not strongly enriched in LREEs (light rare-earth elements) with the result that $La_n/Yb_n<2$ and $Ce_n/Yb_n<2.2$ (subscript "n" indicates a chondrite-normalized value for the element [sample/chondrite] using the average of 10 ordinary chondrite values in Nakamura (1974) as suggested by Rock (1987)). These ratios suggest that complete REE patterns for the BBA suite would be relatively flat if concentrations of all REE were known. Unfortunately, analyses for all REEs in these rocks are not available. However, ratios of some less- to more-incompatible elements are exemplified by non-normalized Zr/Nb ratios in the 10 to 13.5 range for the BBA suite in the northern Havallah Hills (table 9).

Table 10—Normalized chemical analyses, CIPW norms and analytical data of basaltic rocks in Devonian, Mississippian, Pennsylvanian and Permian Havallah sequence, Valmy and North Peak quadrangles, Nevada.†

Analysis No.	1	2	3	4	5	6	7	8	9	10
Field Number 90TT...	284	285	286	308	312	319	314	289	310	328

Chemical analyses, normalized volatile free to 100 percent (in weight percent)										
	Basalt-basaltic andesite			Basaltic trachyandesite-trachyandesite						
SiO_2	51.01	50.68	52.38	54.57	55.6	51.82	53.7	52.59	58.05	58.26
TiO_2	1.63	1.65	1.7	2.07	2.57	3.1	3.2	2.3	1.31	.99
Al_2O_3	15.92	15.42	16.24	16.27	16.37	15.91	17.19	17.04	18.47	16.66
Fe_2O_3	1.57	1.69	1.29	2.1	1.79	2.06	1.64	1.65	1.19	1.85
FeO	7.98	8.62	6.59	10.69	9.15	10.52	8.38	8.43	6.06	9.43
MnO	.18	.22	.24	.17	.15	.18	.15	.2	.14	.45
MgO	8.91	8.64	6.84	2.48	2.09	3.29	2.74	4.59	1.53	.46
CaO	8.50	8.27	10.58	2.94	3.06	4.9	4.23	4.03	2.73	1.45
Na_2O	3.22	3.8	3.37	7.77	8.26	6.96	7.73	6.69	6.51	8.94
K_2O	.36	.35	.35	.17	.31	.14	.41	.81	3.42	.35
P_2O_5	.17	.16	.18	.72	.53	1.07	.79	1.51	.4	1.02
SrO	.03	.03	.03	.04	.04	.04	.04	.07	.04	.03
BaO	.52	.47	.22	.03	.08	.11	.03	.09	.14	.12
Total	100.	100.	100.	100.	100.	100.	100.	100.	100.	100.
mg	62.87	60.29	61.13	25.99	25.68	32.14	33.12	45.2	27.7	6.9

CIPW norms (in weight percent)										
Q	—	—	0.01	—	—	—	—	—	—	—
ne	—	—	—	0.81	2.83	1.88	4.23	—	0.93	—
c	—	—	—	—	—	—	—	1.31	—	1.28
ol	6.6	13.69	—	15.42	10.71	13.92	10.16	15.73	8.97	9.64
(fo)	4.27	8.45	—	4.23	3.14	5.07	4.14	7.75	2.66	0.63
(fa)	2.33	5.24	—	11.19	7.56	8.75	6.02	7.98	6.32	9.02
hy	18.67	8.19	16.45	—	—	—	—	.73	—	3.57
(en)	(12.48)	(5.24)	(10.9)	—	—	—	—	(.38)	—	(.25)
(fs)	(6.19)	(2.95)	(5.55)	—	—	—	—	(.35)	—	(3.32)
di	11.48	13.78	19.2	.84	4.48	4.93	4.25	—	.19	—
(di)	(8.01)	(9.24)	(13.29)	(.27)	(1.54)	(2.07)	(1.97)	—	(.07)	—
(hd)	(3.47)	(4.53)	(5.9)	(.57)	(2.94)	(2.86)	(2.27)	—	(.12)	—
ab	27.31	32.22	28.58	64.23	64.72	55.47	57.64	56.63	53.4	75.66
an	28.04	24.09	28.18	9.03	6.68	11.77	10.96	10.5	11.1	.83
or	2.12	2.05	2.05	.99	1.85	.8	2.44	4.76	20.24	2.04
mt	2.28	2.46	1.88	3.04	2.6	2.99	2.38	2.4	1.73	2.68
il	3.1	3.14	3.23	3.93	4.88	5.71	6.08	4.37	2.49	1.89
ap	.41	.38	.42	1.72	1.26	2.53	1.86	3.57	.95	2.41
Total	100.	100.	100.	100.	100.	100.	100.	100.	100.	100.
norm.Pl(An)	65.9	58.5	65	20.3	14.6	26.6	22.6	25.9	27.1	2
Color index	65.1	62.3	62.6	43.7	42.5	48.	42.3	44	28.8	37

†[Cross, Iddings, Pirsson,, and Washington (CIPW) norms and analytical data calculated from data of table 9 using computer program of M.G. Sawlan (written commun., 1993) after first adjusting weight percent Fe_2O_3 and weight percent FeO so that Fe_2O_3 is equal to 0.15 of total iron oxide (see text);—, not detected]

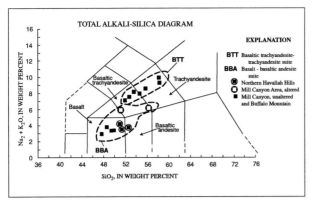

Figure 48 Total alkali versus silica content of chemically analyzed Mississippian BBA–suite rocks and Pennsylvanian BTT–suite rocks of Devonian, Mississippian, Pennsylvanian, and Permian Havallah sequence, as well as volcanic rock classification scheme of Le Bas and others (1986). Calculated volatile free and normalized to 100 percent from data of tables 9–11.

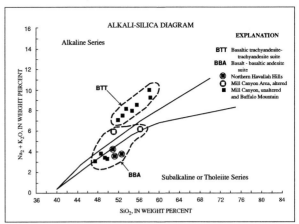

Figure 49 Total alkali content versus silica content of chemically analyzed Mississippian BBA–suite rocks and Pennsylvanian BTT–suite rocks of Devonian, Mississippian, Pennsylvanian, and Permian Havallah sequence, as well as subalkaline fields defined by both McDonald and Katsura (1964) and Irvine and Baragar (1971). Data calculated volatile free and normalized to 100 percent from tables 9–11.

BASALTIC ROCKS AT BUFFALO MOUNTAIN AND IN MILL CANYON AREA (MISSISSIPPIAN)

GEOLOGIC SETTING

Reconnaissance examination and sampling of basaltic rocks in the Havallah sequence at Buffalo Mountain, south of Golconda summit, and in the area of Mill Canyon, just south of the North Peak quadrangle, was undertaken for comparative purposes. The first area examined is on the northwest-facing flank of Buffalo Mountain (fig. 1) in secs. 3 and 4, T. 33 N., R. 41 E., in the Brooks Spring 7–1/2 minute quadrangle, where Marsh and Erickson (1977) show a fairly widespread distribution of greenstone, as much as approximately 2 km² at the surface, that they include with rocks assigned to the Pennsylvanian Pumpernickel Formation. As described previously, the Pumpernickel Formation, which includes pillow lavas and vesicular flows in its greenstone member at Buffalo Mountain, is now (1999) considered to be part of the Havallah sequence. However, it is not entirely clear whether the greenstone at Buffalo Mountain belongs to LU–1 or LU–2 of the Havallah sequence, although the latter is probably the more likely. In addition, greenstone at Buffalo Mountain is within a contact metamorphic aureole surrounding the Jurassic plutonic complex widely exposed at Buffalo Mountain (Marsh and Erickson, 1977). The second area examined is in the Mill Canyon part of the Antler Peak 7–1/2 minute quadrangle, specifically in the bottom of the canyon and approximately 0.8 km southeast from the entrance to the canyon; the area is in W. 1/2 sec. 26, T. 32 N., R. 42 E. Basaltic rock in a small expo-

sure is part of LU–2 in the upper plate of the Willow Creek thrust, and includes a pyrrhotite-bearing volcanogenic massive sulfide occurrence in pillow basalt that has been contact metamorphosed probably both by the Late Cretaceous stock at Trenton Canyon and some other nearby Cretaceous or Tertiary granodiorite porphyry dikes (Doebrich, 1992, 1995). In addition, some basalt at Mill Canyon appears to have been caught up along one of the minor high angle faults that are present in the area.

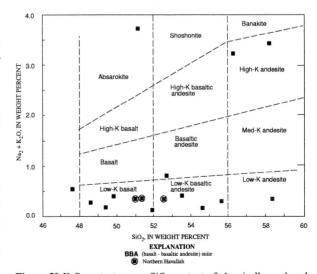

Figure 50 K₂O content versus SiO₂ content of chemically analyzed Mississippian BBA–suite rocks and Pennsylvanian BTT–suite rocks of Devonian, Mississippian, Pennsylvanian, and Permian Havallah sequence, as well as classification scheme of Peccerillo and Taylor (1976). Data calculated volatile free and normalized to 100 percent from tables 9–11.

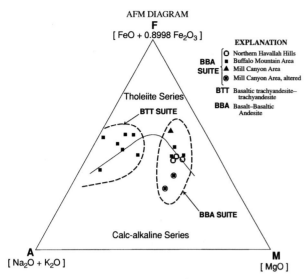

Figure 51 Mississippian BBA–suite rocks and Pennsylvanian BTT–suite rocks of Devonian, Mississippian, Pennsylvanian, and Permian Havallah sequence as well as join between tholeiitic series and calc-alkaline series of Irvine and Baragar (1971) and many others. Data calculated volatile free and normalized to 100 percent from tables 9–11.

CHEMISTRY

Chemical analyses are available for six samples of basaltic rock, three from the Buffalo Mountain area and three from the Mill Canyon area (table 11). The three samples from Buffalo Mountain (analyses 1–3, table 11) are remarkably similar to one another, but the three samples from Mill Canyon (analyses 4–6, table 11) include two samples (analyses 4 and 5) that show highly elevated K_2O contents and some other elements that must result from alteration probably associated with nearby Cretaceous or Tertiary granodiorite porphyry dikes. Barium contents, 3,010 and 4,340 ppm, also are elevated in the two samples that show the highest K_2O contents (table 11). Nonetheless, plots of various analytical data from these six analyses suggest that all six basaltic rocks in the two areas belong to the BBA suite (compare figs. 49–50 and figs. 52–53). The rocks from Buffalo Mountain are definitely sodic ($Na_2O–2 > K_2O$), whereas those from Mill Canyon appear to contain K_2O/Na_2O ratios modified substantially by post–BBA fluids.

CIPW norms appear to show effects of alteration on some of the BBA–suite rocks at Buffalo Mountain and at Mill Creek. Normalized contents of SiO_2, calculated volatile free and using the same Fe^{2+}/Fe total = 0.85 as above, are in the range 48.5 to 49.7 weight percent in BBA samples from Buffalo Mountain and 47.5 to 56 weight percent in BBA samples from Mill Canyon (table 12). The sample with the highest normalized SiO_2 content at Mill Canyon also is obviously altered and one that shows presence of normative quartz in its CIPW norm (analysis 4, table 12).

The two altered samples of BBA at Mill Canyon show approximately 19 to 22 weight percent normative orthoclase in contrast to normative orthoclase roughly in the range 1 to 3 weight percent for the remaining samples (analyses 4, 5, table 12). Two samples from Buffalo Mountain and the apparently least altered sample from Mill Canyon are nepheline normative, suggesting that BBA–suite rocks elsewhere in the Havallah sequence can show moderate amounts of undersaturation with respect to SiO_2 content. Additional implications of various elemental ratios in BBA rocks from both the Buffalo Mountain and Mill Canyon area are included in the immediately preceding section above.

BASALTIC TRACHYANDESITE AND TRACHYANDESITE OF LITHOTECHTONIC UNIT 2 (PENNSYLVANIAN AND [OR] PERMIAN)

GEOLOGIC SETTING

The second occurrence of volcanic rock in LU–2 of the Havallah sequence in the quadrangles is much more widespread than that described in the northern Havallah Hills, and includes presumably Pennsylvanian and (or) Permian basaltic trachyandesite and trachyandesite (BTT), volcaniclastic sedimentary rock, and associated exhalative silica (subunit hba, pl. 2). This subunit crops out in a narrow belt along an approximately 6–km–strike-length south of the Old Marigold Mine. The rocks in the subunit make up tectonically the lowest rocks mapped in LU–2 between the Cottonwood Creek and Oyarbide faults, and the base of the rocks is the Willow Creek thrust as described above.

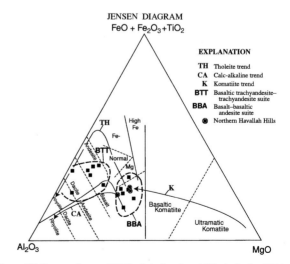

Figure 52 Ternary Jensen (1976) plot of analyzed Mississippian BBA–suite rocks and Pennsylvanian BTT–suite rocks of Devonian, Mississippian, Pennsylvanian, and Permian Havallah sequence. Data calculated volatile free and normalized to 100 percent from tables 9–11.

Age of basaltic trachyandesite and trachyandesite (late Atokan to Wolfcampian Epochs) was established from presence in these rocks of exhalative silica that contains biostratigraphically useable radiolarians assigned to assemblages 6 and 7 of Murchey (1990) and these assemblages are described further by B.L. Murchey (unpub. data, 1999).

In all, 16 standard thin sections and seven polished thin sections of basaltic trachyandesite and trachyandesite first were examined petrographically by transmitted- and reflected-light microscopy. The polished thin sections also were examined subsequently by SEM methods. Volcaniclastic sedimentary rocks in the subunit include some probable manganiferous hyaloclastite breccias. In addition, major- and minor-element chemistry of seven rocks of various types from the subunit are described in this subsection below.

The most common rock in subunit hba is aphyric, fine-grained BTT that is sparsely amygdaloidal, microphenocrystic, and fragmental (fig. 54A, B). As described above, these rocks form part of a magmatic suite that is chemically distinct from BBA occurrences in the northern Havallah Hills and elsewhere throughout the Havallah sequence. Most BTT is relatively fresh, although the rocks apparently have been affected weakly by a subsolidus greenschist metamorphic event, and textural evidence for quenching still is moderately well preserved (fig. 54B). This evidence includes (1) the presence of hollow cores to many acicular crystals of plagioclase thereby producing skeletal outlines, (2) presence of poorly developed swallow-tail and belt-buckle outlines of plagioclase, and (3) dendritic to sheath-like aggregates of fine-grained plagioclase crystals, mostly albite, extending irregularly from edges of some medium-grained crystals of calcium plagioclase that obviously crystallized earlier in the paragenetic history of the rocks. All mineralogic variants in the groundmass of these rocks also are present in textures observed by microscope in submarine basalts along present-day seafloor spreading centers (Bryan, 1972). In contrast to relations noted above in the BBA occurrence in the northern Havallah Hills, calcic plagiolase laths in subunit hba, particularly the larger ones, host inclusions of previously crystallized opaque minerals. Furthermore, many BTT samples have fragmental textures and display highly variable sizes of crystals among quench fabrics in the fragments. Vesicles in amygdaloidal varieties of these rocks are filled with various combinations of calcite, chlorite, biotite, opaque minerals, and, rarely, barite. Some plagioclase in the groundmass of BTT shows conspicuously curved crystal outlines. Opaque minerals also are fairly abundant in the groundmass of all samples examined from subunit hba, perhaps making up as much as 10 volume percent of some samples, and the opaque minerals consist either of scattered equant grains or acicular crystals that form ladder-type textures at high angles to adjoining crys-

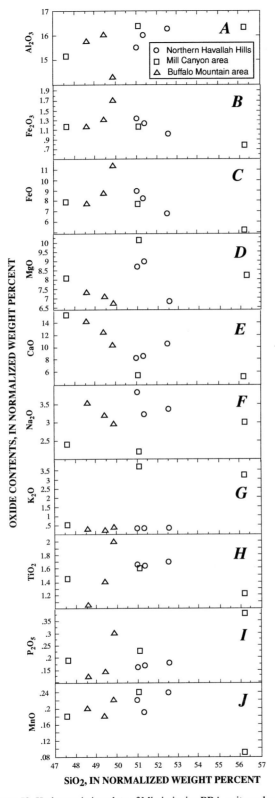

Figure 53 Harker variation plots of Mississippian BBA–suite rocks of Devonian, Mississippian, Pennsylvanian, and Permian Havallah sequence from northern Havallah Hills, Mill Canyon area, and Buffalo Mountain area. Data calculated volatile free and normalized to 100 percent from tables 9–11.

Table 11—Analytical data of Mississippian basalt and basaltic andesite (BBA suite), from Devonian, Mississippian, Pennsylvanian, and Permian Havallah sequence, in Mill Canyon and Buffalo Mountain areas, Humboldt and Lander Counties, Nevada.†

Analysis number	1	2	3	4	5	6
Field number 88TT...	087	088	089	102	103	104

	Chemical analyses (weight percent)					
	Mill Canyon			Buffalo Mountain		
SiO_2	47.2	48.7	48.9	53.8	49.3	44.3
Al_2O_3	15.3	15.8	14.	15.6	15.8	14.1
Fe_2O_3	1.67	2.	2.8	2.29	1.2	1.58
FeO	7.13	8.	10.14	3.68	7.44	6.98
MgO	7.08	6.97	6.59	7.86	9.81	7.53
CaO	13.9	12.2	10.1	5.01	5.27	14.2
Na_2O	3.42	3.15	2.89	2.87	2.13	2.24
K_2O	.27	.18	.39	3.09	3.59	.5
H_2O^+	1.46	1.38	1.83	2.86	2.55	1.86
H_2O^-	.16	.18	.25	.58	.16	.46
TiO_2	1.01	1.37	1.95	1.16	1.54	1.36
P_2O_5	.12	.14	.29	.36	.22	.18
MnO	.19	.18	.22	.09	.23	.17
F	.03	.02	.07	.08	.1	.07
Cl	.233	.043	.042	.015	.086	.018
CO_2	1.6	.13	.09	.19	.25	4.53
Total S	<.01	<.01	.13	<.01	.26	1.42
Subtotal	100.77	100.44	100.68	99.54	99.94	101.5
Less O = F, Cl, S	.06	.02	.07	.03	.13	.37
Total	100.72	100.42	100.61	99.51	99.81	101.13

	Inductively coupled plasma-atomic emission spectroscopy (total) (parts per million)					
Ba	461.0	186.0	626.0	3010.0	4340.0	509.0
Be	—	—	1.	2.	1.	—
Cd	—	—	—	—	—	—
Ce	—	9.	18.	58.	16.	9.
Co	45.	53.	38.	33.	57.	34.
Cr	389.	386.	175.	429.	467.	402.
Cu	61.	46.	127.	36.	11.	61.
Eu	—	—	2.	2.	—	—
Ga	16.	23.	22.	18.	19.	18.
La	5.	9.	14.	34.	11.	8.
Li	18.	8.	26.	54.	66.	21.
Nb	—	—	7.	8.	5.	—
Nd	16.	19.	24.	34.	15.	18.
Ni	98.	128.	51.	153.	235.	198.

	Inductively coupled plasma-atomic emission spectroscopy (total) (parts per million)					
Pb	22.	7.	9.	5.	4.	4.
Sc	41.	50.	40.	20.	31.	35.
Sr	239.	273.	219.	459.	314.	389.
Th	—	—	—	8.	—	—
V	236.	312.	349.	141.	241.	229.

Table 11 – (cont.)

Y	22.	29.	33.	21.	19.	21.
Yb	2.	4.	4.	2.	2.	2.
Zn	96.	88.	84.	56.	105.	59.

Inductively coupled plasma-atomic emission spectroscopy (partial) (parts per million)

Ag	0.1	0.057	0.38	0.14	0.046	0.22
As	35.	4.1	8.7	51.	140.	88.
Bi	.85	—	—	—	—	1.5
Cd	.23	.078	.1	.055	—	.044
Cu	34.	26.	97.	28.	7.6	43.
Mo	.12	—	.43	.58	.18	.09
Pb	15.	2.8	6.9	2.9	.69	2.7
Sb	1.2	.81	1.4	3.	.78	.78
Zn	22.	15.	15.	30.	58.	17.

Energy-dispersive X-ray fluorescence spectrometry (parts per million)

Nb	—	—	—	16.0	—	—
Rb	12.0	—	—	110.	178.0	28.0
Sr	230.	210.0	188.0	355.	285.	330.
Zr	60.	84.	118.	220.	124.	98.
Y	26.	22.	32.	24.	30.	22.

Chemical analyses (parts per billion)

Au	—	—	—	50.0	—	—
Hg	20.	—	—	—	—	—
Pt	0.8	<.5	<.5	0.8	<.5	<.5
Pd	1.7	.9	1.3	<.8	1.3	.9
Rh	<.5	<.5	<.5	<.5	<.5	<.5
Ru	<.5	<.5	<.5	<.5	<.5	<.5
Ir	<.5	<.5	<.5	<.5	<.5	<.5

†[Chemical analyses of major oxides in weight percent by methods of Jackson and others (1987); analysts, J.E. Taggart, A.J. Bartel, D.F. Siems. FeO, H_2O^+, H_2O^-, CO_2 in weght percent by methods of Jackson and others (1987); analyst, T.L. Fries. Total S in weight percent by combustion-infra-red method; analyst, T.L. Fries. F, Cl in weight percent by specific ion electrode method; analyst, T.L. Fries. Inductively coupled plasma-atomic emission spectroscopy (total and partial) in parts per million by methods of Crock and Lichte (1982), Lichte and others (1987), and Motooka (1988); analysts, P.H. Briggs, and J.M. Motooka. Precision for concentration higher than 10 times the detection limit is better than +10–percent relative standard deviation; instrumental precision of the scanning instrument is +2–percent relative standard deviation. Looked for, but not found, at parts-per-million detection levels in parantheses: As (10), Au (8), Bi (10), Cd (2), Ho (4), Sn (5), Ta (40), U (100). Energy-dispersive X–ray fluorescence spectrometry in parts per million by method of Johnson and King (1987); analyst, J. Kent. Chemical analysis for Au by graphite furnace-atomic absorption spectrometry (Meier, 1980); analysts, K.H. Roushey, and P.L. Hageman. Chemical analysis for Hg by cold-vapor atomic absorption spectrometry; analysts, K.H. Roushey, and P.L.Hageman. Chemical analyses for Pt-group elements by NiS fire assay conbined with inductively coupled plasma finish (Meir and others, 1991); —, not detected]

Table 12—Normalized chemical analyses, CIPW norms and analytical data of BBA–suite rocks from Buffalo Mountain and Mill Canyon areas belonging to Devonian, Mississippian, Pennsylvanian, and Permian Havallah sequence, Valmy and North Peak quadrangles, Nevada.†

Analysis number	1	2	3	4	5	6
Field number 88TT..	102	103	104	087	088	089

Chemical analyses normalized volatile free to 100 percent (in weight percent)

	Buffalo Mountain			Mill Canyon		
SiO_2	48.49	49.34	49.75	56.	50.78	47.52
TiO_2	1.04	1.39	1.98	1.21	1.59	1.46
Al_2O_3	15.72	16.01	14.24	16.24	16.27	15.13
Fe_2O_3	1.48	1.66	2.15	1.	1.46	1.5
FeO	7.54	8.44	10.95	5.08	7.46	7.66
MnO	.2	.18	.22	.09	.24	.18
MgO	7.27	7.06	6.70	8.18	10.11	8.08
CaO	14.28	12.36	10.28	5.22	5.43	15.23
Na_2O	3.51	3.19	2.94	2.99	2.19	2.4
K_2O	.28	.18	.4	3.22	3.69	.54
P_2O_5	.12	.14	.3	.38	.23	.19
SrO	.05	.02	.07	.35	.52	.06
BaO	.03	.03	.03	.57	.04	.05
Total	100.	100.	100.	100.	100.	100.
mg	59.38	55.91	48.13	70.94	67.24	61.51

CIPW norms (in weight percent)

Q	—	—	—	1.16	—	—
ne	7.8	.8	—	—	—	5.04
c	—	—	—	1.16	—	—
ol	8.69	12.48	5.14	—	17.62	9.37
(fo)	(5.16)	(7.08)	(2.54)	—	(12.14)	(5.9)
(fa)	(3.54)	(5.4)	(2.6)	—	(5.48)	(3.47)
hy	—	—	15.08	25.94	9.89	—
(en)	—	—	(7.83)	(19.53)	(7.01)	—
(fs)	—	—	(7.25)	(6.41)	(2.87)	—
di	35.81	25.91	20.47	2.46	2.62	37.05
(di)	(23.22)	(16.16)	(11.32)	(1.91)	(1.93)	(25.27)
(hd)	(12.59)	(9.75)	(9.14)	(.55)	(.69)	(11.78)
ab	15.35	25.54	24.89	25.32	18.61	11.05
an	26.31	28.82	24.5	21.44	23.73	28.91
or	1.64	1.08	2.35	19.04	21.85	3.17
mt	2.14	2.4	3.11	1.45	2.13	2.18
il	1.97	2.64	3.77	2.3	3.02	2.77
ap	.29	.34	.7	.89	.54	.46
Total	100.	100.	100.	100.	100.	100.
norm.Pl(An)	61.5	70.6	67.3	55.8	66.	65.
Color index	60.3	60.3	62.6	60.2	58.	67.2

†[Cross, Iddings, Pirsson, and Washington (CIPW) norms and analytical data calculated from data of table 11 using computer program of M.G. Sawlan (written commun., 1993) after first adjusting weight percent Fe_2O_3 and weight percent FeO so that Fe_2O_3 is equal to 0.15 of total iron oxide (see text); —, not detected]

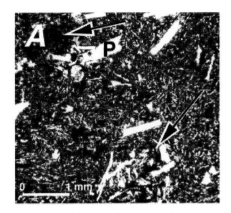

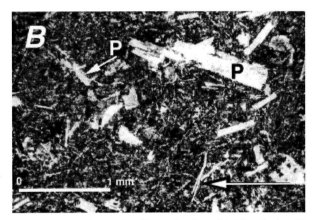

Figure 54 Photomicrographs of Pennsylvanian and (or) Permian basaltic trachyandesite-trachyandesite, subunit hba of LU–2 belonging to Devonian, Mississippian, Pennsylvanian, and Permian Havallah sequence. Area of outcrop is south of Cottonwood Creek fault and north of Oyarbide fault. P, plagioclase. Plane polarized light. *A*, Typical microphenocrystic texture of basaltic trachyandesite-trachyandesite including some fragmental clots of opaque minerals and chlorite (at head of arrow, upper left) and opaque minerals, plagioclase microphenocrysts, and chlorite (at head of arrow, lower right). Note bimodal size distribution of plagioclase with predominant quench-type textures of fine-grained plagioclase in groundmass. Sample 90TT312. *B*, Closeup of *A* showing some plagioclase microlites with swallow-tail shapes (at head of arrow) resulting from quenching

tals of microlitic plagioclase or prismatic clots of fine-grained chlorite. Some equant, generally irregular-shaped opaque grains, mostly ilmenite, in places include extremely fine grained, granular aggregates of apatite that are drab brown in transmitted light. Other equant grains are titaniferous magnetite which seems to be paragenetically early relative to nearby blades of ilmenite, and presence of augite could not be confirmed among plagioclase crystals in groundmass of any of the rocks examined from the subunit. Some small intergranular sphene also is present among the plagioclase crystals as well as chlorite. The latter mineral may reflect devitrification of glass and thereby the previous presence of an intersertal original groundmass texture. Chromite was not found in any of the studied samples. Phenocrysts are rare in BTT and those

phenocrysts that are present consist of stubby laths of plagioclase that have margins that merge gradually into mats of acicular albite in the surrounding groundmass. Plagioclase in all these rocks is only slightly clouded by secondary micas and clays. Thus, overall proportions of modal phenocryst and groundmass minerals in the subunit suggest that the rocks also could be termed highly plagioclase aphyric basalt in the modal classification scheme proposed by Bryan (1983).

A number of the samples from subunit hba also were examined by SEM methods. As with the previously described rocks belonging to the BBA suite, these studies confirm that a probable low-grade greenschist event affected these rocks as well. Furthermore, clinopyroxene could not be found in the groundmass of any of the samples examined by SEM. In addition, some rocks show local remobilization of Fe, probably during Pennsylvanian and (or) Permian time, indicated by wispy Fe–oxide-bearing veins that cut igneous fabrics in the rocks and that are associated with (1) specular hematite, chlorite, and sphene (fig. 55*A*, *B*), (2) magnetite and chlorite (fig. 55*C*), or (3) magnetite-, jasperoid-, barite-, and albite-bearing assemblages (fig. 55*D*). However, albite also is present as the predominant feldspar in that part of BTT showing well-developed quench textures (fig. 55*E*, *F*). Such albite, which has well-developed quench textures in the groundmass, probably is secondary, because it most likely was quenched initially as a calcic plagioclase and then, because of its small grain size, reacted preferentially with Na–enriched fluids under subsolidus conditions either during diagenesis or the subsequent greenschist event.

SEM examination of filled vesicles in BTT reveals complex mineralogic relations in that many of them show vesicles lined by calcite and cores of the vesicles then filled by biotite (fig. 56*A–D*). Contacts between calcite lining the walls of the vesicles and biotite that fills core of the vesicle are quite sharp, but irregular (fig. 56*B*), and alteration of neither mineral on either side of the interface appears to be present. Nonetheless, some calcite may have been dissolved along the highly irregular, stylolitic-like interface between the two minerals at the time of biotite crystallization. Although vesicles in modern submarine basalts have been reported to contain largely primary, magmatic CO_2 (Moore and others, 1977; Sarda and Graham, 1990; Javoy and Pineau, 1991), calcite in the BTT vesicles is probably not the product of coupled chemical reactions involving primary CO_2, sea water, and calcic plagioclase. Some clots and veins of obvious secondary calcite associated with chlorite also are present in BTT, and calcite in the vesicles is probably the same paragenetic age as the definitely secondary carbonate minerals. Apart from the vesicles, biotite is extremely rare in groundmass of these rocks, and biotite in the cores of the vesicles probably is the result of conversion of K–bearing zeolite, pos-

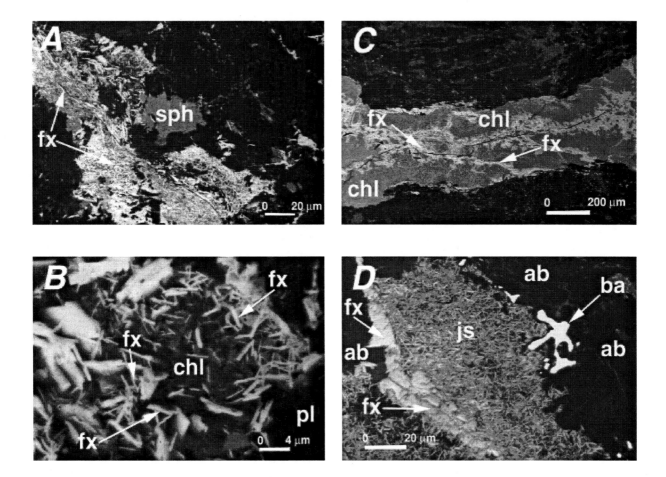

Figure 55 Back-scattered electron micrographs of Pennsylvanian and (or) Permian basaltic trachyandesite-trachyandesite that crops out between Cottonwood Creek and Oyarbide faults in subunit hba of LU–2 of Devonian, Mississippian, Pennsylvanian, and Permian Havallah sequence showing general textural relations. sph, sphene; chl, chlorite; pl, plagioclase; fx, specular hematite and (or) other titaniferous Fe oxide(s). *A*, fragmental basaltic trachyandesite-trachyandesite, including cm–sized, ovoid fragments cut by veins including specular hematite and other Fe oxide(s), and associated with sphene. Locality 92TT048; *B*, closeup of *A* showing extremely small blades of specular hematite tightly intergrown with chlorite; *C*, mat of plagioclase crystals in aphyric basaltic trachyandesite-trachyandesite cut by narrow vein of intergrown titaniferous Fe oxide, probably magnetite, and chlorite. Locality 92TT050; *D*, small clot of jasperoid (js), including barite (ba) and titaniferous Fe oxide surrounded by albite (ab). Locality 92TT050.

sibly mordenite or phillipsite, during the greenschist metamorphic event. Groundmass minerals of BTT are generally reduced in grain size near filled vesicles (fig. 56*C*), and clots of intergrown barite and probable specular hematite are present in calcite lining the walls of some vesicles (fig. 56*D*). This intergrown barite and probable specular hematite may have been among the first minerals to precipitate in the vesicles.

Specific complexities described above in mineralogy of BTT, including presence of an apparently pervasive low-grade greenschist event, must be kept in mind when major-element chemistry is described below because subtle changes in composite mineral assemblages through time can have significant impact on whole-rock chemistry, if some of those changes are metasomatic. However, the relatively low-grade nature of the greenschist event that affected these rocks is further emphasized by the large numbers of samples from which age-diagnostic radiolar

ians have been recovered in sedimentary rocks interlayered with BTT in the Havallah sequence (Tomlinson, 1990; Murchey, 1990).

Volcaniclastic sedimentary rocks in BTT south of the Cottonwood Creek fault (subunit hba, pl. 2), include some unusual microbreccia wherein slightly rounded fragments of monocrystalline and polycrystalline plagioclase, monocrystalline quartz, basaltic rock fragments, sparry micrite, opaque minerals (including manganese oxide), and traces of polycrystalline quartz largely are all supported by a fine-grained plagioclase-chlorite matrix that shows abundant soft-sediment-type deformation features. This microbreccia might be a variety of aquagene tuff as described by Carlisle (1963) who reported that groundmass of the classic aquagene tuffs he studied in British Columbia also include some isolated crystals and clusters of crystals of plagioclase. Morphology of plagioclase fragments, some as much as 1.7 mm long and texturally not at all like

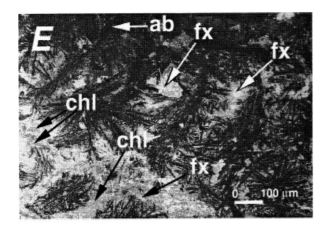

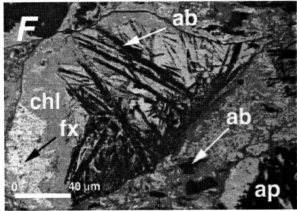

Figure 55 (cont.) *E*, feathery-textured fine-grained albite intergrown with titaniferous magnetite and iron-rich chlorite. Locality 92TT050; *F*, large-scale image of relations same as *E*, but also including a small amount of apatite (A) in field of view.

the plagioclase in associated BTT described above, suggests that plagioclase fragments in microbreccia may have been derived ultimately from a gabbroic protolith, possibly high in a cumulate zone subjacent to the subunit at its site of eruption onto the sea floor (Hess, 1989). Slight-to-moderate rounding of these plagioclase fragments may have occurred during flow of the rocks onto the seafloor. Quartz fragments in the microbreccias are angular and do not include abundant concentrations of fluid inclusions as are common in almost all terrigenous quartz grains described in this report above in calcareous siliciclastic sequences in LU–2. Many quartz fragments are completely devoid of any fluid inclusions, even when viewed at magnifications of 400 X. As such, lack of fluid inclusions in these fragments is similar to absence of fluid inclusions observed by the author in paragenetically late quartz hosted by wairakite-pyrrhotite-sphalerite veins that cut turbidites in some modern seafloor spreading centers. Therefore, quartz fragments contained in microbreccia of subunit hba are probably not terrigenous in origin, but instead are a result of fragmentation subsequent to the hydrothermal

processes that produced exhalative silica on the Pennsylvanian and (or) Permian seafloor.

Sporadic occurrences of massive exhalative silica also are associated with many outcrops of BTT belonging to subunit hba (pl. 2). Generally, exhalative silica in the subunit is blood-red in outcrop and is present as small individual exposures no more than 15 to 20 m^2 in area or as scattered float in colluvium throughout selected areas of BTT. These rocks commonly include some recrystallized radiolarians, in places quite abundant, most of which are too recrystallized to be biostratigraphically significant. Fortunately, some biostratigraphically useful radiolarians are present in some exhalative silica as described above. In addition, under the microscope, exhalative silica typically has a fabric that includes abundant, wavey, discontinous, thin zones of fine-grained silica, and, elsewhere, a brecciated fabric with fractures now filled by Fe oxide and Mn oxide (fig. 57). Narrow veins, including cross-fiber textured quartz and epidote, also are present in some samples of exhalative silica; these veins are discontinuous and, in places, terminate at microfractures that course through the rocks.

CHEMISTRY

Chemical analyses are available for seven samples of BTT collected from representative sites spread almost along the entire exposed strike length of subunit hba, the lowermost mapped subunit in LU–2 south of the Cottonwood Creek fault (pl. 2). These chemical analyses have some similarities and significant differences with the BBA suite (analyses 4–10, table 9). Chemical analyses from subunit hba indicate these rocks to be sodic according to criteria described above, although one of the samples (analysis 6, table 9), contains 3.26 weight percent K$_2$O, a value which suggests the sample is altered. This sample was collected near the northeast corner of sec. 36, T. 33 N., R. 42 E., close to a minor northeast striking fault that cuts subunit hba and its underlying Willow Creek thrust (pl. 2). The anomalously high K$_2$O content of this particular rock relative to the remaining rocks of the BTT suite is readily apparent on a plot of K$_2$O versus SiO$_2$ content (fig. 50) which shows analysis 6 plotting in the high–K andesite field of Peccerillo and Taylor (1976). Petrographic examination of two thin sections cut from this rock reveals that the bulk of the aphyric volcanic rock is essentially unaltered, and that no widespread introduction of secondary K–feldspar or secondary mica into the rock is present. However, a few narrow seams of secondary Fe–oxide minerals are present that probably replace sulfide minerals-some introduction of secondary clay may have occurred along narrow alteration halos adjacent to these seams. These secondary effects may account for the elevated K$_2$O content reported (table 9).

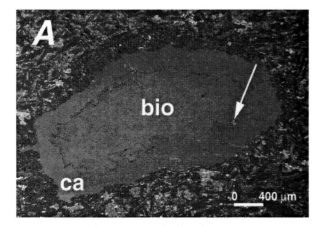

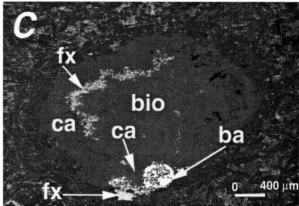

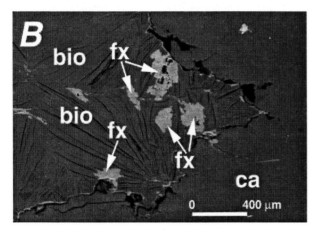

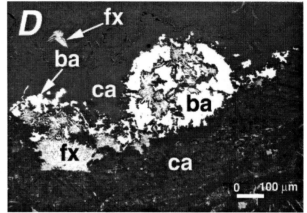

Figure 56 Mineralogy of filled vesicles. bio, biotite; ca, calcite; hm, specular hematite, ba, barite. Locality 92TT049. *A*, Elongate vesicle measuring 1.6 x 3.2 mm lined by approximately 300–μm-wide zone of calcite which has irregular contact with biotite that fills central part of vesicle. Iron oxide, probably specular hematite, at head of arrow. Note decrease in grain size of groundmass microlites near vesicle. *B*, Closeup of relations near head of arrow in *A*. *C*, Ovoid vesicle measuring approximately 2.0 x 2.8 mm lined by calcite, which includes barite and probable specular hematite near wall of vesicle, and containing a core filled by biotite. Iron oxide, probably specular hematite, concentrated near much of irregular interface between calcite and biotite. *D*, Closeup of barite and probable specular hematite textural relations in *C*.

Non-normalized SiO_2 contents in seven analyzed samples of subunit hba are in the range 50.3 to 55.6 weight percent (table 9), and normalized contents of SiO_2 are in the range 51.8 to 58.3 weight percent (table 10). In the classification scheme of Le Bas and others (1986), these seven samples plot in a field that straddles the basaltic trachyandesite and trachyandesite fields (thus defining the BTT–suite of volcanic rocks), and, moreover, rocks of this BTT–volcanic suite are alkalic in contrast to those of the BBA–suite described above (figs. 48, 49). Although discrimination between these two suites of volcanic rock in the Havallah sequence partly involves their strong contrasts in total alkali content, the fact that these contrasts are so marked suggests that even if some local subsolidus intra-suite exchange of alkalis occurred, such exchange cannot account for the wide divergence in total alkali content found in the rocks. Five samples of the BTT–suite contain nepheline in their CIPW norms, also calculated volatile free and normalized to 100 weight percent with Fe^{3+}/Fe (total) = 0.15 as above, and the remaining two

samples are corundum normative (table 10). Contents of normalized Na_2O in the BTT–suite are roughly twice that of the BBA–suite, approximately 6.5 to 8.9 versus 3.2 to 3.8 weight percent (table 10). MgO normalized to 100 weight percent in the BTT–suite of rocks is in the range 0.5 to 4.6 weight percent, significantly less than contents of MgO in the BBA–suite (table 10), resulting in six of the samples of the BTT suite plotting in the tholeiitic field of a Jensen (1976) diagram (fig. 52). The BTT–suite volcanic rocks show diminished values of *mg* relative to the BBA–suite; these values are in the range 6.9 to 45.2 (table 10). In addition, BTT–suite rocks show only moderate to low concentrations of normative hypersthene and normative diopside, together with normative plagioclase approximately in the range 2 to 26 weight percent anorthite, the former being indicative of an overall more highly sodic content in these rocks than the BBA–suite. Thus, rocks of the BTT–volcanic suite apparently are more fractionated or evolved than the BBA–volcanic suite, although their apparent consanguineous evolution from BBA–suite-type

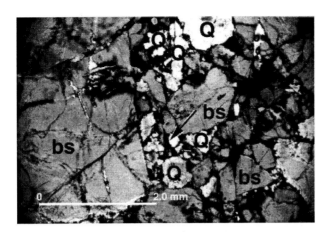

Figure 57 Photomicrograph of microbrecciated exhalative silica associated with Pennsylvanian and (or) Permian basaltic trachyandesite-trachtandesite in subunit hba of Devonian, Mississippian, Pennsylvanian, and Permian Havallah sequence. Locality 90TT313, south of Cottonwood Creek fault and north of Oyarbide fault. Microbrecciated exhalative silica (bs) has open spaces filled partly by manganese oxide (at head of arrow) and coarsely crystalline hydrothermal quartz (Q), some of which also shows effects of brecciation. Plane polarized light.

chemistries must have followed a less Fe oxide- and Ti oxide-enriched path than some classic tholeiitic series described elsewhere, if such an evolution in fact occurred in the Havallah sequence (fig. 52). Various silica variation diagrams, including analyses from both the BTT– and the BBA–suites, appear to show overall chemical trends in these rocks as the marine basalt environment evolved during Mississippian through Pennsylvanian and (or) Permian magmatism in the Havallah basin (fig. 58A–J). The BTT–suite shows a decline in contents of MgO and CaO, and a corresponding increase in content of Na_2O, all with respect to increased contents of SiO_2 (fig. 58D–F).

Minor-element chemistry of the BTT–suite volcanic rock in the Havallah sequence in the Valmy and North Peak quadrangles shows marked differences when compared to the BBA–suite in the northern Havallah Hills. The BTT–suite appears to be more strongly enriched in LREEs than the BBA–suite with the result that average $La_n/Yb_n > 20$ and average $Ce_n/Yb_n > 15$ for the BTT–suite, calculated using the same chondrite-normalizing values as described above. Thus, the REE patterns of the BTT–suite of rocks probably would show convex-upward patterns, if data were available for all REEs in these rocks. As pointed out by Saunders (1984), convex-upward REE patterns probably reflect some kind of fractionation of HREE relative to LREE that is most easily accounted for by association with garnet as a residual phase within which the LREEs are concentrated during the process of melting and generation of the BTT–suite rocks (see also, Sawlan, 1991). La/Nb ratios in the seven analyzed samples from the BTT–suite are less than 1 (table 9); they are in the range 0.34 to 0.73, showing an average value of 0.54. As pointed out by Gill (1981), such La/Nb ratios also are char-

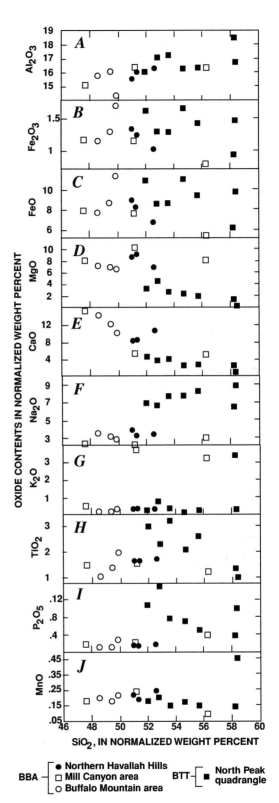

Figure 58 Harker variation diagrams of Mississippian BBA–suite and Pennsylvanian and (or) Permian BTT–suite volcanic rocks of Devonian, Mississippian, Pennsylvanian, and Permian Havallah sequence analyzed from northern Havallah Hills, Mill Canyon area, Buffalo Mountain area, and North Peak quadrangle. Data calculated volatile free and normalized to 100 percent from tables 9, 11.

acteristic of enriched MORB. However, ratios of some less-incompatible to more-incompatible elements in the BTT–suite, exemplified by Zr/Nb ratios in the range 3.2 to 4.65, are approximately three times Zr/Nb ratios in BBA– suite rocks from the northern Havallah Hills (table 9).

BASALT OF LITHOTECTONIC UNIT 1 (PERMIAN)

In the area where argillite, subunit ha2, of LU–1 crops out west-northwest of the abandoned Oyarbide ranch, specifically in W 1/2 sec. 1, T. 32 N., R. 42 E. (pl. 2), the argillite includes some apparently interbedded basaltic flows and cross-cutting basalt (subunit hb, pl. 2) that presumably must be Permian (Leonardian and early Guadalupian) in age, the same apparent age of subunit ha2. These flows and dikes are poorly exposed, and the overall areal extent of basalts in LU–1 is quite limited. Slickensides are prominent on some outcrops of basaltic rock and the slickensides probably are related to the nearby Willow Creek thrust whose trace at the surface has been repeated by displacements along post-Sonoma orogeny normal faults. Much of the microcrystalline and fragmental basaltic rock is dark burgundy-black to maroon-black and it includes abundant Fe oxide and some Mn oxides along wavey fracture surfaces that, in places, are associated with wispy, discontinuous stringers of exhalative silica. Some samples of basaltic rock from the subunit show tabular laths of sodic plagioclase floating in a matrix of Fe and Mn oxides. Murchey (1990) reported greenstone (basalt) to be associated with manganiferous staining and maroon shale in at least two localities in LU–1 in the Willow Creek area of the Battle Mountain Mining District. Greenstone there was presumed by Murchey (1990) to be Early and Middle Pennsylvanian in age because of its spatial association with siliceous argillite that contains some of the oldest faunas recovered from LU–1. However, basaltic rock in subunit ha2 of LU–1 in the North Peak quadrangle is probably significantly younger than Early and Middle Pennsylvanian because this basaltic rock is apparently present at a stratigraphic position well above the unexposed base of the subunit, and its underlying black cherty shale and radiolarians from subunit ha2 are Leonardian to early Guadalupian in age.

CONTINENTAL-SCALE IMPLICATIONS OF ELEMENT RATIOS IN BASALTIC ROCKS OF THE HAVALLAH SEQUENCE

Ratios of some elements in basaltic rocks of the Havallah sequence appear to provide severe constraints on the continental-scale petrotectonic framework prevail-

ing at the time of formation of the basaltic rocks. In two of the more commonly used petrotectonic discrimination diagrams of Meschede (1986) and Pearce and Cann (1973), which primarily utilize various ratios among REE and HFSE—high field strength elements or those elements that are not incorporated appreciably into magmatic minerals (Gill, 1981)—many analyzed rocks of the BBA–suite plot inside various delineated fields, but only some of the BTT–rocks plot inside delineated fields (fig. 59). Basaltic rocks that plot outside delineated fields recently have also been discovered elsewhere and one example is elegantly elaborated upon by Sawlan (1991). Moreover, in all three of these petrotectonic discrimination diagrams, BBA– and BTT–suite rocks plot in separate domains from each other, further emphasizing thereby the sharp geochemical differences in minor element signatures between these two suites of rock. In the Nb–Zr–Y diagram, the BBA–suite appears to show affinities with enriched MORB (that is, mid-ocean-ridge basalt enriched in low field strength elements (LFSE, HFSE and REE), whereas the BTT–suite appears to show its closest affinities with within-plate tholeiites or alkalic rocks (fig. 59A). BBA– and BTT–suite rocks in the Antler Peak 7–1/2 minute quadrangle show some overlap with the domains delineated for these rocks in the Valmy and North Peak quadrangles (fig. 59A). However, the BBA–suite rocks in the Antler Peak quadrangle appear to be slightly more evolved than comparable BBA–suite rocks in the northern part of the Havallah Hills, because the former include higher Zr/(Nb+Y) ratios (Doebrich, 1995).

In the Ti–Zr–Y diagram, all but one of the BBA–suite rocks analyzed plot in a field defined as ocean-floor basalt, low-potassium tholeiite, or alkaline basalt (fig. 59B). The one sample of BBA that plots outside the delineated fields, and well away from all eight of the other samples of BBA, is an altered sample from the Mill Canyon area (see subsection above entitled "Basaltic Rocks at Buffalo Mountain and in Mill Canyon Area"). Four BTT–suite rocks plot either in or near boundaries of the within-plate basalt field delineated on the Ti–Zr–Y diagram, and the three remaining samples of BTT plot outside any delineated fields (fig. 59B). Thus, BBA–suite rocks in tectonic discrimination diagrams using REE and HFSE show affinities with regional tectonic regimes that are less evolved magmatically than regimes with which BTT–suite rocks appear to be allied. Unfortunately, adequate Hf, Ta, and Th data are not available for the BBA– and BTT–suites to apply the tectonic discrimination diagram of Wood (1980). However, in the Zr/Al$_2$O$_3$ ratio versus TiO$_2$/Al$_2$O$_3$ ratio tectonic discrimination diagram of Müller and others (1992) and Müller and Groves (1993), analyses of Mississippian BBA–suite rocks plot in the oceanic arc and continental plus post-collisional arc fields (fig. 59C). Six analyses of Pennsylvanian and (or) Permian BTT–suite rocks plot in

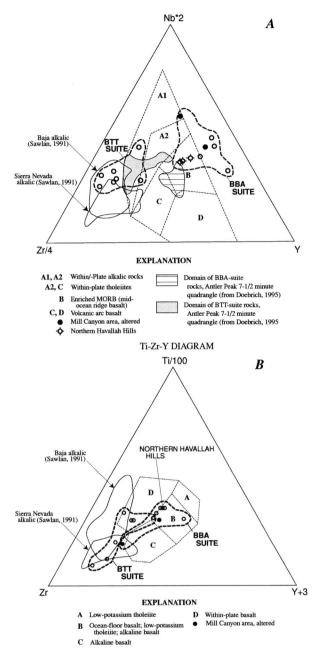

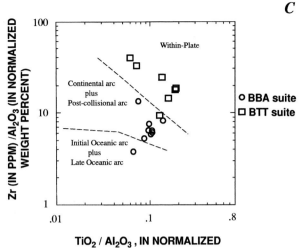

Figure 59 Diagrams commonly used to discriminate among petrotectonic environments of basaltic rocks. Data from tables 10 and 12 using those analytical methods judged to be best for the particular element. *A*, Niobium-zirconium-yttrium petrotectonic classification scheme from Meschede (1986); *B*, Titanium-zirconium-yttrium petrotectonic classification scheme from Pearce and Cann (1973); *C*, Zr/Al_2O_3 ratio versus TiO_2/Al_2O_3 ratios showing classification scheme from Müller and others (1992) and Müller and Groves (1993). Oxides calculated volatile free and normalized to 100 weight percent.

the within-plate field of the diagram, and one analysis of the BTT–suite rock plots just within the continental-plus-post-collisional fields of the diagram. Again, separation between BBA– and BTT–suite rocks is quite good, and, in the latter diagram (fig. 59C), most analyses of BTT–suite rocks show an affinity for within-plate tectonic environments as do some of the analyses of BTT–suite rocks plotted using ternary classification diagrams of Meschede (1986), and Pearce and Cann (1973) (fig. 59A, B). As further pointed out by Müller and others (1992), potassic volcanic rocks that have TiO_2 contents greater than 1.5 weight percent and greater than 350 ppm Zr may be considered "with particular confidence" as belonging to

within-plate petrotectonic environments. Four of the seven analyzed rocks of the BTT–suite contain higher than 1.5 normalized weight percent TiO_2, and three of the BTT–rocks contain more than 350 ppm Zr (tables 10, 12). Thus, both suites of basaltic rock of the Havallah sequence appear to be more evolved when plotted using the tectonic discrimination diagram of Müller and others (1992) and Müller and Groves (1993) than when plotted using the diagrams of Meschede (1986) and Pearce and Cann (1973).

From these various plots of BBA– and BTT–suite rocks, I tentatively infer that basaltic rocks of the Havallah sequence were emplaced as submarine flows and (or) sills and dikes that reflect an evolution of petrotectonic environments with time. These environments appear to be: (1) initially during the Mississippian, an enriched MORB with relatively depleted abundances of HFSE and LREE as the primary magmatic source (here exemplified by BBA–suite rocks), and (2) subsequently during Pennsylvanian and (or) Permian time, either a continental margin regime reflecting effects of some subduction processes or a within-plate regime (exemplified by BTT–suite rocks). The latter rocks in places have a strong enrichment of HFSE mostly through increases in Zr/Nb and Zr/(Ti+Y) ratios, and they also have an enhancement of their HREE abundances (for example, ytterbium) possibly through fractionation processes at depth that may involve garnet. In addition, it does not appear likely that elemental ratios referred to above may have been modified by interactions with continental crust because the pre-Sonoma-orogeny position of the Havallah basin was west of its present position (Miller and others, 1992).

The petrotectonic environment inferred for the Havallah sequence during Pennsylvanian and (or) Permian is based largely on chemical similarity of BTT–suite rocks with Miocene alkali basalt in the Gulf of California (Sawlan, 1991). All basaltic magmatism in the Havallah sequence must have occurred in a relatively confined ocean basin marginal and outboard of the Antler orogenic basin (Murchey, 1990). It is interesting to recall again that Ti–Zr–Y and Hf–Th–Ta plots of basaltic rocks from the Roberts Mountains allochthon show fairly well that emplacement of those rocks occurred in an intracratonic petrotectonic environment (Madrid, 1987). In those data gathered by Madrid (1987), lower Paleozoic basaltic rocks in the Roberts Mountains allochthon appear to show stronger geochemical affinities for some type of crustal involvement in their genesis than either BBA– or BTT–suite rocks in the Golconda allochthon.

Paleozoic, Mesozoic, and Tertiary intrusive igneous rocks

INTRODUCTION

Intrusive igneous rocks in the three quadrangles range in age from Ordovician and (or) Devonian to Tertiary. The Upper Cambrian Harmony Formation in the Snow Gulch quadrangle is intruded by a large number of basaltic dikes and sills that are locally folded along with the surrounding sedimentary rocks of the Harmony Formation (Roberts, 1964), and that are assigned to the Ordovician and (or) Devonian periods because of regionally widespread lower Paleozoic basaltic magmatism of this age (pl. 3). The next younger intrusive igneous activity in the area is represented by small bodies of gabbro and diabase that crop out just to the north of the entrance to Trenton Canyon (pl. 2). The age of these rocks cannot be determined more precisely than either late Paleozoic or early Mesozoic because they intrude the Havallah sequence and are themselves intensely contact metamorphosed by a Late Cretaceous granitoid pluton at Trenton Canyon, which is approximately 89 Ma in age (Theodore and others, 1973).

Following emplacement at 89 Ma of granodiorite, monzogranite, and other associated igneous phases at Trenton Canyon, there was a hiatus of approximately 50 m.y. until emplacement of a small number of granodiorite dikes at roughly 36 to 39 Ma, as documented in the general area of Lone Tree Hill (pl. 1), and by a large number of generally north-trending dikes in the Snow Gulch quadrangle, as well as a number of small, similarly aged stocks near

the southern border of that quadrangle (pl. 3). This age of granitic activity is extremely widespread in the Battle Mountain Mining District (see section below entitled " Potassium-argon Chronology of Cretaceous and Cenozoic Igneous Activity, Hydrothermal Alteration, and Mineralization"). Because many intrusive bodies near the south border of the Snow Gulch quadrangle are described in quite a bit of detail in Theodore and others (1992), those descriptions will not be repeated here. The late Eocene or early Oligocene porphyritic monzogranite of Elder Creek, which is associated with the Elder Creek porphyry copper system in the central part of the Snow Gulch quadrangle, however, will be described in detail below (pl. 3). Some other felsite and altered monzogranite dikes in the Valmy and North Peak quadrangles may be either late Eocene or early Oligocene in age, or they may be Late Cretaceous (pls. 1, 2).

Magmatism in the Great Basin during the Tertiary, however, is reflected primarily in the widespread presence of volcanic rocks (Best and others, 1989). Up until early Miocene, the bulk of the volcanism comprised andesitic to dacitic lavas and flow breccias, and dacite to rhyolite ash-flow tuffs. Basaltic volcanism began sometime after middle Miocene, approximately 17 Ma, and became predominant since about 12 Ma.

Subsequent to emplacement of the small late Eocene or early Oligocene felsic intrusions in the area of Battle Mountain, the next youngest magmatic events are represented by outpouring of the Oligocene (approximately 34.5–Ma) Caetano Tuff and Oligocene augite-olivine basalt, which in the central part of the North Peak quadrangle appears to have a 31.3±0.8–Ma age. Only a small remnant of an outflow facies of the Caetano Tuff crops out near the southeast corner of the Snow Gulch quadrangle (pl. 3). The Oligocene basaltic rocks, which are interbedded with a relatively thick sequence of Tertiary gravel, were followed chronologically by a limited outpouring of 25.5±0.8–Ma calc-alkaline rhyolite tuff and 22.9±0.7–Ma biotite-lithic tuff that are both probably correlative with the late Oligocene and (or) early Miocene Bates Mountain Tuff of Stewart and McKee (1977) and McKee (1994) (pl. 2). Furthermore, some other Tertiary gravels rest depositionally on some of the Miocene calc-alkaline rhyolite tuff at the Eight South deposit. The youngest igneous rocks in the area are the approximately 12–Ma Miocene basaltic rocks that crop out at Treaty Hill (pl. 1).

Basaltic dikes and sills of Ordovician and (or) Devonian age

Ordovician and (or) Devonian basaltic dikes and sills are confined almost entirely to the area of outcrop of Upper Cambrian Harmony Formation in the Snow Gulch quadrangle (pl. 3), and they may be equivalent in age to Paleozoic diabase (unit c d) that crops out in a few localities in the North Peak quadrangle (pl. 2). The dikes in the Harmony Formation must have been emplaced prior to major deformation of the rocks during the Antler orogeny because some dikes are folded together with the surrounding feldspathic arenite and shale of the Harmony Formation (Roberts, 1964; Doebrich, 1994). Assigned provisionally to Ordovician and (or) Devonian periods because of known widespread mafic volcanism of these ages in the region, these dikes and sills are narrow 1– to 8–m-wide, olive-green-to-gray, fine-grained, and dominantly augite-bearing. They are inferred to have the composition roughly of basalt and they are typically as much as several 100 m long in outcrop. They show chilled margins against enclosing sedimentary rocks of the Harmony Formation. Locally, they are well exposed, particularly along ridgelines, but where poorly exposed, their presence is marked by 2– to 4–cm-wide angular basalt chips on colluvium-covered hillsides. In places, basaltic dikes and sills include some sulfide minerals (mostly pyrite, but also including lesser amounts of chalcopyrite and sphalerite) that are generally contemporaneous with development of superposed propylitic or greenschist mineral assemblages in them. Pyrite is commonly present as spongy mm–sized clots (some clots include sphene), as intergrowths among augite and plagioclase, and as tabular infillings among radiating splays of plagioclase. Some plagioclase is altered to albite. Where these basaltic dikes and sills crop out within the outer limit of quartz stockworks associated with the Elder Creek porphyry copper system, they are intensely veined by quartz, and, in places, altered hydrothermally to actinolite±chlorite±secondary biotite mineral assemblages. Rarely, barium-bearing K–feldspar is present in some of the altered domains of the basaltic dikes and sills.

Petrographic examination of 12 thin sections shows most samples to include approximately 10 to 20 volume percent phenocrysts of glomeroporphyritic plagioclase, ovoid clots of olivine (now altered completely to pale apple-green chlorite), and augite that are all as much as 1 to 3 mm wide. Some basaltic dikes and sills within the contact aureole of Tertiary plutons emplaced into the Harmony Formation also show clusters of fine-grained metamorphic biotite developed in them. Some other clots in the dikes and sills are composite and contain mostly chlorite and calcite, thereby giving the appearance that they might be filled vesicles. In places, the ovoid clots of chlorite contain 100– to 175–µm-wide euhedral crystals of spinel that show increased contents of Fe towards texturally complex inclusions made up of augite+albite+chlorite±biotite±sphene composite assemblages, all verified by SEM. The spinel is commonly present in the ovoid clots of chlorite generally as single crystals (fig. 60*A–C*), although some other clots contain a small number of individual crystals of spinel. Some Cr in spinel probably is derived from precursor chromian-olivine magmatic assemblages (M.G. Sawlan, oral commun., 1994). However, the proportion of spinel in a number of the ovoid clots of chlorite suggests that all Cr could not have been present in a chromian olivine of approximately the same size as the clots. Intergranular groundmass of basaltic dikes and sills is mostly plagioclase, augite, and opaque minerals (mostly ilmenite) where unaltered. Roughly 50 percent of the samples examined show extensive alteration of augite to brownish, in thin section, fine-grained intergrowths of amphibole±chlorite even at localities far removed from the widespread secondary alteration associated with Tertiary mineralization. Chalcopyrite, mostly as 10– to 20–µm-wide grains, is preferentially associated with chlorite everywhere it has been identified. In addition, many large glomeroporphyritic clots of plagioclase in these altered samples are replaced by mattes of fine-grained white mica±calcite.

Gabbro and diabase of late Paleozoic or early Mesozoic age

GEOLOGIC RELATIONS AND PETROGRAPHY

Gabbro and diabase of presumed late Paleozoic or early Mesozoic age crop out near the southwest corner of the study area (unit cg g, pl. 2). Two small masses of black hornblende gabbro intrude and contact metamorphose interbedded chert and siltstone of subunit hcs of LU–2 in NE 1/4 sec. 14, T. 32 N., R. 42 E. Diabase is present as a narrow dike emplaced into interbedded gray micrite and black chert, subunit hlc, in SE 1/4 sec. 22, T. 32 N., R. 42 E.; the latter occurrence is about 0.5 km south of the entry to Trenton Canyon (pl. 2). The age of subunit hcs is Mississippian and Pennsylvanian, Chesterian and early Atokan Epochs as described above. Country rock in the immediate area of hornblende gabbro is recrystallized, well-bedded metachert. It is on the basis of presence in hornblende gabbro of well-developed contact metamorphic assemblages undoubtedly associated with Late Cretaceous granodiorite and monzogranite of Trenton Canyon that age of the hornblende gabbro and diabase is assigned to either late Paleozoic or early Mesozoic. These small occurrences of hornblende gabbro probably are related

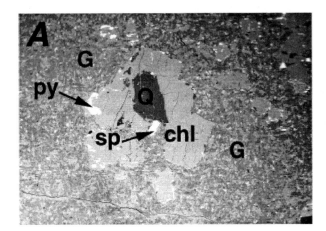

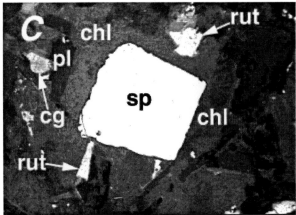

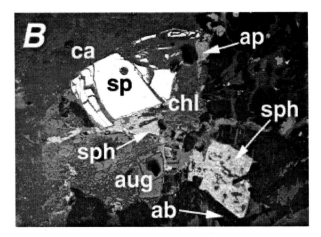

Figure 60 Back-scattered scanning electron micrographs of Ordovician and (or) Devonian basaltic dikes and sills emplaced into Upper Cambrian Harmony Formation in the Snow Gulch quadrangle. chl, chlorite; sp, spinel; sph, sphene; py, pyrite; aug, augite. *A*, Approximately 3–mm-wide clot of chlorite containing spinel, quartz (Q), and pyrite set in a groundmass (G) of mostly altered fine-grained plagioclase; sample 93TT057. *B*, Clot of chlorite including sphene, spinel, and apatite (ap), apparently forming from augite. Plagioclase in groundmass locally altered to albite (ab) calcite (ca) also present in field of view; sample 93TT049. *C*, Euhedral crystal of spinel in chlorite that also forms the matrix for rutile (rut), plagioclase (pl), and a composite grain (cg) consisting of a mixture of ilmenite and chlorite; sample 93TT057.

temporally with some of the basaltic magmatism, Pennsylvanian or Permian in age, described above that is associated with the Havallah sequence in the area. However, the possibility nonetheless remains that the hornblende gabbro also could be Triassic or Jurassic in age. Jurassic gabbro is present near Lovelock, Nev., in the Stillwater Range, approximately 60 km southwest of the study area (Speed, 1963). Some outcrops of hornblende gabbro in the study area appear to have a rare frothy or vesicular, in part amygdaloidal texture in the topographically higher of the two mapped bodies, although the predominant fabric overall both mapped hornblende gabbro bodies is a fairly homogeneous subophitic one. In addition, some outcrops of hornblende gabbro show presence of slickensides similar to those in Mississippian BTT–rocks, suggesting thereby that these rocks have been subjected to deformation associated with emplacement of the Golconda allochthon. Exposures of dense, brownish-black diabase are quite limited, and the diabase, in hand sample, retains only a subtle diabasic fabric through the superposed contact metamorphism.

Petrographic examination of hornblende gabbro reveals that it has been thoroughly contact metamorphosed.

Approximately 3– to 4–mm-long laths of primary hornblende (blue green, optic Z–axis) are in various stages of breakdown to tremolite-actinolite, grayish-brown secondary biotite, chlorite, epidote (trace), calcite, and sphene composite mineral assemblages. In detail, many outlines of former primary hornblende crystals are now represented by patchy concentrations of intergrown fine-grained secondary biotite, chlorite, and granules of sphene that together are commonly laced by needles of tremolite-actinolite or mantled by even somewhat stubby crystals of tremolite-actinolite. Some quartz in the rock appears associated paragenetically with the superposed contact metamorphic event because, in places, quartz is present in narrow, discontinuous veins that contain epidote either in the veins themselves or as alteration halos to the veins. Roughly 50 volume percent hornblende is estimated to have been in the rock prior to contact metamorphism by the nearby Late Cretaceous stock at Trenton Canyon. Primary plagioclase, which is probably more calcic than An_{40} and makes up approximately 40 volume percent of the rocks, is similar in size to primary hornblende, and it is clouded heavily by a pervasive dusting of mostly white mica, as well as some epidote, apatite, and clay minerals. In addition, much plagioclase in the gabbro is replaced partially by generally equant, small crystals of secondary biotite.

Contact metamorphism of the diabase dike resulted in almost complete replacement of plagioclase and primary mafic minerals by felted aggregates of fine-grained biotite, together with lesser concentratons of tremolite-actinolite, sphene, chlorite, and an opaque mineral, probably ilmenite. Somewhat coarser-grained varieties of tremolite-actinolite are also present in the rock as a major constituent of microfolded, several mm–wide veins that include quartz, calcite, and traces of sphene.

CHEMISTRY

Chemical analyses of two samples of hornblende gabbro show that hornblende gabbro is metaluminous, using Shand's (1947) molar index, and chemically somewhat similar to other masses of gabbro in the region, some possibly of the same approximate age, and other gabbros elsewhere that are of entirely different ages and from different tectonic environments (compare analyses 1–2 with analyses 3–6, table 13). Even though only two analyses are available of hornblende gabbro, these two samples (analyses 1–2, table 13) apparently show a covariation in hornblende gabbro of increased MgO content with increased CaO content, as well as three (Ni, Cr, and Sc) of the five compatible elements, the latter of which are concentrated typically in ferromagnesian minerals during hypersolidus magmatic stages. Vanadium contents in hornblende gabbro do not appear to be related directly to MgO content. Silica content in hornblende gabbro apparently is less than the average gabbro of LeMaitre (1976), and hornblende gabbro also contains appreciably higher TiO_2 content than the average gabbro of LeMaitre (1976). Much of the TiO_2 content is in modal sphene and ilmenite, the former mineral present in hornblende gabbro as part of the subsolidus contact metamorphic alteration assemblage that is associated with emplacement of the Late Cretaceous pluton at Trenton Canyon. The remainder of the TiO_2 content in the rock is undoubtedly in secondary biotite that is also related temporally to emplacement of the Late Cretaceous pluton. Data above their respective lower limits of determination are available for three of the titanium-group elements or HFSE (Zr, Nb, and Ti ; see also, Gill, 1981) from hornblende gabbro; they are 178 to 112 ppm Zr, 55 to 27 ppm Nb, and 3.41 to 1.73 weight percent TiO_2 (table 14).

Ratios of Nb/Zr, La/Nb, and Ba/La in hornblende gabbro were compared with similar ratios for BBA– and BTT–suites of volcanic rocks described above in the Havallah sequence to examine minor element signatures that might be suggestive of co-magmatism. The results obtained are not that edifying. Nb/Zr ratios of hornblende gabbro (0.23 to 0.24) are higher than similar ratios in BBA–(0.07 to 0.09) and BTT–(0.07 to 0.16) suites. In addition, La/Nb ratios in hornblende gabbro (0.69 to 0.79) are within the range of ratios for these same elements reported for both BBA (0.66 to 0.71) and BTT (0.64 to 1.78). However, Ba/La ratios for hornblende gabbro (18.5 to 34) are

much closer to Ba/La ratios of BTT (7.1 to 27) than BBA (333 to 820), the latter of which reflect extremely low Nb content (7 to 9 ppm, table 9) in the BBA–suite, particularly in the northern Havallah Hills. In addition, concentrations of the LREE La and Ce in hornblende gabbro are similar to that in BTT–suite rocks (compare tables 9 and 13). If hornblende gabbro is comagmatic with either of the two suites of basaltic rock in the Havallah sequence, it is more likely to be comagmatic with the BTT–suite.

C.I.P.W. norms can represent idealized anhydrous mineral assemblages to which magmas would crystallize stably, provided slow crystallization processes were operative (Gill, 1981). C.I.P.W. norms, calculated volatile free and normalized to 100 weight percent, for hornblende gabbro show that the rocks are nepheline normative, possibly reflecting in these rocks decreased abundances of silica, owing to conversion of some plagioclase to clay minerals, to accomodate normatively the remaining concentrations of Na_2O (table 13). Normative diopside makes up approximately 25 to 29 weight percent of hornblende gabbro.

Potassium-argon chronology of Cretaceous and Cenozoic igneous activity, hydrothermal alteration, and mineralization

by Edwin H. McKee

Fourteen samples from the Battle Mountain area were dated by the K–Ar method for this study. The location of these samples, plus 17 selected published K–Ar or $^{40}Ar/^{39}Ar$ dates, used to evaluate the chronology of Cretaceous and Cenozoic igneous and hydrothermal activity in the region, are shown on fig. 2. A sample of biotite from a drill core from the McCoy Mine, located about 16 km south of the Battle Mountain Mining District, is included because it provides additional information about the age of Au skarn deposits in the region. Analytical data, exact location, and age with its reproducibility (±) of the 14 new age determinations are in table 14. The ages, shown on figures 2 and 61, which summarize the geochronology, are rounded off and their uncertainty is not shown.

The fourteen new K–Ar age determinations include six whole-rock samples and eight purified mineral separates; five biotite, two hornblende and one sanidine. Whole-rock samples were prepared by crushing and sieving from 1– to 0.1–mm size, and treating for 30 minutes in 14 mol percent HNO_3 and for 1 minute in 5 mol percent HF solutions. Mineral phases were separated using magnetic and heavy liquid procedures. Purity of the separate

Table 13—Analytical data on Paleozoic or Mesozoic hornblende gabbro, Humboldt County, Nevada.[†]

Analysis number	1	2	3	4	5	6
Field number 90TT...	446	447				

Chemical analyses (weight percent)						
SiO_2	44.3	43.0	50.14	47.77	47.3	51.3
Al_2O_3	13.5	15.6	15.48	15.	13.8	20.01
Fe_2O_3	3.36	2.95	3.01	2.97	1.8	1.29
FeO	9.49	7.86	7.62	10.54	9.8	2.93
MgO	3.4	6.26	7.59	4.4	7.7	6.11
CaO	10.2	13.5	9.58	9.47	10.2	10.75
Na_2O	3.82	2.04	2.39	4.01	3.	3.63
K_2O	.57	.77	.93	.62	1.	.6
H_2O^+	2.33	2.25	.75	N.d.	2.	1.86
H_2O^-	.34	.13	.11	N.d.	.21	.35
TiO_2	3.41	1.73	1.12	2.86	1.9	.49
P_2O_5	.6	.36	.24	1.33	.7	.14
MnO	.29	.19	.12	.29	N.d.	.05
F	.08	.06	N.d.	N.d.	N.d.	.01
Cl	.073	.044	N.d.	N.d.	N.d.	.05
CO_2	4.52	2.56	.07	N.d.	.14	.15
Total S	<.01	<.01	N.d.	N.d.	N.d.	N.d.
Subtotal	100.28	99.3	99.15	100.7[1]	99.69[2]	99.81
Less O = F, Cl, S	.05	.04	N.d.	N.d.	N.d.	.01
Total	100.23	99.26	99.15	100.7	99.69	99.8

Inductively coupled plasma-atomic emission spectroscopy (total) (parts per million)						
Ba	630.0	620.0	N.d.	187.0	1300.0[3]	N.d.
Be	2.	1.	N.d.	N.d.	N.d.	N.d.
Ce	67.	35.	N.d.	46.	N.d.	N.d.
Co	36.	36.	N.d.	19.	N.d.	N.d.
Cr	2.	240.	N.d.	N.d.	N.d.	N.d.
Cu	38.	28.	N.d.	N.d.	N.d.	N.d.
Eu	2.	—	N.d.	N.d.	N.d.	N.d.
Ga	24.	20.	N.d.	N.d.	N.d.	N.d.
La	34.	18.	N.d.	N.d.	N.d.	N.d.
Li	28.	35.	N.d.	N.d.	N.d.	N.d.
Nb	43.	26.	N.d.	N.d.	N.d.	N.d.
Nd	37.	21.	N.d.	N.d.	N.d.	N.d.
Ni	11.	58.	N.d.	4.	N.d.	N.d.
Pb	4.	14.	N.d.	N.d.	N.d.	N.d.
Sc	24.	31.	N.d.	39.	N.d.	N.d.

Inductively coupled plasma-atomic emission spectroscopy (total) (parts per million)						
Sr	440.	830.	N.d.	409.	N.d.	N.d.
V	320.	220.	N.d.	129.	N.d.	N.d.
Y	30.	18.	N.d.	54.	N.d.	N.d.
Yb	2.	2.	N.d.	5.	N.d.	N.d.
Zn	130.	120.	N.d.	N.d.	N.d.	N.d.

Inductively coupled plasma-atomic emission spectroscopy (partial) (parts per million)						
As	5.9	4.0	N.d.	N.d.	N.d.	N.d.
Cd	.31	.57	N.d.	N.d.	N.d.	N.d.
Cu	40.	30.	N.d.	N.d.	N.d.	N.d.
Mo	.8	.42	N.d.	N.d.	N.d.	N.d.
Pb	11.	17.	N.d.	N.d.	N.d.	N.d.
Sb	1.1	5.	N.d.	N.d.	N.d.	N.d.
Zn	69.	67.	N.d.	N.d.	N.d.	N.d.

Table 13 — (cont.)

	Energy-dispersive X-ray fluorescence spectrometry (parts per million)					
Nb	55.0	27.0	N.d.	N.d.	N.d.	N.d.
Rb	20.	20.	N.d.	N.d.	N.d.	N.d.
Sr	415.	800.	N.d.	N.d.	N.d.	N.d.
Zr	178.	112.	N.d.	116	N.d.	N.d.
Y	28.	19.	N.d.	N.d.	N.d.	N.d.

	Chemical analyses (parts per million)					
As	6.3	4.6	N.d.	N.d.	N.d.	N.d.
Au	—	<.002	N.d.	N.d.	N.d.	N.d.
Hg	—	—	N.d.	N.d.	N.d.	N.d.
W	—	—	N.d.	N.d.	N.d.	N.d.

	C.I.P.W. norms (weight percent)[4]					
Q	—	—	0.73	—	—	—
ne	2.5	3.84	—	—	0.06	—
c	—	—	—	—	—	—
ol	4.82	8.99	—	8.97	16.91	5.66
(fo)	(2.15)	(5.38)	—	(3.94)	(9.56)	(4.49)
(fa)	(2.67)	(3.61)	—	(5.03)	(7.35)	(1.17)
hy	—	—	22.42	4.45	—	5.03
di	25.85	29.22	14.19	14.53	21.05	13.48
(di)	(13.03)	(19.09)	(9.75)	(7.22)	(13.09)	(11.18)
(hd)	(12.82)	(10.13)	(4.44)	(7.31)	(7.96)	(2.3)
ab	30.12	11.21	20.59	34.15	25.98	31.6
an	19.36	33.	29.28	21.23	21.83	37.58
or	3.62	4.82	5.59	3.69	6.07	3.65
mt	5.24	4.53	4.44	4.33	2.68	1.92
il	6.96	3.48	2.17	5.47	3.71	.96
ap	1.53	.9	.58	3.17	1.7	.12
mg^5	32.63	51.49	56.71	37.25	54.59	72.7
norm. pl[6]	49.1	73.	72.8	54.	61.1	69.2
Color index	58.7	58.	64.1	57.	61.7	44.3
D.I.[7]	33.74	16.03	26.91	37.84	32.05	35.25

1. Coarsely crystalline, contact metamorphosed hornblende gabbro (MzPzg, pl. 2).
2. Coarsely crystalline, contact metamorphosed hornblende gabbro (MzPzg, pl. 2).
3. Average of 1,451 analyses of gabbro, from Le Maitre (1976)
4. Hornblende gabbro, from Weibe (1993, table 5, specimen 7).
5. Augite gabbro, from Roberts (1964, table 6, analysis 7), probably part of Mississippian basaltic suite mapped in Devonian, Mississippian, Pennsylvanian, and Permian Havallah sequence (see text; Doebrich, 1992).
6. Hornblende leucogabbro of Humboldt gabbroic complex of Speed (1963), from Speed (1963, table 1, analysis 4).

[1] Includes 1.44 weight percent loss on ignition.
[2] Includes 0.14 weight percent BaO.
[3] Probably determined by emission-spectrographic method.
[4] Normalized to 100 percent and calculated volatile free.

$$^5 mg = \left(\frac{MgO/40.32}{(MgO/40.32 + FeO/71.847 + Fe_2O_3/79.847)} \right) * 100$$

[6] Normative plagioclase.
[7] Differentiation index of Thornton and Tuttle (1960), defined as the total of normative quartz plus normative orthoclase plus normative albite.

†[Chemical analyses of major oxides in weight percent by methods of Jackson and others (1987); analysts, J.S. Mee and D.F.Siems. FeO, H_2O^+, H_2O^- in weght percent by methods of Jackson and others (1987); analyst, N.H. Elsheimer. CO_2 in weight percent by coulometric titration method of Jackson and others (1987); analyst, N.H. Elsheimer. Total S in weight percent by combustion-infra-red method; analyst, N.H. Elsheimer. F, Cl in weight percent by specific ion electrode method; analyst, N.H. Elsheimer. Inductively coupled plasma-atomic emission spectroscopy (total and partial) in parts per million by methods of Crock and Lichte (1982), Lichte and others (1987), and Motooka (1988); analysts, P.H. Briggs and J.M. Motooka. Precision for concentration higher than 10 times the detection limit is better than +10-percent relative standard deviation; instrumental precision of the scanning instrument is +2-percent relative standard deviation. Looked for, but not found, at parts-per-million detection levels in parantheses: Ag (2), As (10), Au (8), Bi (10), Cd (2), Ho (4), Mo (2), Sn (5), Ta (40), Th (4), U (100). Energy-dispersive X-ray fluorescence spectrometry in parts per million by method of Johnson and King (1987); analyst, J. Kent. Chemical analysis for Au by combined chemical separation-atomic absorption method; chemical analysis for As by graphite furnace atomic absorption spectrometry (Wilson and others, 1987; Aruscavage and Crock, 1987); chemical analysis for Hg by cold-vapor atomic absorption spectrometry; chemical analysis for W by spectrophotometric method of Wilson and others (1987); analysts, E.P. Welsch, B.H. Roushey, P.L. Hageman, and L.A. Bradley; —, not detected]

was 99 percent or better, except for the sanidine sample, 93TT063, that contained a significant amount of quartz. The samples were analyzed for potassium by a lithium metaborate-flux fusion-flame-photometry technique (Ingamells, 1970), the argon was analyzed by standard isotope-dilution procedures done by simultaneous measurements of argon isotopic ratios in a five-collector mass spectrometer (Stacy and others, 1981), operated in the static mode. Analytical precision of the reported ages is about 3 percent or less and is based on statistical analysis of a large number of replicated K analyses and duplicate Ar analyses run during the course of earlier studies. The constants used for abundance and radioactive decay of ^{40}K are those recommended by the International Union of Geological Sciences Subcommission on Geochronology (Steiger and Jäger, 1977). The ages shown on figures 2 and 61 from reports published before 1977 (McKee and Silberman, 1970; Theodore and others, 1973) were corrected for constants recommended by the Subcommission on Geochronology.

Twenty six of the radiometric ages currently (1999) available in the Battle Mountain Mining District (fig. 2) are from felsic igneous rocks, or alteration minerals closely associated with igneous rocks, of local origin. Four are from welded tuffs that erupted from sources tens of km from the Battle Mountain region and were dated because they serve as stratigraphic markers that postdate mineralization in the region. These regional welded tuffs include the approximately 34–Ma Caetano Tuff that came from a source about 55 km southeast of the mining district, and three small separate outcrops of a crystal poor calc-alkaline rhyolite tuff tentatively correlated with the approximately 24–Ma Bates Mountains Tuff of unknown, but distant, source.

The ages of locally derived rocks define a temporal pattern of igneous activity (fig. 61). The oldest are of late Cretaceous age, about 85 Ma, and are represented by granitic bodies associated with stockwork molybdenum porphyry systems of the low-fluorine type (Theodore and others, 1992). This includes the Trenton Canyon pluton, the largest igneous body exposed in the area, and several stocks that make up the tectonically segmented Buckingham stockwork molybdenum system. The second, and largest group of ages, is between about 34 and 40 Ma, or, of early Oligocene and late Eocene age. Many small stocks, dikes and other shallow intrusive rocks are this age and many are associated with porphyry copper deposits and their associated gold skarns. In addition, an intensely sericite-altered dike at the recently discovered Trenton Canyon gold deposit has a 34.9±1.0–Ma age (table 14), which indicates superposition of Tertiary mineralization on rocks previously contact metamorphosed by the Late Cretaceous pluton at Trenton Canyon. This age also is roughly the age of alteration and mineralization at a

number of the copper deposits in the central and southern part of the Battle Mountain Mining District and also the gold deposits at Lone Tree in the northern part of the area (that is, north of the Oyarbide fault) that are not clearly associated with any igneous body.

A few small outcrops of basaltic lava are present at Treaty Hill in the northernmost part of the area, and in the southernmost part of the area south of Rocky Canyon and east of Copper Canyon (fig. 2). These lavas are not associated with any known mineral deposits and lie unconformably on many types of older rock. The basalt at Treaty Hill, dated at about 12 Ma, is the same age as basalt that caps the Sheep Creek Range about 25 km to the east of Treaty Hill (Stewart and McKee, 1977). The flows on the southern edge of the area, about 3 Ma, represent the youngest phase of igneous activity in the central part of the Great Basin. Basalts of about this age are present a few km to the south of the Battle Mountain Mining District on the northwest edge of the Fish Creek Mountains. They are not known to be related to mineralization.

Most unaltered igneous rocks in the Battle Mountain Mining District are of early Oligocene or late Eocene age and most hydrothermal alteration and mineralization is the same age. This period of igneous activity has been well documented across a much larger region of central Nevada (McKee and Silberman 1970). It corresponds with the southwestward sweep of calc-alkalic volcanism across the Great Basin that is related to subduction of the Pacific Farallon plate beneath the continental North American plate during Cenozoic time (McKee, 1971; 1996; see also, Dilles and Gans, 1995). The largest volume of igneous rock from this subduction-related event is rhyolitic to dacitic ash flow tuff, but large volumes of andesite and dacite lava were erupted and many small hypabyssal bodies and flow-dome complexes were emplaced as well. All of this igneous activity had a hydrothermal component that produced pervasive alteration and local mineralization throughout much of north-central Nevada. The area of the Battle Mountain Mining District has a relatively small amount of early Oligocene to late Eocene extrusive rock compared to intrusive types—in fact none is known—and the northwestern part of the area has significantly fewer igneous rocks than the southern part of the area. Most igneous bodies of moderate size have been dated radiometrically and are in the southern part of the mining district (fig. 2). The paucity of ages in the northern part of the area reflects the lack of igneous rocks. The ages that have been obtained from north of the Oyarbide fault are on small igneous bodies associated with mineral deposits—the best example being the granodiorite porphyry dike at the south end of Lone Tree Hill. These deposits commonly are probably the distal parts of porphyry systems that are present at depth as explained more fully in the section below entitled "Geology and Geochemistry of Selected Mineralized Areas." A

Table 14 —Postassium-argon ages and analytical data for igneous rocks and alteration material from the Battle Mountain region.

Field number	Location lat.	Location long.	Material dated	K_2O wt percent	$^{40}Ar^{rad}$ mole/g	$^{40}Ar^{rad}$ percent	Apparent age (Ma) ± σ	Rock unit
AP150	40°30'43"	117°12'43"	Whole rock	1.273	5.0527×10^{-12}	15.3	2.75±0.14	Olivine basalt
93TT125	40°51'24"	117°09'36"	Do.	.552	9.5910×10^{-12}	26.9	12.0±0.4	Do.
93TT063	40°40'08"	117°10'32"	Sanidine	7.3	2.0043×10^{-10}	70.	19.0±0.7	Calc-alkaline ryholite tuff
92TT096	40°44'48"	117°09'42"	Biotite	6.8	2.2540×10^{-10}	25.	22.9±0.7	Do.
93TT062	40°40'18"	117°11'10"	Whole rock	4.56	1.6884×10^{-10}	68.	25.5±0.8	Do.
93TT098	40°41'38"	117°12'33"	Do.	2.605	1.8582×10^{-10}	59.9	31.3±0.8	Olivine-augite basaltic andesite
Do.	40°41'38"	117°12'33"	Do.	Do.	1.8748×10^{-10}	27.9	31.4±1.0	Do.
93TT70	40°38'26"	117°10'38"	Whole rock	3.804	1.9264×10^{-10}	55.1	34.9±1.0	Felsite dike, Trenton Canyon deposit
AP148c	40°35'58"	117°14'38"	Biotite	7.09	3.2742×10^{-10}	36.2	31.8±1.0	Altered granodiorite, Buffalo Valley Mine
AP127c	40°35'58"	117°14'48"	Hornblende	.735	3.5955×10^{-11}	31.4	33.7±1.1	Granodiorite stock, Buffalo Valley Mine
BV1	40°35'58"	117°14'45"	Biotite	8.12	4.3594×10^{-10}	46.3	36.9±1.2	Granodiorite dike, Buffalo Valley Mine
Stonehouse	40°49'50"	117°12'08"	Hornblende	1.082	5.7063×10^{-11}	42.4	36.3±1.2	Granodiorite dike, Lone Tree Hill
Do.	40°49'50"	117°12'08"	Biotite	8.73	5.0068×10^{-10}	77.3	39.4±1.3	Do.
92TT119	40°32'30"	117°05'52"	Do.	7.48	4.2048×10^{-10}	57.	38.6±1.2	Granodiorite dike, Iron Canyon Mine
McCoy	40°20'30"	117°14'30"	Do.	8.96	4.9535×10^{-10}	89.2	38.0±1.2	Granodiorite, Brown stock, McCoy Mine

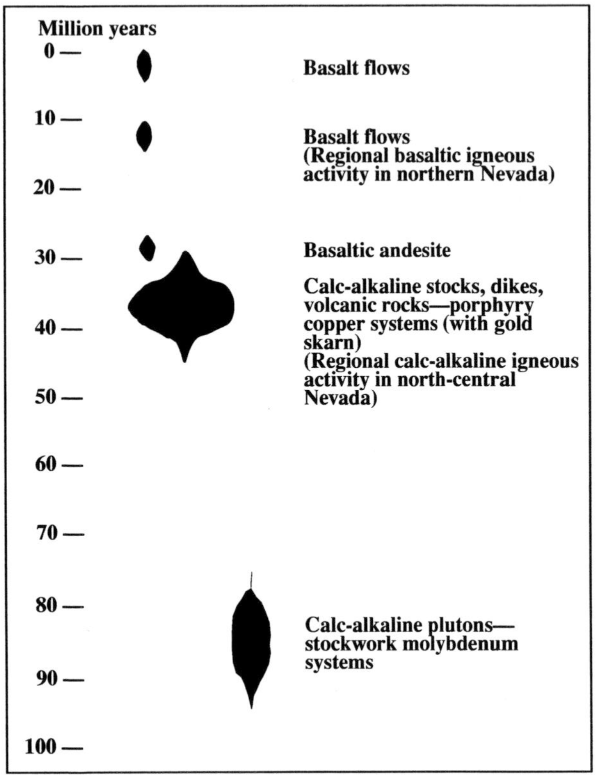

Figure 61 Periods of Cretaceous and Tertiary igneous activity in the Battle Mountain region.

deeper level is exposed in the central and southern part of the mining district south of the Oyarbide fault. Here numerous dikes and small stocks are associated with porphyry copper systems. At this crustal level the upper parts of late Cretaceous molybdenum porphyry bodies are exposed as well.

GRANODIORITE AND MONZOGRANITE OF TRENTON CANYON (CRETACEOUS)

The northernmost lobe, measuring roughly 1.5 km² of an approximately 2.5 km² northwest-southeast elongate composite pluton, crops out in the southern part of the area. This pluton, the Late Cretaceous granodiorite and monzogranite of Trenton Canyon (Roberts, 1964; Theodore and others, 1973), lies astride the southern border of the North Peak quadrangle (unit Kgm, pl. 2). The bulk of the intruded wallrock surrounding the pluton belongs to LU–2 of the Havallah sequence as the pluton is in contact with contact metamorphosed, interbedded chert and siltstone of subunit hcs along almost its entire northern contact. In addition, some metamorphosed black cherty shale, subunit hbc of LU–1 of the Havallah sequence, is in contact

with the pluton in its easternmost parts, also near the southern border of the North Peak quadrangle. Some metamorphosed and, in places, brecciated calcareous sandstones (unit Kbx, pl. 2) are in contact with the pluton near its northwestern border. Compared with exposures of the pluton mapped by Roberts (1964) and Doebrich (1992, 1995) continuing farther to the south, the lobe of the pluton in the North Peak quadrangle also comprises the widest exposed portion of the pluton when measured in an east-west direction, generally along the trend of Trenton Canyon. Along the overall outcrop of exposures of the pluton farther to the south-southeast, the area of outcrop of the pluton decreases dramatically until it eventually pinches out approximately 1.7 km to the south (fig. 9). However, residual magnetics over this entire part of the Battle Mountain Mining District show broad closures of magnetic contours, as much as 300 nT above local background, that are centered on the Late Cretaceous granodiorite and monzogranite of Trenton Canyon (Doebrich, 1995). This relation in turn suggests that additional Cretaceous igneous rock associated temporally with the exposed pluton may underlie an area totaling as much as 25 km². As mapped, the northern contact of the pluton appears to dip to the south as emplacement of this part of the pluton appears to have been controlled partly by a south-dipping thrust fault bounding some of the calcareous siliciclastic subunits of LU–2 in this part of the Havallah sequence (pl. 2). Much of the granodiorite and monzogranite is extensively altered hydrothermally and, in places, is dominated by widespread flooding by quartz veins, the greatest concentrations of these veins being present in two small areas as indicated on the geologic map (pl. 2). In addition, there is a broad contact metamorphic halo surrounding the pluton, and the rocks within this halo are in places cut by wide, throughgoing veins of tremolite-actinolite. Most hydrothermal alteration in the pluton is associated with a deep porphyry Mo–Cu system of the quartz monzonite type (White and others, 1981; Westra and Keith, 1981), also referred to as a low-fluorine Mo type (Theodore and Menzie, 1984), that is described below. A few narrow hornblende granodiorite porphyry dikes, too small to show on the geologic map and presumably Tertiary in age, also cut the granodiorite and monzogranite of Trenton Canyon.

The only metal production recorded up to 1999 from the general area of the pluton at Trenton Canyon, however, is from the patented claim at the Trenton Mine approximately 0.8 km south of the study area (Roberts and Arnold, 1965). This production occurred during 1870, at which time Lott and Co. shipped 9 tonnes of Cu ore then valued at $50 per tonne. Some ore mined from this occurrence is reported to have assayed as high as $157 a tonne Ag and 12 to 14 weight percent Cu (Whitehill, 1873). This mining activity was among the first mineral production recorded from the Battle Mountain Mining District (Roberts and Arnold, 1965; Theodore and others, 1992).

Brief descriptions of the granodiorite and monzogranite of Trenton Canyon are included in earlier published reports by Roberts (1964) and Theodore and others (1973), who both refer to these composite intrusive rocks collectively as the granodiorite of Trenton Canyon. As will be described below, the pluton also includes some more highly evolved igneous phases than these. The former report also describes a rough zonation of igneous fabrics in the pluton, including presence of predominantly porphyritic rocks near the exposed periphery of the pluton and a concentration of generally equigranular hypidiomorphic granular rocks in its central parts. Based on observations during numerous traverses through the pluton, the bulk of the exposed porphyritic facies of the pluton appears to be present in the general areas of the concentrations of vein quartz (pl. 2). It is not clear, however, whether these porphyritic rocks are gradational with equigranular ones, or whether they are the result of emplacement of another distinct pulse or pulses of magma that cut equigranular phases. As will be described shortly, a complex assortment of igneous rock is present throughout the pluton, partly resulting from widespread effects of a subsolidus history of alteration.

The report by Theodore and others (1973) includes two K–Ar ages determined on primary biotite separates, each approximately 90 Ma (recalculated), and a number of minor-element analyses of selected rocks from across an approximately 1–km-long geochemical traverse along the bottom of Trenton Canyon. The samples from which the radiometric ages were obtained are largely equigranular granodiorite collected from outcrops near the bottom of Trenton Canyon (Theodore and others, 1973). In addition, Roberts (1964) describes an augite granodiorite phase of the pluton near the south fork of Trenton Canyon that may owe its origins to assimilation of Ca from some of the nearby intruded sequences of calcareous rock in the Havallah sequence. Thus, the pluton at Trenton Canyon was emplaced coincident with magmatic culmination along the Cordillera of western North America during the Cretaceous (Armstrong and Ward, 1993).

Although a wide variety of rocks crop out in the pluton of Trenton Canyon, the most common type is probably granodiorite. Although rocks throughout much of the pluton weather deeply to grus-covered slopes, the presence of highly incised Trenton Canyon through the pluton provides good exposures, particularly along bottoms of major drainages (fig. 62*A, B*). Silicified parts and quartz stockwork-veined parts of the pluton typically are resistant to weathering and form topographic highs, as does some of the breccia, in places near the border of the pluton (fig. 62*C*). Fragments in the breccia show sucrosic fabrics, some staining by secondary Cu minerals, and pres-

ence of bright yellow-green colors on some weathered surfaces probably indicative of ferrimolybdite that replaces molybdenite. Molybdenite also is present as rare extremely small grains in some breccia fragments. Bleached rocks resulting from oxidation of phyllic-altered quartz-K–feldspar pegmatoid zones are common in areas affected most strongly by flooding of quartz veins (fig. 62*D*). In some areas of the pluton that were prospected in the past for turquoise, as in SW 1/4 sec. 24, T. 32 N., R. 42 E., fairly large clots of Fe oxide, measuring as much as 5 cm wide, replace previously crystallized concentrations of pyrite. The amount, if any, of turquoise mined in the past from these occurrences is not reported (Roberts and Arnold, 1965). In addition, some shear zones, defined by planar sets of intense fracturing and concentrations of quartz stockworks, extend irregularly through the pluton and appear to be cut by an opaline silica-bearing set of veins that definitely truncates the stockwork-type quartz veins. These opaline veins may be related to a Tertiary magmatic event subsequent to emplacement of the pluton during the Late Cretaceous.

PETROGRAPHY

Petrographic examination of 23 thin-sectioned rocks from the Late Cretaceous pluton, including relatively unaltered and intensely altered igneous rocks and a small number of quartz veins, confirms an overall areal heterogeneity of magmatic phases that make up the mostly granodiorite and monzogranite pluton of Trenton Canyon. Equigranular and porphyritic varieties of hornblende-biotite granodiorite and monzogranite probably are the most widespread exposed phases of the pluton, but other phases, including biotite granite and altered porphyry, also are present. No attempt is made here to map these phases for the present investigation because of scale and time limitations. Under the microscope, unaltered equigranular hornblende-biotite monzogranite includes fresh, normally zoned plagioclase crystals, as calcic as An_{45-50}, K–feldspar, blue-green (optic Z–axis) hornblende, brown (optic Z–axis) stubby books of biotite, and small dispersed masses of quartz (fig. 63*A*). Mafic minerals make up about 15 volume percent of these rocks, and accessory minerals include sphene and apatite, and, in some samples, allanite. The quartz crystals do not show well-developed crystal outlines, and concentrations of fluid inclusions are overall quite sparse relative to concentrations associated with many other porphyry Cu and Mo systems. Porphyritic facies of intermediate-argillic altered monzogranite, however, include a well-developed aplitic groundmass largely including quartz and K–feldspar grains showing average grain sizes of about 0.06 mm and abundant 120° dihedral angles at grain boundaries. In addition, along some margins of

Figure 62 Typical exposures and rocks associated with Late Cretaceous monzogranite and granodiorite pluton of Trenton Canyon. *A*, General overview to southwest across the monzogranite and granodiorite of Trenton Canyon showing grus-covered slopes characteristic of the pluton. Trenton Canyon in central part of photograph, and ridge-forming wallrock breccia developed near northwest border of pluton at head of arrow (unit Kbx, pl. 2). *B*, Closeup view of equigranular hornblende biotite monzogranite facies common throughout much of pluton. *C*, Silicified rib of massive bull quartz, as much as 3 m wide, emplaced into 30–to 50–m-wide shattered fault zone in southwest part of pluton. *D*, Closeup view of pegmatoid quartz, K–feldspar, and white mica in central part of zone of abundant quartz veins.

relatively unaltered laths of K–feldspar phenocrysts, small inclusions of quartz are present approximately the same size as the quartz crystals in the adjoining groundmass (fig. 63*B*). No optical discontinuity is present between cores and rims of the K–feldspar phenocrysts. Such textural relations are interpreted to reflect continued growth of phenocrystic K–feldspar after quenching of groundmass, which, in turn, suggests that development of groundmass in these rocks is (1) primarily a chilling event during the latter stages of magmatism, and is (2) not entirely the result of alteration associated with circulation of subsolidus fluids. Generally, former phenocrysts of biotite, hornblende, and plagioclase, all approximately 2 to 4 mm wide but including some phenocrysts as much as 7 mm wide, show complete alteration to white mica. Other samples include variable concentrations of chlorite, epidote, secondary sphene, and pyrite or iron oxide that replace py-

rite. It appears from relict plagioclase that survived the intermediate-argillic event in these porphyritic rocks that the plagioclase is somewhat less calcic, An_{35-45}, than that in the equigranular facies. Secondary fluid inclusions are much more abundant in magmatic quartz crystals of these altered porphyritic rocks than in the quartz of unaltered equigranular facies described above.

Some altered porphyry crops out close to the breccia mapped near the northwest margin of the pluton and appears to be depleted significantly in abundance of mafic minerals in comparison to much of the surrounding porphyritic and equigranular granodiorite and monzogranite (K. Ratajeski, written commun., 1993; Ratajeski, 1995). The altered porphyry shows a well-developed intermediate argillic alteration assemblage wherein all mafic minerals and plagioclase are altered to white mica assemblages (±sphene±apatite±opaque minerals), and the phenocrysts of K–feldspar show only a moderate clouding to micas and clay minerals. Quartz phenocrysts are either euhedral or are partially resorbed, and they also show exceptionally low concentrations of fluid inclusions, especially when one considers that they are in the central core area of a weakly developed porphyry system. Average grain size of the matrix is approximately 0.01 mm, and the matrix makes up approximately 50 volume percent of the rock. The relative age between altered porphyry and nearby less altered phases of the pluton is unknown.

Miarolitic cavities are present in some intensely phyllitic-altered porphyritic facies of the pluton that, before alteration, may have approached the composition of granite, sensu stricto, based on visually estimated initial hypersolidus proportions of plagioclase and K–feldspar. Such cavities were first recognized in outcrops of the pluton near the bottom of Trenton Canyon by J.M. Hammarstrom (oral commun., 1985). The pre-alteration chemistry of this granite is probably equivalent to the chemistry of a dike of granophyric microgranite that crops out just beyond the southeast border of the pluton and described below. The phyllic-altered porphyritic granite is present in the general area of intensely quartz-veined rocks of the pluton (pl. 2). Miarolitic cavities in these rocks are defined by euhedral terminations of K–feldspar projecting into irregularly-shaped, mm-sized domains now filled by quartz and some minor amounts of white mica (fig. 63C). Thus, K–feldspar crystal terminations show angular jagged outlines, quite conspicuous in thin section, along margins of quartz-cored domains, which domains surprisingly, in light of the intensity of alteration effects throughout these rocks, also do not host concentrations of fluid inclusions comparable to those in similarly altered rocks in the Buckingham stockwork molybdenum system (Theodore and others, 1992). Those fluid inclusions that are present at room temperature in quartz at Trenton Canyon are mostly of the two-phase variety (liquid plus vapor), and they show a much higher proportion of liquid than vapor. Presence of miarolitic cavities in these rocks is only one manifestation of emplacement of these igneous rocks to high levels in the crust (Cox and others, 1979). However, preliminary evaluations using the aluminum-in-hornblende geobarometer methodology of Hammarstrom and Zen (1986) resulted in an extremely broad range of emplacement pressures for these igneous rocks on the basis of analyses of an actinolite-bearing sample (Ratajeski, 1995): 0.1±0.5 to as much as 6.3±0.6 kbar. These data were all calculated using calibrations by Johnson and Rutherford (1989) and Schmidt (1992) to the original geobarometric equation of Hammarstrom and Zen (1986). The presence of secondary actinolite in the sample analyzed from Trenton Canyon, however, may have introduced some unknown variable into proper determination of the level of emplacement. Shallow levels of emplacement for Cretaceous and Tertiary plutons in the Battle Mountain Mining District have been proposed previously, and their

levels of emplacement defined quite precisely on the basis of fluid-inclusion studies that quantified temperatures at which boiling, magmatically derived fluids circulated at Copper Canyon and at Buckingham (Nash and Theodore, 1973; Theodore and others, 1992; Myers, 1994). Ratajeski (1995) further suggests that the minimum initial H_2O content of the magma(s) at Trenton Canyon were greater than 4 weight percent. However, presence of resorption textures involving primary quartz and K–feldspar phenocrysts suggests that the magma was still undersaturated with respect to H_2O even at high crystallinities (Ratajeski, 1995).

Some of the massive quartz veins associated with emplacement of stockworks at Trenton Canyon show apparently cogenetic, liquid-rich and vapor-rich, two-phase fluid inclusions that may be indicative of boiling, although the liquid-rich varieties are overwhelmingly the most common type of fluid inclusion in the veins. Some liquid CO_2-bearing fluid inclusions also are part of the fluid-inclusion signature of these quartz veins, a relationship that this system shares with many other stockwork molybdenum systems of the quartz monzonite type (Theodore and Menzie, 1984). The absence of halite-bearing, extremely saline fluid inclusions in most exposed vein quartz associated with the pluton at Trenton Canyon also is one of the characteristics of this particular magmatic-hydrothermal system—a characteristic which this system shares with the one at Buckingham (Theodore and others, 1992). Some vein quartz from the southwest border of the mapped concentration of quartz veins in SW 1/4 sec. 24, T. 32 N., R. 42 E. (pl. 2), however, shows relatively large numbers of fluid inclusions characterized by presence of at least one non opaque daughter mineral, including common halite. In addition, the vapor proportions in these fluid inclusions suggest trapping at some of the highest subsolidus temperatures of the entire system. Nonetheless, there may be deep within the system additional and more widespread concentrations of extremely saline fluids, because of the common ponding at depth as a result of the increased density of such hydrothermal fluids in many other porphyry systems (Roedder, 1984).

Small exposures of originally sulfide mineral-poor, andradite-diopside skarn, only as much as 3– to 5–m wide, are present in some metasomatized rocks of the Battle Formation where they crop out adjoining the pluton of Trenton Canyon just to the south of the southern border of the North Peak quadrangle. Diopside makes up less than 5 volume percent of the skarn, and iron oxides that replace iron sulfide minerals along grain boundaries perhaps make up as much as approximately 2 volume percent of the skarn. Under the microscope, andradite in this skarn body shows complex zoning patterns among isotropic and anisotropic growth bands, commonly with cores of most crystals being isotropic and sector twinning being well

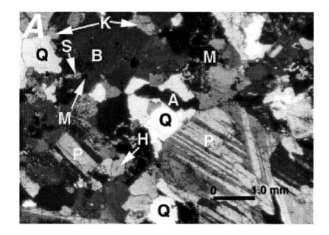

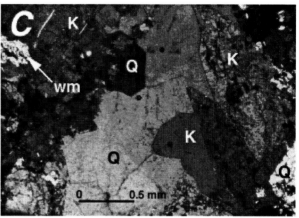

Figure 63. Late Cretaceous monzogranite and granite from the pluton at Trenton Canyon . H, hornblende; B, biotite; Q, quartz; P, plagioclase; K, K–feldspar; S, sphene; M, magnetite. *A*, Fabric of essentially unaltered equigranular hornblende-biotite monzogranite. Locality 90TT452, crossed nicols; *B*, porphyritic monzogranite showing K–feldspar phenocryst with quartz grains (at head of arrow) from groundmass (G) included near its margins. Locality 93TT127, crossed nicols; *C*, miarolitic cavity showing euhedral terminations of K–feldspar projecting into irregularly-shaped domains filled by quartz in phyllic-altered porphyritic granite. Locality 84BF062, partly crossed nicols.

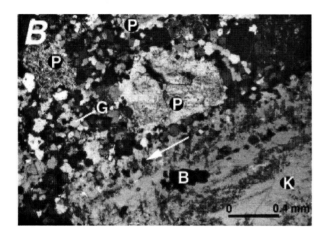

developed. However, this skarn also lacks development of yellowish, apparently resorbed isotropic cores common in many of the productive gold-bearing skarns elsewhere in the Battle Mountain Mining District (Theodore and Hammarstrom, 1991; Theodore and others, 1992). In andradite from the productive skarns, the resorbed cores mark a hiatus or transition from a regime in which highly iron-enriched fluids circulated to one less enriched in overall iron during prograde crystallization, but also including notable fluctuations in iron during the final stages of prograde crystallization of the garnets there.

Metamorphosed rocks of the Battle Formation in this general area also include some narrow medium-grained dikes of granophyric microgranite that have been emplaced into the formation and appear to be related to the nearby Cretaceous-age pluton at Trenton Canyon (K. Ratajeski, written commun., 1993; see also the next subsection below entitled "Chemistry"). These dikes have sharp contacts with adjoining rocks of the Battle Formation and the contacts are commonly marked by presence of stout crystals of K–feldspar oriented at right angles to the contact. One of these dikes, located approximately 0.1 km from the pluton, has well formed granophyric textures of several different types including radiating fringe, vermicular,

and cuneiform (Smith, 1974). The granophyric microgranite dike does not contain myrmekite as defined by Castle and Lindsley (1993). However, the granophyric microgranite dike is quite leucocratic, and contains optically homogeneous original plagioclase (sodic oligoclase or calcic albite, but most likely calcic albite) , K–feldspar (modally much more abundant than plagioclase), quartz, and much lesser concentrations of primary amphibole, sphene, zircon, and allanite. The K–feldspar is generally euhedral near the border of the dike, and elongate crystals of K–feldspar are capped towards the dike by a narrow zone, as much as 5 to 10 mm wide, of granophyre. Some textural types of granophyre in the zone appear largely to be developing as replacement phenomena either at the expense of earlier crystallized essential quartz, or at the expense of K–feldspar. In addition, presence of radiating fringe varieties of granophyre developed on euhedrally-terminated crystal faces of K–feldspar suggests a rapid change of primary crystallization conditions, possibly involving cotectic development of the radiating fringe granophyre near final stages of crystallization of those K–feldspar crystals that are oriented at right angles to the contact of the dike (Smith, 1974). Some granophyre seems to be concentrated near grain boundaries between equivalently sized crystals of essential quartz and K–feldspar. Another textural variety of granophyre seems to be one wherein vermicules of quartz are confined entirely within ovoid to lobate zones formed entirely within large crystals of K–feldspar, in places even preserving the overall outlines of the K–feldspar. The latter type of granophyre appears without question to have formed at the expense of previously crystallized K–feldspar and suggests intergranular mobility of SiO_2 and (or) K (Castle and Lindsley, 1993). Smith (1974) also notes that presence of granophyric textures is almost universally confined to felsic high-level granitoids which is compatible with the environment of emplacement of the pluton at Trenton Canyon.

CHEMISTRY

A small number of chemical analyses for major elements of rocks from the pluton at Trenton Canyon are available (table 15). Most chemical analyses are from granodioritic phases that appear to make up a significant proportion of the outer marginal parts of the exposed pluton. Five chemical analyses of select samples of hornblende biotite granodiorite from the main mass of the pluton at Trenton Canyon and one sample of the 5–cm-wide granophyric microgranite emplaced into the Battle Formation are listed in table 15 (analyses 1–6). The granophyric microgranite dike can only be inferred to be approximately the same age as the main body of the pluton at Trenton Canyon. Total alkalis in the five samples of analyzed hornblende biotite granodiorite are in the range 6.76 to 7.44 weight percent, and the analyzed sample of granophyric microgranite shows a total alkali content of 8.83 weight percent (table 15). The latter analysis also shows a 75.9 weight percent SiO_2 content, which is significantly higher than the SiO_2 contents of analyzed samples of hornblende biotite granodiorite from the pluton, and it is probably representative of SiO_2 content of unanalyzed biotite granite phases that are present near quartz vein-flooded areas of the pluton. The biotite granite probably is associated with development of the Mo–enriched parts of the composite pluton. The sample of granophyric microgranite analyzed includes the particularly well-developed narrow zone of granophyre described previously. The chemical analysis of rock previously reported by Roberts (1964) from the pluton shows total alkalis of 7.0 weight percent (analysis 1, table 15), and all oxide values of this analysis, and as well as those from the other four analyses of hornblende biotite granodiorite, are quite close to the analysis reported for the "average" granodiorite by LeMaitre (1976) (analysis 7, table 15). However, the "average" granodiorite of LeMaitre (1976) shows somewhat depleted K_2O and Al_2O_3 content, and elevated MgO content when compared to the five analyzed samples of hornblende biotite granodiorite from the stock at Trenton Canyon (table 15). In addition, three of the five analyzed samples of hornblende biotite granodiorite from the stock are slightly corundum normative. The differentiation indices of all five of these samples of hornblende biotite granodiorite are within a relatively narrow spread of values in the range 72.49 to 75.92, which values are somewhat more evolved than the 70.83 differentiation index calculated for the "average" granodiorite

of LeMaitre (1976) (table 15). The five analyses of these samples also plot close to the join between the metaluminous and peraluminous fields (fig. 64), and close to the join between the tonalite and granodiorite fields of the R_1R_2 classification diagram of De la Roche and others (1980), where $R_1 = [4000\ Si–11,000\ (Na+K)–2,000\ (Fe^{2+}+Fe^{3+}+Ti)]$, and $R_2 = [6,000\ Ca+2,000\ Mg+1,000\ Al]$ (fig. 65). On the latter diagram, the "average" granodiorite of Le Maitre (1976) plots well into the tonalite field and the analyzed sample of granophyric microgranite plots near the join between the granite and alkali granite fields.

Available whole-rock analyses of igneous rocks from some other low-fluorine stockwork Mo systems, or calc-alkaline Mo systems as they are referred to by other authors, show an overwhelming preponderance of analyses of supposedly unaltered associated igneous rocks that plot in the peraluminous field (Westra and Keith, 1981; John and others, 1993). Thirty four of 35 data points plotted by these authors plot in the peraluminous field, and that one other data point barely plots in the metaluminous field. In fact, some of their samples show values of $Al_2O_3/(K_2O+Na_2O+CaO)$ ratios, in molecular percent, that are as high as 1.35. However, many samples described as unaltered by Westra and Keith (1981) may in fact be altered heavily by subsolidus hydrothermal fliuds. The least altered sample of the igneous rocks associated with the Buckingham stockwork Mo system (Loucks and Johnson, 1992, table 17, analysis 1) is metaluminous because it shows a value of 0.97 for its molar $Al_2O_3/(K_2O+Na_2O+CaO)$ ratio, calculated volatile free and normalized to 100 percent. Thus, unaltered igneous rocks associated genetically with low-fluorine stockwork Mo systems should not be characterized as being restricted only to suites of igneous rock that are peraluminous.

Only two of 11 analyses for minor metals of rocks from the pluton reported by Theodore and others (1973) contain as much as 70 ppm Mo. However, these 11 analyses are of rocks collected generally near the margin of the pluton and well away from some of the most intensely altered and quartz-veined areas in the central parts of the pluton. For additional detailed discussion of Mo contents of rocks in the mineralized parts of the pluton at Trenton Canyon see the subsection below entitled "Trenton Canyon Stockwork Molybdenum System." In comparison, analyzed quartz-stockwork-veined rocks from the surface of the Buckingham stockwork Mo system generally have concentrations higher than 50 ppm Mo, and distribution of these Mo–enriched samples define the surface projection of the steeply plunging Mo–enriched shells at depth quite well (Theodore and others, 1992).

Table 15—Chemical analyses, CIPW norms, and analytical data of Late Cretaceous intrusive rocks from the composite pluton at Trenton Canyon, Lander County, Nevada.†

Analysis number	[1]1	2	3	4	[2]5	6	[3]7
Field number		55-AP-5	TC-25	TC-35	TC-41	AP-108	TC-7.C
Chemical analyses (weight percent)							
SiO_2	66.1	65.5	66.8	65.1	66.5	75.9	66.09
Al_2O_3	16.9	16.6	16.2	16.5	16.1	12.9	15.73
Fe_2O_3	1.8	1.58	1.34	1.07	1.44	.09	1.38
FeO	1.7	1.76	1.56	1.96	1.79	.15	2.73
MgO	1.1	1.27	1.16	1.36	1.11	.13	1.74
CaO	3.9	4.08	3.7	4.37	3.82	.75	3.83
Na_2O	4.1	3.79	3.99	3.93	3.75	2.16	3.75
K_2O	2.9	2.97	3.45	3.35	3.19	6.67	2.73
H_2O^+	.72	.85	.78	.69	.58	.33	.85
H_2O^-	—	.22	.22	.22	.16	.21	.19
TiO_2	.45	.52	.46	.49	.46	.05	.54
P_2O_5	.19	.2	.19	.23	.23	.03	.18
MnO	.05	.09	.08	.09	.07	.02	.08
F	—	.05	.05	.06	—	.02	N.d.
Cl	—	.01	.01	.01	—	<.01	N.d.
CO_2	.05	.25	.19	<.01	<.01	.32	.08
Total S	—	.04	.06	.04	—	<.01	N.d.
Subtotal	99.96	99.78	100.24	99.47	99.2	99.73	99.9
Less O = F, Cl, S	—	.03	.04	.04	—	.01	N.d.
Total	99.96	99.75	100.2	99.43	99.2	99.72	99.9
Energy-dispersive X–ray fluorescence spectrometry (parts per million)							
Ba	N.d.	1300.0	1200.0	1600.0	1200.0	4400.0	N.d.
Nb	N.d.	30.	30.	20.	11.	20.	N.d.
Rb	N.d.	80.	130.	110.	118.	220.	N.d.
Sr	N.d.	850.	820.	920.	650.	270.	N.d.
Y	N.d.	20.	<10.	20.	14.	<10.	N.d.
Zr	N.d.	120.	110.	120.	128.	70.	N.d.
[4]CIPW norms (in weight percent)							
Q	21.41	21.81	21.18	18.6	22.97	36.44	22.38
c	.38	.22	—	—	.09	.85	.07
hy	3.75	4.47	3.68	4.46	4.3	.49	7.56
(en)	(2.76)	(3.22)	(2.64)	(2.77)	(2.81)	(.33)	(4.39)
(fs)	(.99)	(1.25)	(1.04)	(1.69)	(1.49)	(.16)	(3.17)
di	—	—	.82	2.22	—	—	—
(di)	—	—	(.61)	(1.45)	—	—	—
(hd)	—	—	(.21)	(.77)	—	—	—

Table 15 —(cont.)

ab	34.97	32.6	34.13	33.78	32.22	18.49	32.12
an	18.25	19.25	16.28	17.76	17.72	3.57	18.04
or	17.28	17.84	20.61	20.11	19.14	39.88	16.33
mt	2.63	2.33	1.96	1.58	2.12	.13	2.03
il	.86	1.	.88	.95	.89	.1	1.04
ap	.45	.48	.45	.55	.55	.07	.43
Total	100.	100.	100.	100.	100.	100.	100.
[5] D.I.	73.66	72.25	75.92	72.49	74.33	94.81	70.83

[1]From Roberts (1964, table 6).

[2]From Doebrich (1995).

[3]Average granodiorite from LeMaitre (1976).

[4]Calculated volatile free and normalized to 100 weight percent using program of M.G. Sawlan (written commun., 1993).

[5]Differentiation index of Thornton and Tuttle (1960) defined as the sum normative quartz and normative albite and normative orthoclase.

1.–5. Hornblende biotite granodiorite.

6.   Granophyric microgranite.

[†][Contents of SiO_2, Al_2O_3, Fe_2O_3, MgO, CaO, Na_2O, K_2O, TiO_2, P_2O_5, and MnO and energy-dispersive X–ray fluorescence spectrometry of analysis 2-4 and analyses 6 by XRAL Activation Services, Inc. (K. Ratajeski, written commun., 1994). FeO, H_2O^+, H_2O^-, and CO_2 in analyses 2–4 and analysis 6 in weight percent by methods of Jackson and others (1987); analyst, N.H. Elsheimer Total S in weight percent by combustion-infra-red method in analyses 2–4 and analysis 6, analyst, N.H. Elsheimer F, Cl in analyses 2-4 and analysis 6 by specific ion electrode method; analyst, N.H. Elsheimer; —, not detected; N.d., not determined]

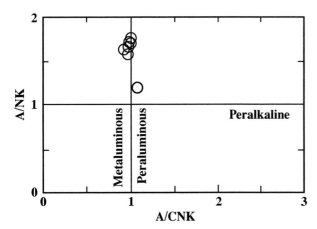

Figure 64 Plot of $Al_2O_3/(Na_2O+K_2O)$ ratio versus $Al_2O_3/(CaO+Na_2O+K_2O)$ ratio, in normalized molecular percent, for samples of Late Cretaceous hornblende biotite granodiorite at Trenton Canyon, probable Late Cretaceous granophyric microgranite at Trenton Canyon, and "average" granodiorite of LeMaitre (1976). Metaluminous, peraluminous, and peralkaline fields also shown. Data from table 15.

CRETACEOUS OR TERTIARY ROCKS

A number of felsic narrow dikes and small stocks that were emplaced presumably during either the Cretaceous or the Tertiary crop out at various places in the Valmy and North Peak quadrangles (pls. 1, 2). These dikes and stocks provisionally are postulated to have such a range in their ages of emplacement because no radiometric ages

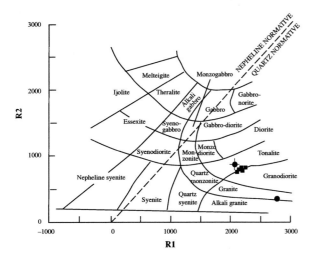

EXPLANATION

■   Hornblende biotite granodiorite from pluton at Trenton Canyon

●   Granophyric microgranite at Trenton Canyon

✦   "Average" granodiorite of LeMaitre (1976)

Figure 65 Values of R_1 versus R_2 for Late Cretaceous hornblende biotite granodiorite at Trenton Canyon, probable Late Cretaceous granophyric microgranite at Trenton Canyon, and "average" granodiorite of LeMaitre (1976). $R_1 = [4,000\ Si - 11,000\ (Na+K) - 2,000\ (Fe^{2+}+Fe^{3+}+Ti)]$, and $R_2 = (6,000\ Ca + 2,000\ Mg + 1,000\ Al)$ from De la Roche and others (1980). Analyses from table 16.

are available from any of them to refine their time of emplacement more precisely, and because of their general lithologic similarities to intrusive rocks elsewhere in the mining district known to be either Cretaceous or Tertiary in age.

There are six occurrences of Cretaceous or Tertiary felsic dikes in the Valmy quadrangle, and these widely-spaced dikes span almost the entire north-south length of the Havallah Hills (pl. 1). The dikes generally show north-south strikes and some of them appear to have been emplaced along faults of a similar orientation. The dikes are altered porphyritic hornblende biotite monzogranite, have minimal amounts of quartz phenocrysts (0 to 5 volume percent), and generally have intense alteration of their primary mafic minerals to chlorite-calcite-Fe oxides (replacing pyrite)±white mica±epidote composite assemblages. Alteration is confined largely to the felsic dikes, and only outlines of magmatic hornblende and biotite phenocrysts are preserved in hand samples. Aplitic groundmass (average grain size approximately 0.04 mm) makes up 20 to 40 volume percent of the rocks, and phenocrysts, mostly feldspar, are medium grained (2 to 4 mm). Only the altered hornblende biotite monzogranite dike mapped in the NE 1/4 sec. 35, T. 34 N., R. 42 E., which includes some well-preserved pink K–feldspar phenocrysts, shows some nearby alteration in its wallrock that may be related to the dike (pl. 1). Approximately 0.2 km to the southeast of this dike and along a north-northwest striking fault, ochre-brown to reddish-brown jasperoid, approximately 3 to 5 m wide in outcrop and cropping out along the trace of the fault, shows multiple generations of brecciation and silicification in thin section. In addition, jasperoid from this locality contains hydrothermal barite and Fe–oxide minerals that replace pyrite.

Cretaceous or Tertiary porphyritic monzogranite dikes and small intrusive bodies also are present in a number of areas in the North Peak quadrangle (pl. 2). The largest of these bodies crops out in the southeastern part of the Havallah Hills where a 0.4–km– long mass, elongate in a northeasterly direction, is present in NE 1/4 sec. 13, T. 33 N., R. 42 E. (pl. 2). This body is composite and includes both equigranular and porphyritic phases, the latter of which contains intensely propylitically altered biotite and hornblende phenocrysts and a sparse concentration of partially resorbed quartz phenocrysts—possibly indicative of H_2O undersaturation late in the process of crystallization. Although alteration of the igneous body itself is quite pronounced, no visible alteration is present in thin sequences of nearby gray micrite that are part of subunit hss that butts up against the composite igneous body. A small number of Cretaceous or Tertiary porphyritic monzogranite dikes, individually of quite limited areas of exposure, also are clustered in the north-central part of sec. 31, T. 33 N., R. 43 E., where they are emplaced into rocks of the Battle and Edna Mountain Formations (pl. 2). Quartz phenocrysts are apparently not present in these rocks. This area was explored extensively in the past, and a number of holes were drilled. The porphyritic dike showing an east-west strike at the locality, and thus almost at right angles to the strike of the normal faults bounding the Antler sequence here, is altered intensely to a phyllic alteration assemblage. Some jasperoid also is present along normal faults in this area, which, in a few places, are marked as well by clay-rich gouge zones approximately 3 m wide. The other small intrusive body just to the east of the normal fault bounding the Edna Mountain and Battle Formations includes wispy hairline microveins of secondary biotite, now converted almost entirely to chlorite, the secondary biotite being relict from earlier-stage potassic alteration. Plagioclase phenocrysts are grayish-brown in thin section as a result of their conversion to micas and clays during alteration.

A small number of narrow Cretaceous or Tertiary dikes also crop out southwest of North Peak where they have been emplaced into rocks of the Valmy Formation (pl. 2). These dikes are quite inconspicuous even in outcrop, and they show no visible signs of alteration in their adjoining wall rocks.

Tertiary rocks

Igneous rocks, including those both known to be and those presumed to be, emplaced or extruded during Tertiary time locally include rocks as old as approximately 39 Ma and as young as approximately 12 Ma in the Valmy, North Peak, and Snow Gulch quadrangles (see subsection above entitled "Potassium-argon Chronology of Cretaceous and Cenozoic Igneous Activity, Hydrothermal Alteration, and Mineralization"). These volumetrically sparse rocks crop out intermittently throughout the quadrangles, and they include granodiorite porphyry dikes and small stocks, composite leucogranite (*sensu lato*) made up of tonalite and monzogranite phases, felsite dikes, calc-alkaline rhyolite tuff, biotite crystal tuff and lacustrine tuff, and basalt, as well as miscellaneous other minor occurrences of igneous rock (pls. 1–3). In all, these rocks probably amount to no more than a few areal percent of the entire bedrock area of the quadrangles. Nonetheless, they are, in places, extremely important with regards to their association with base- and precious-metal mineralization, and with regards to their providing geologic markers that can be used to establish separations across some of the more regionally prominent faults.

Granodiorite porphyry dikes

Narrow granodiorite porphyry dikes crop out in two areas in the Valmy and North Peak quadrangles, both of which involve emplacement of granodiorite porphyry into rocks of the Valmy Formation: the first is at the south end of Lone Tree Hill (pl. 1), and the second is in NE 1/4 sec. 28, T. 33 N., R. 43 E. in the northern part of the North Peak quadrangle near its eastern border (pl. 2). The latter occurrence is east-southeast of Mud Spring. Primary biotite from the dike at the south end of Lone Tree Hill has yielded a 39.4±1.2–Ma age and primary hornblende from the same sample has yielded a 36.3±1.3–Ma age (see subsection above entitled "Potassium-argon Chronology of Cretaceous and Cenozoic Igneous Activity, Hydrothermal Alteration, and Mineralization"). The granodiorite porphyry at the south end of Lone Tree Hill typically is grayish brown in fresh exposures that, in turn, weather to reddish gray-brown. Petrographic examination of the granodiorite porphyry dike reveals that it is made up of approximately 40 to 45 volume percent groundmass, average grain size 0.06 mm, that is extremely poor in quartz and that is not of an aplitic-type fabric (fig. 66*A*). Groundmass of the dike contains a relatively low percentage of quartz, and a high percentage of plagioclase, together with variable amounts of biotite, some of which is obviously secondary, and igneous hornblende. In fact, fresh hornblende (blue green, optic Z–axis) in the rock shows a seriate-type textural habit ranging from small microphenocrysts distributed throughout the groundmass to large ones as wide as 2 mm. In addition, some clots are comprised of fine-grained crystals of hornblende that are distributed randomly in granodiorite porphyry. Secondary biotite is present in granodiorite porphyry as partial replacement of hornblende and as hairline microveins that cut highly oscillatory-zoned plagioclase crystals, which are as calcic as An_{40}. The secondary biotite is probably related to potassic alteration present in the nearby Lone Tree deposit, and which is clearly associated there with introduction of Au (see subsection below entitled "Lone Tree Gold Deposit"). Phenocrysts of biotite (dark red brown, optic Z–axis) in granodiorite porphyry apparently are visibly not affected by potassic alteration, and some phenocrysts are in the form of euhedral laths as much as 4 mm wide. Quartz phenocrysts are either well rounded due to resorption or they are euhedral; they also commonly contain moderate concentrations of apparently cogenetic liquid-rich and vapor-rich, two-phase secondary fluid inclusions suggestive of circulation of boiling fluids some time during the subsolidus history of the dike. Generally, concentrations of fluid inclusions in these rocks are less than those reported previously in some of the porphyry systems in the Battle Mountain Mining District (Nash and Theodore, 1973; Theodore and others, 1992; Myers, 1994).

Table 16 — Analytical data on Tertiary granodiorite porphyry at south end of Lone Tree Hill compared with analytical data from some other Late Cretaceous and Tertiary intrusive rocks in the Battle Mountain Mining District.†

Analysis No.	1	2	3	4	5	6	7	8	9	10	11	12
Field No.	90TT174	B24-1419	—	[1]85TT125	[2]85TT126	[3]KSO516	[4]KSO517	48-R-112	68-259	—	MB–12	MB–11

Chemical analyses (weight percent)												
SiO_2	64.9	69.4	66.9	63.5	64.0	64.8	63.2	70.25	66.43	66.15	66.3	70.5
Al_2O_3	15.5	14.2	14.3	15.4	15.3	17.8	14.8	13.46	14.47	14.39	15.2	14.5
Fe_2O_3	.08	2.1	.7	1.21	.86	2.91	.18	[5]3.42	[5]4.29	3.06	.62	.9
FeO	2.58	1.8	1.7	1.47	1.6	.64	.91	N.d.	N.d.	2.9	2.8	1.
MgO	2.71	1.7	2.	2.9	2.6	1.69	2.67	1.46	2.7	4.15	3.	1.5
CaO	4.6	2.9	3.4	5.31	5.11	.57	6.46	2.32	2.52	2.28	4.5	2.2
Na_2O	3.48	2.9	2.57	2.7	2.6	<.15	.94	2.02	2.52	1.25	3.	2.1
K_2O	2.51	4.1	3.7	4.1	4.27	6.05	7.32	4.88	4.22	4.73	2.6	5.6
H_2O^+	.78	N.d.	1.5	.9	.99	3.04	.62	.45	.3	N.d.	1.	.84
H_2O^-	.17	N.d.	.45	.64	.68	.82	.74	.9	1.12	N.d.	.09	.26
TiO_2	.4	N.d.	.29	.52	.47	.76	.39	.36	.42	.38	.54	.43
P_2O_5	.17	N.d.	.08	.18	.16	.37	.15	.15	.3	.1	.1	.07
MnO	.06	N.d.	.07	.08	.06	<.02	<.02	.02	.03	.05	.05	—
F	.04	.052	.035	.06	.06	.09	.04	N.d.	N.d.	N.d.	N.d.	N.d.
Cl	.013	N.d.	.022	.015	.15	.005	.049	N.d.	N.d.	N.d.	N.d.	N.d.
CO_2	.26	N.d.	2.	.59	.59	.02	.33	.08	.12	N.d.	<.05	<.05
Total S	<.01	.6	.31	<.01	<.01	<.01	<.01	.51	.81	N.d.	N.d.	N.d.
Subtotal	98.25	99.75	100.03	98.58	98.50	99.57	98.8	100.28	100.08	99.44	100.	100.
Less O = F, Cl, S	.02	.17	.11	.03	.03	.04	.03	.13	.2	N.d.	N.d.	N.d.
Total	98.23	99.58	99.52	98.55	98.47	99.53	98.77	100.15	99.88	99.44	100.	100.

Inductively coupled plasma-atomic emission spectroscopy (total) (parts per million)												
Ba	1200.0	N.d.	N.d.	1200.0	1300.0	829.0	3200.0	N.d.	N.d.	N.d.	N.d.	N.d.
Be	2.	N.d.	N.d.	.2	.2	.2	2.	N.d.	N.d.	N.d.	N.d.	N.d.
Ce	34.	N.d.	N.d.	33.	47.	63.	22.	N.d.	N.d.	N.d.	N.d.	N.d.
Co	13.	N.d.	N.d.	13.	12.	6.	6.	N.d.	N.d.	N.d.	N.d.	N.d.
Cr	170.	N.d.	N.d.	110.	110.	9.	126.	N.d.	N.d.	N.d.	N.d.	N.d.
Cu	73.	N.d.	N.d.	5.	3.	35.	2.	N.d.	N.d.	N.d.	N.d.	N.d.
Ga	18.	N.d.	N.d.	16.	16.	21.	15.	N.d.	N.d.	N.d.	N.d.	N.d.
La	19.	N.d.	N.d.	15.	22.	35.	14.	N.d.	N.d.	N.d.	N.d.	N.d.
Li	17.	N.d.	N.d.	17.	17.	31.	17.	N.d.	N.d.	N.d.	N.d.	N.d.
Nb	9.	N.d.	N.d.	—	—	7.	—	N.d.	N.d.	N.d.	N.d.	N.d.
Nd	18.	N.d.	N.d.	19.	24.	33.	14.	N.d.	N.d.	N.d.	N.d.	N.d.
Ni	41.	N.d.	N.d.	23.	22.	6.	21.	N.d.	N.d.	N.d.	N.d.	N.d.
Pb	14.	N.d.	N.d.	18.	18.	98.	49.	N.d.	N.d.	N.d.	N.d.	N.d.
Sc	12.	N.d.	N.d.	12.	12.	9.	10.	N.d.	N.d.	N.d.	N.d.	N.d.
Sr	650.	N.d.	N.d.	560.	560.	56.	396.	N.d.	N.d.	N.d.	N.d.	N.d.

Inductively coupled plasma-atomic emission spectroscopy (total) (parts per million)												
Th	4.	N.d.	N.d.	7.	6.	10.	7.	N.d.	N.d.	N.d.	N.d.	N.d.
V	76.	N.d.	N.d.	82.	81.	103.	66.	N.d.	N.d.	N.d.	N.d.	N.d.
Y	17.	N.d.	N.d.	14.	14.	11.	13.	N.d.	N.d.	N.d.	N.d.	N.d.
Yb	2.	N.d.	N.d.	2.	2.	1.	2.	N.d.	N.d.	N.d.	N.d.	N.d.
Zn	39.	N.d.	N.d.	2900.	2200.	165.	79.	N.d.	N.d.	N.d.	N.d.	N.d.

Inductively coupled plasma-atomic emission spectroscopy (partial) (parts per million)												
Ag	0.13	N.d.	N.d.	N.d.	N.d.	0.66	0.42	N.d.	N.d.			
As	4.1	N.d.	N.d.	N.d.	N.d.	170.	4.8	N.d.	N.d.			
Bi	—	N.d.	N.d.	N.d.	N.d.	—	—	N.d.	N.d.			
Cd	.12	N.d.	N.d.	N.d.	N.d.	2.2	.8	N.d.	N.d.			
Cu	75.	N.d.	N.d.	N.d.	N.d.	33.	1.9	N.d.	N.d.			

Table 16 — (cont.)

Mo	.28	N.d.	N.d.	N.d.	N.d.	.99	.23	N.d.	N.d.
Pb	3.7	N.d.	N.d.	N.d.	N.d.	150.	58.	N.d.	N.d.
Sb	<.67	N.d.	N.d.	N.d.	N.d.	1.7	—	N.d.	N.d.
Zn	29.	N.d.	N.d.	N.d.	N.d.	110.	43.	N.d.	N.d.

Energy-dispersive X-ray fluorescence spectrometry (parts per million)

Nb	—	N.d.	N.d.	N.d.	N.d.	N.d.	N.d.	N.d.	N.d.
Rb	75.0	190.0	N.d.	N.d.	N.d.	N.d.	N.d.	N.d.	N.d.
Sr	620.	N.d.	N.d.	N.d.	N.d.	N.d.	N.d.	N.d.	N.d.

Energy-dispersive X-ray fluorescence spectrometry (parts per million)

Zr	122.	N.d.	N.d.	N.d.	N.d.	N.d.	N.d.	N.d.	N.d.
Y	21.	N.d.	N.d.	N.d.	N.d.	N.d.	N.d.	N.d.	N.d.

Chemical analyses (parts per million)

As	5.3	N.d.	N.d.	18.0	33.0	N.d.	N.d.	N.d.	N.d.
Au	<.002	N.d.	N.d.	N.d.	N.d.	.004	.004	N.d.	N.d.
Hg	<.02	N.d.	N.d.	N.d.	N.d.	.1	—	N.d.	N.d.
W	<1.	N.d.	N.d.	N.d.	N.d.	N.d.	N.d.	N.d.	N.d.

C.I.P.W. norms (weight percent)

Q	20.91	27.79	27.94	18.53	19.5	40.28	15.96	33.65	27.18	29.15	23.92	30.11
c	—	—	.08	—	—	—	—	—	—	3.33	—	1.
or	15.29	24.45	22.85	24.88	26.	37.37	44.58	29.32	25.52	28.09	15.57	33.52
ab	30.36	24.76	22.72	23.46	22.67	.62	8.2	17.38	21.82	10.63	25.72	18.
an	19.86	13.74	17.08	18.27	18.	.43	14.99	10.71	11.92	10.71	20.6	11.05
hy	10.21	5.56	7.5	5.62	5.62	4.4	1.44	3.7	6.88	12.66	10.92	4.17
(en)	(6.34)	(4.06)	(5.21)	(4.94)	(4.53)	(4.4)	(1.27)	(3.7)	(6.88)	—	—	—
(fs)	(3.87)	(1.5)	(2.29)	(.68)	(1.09)	—	(.17)	—	—	—	—	—
di	2.06	.62	—	5.98	5.61	—	13.43	—	—	—	1.1	—
(di)	(1.34)	(.47)	—	(5.34)	(4.63)	—	(12.04)	—	—	—	—	—
(hd)	(.71)	(.15)	—	(.64)	(.98)	—	(1.39)	—	—	—	—	—
mt	.12	3.07	1.06	1.8	1.28	—	.27	—	—	4.46	.91	1.32
il	.78	—	.58	1.01	.92	1.41	.76	.04	.07	.73	1.04	.83
hm	—	—	—	—	—	3.04	—	3.48	4.39	—	—	—
ap	.42	—	.2	.44	.39	.92	.37	.36	.32	.24	.24	—
ru	—	—	—	—	—	.05	—	.34	.4	—	—	—
Total	100.	100.	100.	100.	100.	100.	100.	100.	100.	100.	100.	100.

[6]D.I.	66.56	77.	73.51	66.87	68.17	78.27	68.74	80.35	74.52

[1]Includes: 0.2 ppm Be; 130 ppm Cd; 18 ppm As; 4.4 ppm Sb; 1.1 ppm Tl. [2]Includes: 0.2 ppm Be; 110 ppm Cd; 33 ppm As; 6.7 ppm Sb; 1.1 ppm Tl.
[3]Includes: 2 ppm Cd. Concentration of 0.07 weight percent Na_2O assumed in calculation of C.I.P.W. norm. Pt, Pd, Rh, Ru, and Ir not detected at etection thresholds of 5 ppb (parts per billion).
[4]Pt, Pd, Rh, Ru, and Ir not detected at detection thresholds of 5 ppb. [5]Total iron as Fe_2O_3.
[6]Differentiation index of Thornton and Tuttle (1960) defined as the sum normative quartz and normative albite and normative orthoclase.

1. Granodiorite porphyry dike from south end of Lone Tree Hill (pl. 1).
2. Quartz latite porphyry from Loucks and Johnson (1992, table 16, analysis 3).
3. Granodiorite porphyry, average of six analyses from Theodore and others (1992, table 11).
4. Granodiorite porphyry, emplaced in middle member of Middle Pennsylvanian Battle Formation and the Upper Pennsylvanian and Lower Permian Antler Peak Limestone approximately 250 m west-northwest of Surprise Mine. From Theodore and others (1992, table 12).
5. Do.
6. Augite granodiorite porphyry phase, at Labrador Mine, of late Eocene or early Oligocene porphyritic leucogranite of Theodore and others (1992). K–feldspar makes up bulk of groundmass mica.
7. Phyllic-altered biotite monzogranite porphyry, at Surprise Mine, presumably of Late Cretaceous age. Phenocrystic biotite converted to white.
8. Monzogranite phase of granodiorite of Copper Canyon. From Roberts (1964, table 6).
9. Monzogranite porphyry phase of granodiorite of Copper Canyon. From Roberts (1964, table 6).
10. Average of 15 analyses of granodiorite of Copper Canyon. From Meyers (1994, Appendix A). Most of the 15 samples are altered variably to propylitic, phyllic, and potassic secondary assemblages.
11. Porphyritic monzogranite of Elder Creek. From Theodore and others (1973).
12. Do.

†[Chemical analyses of major oxides in weight percent by methods of Jackson and others (1987); analysts, J.S. Mee and D.F. Siems. FeO, H_2O^+, H_2O^-, CO_2 in weght percent by methods of Jackson and others (1987); analyst, N.H. Elsheimer. Total S in weight percent by combustion-infra-red method; analyst, N.H. Elsheimer. F, Cl in weight percent by specific ion electrode method; analyst, N.H. Elsheimer. Inductively coupled plasma-atomic emission spectroscopy (total and partial) in parts per million by methods of Crock and Lichte (1982), Lichte and others (1987), and Motooka (1988); analysts, P.H. Briggs, and J.M. Motooka. Precision for concentration higher than 10 times the detection limit is better than +10-percent relative standard deviation; instrumental precision of the scanning instrument is +2-percent relative standard deviation. Looked for, but not found, at parts-per-million detection levels in parantheses: As (10), Au (8), Bi (10), Cd (2), Eu (2), Ho (4), Sn (5), Ta (40), U (100). Energy-dispersive X-ray fluorescence spectrometry in parts per million by method of Johnson and King (1987); analyst, J. Kent. Energy-dispersive X-ray fluorescence spectrometry (Meier, 1980); analysts, E.P. Welsch, B.H. Roushey, and P.L. Hageman. Chemical analysis for As by graphite furnace atomic absorption spectrometry (Wilson and others, 1987; Aruscavage and Crock, 1987); chemical analysis for Hg by cold-vapor atomic absorption spectrometry; chemical analysis for W by spectrophotometric method of Wilson and others (1987); analysts, E.P. Welsch, B.H. Roushey, and P.L.Hageman; —, not detected]

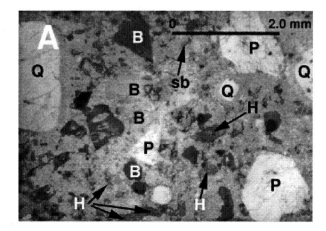

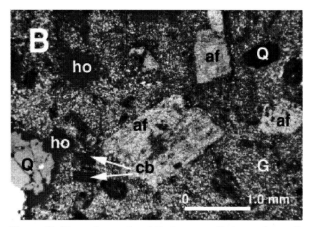

Figure 66 Photomicrographs of Tertiary granodiorite porphyry. H, hornblende; B, biotite; Q, quartz; P, plagioclase. *A*, Fabric of relatively fresh dike that crops out at south end of Lone Tree Hill, and which shows presence of some secondary biotite (sb) in groundmass and in microveins that cut laths of plagioclase. Partly crossed nicols. Sample 90TT174; *B*, Phyllic-altered granodiorite porphyry near outer limit of alteration halo that surrounds Elder Creek porphyry copper system. Shows altered feldspar (af) and chloritized biotite (cb) phenocrysts set in quartz-feldspar groundmass (G); ho, hole in thin section. Sample 93TT052, partly crossed nicols.

Only one chemical analysis is available from the granodiorite porphyry dike that crops out at the south end of Lone Tree Hill (analysis 1, table 16). Also, for comparative purposes in this table, four additional analyses of granodiorite porphyry are compiled from elsewhere in the Battle Mountain Mining District (analyses 2–5, table 16). Although one of these (analysis 2, table 16) is referred to as quartz latite porphyry in the original source document by Loucks and Johnson (1992), the rock in fact consists of the same lithology and fabric as other rocks in the mining district referred to as granodiorite porphyry. In addition, one of the listed analyses of granodiorite porphyry (analysis 3, table 16) reflects average oxide contents from six analyses of Tertiary granodiorite porphyry dikes emplaced into the Late Cretaceous Buckingham stockwork molybdenum system (Theodore and others, 1992). Each of the

six analyses incorporated into this average is plotted individually in the diagrams below graphically showing chemical variations for this suite of igneous rocks. Finally, table 16 also includes, for comparative purposes, analyses of intrusive rocks apparently associated genetically with the Tertiary Labrador and apparently Late Cretaceous Surprise gold deposits, which are located in the east-central part of the mining district (Theodore and others, 1992).

The chemically analyzed sample of granodiorite porphyry at the south end of Lone Tree Hill is just barely metaluminous and plots in the tonalite field of the R_1R_2 classification diagram of De la Roche and others (1980) (fig. 67). Nine samples of granodiorite porphyry or other related igneous rocks from the mining district, however, plot in the granodiorite field of the diagram, which data also show that the analysis of an augite monzogranite porphyry from the Labrador Mine (Theodore and others, 1992) plots close to the border between the granodiorite and granite fields. The contrast in terms of $K_2O–Na_2O–CaO$ content between granodiorite porphyry at the south end of Lone Tree Hill and augite monzogranite porphyry at the Labrador Mine is depicted dramatically in a ternary diagram for these oxides wherein the former rock plots near the $Na_2O–CaO$ sideline of the diagram, and the latter plots near the K_2O corner (fig. 68). The plotted position of the sample from the south end of Lone Tree Hill and the known intense potassic alteration at the Labrador and Lone Tree Mines emphasizes the remarkably unaltered state of the sample analyzed from the south end of Lone Tree Hill. Silica variation diagrams provide some chemical comparisons of the granodiorite porphyry dike at the south end of Lone Tree Hill with other analyzed samples of granodiorite porphyry in the database available from the Battle Mountain Mining District (fig. 69). These silica variation diagrams appear to show a consanguinity of granodiorite porphyry at the south end of Lone Tree Hill with many other intrusive rocks in the Battle Mountain Mining District with the possible exception of the analyzed sample of augite monzogranite porphyry at the Labrador Mine. The data base used to construct these silica variation diagrams has been augmented significantly by recent availability of 15 additional whole-rock analyses of the stock at Copper Canyon (Myers, 1994). However, as pointed out by Myers (1994), most igneous rocks analyzed from the Copper Canyon area are affected by variable concentrations of subsolidus alteration assemblages, including potassic, phyllic, and argillic. Nonetheless, the elevated MgO contents of these rocks at Copper Canyon relative to many of the other Tertiary igneous rocks from the mining district is quite evident (fig. 69).

Tertiary granodiorite porphyry dikes are the most widely exposed of the various intrusive units in the Snow Gulch quadrangle, and generally they are best exposed as rounded knobby hill-forming dikes and elongate masses

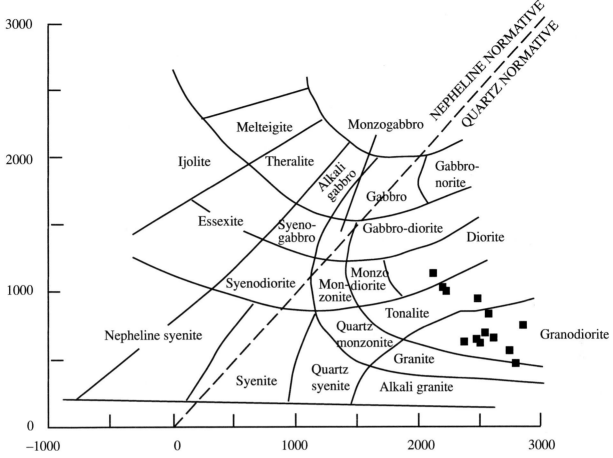

Figure 67 Values of R_1 versus R_2 for Tertiary granodiorite porphyry at south end of Lone Tree Hill with other analyses of Tertiary granodiorite porphyry and other intrusive rocks from the Battle Mountain Mining District. $R_1 = [4,000\ Si–11,000\ (Na+K)–2,000\ (Fe^{2+}+Fe^{3+}+Ti)]$, and $R_2 = (6,000\ Ca+2,000\ Mg+1,000\ Al)$ from De la Roche and others (1980). Analyses from table 16.

along ridgetops (pl. 3). Many of them are emplaced along north-striking faults. Commonly, granodiorite porphyry dikes in this part of the area are gray on fresh surfaces, but, in places, they are deeply weathered to various shades of drab pale orange. Locally, rocks of this unit contain conspicuous 2– to 4–cm-wide euhedral K–feldspar phenocrysts and subrounded "eyes" of quartz bipyramids, together making up as much as 15 volume percent of the rocks. Large phenocrysts of primary biotite are partly altered to chlorite. Groundmass is mostly aphanitic, but may contain fairly abundant partially chloritized needles of hornblende. Some dikes show contact metamorphic aureoles at least 30 to 50 m wide. Primary hornblende in granodiorite porphyry from SW 1/4 sec. 21, T. 32 N., R. 44 E., near the south edge of the quadrangle, yielded 35.4±1.1–Ma age by the potassium-argon method (McKee, 1992). A number of granodiorite porphyry dikes cut the Buckingham stockwork molybdenum system, and their chemistry shows variations similar to other Tertiary igneous rocks in the mining district (fig. 69).

PORPHYRITIC MONZOGRANITE OF ELDER CREEK

Porphyritic monzogranite of Elder Creek makes up a large number of small, generally potassic-altered intrusive bodies that are clustered in general area of Elder Creek and associated genetically with emplacement of the approximately 5–km-wide Elder Creek porphyry Cu system, the broad core of which is probably centered approximately in NW 1/4 sec. 1, T. 32 N., R. 43 E. (pl. 3). Most porphyritic monzogranite of Elder Creek is present within the area of outcrop of quartz stockwork veins associated with the system. The rocks of this unit include undivided hornblende-biotite and biotite-only phases, both showing highly variable concentrations of phenocrystic quartz bipyramids. Presence of numerous pendants in porphyritic monzogranite of Elder Creek suggests that current overall level of exposure of the system probably is quite shallow. Locally, secondary Cu minerals are abundant, and in places intense flooding and stockwork veining by quartz result in almost complete obliteration of igneous texture of the porphyritic monzogranite of Elder Creek. Quartz stockworks in some places, as near the portal to the Morn-

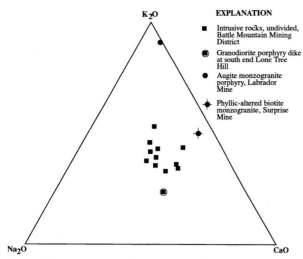

Figure 68 K₂O–Na₂O–CaO for Tertiary granodiorite porphyry at south end of Lone Tree Hill, other analyses of Tertiary granodiorite porphyry, and other Tertiary and Late Cretaceous intrusive rocks from selected sites in the Battle Mountain Mining District. Analyses from table 16.

ing Star Mine (pl. 3), define well-developed "stars" (intersections of planar sets of veins having variable strikes along a common line) in outcrops indicative of their being close to the center of the porphyry system. Many outcrops near the center of the system also show abundant concentrations of secondary Cu minerals, including blue and green chrysocolla, and secondary Cu oxides—all formed in porphyritic monzogranite of Elder Creek and its adjoining wallrocks. Some outcrops of brecciated rock show development of slickensides on secondary-Cu–bearing fractures indicative of post-mineral movement along prominent N. 45° W.–striking joints that cut porphyritic monzogranite of Elder Creek. In addition, some exposures of brecciated rock include well-rounded fragments of biotite hornfels derived from protoliths of the Harmony Formation, and these brecciated rocks show infilling by porphyritic monzogranite of Elder Creek and subsequent veining by additional Cu–stained quartz veins. These relations suggest that some brecciation near the core of this porphyry Cu system preceeded final circulation of fluids associated with development of the system.

A sample of primary biotite from porphyritic monzogranite of Elder Creek obtained from NE 1/4 sec.2, T. 32 N., R. 43 E., yields 37.3±0.7–Ma age by the potassium-argon method (Theodore and others, 1973).

Figure 69 Harker variation plots comparing Tertiary granodiorite porphyry at south end of Lone Tree Hill, other analyses of Tertiary granodiorite porphyry, and other Tertiary and selected Late Cretaceous intrusive rocks from the Battle Mountain Mining District. Analyses from table 17 (see text). A, Al₂O₃ versus SiO₂ content; B, Fe₂O₃ versus SiO₂ content; C, FeO versus SiO₂ content; D, MgO versus SiO₂ content; E, CaO versus SiO₂ content; F, Na₂O versus SiO₂ content; G, K₂O versus SiO₂ content; H, TiO₂ versus SiO₂ content; I, P₂O₅ versus SiO₂ content; J, MnO versus SiO₂ content.

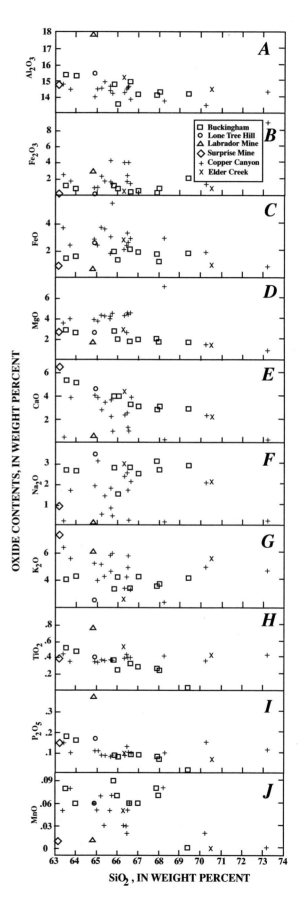

Petrographic examination of approximately 20 samples of porphyritic monzogranite of Elder Creek indicate quenched aplitic groundmass (average grain size in the range 0.06 to 0.12 mm) throughout the unit, which groundmass includes mostly quartz and K–feldspar, and minor amounts of biotite and hornblende. Plagioclase ($An_{35–45}$) is exclusively phenocrystic and K–feldspar is almost entirely confined to the groundmass suggesting that bulk of the quenching occurred just before evolution of the magma(s) reached the cotectic between plagioclase solid solution and K–feldspar solid solution. Those samples containing phenocrystic K–feldspar show the K–feldspar to be approximately one-third to one-tenth the size of the plagioclase and also show a corresponding decrease in the amount of K–feldspar present in the groundmass. However, other samples showing comparatively small volumes of groundmass (roughly 10 volume percent) do not show an increased presence of phenocrystic K–feldspar, and may be indicative of saturation with regards to H_2O at relatively high crystallinities. Crystallization of primary biotite appears to have preceded that of hornblende except in those rocks showing relatively high concentrations of hornblende and, in places, hornblende and biotite continued to crystallize synchronous with development of groundmass. Porphyritic monzogranite of Elder Creek that shows abundant modal hornblende shows multiple generations of hornblende, including some that crystallized prior to final crystallization of biotite phenocrysts and some that is definitely secondary, as indicated by its presence along and within veins that cut the igneous fabric of the rocks.

Classic indications of potassic alteration in porphyritic monzogranite of Elder Creek are reflected in a dominant presence of feldspar-stable alteration assemblages indicated by (1) secondary biotite replacing primary hornblende and primary biotite, (2) quartz-K–feldspar±biotite±hornblende±sphene in veins cross-cutting igneous fabric of porphyritic monzogranite of Elder Creek, and (3) widespread presence of K–feldspar in alteration selvages along margins of quartz±K–feldspar±sulfide mineral veins. Minor accessories include sphene (primary and secondary), apatite, and traces of allanite. Overall intensity of potassic alteration (mostly as replacement of hornblende by shreddy secondary biotite) decreases in narrow dikes of porphyritic monzogranite of Elder Creek that crops out near the outer limit of quartz stockwork veins. Late-stage chlorite, primarily as a marginal alteration of primary and secondary biotite is generally sparse, although some rocks show complete replacement of primary biotite by chlorite, and a few other samples of porphyritic monzogranite of Elder Creek show chlorite+K–feldspar+epidote+sphene veins. Overall the porphyritic monzogranite of Elder Creek, effects of intermediate argillic alteration (white mica+clay mineral alteration of

plagioclase and continued stability of K–feldspar) appear to be confined to areas close to some of the major post-mineral faults that cut the porphyry system, such as the Elder Creek fault (pl. 3). Additional details about the porphyry system are included in the section below entitled "The Elder Creek Porphyry Copper System."

Two chemical analyses are available of porphyritic monzogranite of Elder Creek (table 16). Elevated K_2O content in one of these analyses (5.6 weight percent, analysis 12) is primarily the result of the superposed potassic alteration that has affected the rocks. The chemistry of these two analyses also is compared with that from other Tertiary igneous rocks in the mining district in figure 69. These limited chemical data from porphyritic monzogranite of Elder Creek appear to confirm a general compatibility among all of these rocks in the mining district. However, based on these limited data, porphyritic monzogranite at Elder Creek does not appear to be as enriched in MgO content as the igneous rocks in the Copper Canyon area (fig. 69).

FELSITE DIKES

Felsite dikes, presumably Tertiary in age, crop out in four localities in the North Peak quadrangle (pl. 2). Three of these relatively minor occurrences are fairly close to one another. They are southwest of North Peak, and are near the major drainage through Trout Creek, where they are emplaced into rocks of the Valmy Formation. The fourth and largest occurrence is near the south border of the North Peak quadrangle, in SW sec. 19, T. 32 N., R. 43 E., where an approximately 0.3–km-long west-northwest striking dike is emplaced into metamorphosed calcareous siltstone and chert belonging to subunit hcs of lithotectonic unit LU–2 of the Havallah sequence. At this latter locality, marked contrast in igneous fabric of the felsite dike with nearby exposed rocks of the Late Cretaceous pluton in Trenton Canyon, approximately 0.7 km to the west, suggests quite strongly that the dike is not part of the Late Cretaceous igneous event. The felsite dike is therefore assigned a Tertiary age, and is inferred to be either late Eocene or early Oligocene in age similar to many other dikes and small stocks in the mining district (Theodore and others, 1973; Theodore and others, 1992).

Petrographic examination of a sample from the 0.3–km-long felsite dike indicates that the largely cryptocrystalline quartz-feldspar rock, average grain size 0.1 mm, is phyllic altered intensely, and includes approximately 2 to 3 volume percent Fe oxide that replaces pyrite. Nearby contact metamorphosed siltstones are veined by narrow quartz-pyrite-white mica-barite-chlorite veinlets that show development of irregular, mm–sized bleached margins with the surrounding biotite hornfels. These veinlets most likely are associated with emplacement of the felsite dike. SEM examination of one of these veinlets confirmed extensive intergrowths between barite and Fe oxide minerals that previously replaced pyrite, but this examination by SEM failed to detect the presence of any precious metals in the veinlet.

MISCELLANEOUS INTRUSIVE ROCKS

A large number of Tertiary intrusive rocks of various compositions are present near the south-central border of the Snow Gulch quadrangle (pl. 3), and, for discussion purposes, are collectively considered here as miscellaneous intrusive rocks that need not be described in detail because they have been adequately described previously (Theodore and others, 1992; Theodore, 1994; Ivosevic and Theodore, 1995, 1996).

A rare lamprophyre dike also is present in the Elder Creek area of the Snow Gulch quadrangle (pl.3), specifically in SW 1/4 sec. 36, T. 33 N., R. 43 E., where it is present as an approximately 2– to 3–m–wide, brownish-black fine-grained dike emplaced into contact metamorphosed rocks of the Upper Cambrian Harmony Formation. The dike is cut by quartz-stockwork veins associated with the Eocene or Oligocene Elder Creek porphyry copper system, and it is relatively well exposed in a shallow bulldozer cut. In thin section, the lamprophyre essentially shows apparently compatible euhedral blue-green (optic Z–axis) hornblende (average grain size 0.1 mm), clouded plagioclase (approximately An$_{60}$), and dark brown (optic Z–axis) primary biotite—listed in order of declining abundance. Hornblende probably makes up about 60 volume percent of the lamprophyre. Biotite includes 10– to 20– µm anhedral blebs of ilmenite, which in turn are mantled by thin rims of sphene. Biotite also includes some uncommonly small grains of zircon, approximately 0.2 µm wide. Extremely sparse K–feldspar in lamprophyre is associated with late-stage hornblende-quartz veins. The lamprophyre falls into the calc-alkaline branch of the lamprophyre clan of Rock (1991), and should best be termed spessartite according to his classification scheme. The lamprophyre is assumed to be early Tertiary in age on the basis of lithologic similarity with some of the potassium-argon-dated Tertiary biotite and hornblende-biotite lamprophyre dikes present in general area of Hancock Canyon in the Shoshone Range, approximately 25 km to the southeast of the quadrangle (E.H. McKee, oral commun., 1994). Gilluly and Gates (1965) also show the Tertiary granitic body at Granite Mountain to be intruded by a number of lamprophyre dikes.

OLIGOCENE OLIVINE AND AUGITE BASALTIC ANDESITE

Thin flows of Oligocene olivine augite basaltic andesite and augite basaltic andesite crop out in two localities in the North Peak quadrangle (unit Tba, pl. 2). These localities include (1) olivine augite basaltic andesite in the general area of the middle stretches of Cottonwood Creek, roughly in the central part of the quadrangle, and (2) presumably Oligocene augite basaltic andesite in the central part of sec. 11, T. 32 N., R. 42 E, approximately 5 km south of the first locality. In the first area, olivine augite basaltic andesite unconformably overlies rocks belonging to LU–1 and LU–2 of the Havallah sequence, and some exposures of this olivine augite basaltic andesite also are interbedded with Tertiary gravel (unit Tg, pl. 2). Two replicate determinations of a sample of olivine augite basaltic andesite at the latter occurrence in NE 1/4 sec. 35, T. 33 N., R. 42 E. yielded whole-rock potassium-argon ages of 31.3±0.8 and 31.4±1.0 Ma (see section above entitled "Potassium-argon Chronology of Cretaceous and Cenozoic Igneous Activity, Hydrothermal Alteration, and Mineralization"). Total area of exposure of olivine augite basaltic andesite at the first locality is quite minimal, however, amounting, in all, to approximately 0.1 km^2. The second locality is an even smaller, isolated occurrence of augite basaltic andesite that rests on subunit hss of LU–2, and it is assumed to have the same Oligocene age as the radiometrically dated olivine augite basaltic andesite at the first locality (pl. 2).

Petrographic examination of a small number of samples of Oligocene basaltic andesite reveals that several different textural varieties of basaltic andesite are present in the two localities. Oligocene basaltic andesite in the general area of the middle stretches of Cottonwood Creek is phyric olivine augite basaltic andesite, which includes some notable petrographic differences when compared with Miocene basalt at Treaty Hill to be described in a subsection below entitled "Miocene Olivine Basalt." At Cottonwood Creek, the predominant phenocrysts, in terms of size and volume, are augite, although microphenocrysts of olivine, heavily altered to iddingsite, are relatively abundant throughout the rocks. Generally similar textural varieties of basaltic andesite are present at the second locality also. However, the rock at this latter locality is a phyric augite basaltic andesite containing abundant glomeroporphyric, 1– to 2–mm–wide clots of augite, some individual crystals in the clots have well-developed sector twinning, and extremely sparse concentrations of microphenocrysts of olivine in the groundmass. Plagioclase phenocrysts are not present, and plagioclase microlites are as calcic as An$_{70}$. Thus, there appears to be a north-to-south decline in proportion of olivine in Oligocene basaltic andesite together with a proportionate increase toward the south in the amount of phenocrystic augite in the rocks. Fabrics in basaltic andesite are the result of distribution and density of precrystallization nucleation sites, cooling rates, time of residency at liquidus temperatures, crystallization kinetics, and temperature of the magma (Lofgren, 1983; Cashman, 1990). Studies by these two authors further suggest that size of augite phenocrysts largely may be a function of cooling rates in basaltic rocks. Additional information concerning chemical zoning in augite phenocrysts from Tertiary phyric augite basalt in the North Peak quadrangle is contained in the following subsection.

Mineral chemistry of augite in basaltic andesite

by Robert L. Oscarson and Ted G. Theodore

An electron microprobe was used to examine mineral chemistry and cation zoning of augite in one sample of Oligocene olivine augite basaltic andesite from the middle stretches of Cottonwood Creek. As described above, the specific site from which the sample was collected is an outcrop of approximately 31–Ma olivine augite basaltic andesite that is interbedded with Tertiary gravel (pl. 2). Mineral assemblage of the sample analyzed by microprobe is olivine, augite, ilmenite, and calcic plagioclase. In addition, major- and minor-element data for this particular sample (analysis 3, table 17) are described in the immediately succeeding subsection below entitled "Chemistry." Mineral chemistry of augite in the selected polished thin section was studied using a Jeol electron microprobe, model JXA 8800, equipped with five wavelength dispersive spectrometers. Analysis conditions for standards and unknowns were the same as those employed during microprobe investigation of augite in BBA–suite rocks of the Havallah sequence described above. The nomenclature followed for classification of pyroxene is that of Morimoto and others (1988), and tests of acceptability of microprobe data generally complied with those suggested by Robinson (1980). As described previously, to preserve compositional differences recorded by low-level differences in chemical compositions of core-, intermediate-, and rim-domains of individual augite crystals, all analyses reported represent single points rather than averages. In the first part of the study by microprobe, three points were analyzed on each of ten representative well-formed augite crystals; these points were chosen to characterize the three domains of the augite crystals, which crystals overall are in the range 200 to 400 μm in long

dimension. Subsequently, an additional three points were analyzed on six of the selected ten grains in order to confirm subtle differences in composition among the domains. In the second part of the microprobe study, 38 points were analyzed along a traverse across one of the ten augite crystals that shows well-developed Fe–Mg zoning when viewed by microprobe and SEM using back-scattered electrons. Approximately 22–μm distances were present between most points analyzed along the traverse. Representative microprobe analyses are presented in table 18, and all 86 microprobe analyses of augite from the Oligoclase augite basaltic andesite are plotted as individual points in synoptic diagrams to follow.

Some assumptions are required for conversion of microprobe analyses reported as oxide weight percentages, with total Fe reported as FeO and no analyses for H_2O available by microprobe, to formulas based on a fixed number of oxygens per formula unit. As was done above, we calculated formulas on the basis of four cations and six oxygens (Vieten and Hamm, 1978), and adjusted the Fe^{3+}/Fe^{2+} ratio in the standard $(M2)(M1)T_2O_6$ standard pyroxene formula as follows: (1) initially to sum the tetrahedral (T) site to 2.00 cations; (2) then to sum the M1 site to 1.00 using the required amount of Fe^{3+} cations; and (3), finally, to assign all remaining Fe to Fe^{2+} in the M2 site. As shown in table 18, this results in the M2–group cations totaling at most five percent more than the 1.00 available positions in the M2 site, and a slight amount of Al, 0.01–0.08 cations, being assigned to the M1 site. Approximately 77 to 83 percent of the M2–group site apparently is filled by Ca, and approximately 2 to 3 percent of the site is also occupied by Na (table 18), of which the latter element does not calculate to an amount larger than the number of cations (0.1 cations of Na) requiring the "sodian" adjectival modifier to the mineral name (Morimoto and others, 1988). However, Cr_2O_3 content mostly in the cores of some of the augite crystals is in concentrations such that Cr^{3+} cations make up more than 0.01 cations, actually as many as 0.02 cations, and thus meeting concentration levels that could be then designated as "chromian."

All 86 analyses of pyroxene in the sample of the Oligocene olivine augite basaltic andesite plot in the augite compositional field of Morimoto and others (1988) near the Wo-En sideline (fig. 70). In addition, only some of the ten augite crystals examined in the olivine augite basalt show moderate differences in chemical composition between rim- and core-domains of individual crystals analyzed by microprobe (table 18). In some augite crystals, a general low-level enrichment of Cr_2O_3 and Al_2O_3 contents is present in core domains of the crystals, together with a corresponding enrichment of MgO content (grain A, table 18). However, many rim domains of these same augite crystals have enrichments in contents of Fe, TiO_2, and possibly Na_2O, the latter being present in concentrations

Table 17—Analytical data on Oligocene augite basaltic andesite from area of middle stretches of Cottonwood Creek, and on Miocene olivine basalt from area of Treaty Hill, Humboldt County, Nevada.†

Analysis number	1	2	3	4	5
Field number	90TT320	90TT321	91TT098	90TT169	90TT170

Chemical analyses (weight percent)

	Cottonwood Creek area			Treaty Hill area	
SiO_2	53.7	53.3	53.6	51.4	47.5
Al_2O_3	14.6	15.2	15.5	16.2	16.2
Fe_2O_3	4.14	4.49	3.86	4.97	7.26
FeO	3.76	4.16	4.55	5.94	4.4
MgO	5.83	5.3	5.6	4.97	6.02
CaO	9.1	8.68	8.38	7.98	9.92
Na_2O	2.28	2.31	2.37	3.28	2.97
K_2O	2.79	2.68	2.72	1.42	.82
H_2O^+	.55	.73	.61	.45	.51
H_2O^-	.77	.81	.79	.61	.63
TiO_2	.82	.92	.93	1.78	1.88
P_2O_5	.42	.42	.42	.53	.55
MnO	.14	.14	.14	.17	.15
F	.1	.08	.08	.06	.04
Cl	.009	.01	.008	.009	.012
CO_2	.74	.18	.04	.22	.73
Total S	.01	.03	<.01	<.01	.13
Subtotal	99.76	99.44	99.6	99.99	99.72
Less O = F, Cl, S	.04	.03	.03	.03	.05
Total	99.72	99.41	99.57	99.96	99.67

Inductively coupled plasma-atomic emission spectroscopy (total) (parts per million)

Ba	1300.0	2700.0	1600.	720.0	500.0
Be	2.	2.	2.	1.	1.
Ce	49.	48.	49.	50.	41.
Co	34.	37.	36.	34.	40.
Cr	310.	300.	250.	63.	110.

Inductively coupled plasma-atomic emission spectroscopy (total) (parts per million)

Cu	50.	45.	43.	27.	41.
Ga	16.	17.	18.	21.	20.
La	26	24.	24.	23.	19.
Li	13.	13.	14.	16.	18.
Nb	10.	11.	11.	11.	10.
Nd	28.	27.	27.	29.	26.
Ni	60.	40.	40.	38.	59.
Pb	8.	7.	9.	4.	—
Sc	28.	31.	30.	25.	28.
Sr	620.	640.	620.	560.	600.

Table 17 —(cont.)

Analysis number	1	2	3	4	5
Field number	90TT320	90TT321	91TT098	90TT169	90TT170
Th	6.	7.	6.	—	—
V	210.	240.	230.	260.	260.
Y	21.	20.	21.	30.	27.
Yb	2.	2.	2.	3.	3.
Zn	74.	71.	72.	100.	100.

Inductively coupled plasma-atomic emission spectroscopy (partial) (parts per million)					
Ag	—	—	—	0.16	0.7
As	5.0	11.0	—	3.7	3.7
Cd	.09	.087	—	.098	.064
Cu	43.	43.	41.0	29.	23.
Mo	.38	.7	.44	.57	.73
Pb	1.9	2.5	2.3	2.2	2.4
Zn	61.	54.	56.	55.	74.

Energy-dispersive X-ray fluorescence spectrometry (parts per million)					
Nb	—	—	—	16.0	12.0
Rb	64.0	80.0	69.0	27.	18.
Sr	580.	600.	570.	540.	580.
Zr	144.	146.	138.	142.	124.
Y	26.	26.	18.	32.	27.

Chemical Analyses (parts per million)					
As	5.3	11.0	2.3	5.4	5.3
Au	—	—	—	—	—
Hg	—	.02	.02	—	.02
W	—	—	—	—	—

†[Chemical analyses of major oxides in weight percent by methods of Jackson and others (1987); analysts, J.S. Mee and D.F. Siems. FeO, H_2O^+, H_2O^-, CO_2 in weght percent by methods of Jackson and others (1987); analyst, N.H. Elsheimer. Total S in weight percent by combustion-infra-red method; analyst, N.H. Elsheimer. F, Cl in weight percent by specific ion electrode method; analyst, N.H. Elsheimer. Inductively coupled plasma-atomic emission spectroscopy (total and partial) in parts per million by methods of Crock and Lichte (1982), Lichte and others (1987), and Motooka (1988); analysts, P.H. Briggs, and J.M. Motooka. Precision for concentration higher than 10 times the detection limit is better than +10-percent relative standard deviation; instrumental precision of the scanning instrument is +2-percent relative standard deviation. Looked for, but not found, at parts-per-million detection levels in parantheses: As (10), Au (8), Bi (10), Cd (2), Eu (2), Ho (4), Sn (5), Ta (40), U (100). Energy-dispersive X–ray fluorescence spectrometry in parts per million by method of Johnson and King (1987); analyst, J. Kent. Chemical analysis for Au by graphite furnace-atomic absorption spectrometry (Meier, 1980); analysts, E.P. Welsch, B.H. Roushey, and P.L. Hageman. Chemical analysis for As by graphite furnace atomic absorption spectrometry (Wilson and others, 1987; Aruscavage and Crock, 1987); chemical analysis for Hg by cold-vapor atomic absorption spectrometry; chemical analysis for W by spectrophotometric method of Wilson and others (1987); analysts, E.P. Welsch, B.H. Roushey, and P.L.Hageman; —, not detected].

that yield extremely low-levels of elemental contrasts between cores and rims (table 18). However, many other augite crystals in the sample of olivine augite studied do not show enrichments of Cr in their cores. Nonetheless, systematic, generally concentric growth zones primarily reflecting differing cationic ratios of Mg and Fe are present in many crystals and these growth zones are best displayed in back-scatter electron images (fig. 71). As shown on these images, several types of concentric chemical zonation are present in the augite grains. The more common type of zonation consists of a light band, or higher effective atomic number, present in the rim region of the augite grains (fig. 71A). However, some other grains of augite show in the areas of their rims fairly dark bands in their back-scatter images, which bands are bounded sharply on the side of the core by a narrow zone that has a slightly higher effective atomic number (fig. 71B). These narrow zones are then succeeded toward the core of the augite crystals by gradually decreasing effective atomic numbers which then become uniform over relatively broad core areas of crystal grains.

Microprobe analyses along a traverse at high angles to growth zones across a selected augite crystal (fig. 72A–I) show changes in chemistry compatible with the sequence of the bands in the back-scatter electron images, including increased TiO_2, Al_2O_3, FeO, and MnO contents in the area of the rim of the crystal which thus has a higher effective atomic number to the back-scatter electrons. The point plotted at a distance of zero μm along the depicted traverse across the augite crystal may include some elemental effects from the surrounding matrix in the sample, and should not be considered representative of compositions at the very edge of the selected augite grain. These chemical changes across the augite grain contrast with reported common occurrences of core-to-rim depletions of Al_2O_3 and TiO_2 contents in augite of many basaltic rocks elsewhere (Deer and others, 1978). However, these same authors

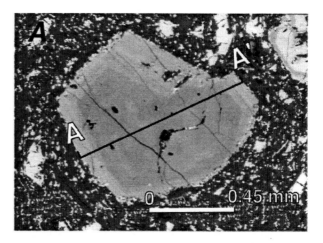

Figure 71 Back-scatter electron images showing growth zones in select grains of augite from Oligocene olivine augite basaltic andesite near middle stretches of Cottonwood Creek. Sample 91TT098. *A*, Grain D showing a relatively large number of growth bands and trace of 800-μm-long traverse (A–A') along which spot microprobe analyses depicted in figure 72 were made; *B*, grain I showing a small number of growth bands confined near the borders of augite crystal and a wide region in the core of the crystal showing apparently uniform Fe and Mg composition.

further report that some phenocrystic augites in quenched alkali basaltic rocks in New South Wales do show core-to-rim enrichments of Al_2O_3, TiO_2, and FeO contents. In addition, Aurisicchio and others (1988) refer to core-to-rim evolution trends in augites at Alban Hills, Italy, similar to those that we have found as being "normal." They proceed to infer that "reverse" core-to-rim compositional trends are the result of changes of H_2O pressures during crystallization of augite. All of the dramatic chemical changes recorded in the augite grains during our investigation, moreover, suggest that augite in the Oligocene olivine augite basaltic andesite must have crystallized under fluctuating P, T, and X conditions, the latter of which may also have been partly controlled by partition coefficients between crystals and magma and activity of silica.

Many differences in oxide components of augite crystals among rim-, intermediate-, and core domains are depicted graphically in figure 73A–H. As shown in these

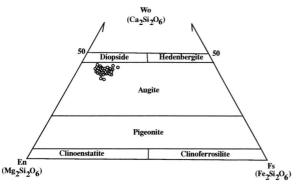

Figure 70 Quadrilateral diagram showing 86 microprobe analyses of augite from a select sample of Oligocene basaltic andesite from area of middle stretches of Cottonwood Creek plotted using ternary molecular proportions of end-member pyroxenes Wo (Ca_2Si_2O6), En ($Mg_2Si_2O_6$), and Fs ($Fe_2Si_2O_6$). Classification scheme of pyroxenes from Morimoto and others (1988).

Table 18—Representative microprobe analyses of augite from Oligocene olivine augite basaltic andesite cropping out in area of middle stretches of Cottonwood Creek, Humboldt County Nevada.†

Grain	B			C			E		
[1]Domain	rim	inter.	core	rim	inter.	core	rim	inter.	core

Microprobe analyses (weight percent)

SiO_2	52.17	51.81	52.02	50.25	52.15	51.05	51.2	52.72	49.91
Al_2O_3	3.19	2.94	2.85	5.3	2.42	2.63	3.79	2.67	4.95
FeO	6.53	6.49	5.25	7.45	4.49	9.1	5.8	5.07	8.67
MgO	17.38	17.03	17.71	16.16	17.9	15.82	16.86	17.84	15.74
CaO	19.64	20.21	20.72	17.21	21.02	19.22	20.46	20.6	19.45
Na_2O	.33	.34	.26	.35	.3	.25	.35	.29	.32
TiO_2	.42	.31	.26	.6	.25	.62	.51	.28	.66
MnO	.16	.15	.12	.17	.13	.25	.17	.12	.18
Cr_2O_3	.26	.3	.57	.17	.47	.11	.24	.54	.05
Total	100.08	99.58	99.76	99.66	99.13	99.05	99.38	100.13	99.93

Cations (normalized to 4)

Si (T)	1.9	1.9	1.9	1.85	1.91	1.9	1.88	1.92	1.84
Al^{IV}	.1	.1	.1	.15	.09	.1	.12	.08	.16
Al^{VI}	.04	.03	.02	.08	.01	.02	.04	.03	.05
Fe^{3+}	—	.02	—	—	—	.08	.01	—	.07
Ti (M1)	.01	.01.	.01	.02	.01	.02	.01	.01	.02
Cr	.01	.01	.02	.01	.01	—	.01	.02	—
Mg	.94	.93	.95	.89	.97	.88	.92	.94	.86
Mg	.01	—	.01	—	.01	—	—	.03	—
Fe^{2+} (M2)	.2	.18	.16	.23	.14	.26	.17	.15	.2
Ca	.77	.8	.81	.76	.83	.77	.81	.8	.77
Na	.02	.02	.02	.03	.02	.02	.03	.02	.02

Molecular proportions of end-member pyroxene

En	49.4	48.4	49.8	47.3	50.4	45.5	48.4	50.3	45.5
Wo	40.2	41.3	41.9	40.5	42.5	39.8	42.3	41.7	40.4
Fs	10.4	10.3	8.3	12.2	7.1	14.7	9.3	8.	14.1

[1]Domain category of analyzed crystal grain; inter., intermediate.

†[Field number 91TT098; same as analysis 3, table 17;—not detected]

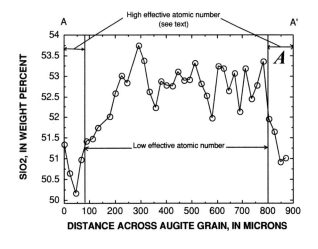

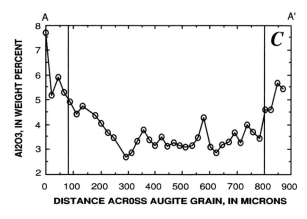

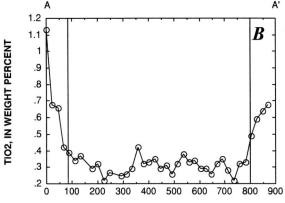

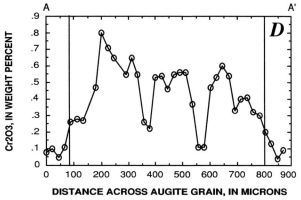

Figure 72 Microprobe traverse across augite from Oligocene olivine augite basaltic andesite near middle stretches of Cottonwood Creek. Trace of traverse shown in figure 71*A*. Distance versus SiO_2 content (*A*), versus TiO_2 content (*B*), versus Al_2O_3 content (*C*), versus Cr_2O_3 content (*D*), versus FeO content (*E*), versus MnO content (*F*), versus MgO content (*G*), versus CaO (*H*), and versus Na_2O content (*I*).

diagrams, cation concentrations at sites categorized as belonging to all three domains span the entire range of reported oxide compositions. In addition, significant overlap apparently is present in measured concentrations of these oxides in all domains of the augite crystals—including a relatively strong inverse relation between FeO and TiO_2 contents versus Cr_2O_3 content overall the augite grains; many analyses at the cores of the augite crystals are elevated in Cr_2O_3 content (fig. 73*F–G*). However, the highest Cr_2O_3 content detected, approximately 0.8 weight percent, is present in one of the intermediate domains analyzed by microprobe. Chrome content of rock sample 91TT098 which hosts the microprobe-analyzed grains is 250 ppm (analysis 3, table 17). In addition, considerably more scatter is present in these microprobe data when classified by domain than there was detected previously above in augite crystals of the BBA–suite rocks of the Havallah sequence (compare figs. 45 and 73).

Qualitative distribution of a number of cations in two augite crystals of the sample of Oligocene olivine augite basaltic andesite studied quantitatively above by microprobe (91TT098) was determined by preparation of elemental X–ray intensity maps also using the microprobe. Figure 74*A–F* shows distribution of Mg, Cr, Na, Mn, and Ca in a representative augite crystal (grain D, same as fig. 71*A*), and figure 75*A–F* shows distribution of those same five elements in augite grain F. The X–ray intensity maps reveal depletion of Mg in the rim domains of both augite crystals as discussed above. Furthermore, a marked difference in distribution of Cr is present in these two augite crystals. Grain D shows a splotchy, irregular distribution of increased concentrations of Cr in its core domain (fig. 74*B*), whereas grain F shows increased concentrations of Cr to be present narrowly in its rim domains (fig. 75*B*). Again, these data suggest that many crystals of augite now residing within millimeters of one another in the Oligocene basaltic andesite must have had radically different hypersolidus histories prior to their eventual extrusion onto the Oligocene surface.

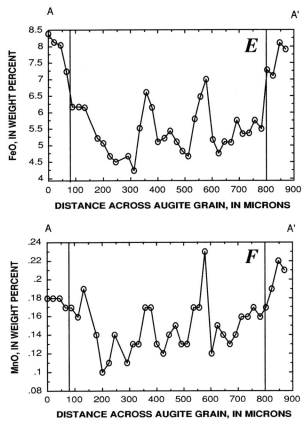

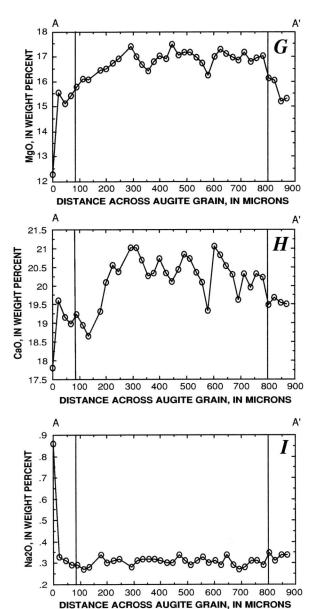

Figure 72 (cont.)

CHEMISTRY

Three chemical analyses of Oligocene olivine augite basaltic andesite are available (analyses 1–3, table 17). All three of these analyses are from the same outcrop in NE 1/4 sec. 35, T. 32 N., R. 42 E., from which outcrop the two potassium-argon ages also were obtained. One of the analyzed samples (analysis 3, table 17) represents an aliquot of the rock from which both potassium-argon ages were determined (see section above entitled "Potassium-argon Chronology of Cretaceous and Cenozoic Igneous Activity, Hydrothermal Alteration, and Mineralization"). Non-normalized contents of SiO_2 in these rocks are in the range 53.3 to 53.7 weight percent (table 17), and normalized contents, calculated volatile free, are in the range 54.79 to 55.2 weight percent (table 19). In the classification scheme of Le Bas and others (1986), these three samples plot in the basaltic andesite field (fig. 76). All three samples of Oligocene olivine augite basaltic andesite contain normative quartz in their CIPW norms, also calculated volatile free and normalized to 100 weight percent with Fe^{3+}/Fe (total) = 0.15 as above. The three chemically analyzed samples contain no normative olivine although they do contain small amounts of modal olivine in their fabrics as described previously (table 19). Contents of normalized Na_2O in basaltic andesite are constrained tightly to the range 2.34 to 2.42 weight percent, and normalized K_2O contents are in the range 2.76 to 2.87 weight percent (table 19). In addition, basaltic andesite shows moderate concentrations of normative hypersthene and normative diopside, these two minerals together accounting for approximately 30 normative weight percent of total normative makeup of the rocks. Overall chemistry of the Oligocene olivine augite basaltic andesite is markedly different from Miocene basalt cropping out in the area of Treaty Hill (see subsection below entitled "Miocene Olivine Basalt").

Oligocene basaltic andesite along the middle stretches of Cottonwood Creek shows high values of *mg* (100*atomic Mg/(Mg+Fe^{2+}), again setting Fe^{2+}/Fe^{total} =0.85 as we did above) that are in the range 53.54 to 58.13, which

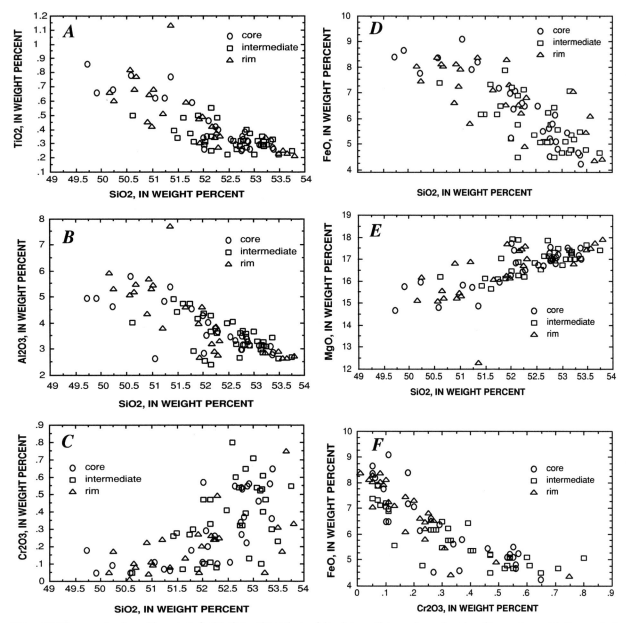

Figure 73 Diagrams showing oxide contents for 86 microprobe analyses of rim-, intermediate-, and core-domains of ten augite crystals from a select sample of Oligocene olivine augite basaltic andesite from the area of middle stretches of Cottonwood Creek. *A*, TiO_2 versus SiO_2 content; *B*, Al_2O_3 versus SiO_2 content; *C*, Cr_2O_3 versus SiO_2 content; *D*, FeO versus SiO_2 content; *E*, MgO versus SiO_2 content; *F*, FeO versus Cr_2O_3 content; *G*, TiO_2 versus Cr_2O_3 content; *H*, FeO versus MgO content.

values suggest that the rocks are magnesian (table 19). In addition, basaltic andesite is enriched in LREEs with the result that $La_n/Yb_n > 8$ and $Ce_n/Yb_n > 6$ (subscript "n" indicates a chondrite-normalized value for the element [sample/chondrite] using the average of ten ordinary chondrite values in Nakamura (1974) as suggested by Rock (1987)). These ratios are calculated from data in table 17, and they suggest that complete REE patterns for basaltic andesite would slope significantly downwards toward HREEs if concentrations of all REE were known. That is, the REE patterns may be convex-upwards in shape, which shape indicates a fractionation of LREE from HREE that has

been inferred to involve garnet at depth during genesis of magmas associated with these types of REE patterns (Saunders, 1984). Unfortunately, analyses for all REEs in these rocks are not available. However, ratios of some less- to more-incompatible elements also are exemplified by Zr/Nb ratios approximately of 14 for analyses of basaltic andesite (table 17).

Chemically analyzed samples of Oligocene olivine augite basaltic andesite also meet potassic volcanic rock criteria of Müller and others (1992) and Müller and Groves (1993). However, the apparent within-plate extensional tectonic setting of these rocks apparently is not identified

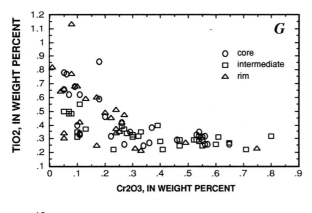

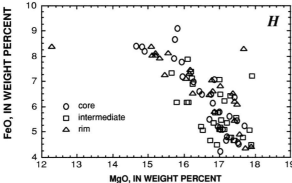

Figure 73 (cont.)

correctly on the Zr (in ppm)/Al_2O_3 (in weight percent) ratio versus TiO_2/Al_2O_3 (in weight percent) ratio discrimination diagram of Müller and others (1992) (fig. 77A). As shown on this figure, three available analyses of Oligocene olivine augite basaltic andesite plot in the continental arc plus post-collisional arc field of their discrimination diagram. Further examination of available Zr, Nb, Ce, and P_2O_5 content of these three rocks shows that the data plot in the post-collisional-arc field of the ternary diagram used by Müller and others (1992) to discriminate between continental arc and post-collisional arc tectonic settings (fig. 77B). However, as pointed out by Sawlan (1992) and M.G. Sawlan (oral commun., 1993), it is possible to show inherited arclike geochemical signatures in subsequent largely extensional tectonic settings within a geologic province well after cessation of prior subduction-related processes. In this part of Nevada during the Oligocene, approximately 30 m.y. ago, limited outpouring of basaltic andesite thus appears to have occurred approximately at the same time as initial intersection of the East Pacicic Rise with the American Plate (Christiansen and Lipman, 1972). However, at approximate longitudes of Battle Mountain, onset of regional extensional breakup appears locally to have occurred somewhat earlier, possibly beginning at approximately 40 Ma (Theodore and others, 1973; Seedorff, 1991; Doebrich and others, 1995). In addition, elemental ratios in magmas associated with Oligocene basaltic andesite may have been affected

dramatically by interaction(s) of magma with Proterozoic continental crust, the westernmost edge of which is inferred to be present at depth west of the general area the Battle Mountain Mining District (Kistler, 1983, 1991; Robinson, 1993, 1994; Kistler and Fleck, 1994). Therefore, the roughly 30–Ma basaltic andesite, some of the oldest volcanic rocks in the general area of the mining districts, may mark a lithologic transition between predominantly arc-related andesitic magmatism and subsequent extensional bimodal volcanic rocks that are extremely widespread in the region (Christiansen and Lipman, 1972; Lipman and others, 1972).

OLIGOCENE OR MIOCENE CALC-ALKALINE RHYOLITE TUFF

Late Oligocene and (or) early Miocene calc-alkaline rhyolite tuff, no more than a few tens of meters thick, crops out discontinuously along a northeast trending zone, approximately 6 km long and at most 2 km wide, in the hanging wall of the Oyarbide fault, where the tuff rests unconformably both on rocks of the Valmy Formation and on rocks of the Havallah sequence (unit Tt, pl. 2). Potassium-argon whole rock age of a sample of calc-alkaline rhyolite from the Trout Creek area is 25.5±0.8 Ma (see section above entitled "Potassium-argon Chronology of Cretaceous and Cenozoic Igneous Activity, Hydrothermal Alteration, and Mineralization"). Similar tuffs are present in the general area of the Buffalo Valley Au mine, approximately 5 km south of the North Peak quadrangle (Roberts, 1964; Doebrich, 1992, 1995), and a sample of a probably correlative biotite crystal tuff described below in the section titled "Geology of the Marigold Mine Area" has been dated at 22.9±0.7 Ma. On the basis of these ages and the chemistry described below, the calc-alkaline rhyolite tuff is probably part of the Bates Mountain Tuff whose reported age range is 22.2 to 24.7 Ma (Stewart and McKee, 1968; McKee, 1968; Stewart and McKee, 1977; McKee, 1994), which age range straddles the boundary between the Oligocene and Miocene Epochs. If these correlations are correct, calc-alkaline rhyolite tuff in the North Peak quadrangle represents the northernmost exposures of Bates Mountain Tuff known to 1999 (see also, fig. 15 of Stewart and McKee, 1977). The Bates Mountain Tuff is one of the most widespread ash-flows in the region, showing an overall area of distribution that is more than 9,000 km².

The striking yellowish-white calc-alkaline rhyolite tuff, which is quite prominent relative to many of the surrounding drab-colored rocks and unconsolidated deposits in the North Peak quadrangle, is overlain by Tertiary gravels and unconsolidated Quaternary deposits (pl. 2). Some Quaternary deposits near outcrops of this tuff also include well-rounded boulders of the tuff. In this area, the tuff

Table 19—Normalized chemical analyses, CIPW norms, and analytical data of Oligocene augite basaltic andesite cropping out in area of middle stretches of Cottonwood creek, and Miocene olivine basalt from area of Treaty Hill, Humboldt County, Nevada.†

Analysis number	1	2	3	4	5
Field Number	90TT320	90TT321	91TT098	90TT169	90TT170

Chemical analyses (normalized volatile free to 100 weight percent)

	Cottonwood Creek area			Treaty Hill area	
SiO_2	55.2	54.79	54.8	52.28	48.91
Al_2O_3	15.01	15.62	15.85	16.48	16.68
Fe_2O_3	1.28	1.41	1.37	1.77	1.88
FeO	6.54	7.16	6.97	9.	9.57
MgO	5.99	5.45	5.73	5.06	6.2
CaO	9.35	8.92	8.57	8.12	10.21
Na_2O	2.34	2.37	2.42	3.34	3.06
K_2O	2.87	2.76	2.78	1.44	.84
TiO_2	.84	.95	.95	1.81	1.94
P_2O_5	.43	.43	.43	.54	.57
MnO	.14	.14	.14	.17	.15
[1]Total	99.99	100.	100.01	100.01	100.01

[2]C.I.P.W. norms (in weight percent)

Q	3.02	3.33	3.01	0.67	—
ol	—	—	—	—	10.46
(fo)	—	—	—	—	5.4
(fa)	—	—	—	—	5.05
hy	16.13	17.17	18.39	20.48	7.13
(en)	(9.73)	(9.6)	(10.65)	(10.32)	(3.85)
(fs)	(6.4)	(7.57)	(7.74)	(10.16)	(3.27)
di	17.65	14.45	12.73	9.11	14.54
(di)	(11.21)	(8.56)	(7.79)	(4.9)	(8.36)
(hd)	(6.43)	(5.89)	(4.94)	(4.21)	(6.19)
ab	19.83	20.09	20.5	28.22	25.87
an	21.95	23.83	24.14	25.71	29.28
or	16.94	16.28	16.43	8.53	4.99
mt	1.86	2.04	1.98	2.56	2.72
il	1.6	1.8	1.81	3.44	3.68
ap	1.02	1.02	1.02	1.28	1.34
Total	100.	100.	100.	100.	100.
[3]D.I.	39.79	39.7	39.94	37.42	30.86

[1]May deviate slightly from 100 weight percent because of independent rounding.

[2]Calculated using program of M.G. Sawlan (written commun., 1993).

[3]Differentiation index of Thornton and Tuttle (1960), defined as the total of normative quartz plus normative orthoclase plus normative albite.

†[Chemical analyses of major oxides in weight percent by methods of Jackson and others (1987); analysts, J.S. Mee and D.F. Siems. FeO, H_2O^+, H_2O^-, CO_2 in weght percent by methods of Jackson and others (1987); analyst, N.H. Elsheimer. Total S in weight percent by combustion-infra-red method; analyst, N.H. Elsheimer. F, Cl in weight percent by specific ion electrode method; analyst, N.H. Elsheimer. Inductively coupled plasma-atomic emission spectroscopy (total and partial) in parts per million by methods of Crock and Lichte (1982), Lichte and others (1987), and Motooka (1988); analysts, P.H. Briggs, and J.M. Motooka. Precision for concentration higher than 10 times the detection limit is better than +10–percent relative standard deviation; instrumental precision of the scanning instrument is +2–percent relative standard deviation. Looked for, but not found, at parts-per-million detection levels in parantheses: As (10), Au (8), Bi (10), Cd (2), Eu (2), Ho (4), Sn (5), Ta (40), U (100). Energy-dispersive X–ray fluorescence spectrometry in parts per million by method of Johnson and King (1987); analyst, J. Kent. Chemical analysis for Au by graphite furnace-atomic absorption spectrometry (Meier, 1980); analysts, E.P. Welsch, B.H. Roushey, and P.L. Hageman. Chemical analysis for As by graphite furnace atomic absorption spectrometry (Wilson and others, 1987; Aruscavage and Crock, 1987); chemical analysis for Hg by cold-vapor atomic absorption spectrometry; chemical analysis for W by spectrophotometric method of Wilson and others (1987); analysts, E.P. Welsch, B.H. Roushey, and P.L.Hageman; —, not detected]

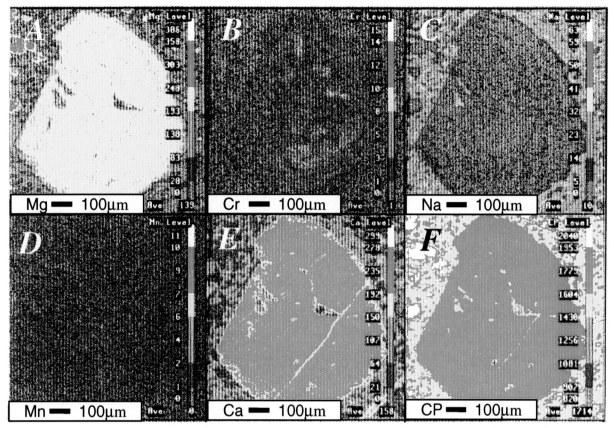

Figure 74 X-ray intensity maps showing distribution of elements in select augite crystals of Oligocene olivine augite basaltic andesite near middle stretches of Cottonwood Creek. Sample 91TT098 (same as analysis 3, table 17). *A*, magnesium; *B*, chromium; *C*, sodium; *D*, manganese; *E*, calcium; *F*, total X-ray counts showing compositional image. Grain D; same as grain D (fig. 71*A*).

strikes to the northeast and either dips shallowly to the southeast into the projected plane of the Oyarbide fault at depth, or the tuff is horizontal in attitude. Fresh surfaces of outcrop of calc-alkaline rhyolite tuff commonly are pinkish gray to grayish pink and are, in places, either crystal- or lithic-rich; the best exposures of the tuff are present in SW 1/4 sec. 6, T. 32 N., R. 43 E., approximately 0.8 km west of the abandoned Oyarbide ranch. Flow banding and collapsed pumice shards are common in many of these variably welded rocks (fig. 78*A*). In addition, some of the tuffs are quite glassy (fig. 78*B*). Microphenocrysts of sanidine and quartz probably make up less than 10 volume percent of most of the 12 samples of tuff examined petrographically. Lithics in some samples of tuff also include small quartzarenite fragments derived from underlying rocks of the Valmy Formation.

Chemical analyses of two samples of calc-alkaline rhyolite tuff from the North Peak quadrangle are available (analyses 1–2, table 20), and these two analyses are compared in this table with five analyses of Oligocene or Miocene Bates Mountain Tuff from Stewart and McKee (1977). The second of the samples apparently is metaluminous and the first is just barely peraluminous. However, the second sample may have a slightly distorted

A/CNK ratio owing to presence of secondary calcite in the rock. Many analyzed samples of Bates Mountain Tuff also contain more H_2O^+, possibly indicative of elevated abundances of clay minerals and (or) white mica in these rocks. Nonetheless, all seven samples plot in a relatively tight domain in the rhyolite field of the total alkali versus silica diagram (fig. 79). However, six analyses plot either in the alkali-rhyolite field or on the join between the alkali-rhyolite and rhyolite fields of the R_1R_2 classification diagram of De la Roche and others (1980) (fig. 80). The position of that one sample that plots in the rhyolite field of the R_1R_2 classification diagram is somewhat separated from the overall trend of the remaining six samples and this separation is probably also a reflection of increased presence of Ca in the rock because of some included secondary calcite. Silica variation diagrams appear to show a consanguinity of Oligocene or Miocene calc-alkaline rhyolite tuff in the North Peak quadrangle and Oligocene or Miocene Bates Mountain Tuff with a possible exception of Al_2O_3 and Na_2O contents (fig. 81*A–J*).

A small number of minor-element analyses are available for cooling unit D of Grommé and others (1972), the most widespread of the four cooling units that make up the Bates Mountain Tuff (Deino, 1985). These analyses

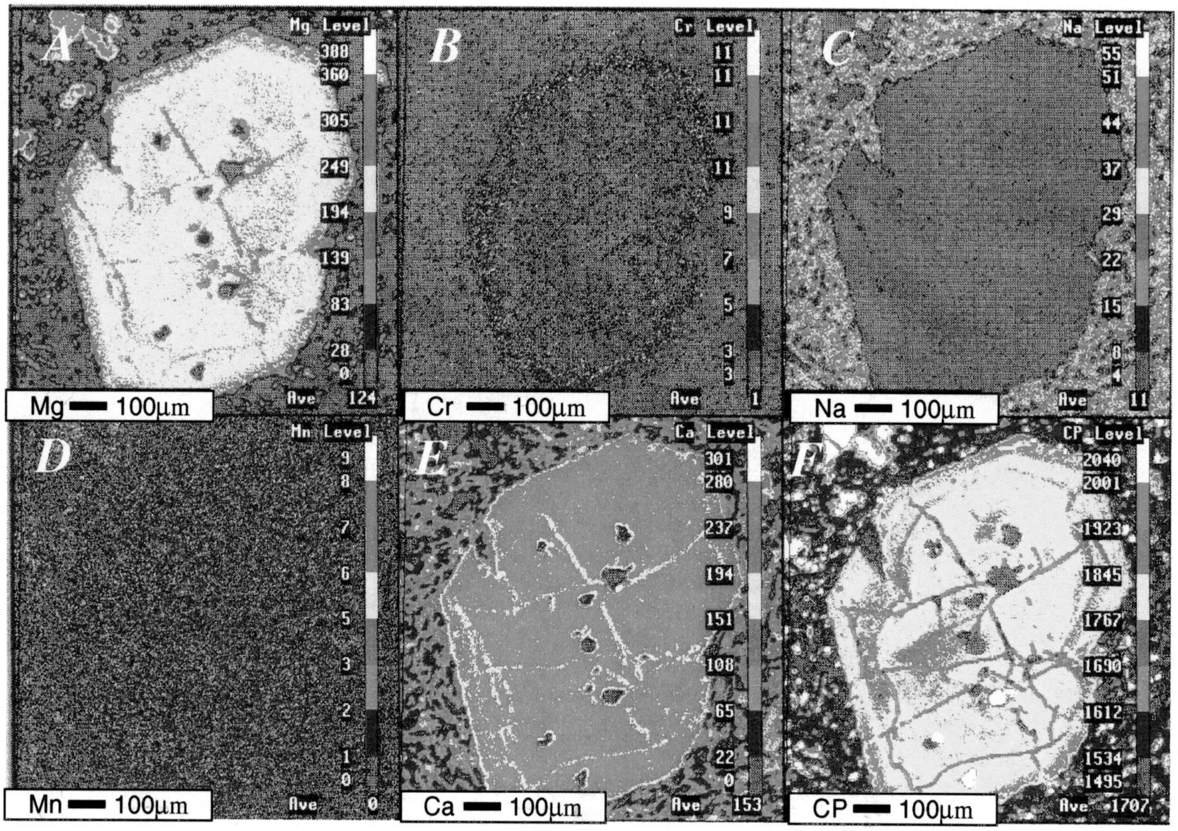

Figure 75 Grain F (see Figure 74).

can be used to test further whether calc-alkaline ryholite tuff in the North Peak quadrangle may be comagmatic with Bates Mountain Tuff, which crops out far to the south of the Battle Mountain Mining District. Cooling unit D of the Bates Mountain Tuff contains 406 ppm Zr and 30.5 ppm Nb, average of nine analyses (Deino, 1985; Best and others, 1989). These values are remarkably similar to the average Zr and Nb contents, 395 ppm and 36 ppm respectively (table 20), of the two analyses available of calc-alkaline rhyolite tuff, even though one of the two analyses (analysis 1, table 20) is much fresher than the other—analysis 2 contains secondary calcite (1.9 weight percant CO_2). In addition, the TiO_2, Rb, Sr, Th, and Y contents of analysis 1 all compare unusually well with the average of nine analyses of unit D of the Bates Mountain Tuff (Deino, 1985). Therefore, trace element data—although admittedly sparse in terms of the numbers of available analyses—add credence to the hypothesis that calc-alkaline rhyolite tuff in the Battle Mountain Mining District and Bates Mountain Tuff were comagmatic.

OLIGOCENE OR MIOCENE BIOTITE CRYSTAL TUFF

Although Oligocene or Miocene biotite crystal tuff is exposed only in the open pit of the Eight South deposit,

this type of rock, probably coeval with Oligocene or Miocene calc-alkaline rhyolite tuff just described above, is assumed to be much more widespread under Tertiary and Quaternary gravels in the quadrangles. A small 20–m-wide mass of light grayish-white biotite crystal tuff crops out near the then (1992) northernmost bottom (4,620–ft elevation) of the open pit at the Eight South deposit. In this area, biotite crystal tuff ranges in thickness from 5 cm to approximately 2 m. Euhedral crystals of 1–mm-wide biotite make up 15 to 20 volume percent of the tuff and this primary biotite has yielded a 22.9±0.7–Ma age by the potassium-argon method (see section above entitled "Potassium-argon Chronology of Cretaceous and Cenozoic Igneous Activity, Hydrothermal Alteration, and Mineralization"). In the small area of the Eight South deposit where biotite crystal tuff crops out, biotite crystal tuff is partly in direct contact with unconformably underlying, mineralized, debris-flow subunits of the Edna Mountain Formation (see section below entitled "Geology of the Marigold Mine Area"). Some parts of biotite crystal tuff also are interbedded with valley-fill gravel that also includes some mineralized rounded cobbles derived from underlying ore. In addition, biotite crystal tuff is, in places, cut by steeply-dipping normal faults exposed in the walls of the open pit. These faults do not have much displacement along them. Much of the gravel that immediately

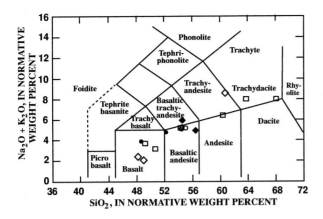

EXPLANATION

o   Oligocene olivine augite basaltic andesite, Cottonwood Creek area (table 17)

◇   Miocene basalt-trachyandesite, Sheep Creek Range (from McKee and Mark, 1971; Stewart and McKee, 1977)

◆   Miocene basaltic trachyandesite and andesite, Shoshone Range (from Gilluly and Gates, 1965)

□   Miocene basalt-trachydacite, Beowawe Geysers area (Struhsacker, 1980)

●   Miocene olivine basalt, Treaty Hill area (table 17)

Figure 76 Total alkali content versus silica content diagram of three samples of chemically analyzed Oligocene olivine augite basaltic andesite from middle stretches of Cottonwood Creek and two samples of Miocene basalt from Treaty Hill compared with available analyses of basaltic rocks from Sheep Creek Range and Shoshone Range, Nevada. Volcanic rock classification scheme of Le Bas and others (1986). Calculated volatile free and normalized to 100 weight percent from data of table 17.

overlies biotite crystal tuff in the Eight South open pit also is probably Tertiary in age.

The biotite crystal tuff exposed by mining at the Eight South deposit is probably comagmatic and coeval with a thin rhyolite encountered at depth in gravels approximately 1.5 km west of the Lone Tree deposit (see subsection below entitled "Lone Tree Gold Deposit"). Rhyolite at this latter occurrence is at a depth of approximately 70 to 85 m below the present erosion surface and it, and the enclosing gravels, are cut by northeast-striking faults probably associated with displacements that are roughly parallel to post-9–Ma, north-northwest extension along the Midas trend (Zoback and Thompson, 1978; Wallace, 1991).

Similar tuff also has been encountered at depth under gravels by drill holes put down in support of mineral-exploration activities in Buffalo Valley along the range front to the southwest of the study area (J.L. Doebrich, oral commun., 1992). These tuffs undoubtedly continue north along Buffalo Valley well into the study area and sporadic occurrences of them are probably present in the subsurface in the northern parts of the valley near and along the west-facing front of the mountain range (pl. 2).

MIOCENE OLIVINE BASALT

Thin flows of approximately 12–Ma (see section above entitled "Potassium-argon Chronology of Cretaceous and Cenozoic Igneous Activity, Hydrothermal Alteration, and Mineralization") Miocene olivine basalt crop out in the Valmy quadrangle north of the Humboldt River in the general area of Treaty Hill (unit Tb, pl. 1). Olivine basalt at Treaty Hill is the most widely exposed Tertiary basalt in the quadrangles, but the exposures at Treaty Hill are nevertheless limited only to a total area of approximately 1 km², where olivine basalt unconformably over-

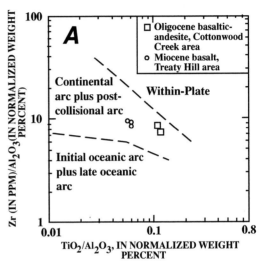

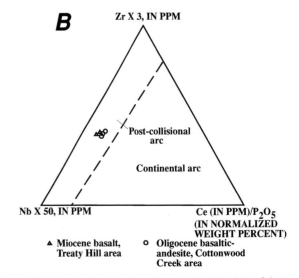

Figure 77 Tectonic discrimination diagrams showing plots of three samples of chemically analyzed Oligocene olivine augite basaltic andesite from middle stretches of Cottonwood Creek and two samples of Miocene basalt from Treaty Hill. Classification of tectonic settings from Müller and others (1992) and Müller and Groves (1993). *A*, Zr (in ppm)/Al₂O₃ (in normalized weight percent) ratio versus TiO₂/Al₂O₃ (in normalized weight percent) ratio diagram; *B*, Zr*3 (in ppm)–Nb*50 (in ppm)–Ce (in ppm)/P₂O₅ (in normalized weight percent) ternary diagram.

Table 20—Analytical data on late Oligocene or early Miocene calc-alkaline rhyolite tuff, Humboldt County, and densely welded or vitrophyre parts of late Eocene and (or) early Miocene Bates Mountain Tuff, Lander County, Nevada.†

Analysis No.	1	2	3	4	5	6	7
Field No.	90TT329	91TT099	15592-1	CC-3A	15978-1	15891-5	15891-5A

Chemical analyses (weight percent)							
SiO_2	74.2	69.4	71.0	73.6	74.0	73.4	73.8
Al_2O_3	12.6	12.8	12.9	12.9	13.	12.7	13.
Fe_2O_3	1.33	2.44	1.4	.85	.47	.56	1.6
FeO	.18	.05	.6	.32	.48	.4	.24
MgO	.19	.28	.23	.16	.1	.2	—
CaO	.38	2.46	.5	.5	.72	.91	.26
Na_2O	3.86	3.74	2.9	3.	3.	3.4	4.2
K_2O	5.02	5.01	5.8	4.8	5.	5.6	5.3
H_2O^+	.56	.6	3.5	2.8	2.5	2.2	.44
H_2O^-	.82	.34	.53	.29	.16	.27	.3
TiO_2	.17	.26	.16	.11	.08	.08	.13
P_2O_5	.06	.17	—	—	—	—	.05
MnO	.02	.15	.2	.06	.05	.06	.07
F	.02	.02	N.d.	N.d.	N.d.	N.d.	N.d.
Cl	.015	.003	N.d.	N.d.	N.d.	N.d.	N.d.
CO_2	.39	1.9	<.05	<.05	<.05	<.05	<.05
Total S	<.01	.01	N.d.	N.d.	N.d.	N.d.	N.d.
Subtotal	99.82	99.63	99.72	99.39	99.56	99.78	99.39
Less O = F, Cl, S	.01	.01	N.d.	N.d.	N.d.	N.d.	N.d.
Total	99.81	99.62	99.72	99.39	99.56	99.78	99.39

Inductively coupled plasma-atomic emission spectroscopy (total) (parts per million)							
Ba	290.0	1200.0	N.d.	N.d.	N.d.	N.d.	N.d.
Be	3.	2.	N.d.	N.d.	N.d.	N.d.	N.d.
Ce	90.	100.	N.d.	N.d.	N.d.	N.d.	N.d.
Co	—	3.	N.d.	N.d.	N.d.	N.d.	N.d.
Cr	2.	2.	N.d.	N.d.	N.d.	N.d.	N.d.
Cu	5.	13.	N.d.	N.d.	N.d.	N.d.	N.d.
Ga	20.	20.	N.d.	N.d.	N.d.	N.d.	N.d.
La	70.	57.	N.d.	N.d.	N.d.	N.d.	N.d.
Li	11.	20.	N.d.	N.d.	N.d.	N.d.	N.d.
Nb	26.	17.	N.d.	N.d.	N.d.	N.d.	N.d.
Nd	53.	44.	N.d.	N.d.	N.d.	N.d.	N.d.
Ni	—	5.	N.d.	N.d.	N.d.	N.d.	N.d.
Pb	28.	26.	N.d.	N.d.	N.d.	N.d.	N.d.
Sc	5.	5.	N.d.	N.d.	N.d.	N.d.	N.d.
Sr	36.	100.	N.d.	N.d.	N.d.	N.d.	N.d.
Th	28.	23.	N.d.	N.d.	N.d.	N.d.	N.d.
V	4.	15.	N.d.	N.d.	N.d.	N.d.	N.d.
Y	31.	32.	N.d.	N.d.	N.d.	N.d.	N.d.
Yb	3.	3.	N.d.	N.d.	N.d.	N.d.	N.d.
Zn	78.	55.	N.d.	N.d.	N.d.	N.d.	N.d.

Inductively coupled plasma-atomic emission spectroscopy (partial) (parts per million)							
Ag	—	—	N.d.	N.d.	N.d.	N.d.	N.d.
As	—	3.3	N.d.	N.d.	N.d.	N.d.	N.d.
Au	—	—	N.d.	N.d.	N.d.	N.d.	N.d.
Bi	—	—	N.d.	N.d.	N.d.	N.d.	N.d.
Cd	—	.19	N.d.	N.d.	N.d.	N.d.	N.d.

Analysis No.	1	2	3	4	5	6	7
Field No.	90TT329	91TT099	15592-1	CC-3A	15978-1	15891-5	15891-5A
Mo	0.24	.4	N.d.	N.d.	N.d.	N.d.	N.d.
Pb	6.9	2.7	N.d.	N.d.	N.d.	N.d.	N.d.
Sb	—	4.9	N.d.	N.d.	N.d.	N.d.	N.d.
Zn	54.	45.	N.d.	N.d.	N.d.	N.d.	N.d.

Energy-dispersive X-ray fluorescence spectrometry (parts per million)

Nb	40.0	32.0	N.d.	N.d.	N.d.	N.d.	N.d.
Rb	205.	166.	N.d.	N.d.	N.d.	N.d.	N.d.
Sr	36.	102.	N.d.	N.d.	N.d.	N.d.	N.d.
Zr	440.	350.	N.d.	N.d.	N.d.	N.d.	N.d.
Y	44.	46.	N.d.	N.d.	N.d.	N.d.	N.d.

Chemical analyses (parts per million)

As	2.9	5.8	N.d.	N.d.	N.d.	N.d.	N.d.
Au	—	—	N.d.	N.d.	N.d.	N.d.	N.d.
Hg	.02	—	N.d.	N.d.	N.d.	N.d.	N.d.
W	—	7.	N.d.	N.d.	N.d.	N.d.	N.d.

[2]C.I.P.W. norms (weight percent)

Q	32.25	26.41	31.87	37.87	36.65	31.02	29.06
c	.27	—	.98	1.93	1.39	—	—
hy	.48	—	.65	.41	.73	.1	—
(en)	(.48)	—	(.6)	(.41)	(.26)	(.07)	—
(fs)	—	—	(.05)	—	(.47)	(.03)	—
di	—	1.59	—	—	—	1.38	—
(di)	—	(1.59)	—	—	—	(.96)	—
(hd)	—	—	—	—	—	(.42)	—
ab	33.33	33.53	25.64	26.36	26.2	29.57	36.03
an	1.52	3.54	2.59	2.58	3.69	2.93	.98
or	30.27	31.37	35.82	29.46	30.49	34.01	31.75
mt	.16	—	2.12	.94	.7	.83	.63
il	.33	.45	.32	.22	.16	.16	.25
hm	1.25	2.58	—	.23	—	—	1.18
ap	.14	.43	—	—	—	—	.12
tn	—	.09	—	—	—	—	—
Total	100.	100.	100.	100.	100.	100.	100.
[1]D.I.	95.85	91.31	93.33	93.69	93.34	94.6	96.84

[1]Differentiation index of Thornton and Tuttle (1960) defined as the sum normative quartz and normative albite and normative orthoclase.
[2]Calculated volatile-free using normalized chemical data.

1. Calc-alkaline rhyolite tuff, SW 1/4 sec. 31, T. 33 N., R. 43 E., North Peak quadrangle.
2. Calc-alkaline rhyolite tuff, SW 1/4 sec. 6 , T. 32 N., R. 43 E., North Peak quadrangle.
3.–7. Bates Mountain Tuff from Stewart and McKee (1977, table 5).

[†][Chemical analyses of major oxides in weight percent by methods of Jackson and others (1987); analysts, J.S. Mee and D.F. Siems. FeO, H_2O^+, H_2O^-, CO_2 in weght percent by methods of Jackson and others (1987); analyst, N.H. Elsheimer. Total S in weight percent by combustion-infra-red method; analyst, N.H. Elsheimer. F, Cl in weight percent by specific ion electrode method; analyst, N.H. Elsheimer. Inductively coupled plasma-atomic emission spectroscopy (total and partial) in parts per million by methods of Crock and Lichte (1982), Lichte and others (1987), and Motooka (1988); analysts, P.H. Briggs, and J.M. Motooka. Precision for concentration higher than 10 times the detection limit is better than +10–percent relative standard deviation; instrumental precision of the scanning instrument is +2–percent relative standard deviation. Looked for, but not found, at parts-per-million detection levels in parantheses: As (10), Au (8), Bi (10), Cd (2), Eu (2), Ho (4), Sn (5), Ta (40), U (100). Energy-dispersive X–ray fluorescence spectrometry in parts per million by method of Johnson and King (1987); analyst, J. Kent. Chemical analysis for Au by graphite furnace-atomic absorption spectrometry (Meier, 1980); analysts, E.P. Welsch, B.H. Roushey, and P.L. Hageman. Chemical analysis for As by graphite furnace atomic absorption spectrometry (Wilson and others, 1987; Aruscavage and Crock, 1987); chemical analysis for Hg by cold-vapor atomic absorption spectrometry; chemical analysis for W by spectrophotometric method of Wilson and others (1987); analysts, E.P. Welsch, B.H. Roushey, and P.L.Hageman; N.d., not determined; —, not detected]

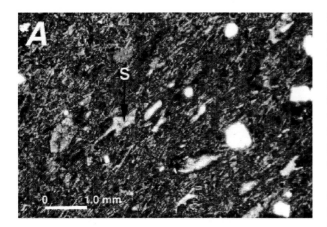

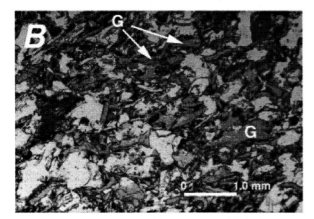

Figure 78 Photomicrographs of late Oligocene or early Miocene calc-alkaline rhyolite tuff. Plane polarized light. *A*, Collapsed pumice shards (S) and well-developed flow banding. Sample 90TT376; *B*, extremely glass-rich (G) crystal-lithic welded tuff showing discontinuous laminae. Clear areas are mostly holes, but other domains of sample include minor amounts of quartz, sanidine, plagioclase, and amphibole. Sample 90TT384.

lies rocks belonging to both the Valmy Formation and the Edna Mountain Formation.

In that part of the Treaty Hill area underlain by the olivine basalt, topography is generally subdued and consists of low rounded hills resulting largely from deep weathering of olivine basalt enhanced by closely-spaced fractures throughout much of the unit. Flow banding is poorly preserved in olivine basalt, and some of the better exposures are made up mostly of a small number of 3– to 4–m-thick individual flows that weather to blocky, boulder-strewn slopes (fig. 82*A*). Fresh olivine basalt is grayish black, and it weathers to dark reddish black. Much of the olivine basalt is amygdaloidal and many of the 3– to 10–mm-wide vesicles are filled with waxy-appearing quartz (fig. 82*B*).

Petrographic examination of a small number of samples reveals that basaltic rocks in the Treaty Hill area are olivine basalt, including phyric vesicular and non-vesicular types—latter includes phenocrysts of olivine (0.4 mm average size; 5 volume percent), augite (0.8 mm; 1 to

2 volume percent), and plagioclase (0.8 mm; 10 volume percent) set in an augite-rich, ilmenite-bearing 0.1– to 0.2–mm-sized groundmass of mostly plagioclase microlites. Sparse exotic fragments of recrystallized calcite include numerous blades of intergrown diopside. Plagioclase phenocrysts are as calcic as labradorite, An_{60}. The vesicular varieties commonly show irregularly shaped crystals of augite that have incorporated numerous small laths of groundmass plagioclase in a subophitic fabric. The walls of many vesicles are also lined with late calcite.

Chemical analyses of two samples of Miocene basalt from the area of Treaty Hill show that these rocks are tholeiitic basalt (analyses 4–5, tables 17, 19). Non-normalized contents of SiO_2 in these rocks are in the range 47.5 to 51.4 weight percent (table 17), and normalized contents, calculated volatile free are in the range 48.91 to 52.28 weight percent (table 19). In the classification scheme of Le Bas and others (1986), one of these two samples plots in the basalt field and the other plots in the basaltic andesite field, just barely beyond the join between the basalt and the basaltic andesite fields (fig. 76). The more siliceous of the two samples contains 0.67 weight percent normative quartz in its CIPW norm, also calculated volatile free and normalized to 100 weight percent with Fe^{3+}/Fe (total) = 0.15 as above, and no normative olivine, whereas the other sample contains approximately 10.5 weight percent normative olivine (table 19). Normalized Na_2O contents in basaltic andesite are constrained tightly to the range 3.06 to 3.34 weight percent, and normalized K_2O contents are in the range 0.84 to 1.44 weight percent, both of which differ significantly from their respective contents in Oligocene basaltic andesite described above (table 20). MgO contents normalized to 100 weight percent in the Miocene olivine basalt is in the range 5.06 to 6.2 weight percent, and normalized TiO_2 contents are in the range 1.81 to 1.94 weight percent. The TiO_2 contents of Miocene olivine basalt, thus, are approximately twice that in Oligocene basaltic andesite. In addition, Miocene

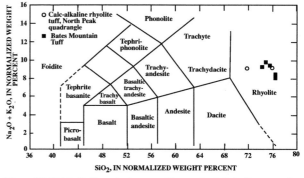

Figure 79 Total alkali content versus silica content plot of two samples of chemically analyzed late Oligocene or early Miocene calc-alkaline rhyolite tuff from North Peak quadrangle and five samples of Oligocene or Miocene Bates Mountain Tuff showing volcanic rock classification scheme of Le Bas and others (1986). Calculated volatile free and normalized to 100 percent from data of table 20.

159

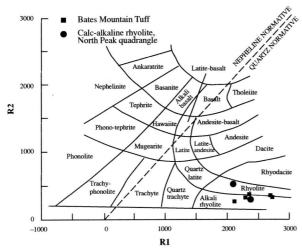

Figure 80 Values of R1 versus R2 for two analyses of late Oligocene or early Miocene calc-alkaline rhyolite tuff from North Peak quadrangle with five analyses of Oligocene or Miocene Bates Mountain Tuff from Stewart and McKee (1977). R1 = [4000Si-11,000(Na+K)-2,000(Fe^{2+} Fe^{3+}+Ti)], and R2 = [6,000Ca+2,000Mg+1,000Al]. Analyses from table 20.

olivine basalt has moderate concentrations of normative hypersthene and normative diopside, generally somewhat less than their calculated contents in Oligocene basaltic andesite. Finally, a marked difference in Cr and Rb contents is present in Miocene olivine basalt when compared to available analyses of Oligocene basaltic andesite; Miocene olivine basalt appears to contain approximately one-third of the amounts of these two elements in the Oligocene basaltic andesite (table 17).

Forty three additional chemical analyses of Miocene, 14– to 17–Ma, basaltic rock are available for comparative purposes from various sites along the main trend of the northern Nevada rift (Struhsacker, 1980; Zoback and others, 1994), approximately 50 to 70 km east-southeast of Treaty Hill—three are from the Sheep Creek Range (McKee and Mark, 1971; Stewart and McKee, 1977) and two from the Shoshone Range (Gilluly and Gates, 1965). As noted by Zoback and others (1994), 14 of 17 basaltic rock samples from the northern Nevada rift dated by the potassium-argon method fall between 14 and 17 Ma. Struhsacker (1980) includes an additional six potassium-argon age determinations of basaltic rock from the Beowawe Geysers portion of the rift (fig. 1) that range from 16.1±0.6 Ma to 16.7±0.7 Ma. Thus, the 12–Ma age of olivine basalt at Treaty Hill is approximately 2 m.y. younger than the youngest age reported to date for lavas

Figure 81 Harker variation plots comparing two analyses of late Oligocene or early Miocene calc-alkaline rhyolite tuff from North Peak quadrangle with five analyses of late Oligocene or early Miocene Bates Mountain Tuff from Stewart and McKee (1977). Analyses from table 20. A, Al$_2$O$_3$ versus SiO$_2$ content; B, Fe$_2$O$_3$ versus SiO$_2$ content; C, FeO versus SiO$_2$ content; D, MgO versus SiO$_2$ content; E, CaO versus SiO$_2$ content; F, Na$_2$O versus SiO$_2$ content; G, K$_2$O versus SiO$_2$ content; H, TiO$_2$ versus SiO$_2$ content; I, P$_2$O$_5$ versus SiO$_2$ content; J, MnO versus SiO$_2$ content.

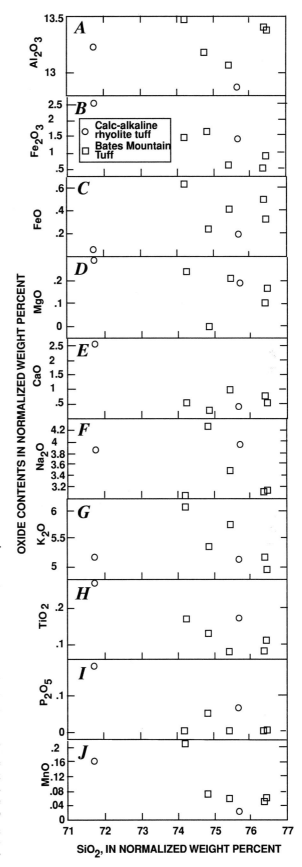

Figure 82 Photographs showing characteristics of Miocene approximately 12–Ma basalt in southeastern part of Treaty Hill, Humboldt County, Nevada. *A*, Thin basalt flows with weathered-out blocks measuring 0.5 to 1.0 m on a side; *B*, close-up view of vesicular basalt.

close to the northern Nevada rift. However, there are two approximately 12–Ma ages of basaltic rock from the Sheep Creek Range several km east of the rift (Stewart and Mckee, 1977).

Two samples from the Sheep Creek Range plot in the basalt field, and the other sample from the range plots in the trachyandesite field (fig. 76). Six analyses of Miocene volcanic rocks from the general area of Beowawe Geysers are included in the report by Struhsacker (1980, table 2), and these analyses define a spread of compositions that ranges from basalt to trachydacite (fig. 76). These samples, together with the remaining analyzed samples of Miocene volcanic rocks in the region, define an extraordinarily broad extent of composition for mid-Miocene basaltic volcanism along the northern Nevada rift. Zoback and others (1994) suggest that varying amounts of crustal contamination may have been important in development of such wide ranges of composition.

The two chemically analyzed samples of Miocene olivine basalt at Treaty Hill also meet potassic volcanic rock criteria of Müller and others (1992) and Müller and Groves (1993). However, the within-plate extensional tectonic setting of these rocks apparently is not identified

correctly on the Zr (in ppm)/Al_2O_3 (in weight percent) ratio versus TiO_2/Al_2O_3 (in weight percent) ratio discrimination diagram of Müller and others (1992) (fig. 77*A*). As shown, the two available analyses of Miocene olivine basalt plot in the continental arc plus post-collisional arc field of their discrimination diagram. However, these two analyses plot closer to the within-plate field of the classification tectonic classification diagram than the analyses of Oligocene basaltic andesite from the Cottonwood Creek area. In addition, Zr, Nb, Ce, and P_2O_5 contents of Miocene olivine basalt from Treaty Hill show that the data plot in the post-collisional-arc field of the ternary diagram used by Müller and others (1992) to discriminate between continental arc and post-collisional arc tectonic settings (fig. 77*B*). As described above, these arclike geochemical signatures in an unquestionably extensional tectonic setting may be inherited from prior subduction-related processes, or these geochemical signatures may reflect interaction with other crustal materials during their ascent to the surface as suggested previously by Zoback and others (1994).

Unconsolidated deposits

Tertiary system

Unconsolidated Tertiary gravel deposits are present at the surface only in the North Peak quadrangle where they crop out in four areas as elevated preserved remnants of former channels that now either form low-lying rounded hills somewhat above present-day stream beds, or are present as small patches almost 1 km above present-day stream beds (pl. 2). Boulders of mostly chert, quartzarenite, and conglomerate in these deposits are well-rounded and as much as 0.7 m wide. They are reported by Roberts (1964) to be as much as 1 m wide in one of the elevated remnants approximately 1.5 km west of Antler Peak. The largest area of outcrop of unconsolidated Tertiary gravel deposits, a north-south elongate body of boulders and cobbles making up low-lying hills approximately 1.5 km² in total area, is on the west side of Cottonwood Creek, near the point of entry of Cottonwood Creek into Buffalo Valley. The gravel deposits in this particular area include some small masses of Oligocene augite basalt in NE 1/4 sec. 35, T. 33 N., R. 42 E. that has been dated at approximately 31 Ma (see section above entitled "Potassium-argon Chronology of Cretaceous and Cenozoic Igneous Activity, Hydrothermal Alteration, and Mineralization"). Presumably, these gravel deposits are interbedded with the augite basalt and the gravel deposits must also be Oligocene in age. Approximately 0.5 km to the east across Cotton-wood Creek, another area also underlain by similar Tertiary gravel deposits must, at one time, have been contiguous with the first occurrence, but these latter deposits subsequently must have been dissected by Quaternary younger alluvium (unit Qya, pl. 2) along Cottonwood Creek owing to later uplift. Presumably Tertiary gravel deposits are also present in the general area of the site of the abandoned Oyarbide Ranch, and in this area they are in the hanging wall of the Oyarbide fault and are also cut by the Oyarbide fault (pl. 2). These latter gravel deposits, in places, have been deposited on Tertiary calc-alkaline rhyolite tuff, and the basal contact of the gravel deposits is at the 6,000-ft elevation. The last of the four occurrences of Tertiary gravel deposits is a small remnant present on the ridge north of Trenton Canyon at the 7,956-ft elevation; implications of these vertical offsets are discussed in the "Tertiary Deformation" subsection below.

Although unconsolidated Tertiary gravel deposits are not present at the natural present-day erosion surface in the Valmy quadrangle (pl. 1), they nonetheless are exposed in open pits at the Lone Tree and Eight South Mines where, especially at the latter mine, some gravel deposits are cut by minor, generally north-trending normal faults. Some gravel deposits at the Eight South deposit are interbedded with biotite crystal tuff at the 4,620-ft elevation, which tuff has been dated at 22.9±0.7 Ma or Early Miocene (see section above entitled "Potassium-argon Chronology of Cretaceous and Cenozoic Igneous Activity, Hydrothermal Alteration, and Mineralization"). Thus, fairly thick gravel deposits as old as Early Miocene, and possibly even as old as Oligocene, appear to make up a significant proportion of the unconsolidated deposits of the valleys in the region.

QUATERNARY SYSTEM

Older alluvium, older fanglomerate, terrace and flood plain deposits of the Humboldt River, slope wash, talus, and younger alluvium and fanglomerate make up the surficial deposits of the Quaternary system in the quadrangles (pls. 1–3). Older alluvium was distinguished from Tertiary gravel deposits largely on the basis of respective relations to present-day geomorphology of the area, but also including geologic relations with other map units. In addition, as mapped, younger alluvium and fanglomerate includes shoreline gravel deposits upon which lakes were formed subsequent to maximum development of the fans that surround many of the mountain ranges in the region (Roberts, 1964). As pointed out by Roberts, one of these lakes extended from a point 8 km east of Golconda along the main Humboldt River drainage to the town of Battle Mountain, and another one partly filled Buffalo Valley, just west of the Valmy and North Peak quadrangles, as

well as an outlet to the Humboldt River drainage through the Reese River valley. There apparently is a wave-cut bench formed in bedrock at the north end of Lone Tree Hill (R.J. Roberts, oral commun., 1967). Although older alluvium and some of the younger alluvium deposits in some of the drainages elsewhere in the Battle Mountain Mining District were extensively worked for placer gold in the past (Roberts and Arnold, 1965; Theodore and others, 1992), none of these deposits in the Valmy and North Peak quadrangles apparently show evidence of such prior exploitation. Some of the older alluvium at the south end of Lone Tree Hill (pl. 1) was found to contain extensive manganese oxide minerals that cement the gravels and some siliceous sinter, when the older alluvium was cut by roads associated with the mining operations at the Lone Tree Mine. The age of this hot spring activity is unknown, but it probably is the same as that in the Golconda area (fig. 1), approximately 30 km to the northwest where Mn–W mineralized rocks are present as replacements in lakebed deposits of Pleistocene Lake Lahontan (Peters and others, 1996). Some relatively minor Au placer mining operations were conducted in the Snow Gulch quadrangle as recently as the 1980s (pl. 3).

Structural geology

ANTLER OROGENY DEFORMATION

Structures in the North Peak and Snow Gulch quadrangles associated with emplacement of the Roberts Mountains allochthon during the Antler orogeny include the Dewitt thrust at the base of the Harmony Formation and several other intraformational low-angle faults bounding units that make up the Valmy Formation (pls. 2, 3; see also, Roberts, 1964). Low-angle structures associated with the Roberts Mountains allochthon do not crop out in the Valmy quadrangle part of the study area (pl. 1). However, rocks of the Scott Canyon Formation in the Snow Gulch quadrangle—irrespective of the fact that they make up an extremely small percentage of the total area of bedrock in the area—are tectonically the lowest rocks exposed in the Roberts Mountains allochthon overall, and they comprise a package of rocks below the Valmy Formation (Roberts, 1964). Elsewhere in the southeastern part of the Battle Mountain Mining District, the Scott Canyon Formation is fairly widespread, and rocks of the formation are present below rocks of the overthrust Valmy Formation (see also, Doebrich, 1994). The Dewitt thrust apparently is structurally the highest interformational low-angle fault asso-

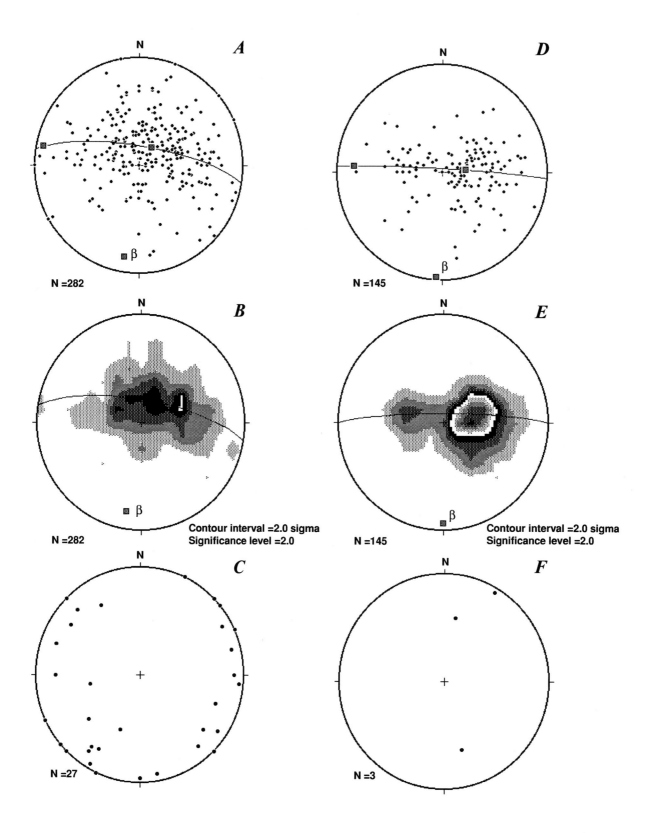

Figure 83 Poles to bedding, contoured poles to bedding, and minor fold axes from most of Ordovician Valmy Formation. Contoured in 2σ intervals where 2σ is the expected number of data points within a counting area for a uniform distribution across the entire stereogram (Kamb, 1959). N, number of points. *A–C,* From footwall of Oyarbide fault; *D–F,* From hanging wall of Oyarbide fault.

ciated with the Roberts Mountains allochthon in the area. Absence of suitable sequences of rock to act as markers in the Harmony Formation, however, hindered recognition of any intraformational low-angle faults even in the wide area of its outcrop in the Snow Gulch quadrangle (pl. 3). Moreover, the surface of the Dewitt thrust itself, which marks a significant tectonic boundary between rocks of the Harmony Formation and those of the Valmy and Scott Canyon Formations, is fairly well exposed along much of the western border of the Snow Gulch quadrangle (pl. 3; Roberts, 1964). Nonetheless, the map pattern traced by the Dewitt thrust in the North Peak quadrangle suggests quite convincingly that rocks of the Valmy Formation project down dip below rocks of the Harmony Formation (see also, Madrid, 1987). As pointed out by Roberts (1964), no direct evidence is present in the mining district that rocks of the Valmy and Harmony Formations are allochthonous. However, from previously recognized regional relations of rocks belonging to these formations and now (1999) reinforced by (1) recent recognition of rocks of lower Paleozoic-age carbonate assemblage in the East Range well to the west of the study area (Peters and others, 1996), and by (2) exposure of the Roberts Mountains thrust in many of the mountain ranges to the east, Roberts and others (1958) and Roberts (1964) previously concluded that rocks of the Valmy and Harmony Formations were transported generally to the east during continental-scale shortening associated with the Antler orogeny.

Several low-angle intraformational thrust faults of Antler orogeny age, at this time all dipping shallowly to the southeast, are present in the Valmy Formation. Strains along these faults undoubtedly contributed to present geometries of bedding and minor folds in the Valmy Formation (pl. 2). At least two of these thrust faults, one at the base of unit 4 and the other at the base of subunit Ov_{3sc} of the Valmy Formation, appear to be fairly major structures because their traces can be mapped continuously for a number of kilometers in the general area of Trout Creek. Attitudes of poles to bedding of the Valmy Formation in the footwall block of the Oyarbide fault show quite a bit of scatter (fig. 83A), but nonetheless define quite well a statistically significant girdle, the axis of which (geometric fold axis or ß axis) plunges about 15° to S. 9° W. (fig. 83B). Minor folds in the block of the Valmy Formation making up the footwall of the Oyarbide fault, however, show significant scatter in their trends, although the plunge of most minor folds is quite shallow (fig. 83C). Moreover, no strong preferred concentration of measured minor fold axes is coaxial with the statistically determined geometric fold axis. All of these data were gathered during regional geologic mapping of the Valmy and North Peak quadrangles, and they do not incorporate data obtained from locality 92TT046 in the Valmy Formation to be described in this section below.

In order to evaluate the possibility of reorientation of minor structures of the Roberts Mountains allochthon where these structures crop out in close proximity to faults that are post-Antler in age, large numbers of structural data were obtained from a small area in the Valmy Formation close to the Miocene-age Oyarbide fault. The rocks selected include some sequences of ribbon chert that are well-exposed at locality 92TT046, which comprises a series of outcrops along a ridge in SE 1/4 sec. 33, T. 33 N., R. 43 E. (pl. 2). This particular ridge is in the footwall of the Oyarbide fault, and all structural data from the locality were obtained within approximately 0.4 km of the trace of the fault (fig. 84). A typical example of some outcrop-scale folds in chert of subunit Ov_{2c} exposed along the ridge is shown above (fig. 18B). Poles to bedding at locality 92TT046 show some scatter mostly indicating variations in dip, but their attitudes nonetheless establish fairly well a statistically defined girdle whose axis plunges 36° to S. 59° W., an orientation significantly different from that of the structural data gathered throughout the remaining footwall of the Oyarbide fault (figs. 84A, B; compare with figs. 83A–C). The expected number for a uniform distribution of poles to bedding across the entire stereonet of the pole-to-bedding data at locality 92TT046 is 3.86 using a significance level of 2.0 (see Kamb (1959) for an explanation of the contouring procedure used).

Most data in the latter figures were obtained from outcrops far removed from the trace of the Oyarbide fault. The statistical differences in orientation of bedding between locality 92TT046 and the remaining rocks of the Valmy Formation in the footwall of the Oyarbide fault most conveniently are expressed through comparisons of differences in orientation of their geometric fold axes. These differences amount to a clockwise rotation of 49°, measured along a great circle, and the clockwise sense of rotation established from a comparison of the regional geometric fold axis with the geometric fold axis present at locality 92TT046. This sense of rotation is compatible with the inferred sense of drag that would accompany the combined dip-slip, northwest block down, and right-lateral components of Miocene offset(s) along the Oyarbide fault. Thus, present orientation of bedding and other Antler-age structures near major post-Antler faults in the Valmy Formation may reflect some reorientation due to drag.

Most minor folds measured at locality 92TT046 trend and plunge shallowly to the southwest (fig. 84C, D). The center of maximum defined by the 8σ contour for the fold axes at this locality plunges 30° to S. 44 W., which attitude diverges approximately 14° from orientation of the geometric fold axis defined by pole-to-bedding data from the same locality (see this section above). The measured angular difference between the center of maximum for minor fold axes at locality 92TT046 and the geometric fold axis for the remaining parts of the Valmy Formation

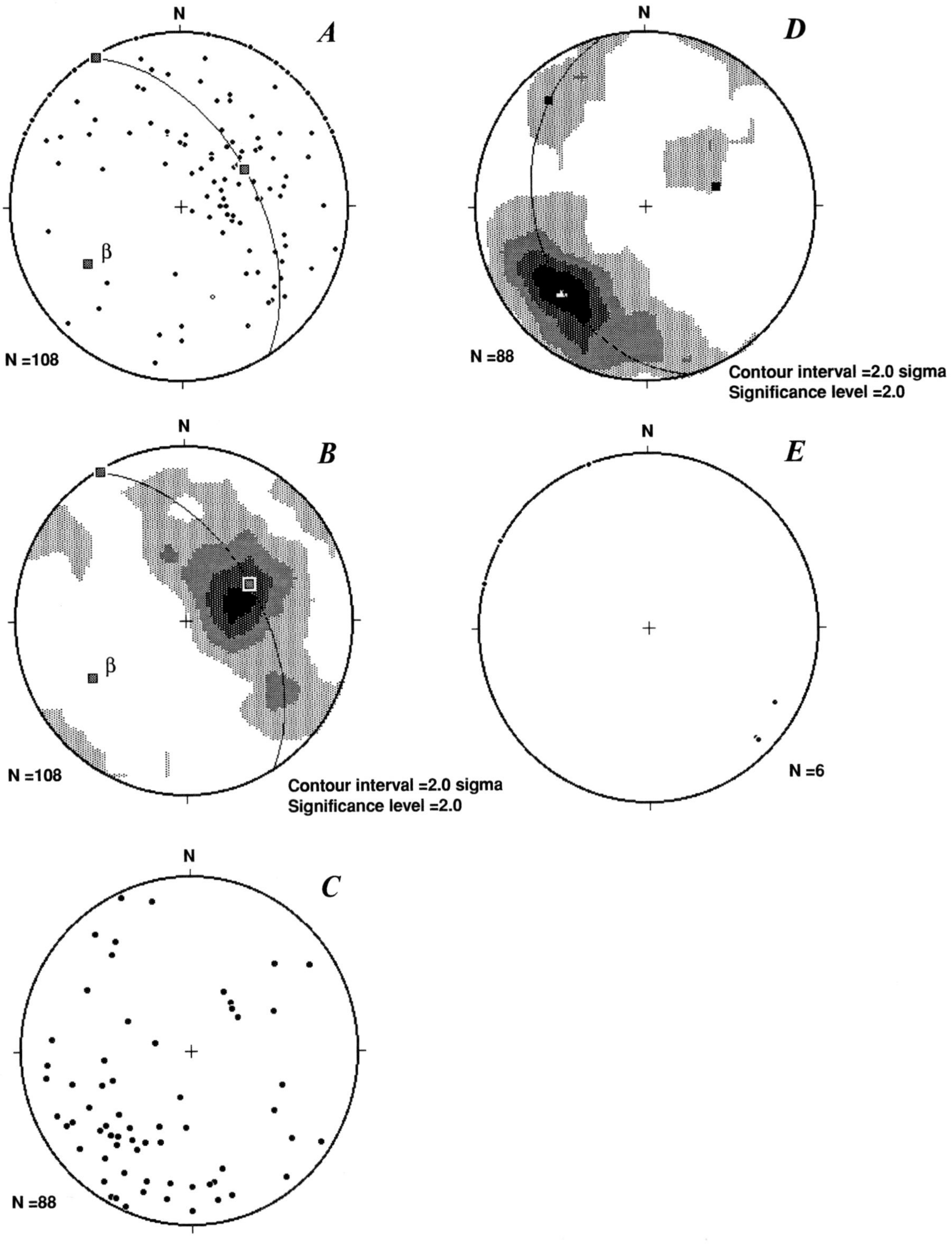

Figure 84 Poles to bedding (*A*), contoured poles to bedding (*B*), minor fold axes (*C*), contoured minor fold axes (*D*), and poles to axial planes (*E*) in subunit Ov$_2$c of Ordovician Valmy Formation at locality 92TT046 in SE 1/4 sec. 33, T. 33 N., R. 43 E. Contouring procedures same as figure 83.

in the footwall of the Oyarbide fault is approximately 36°. This value is somewhat smaller than the difference (49°, see this section above) measured using geometric fold axes from the two data sets, but the probable clockwise sense of drag on the rocks at locality 92TT046 by movements along the Oyarbide fault is still the same. In addition, concentration of fold axes at locality 92TT046 are elongated largely along a girdle, defined by great-circle elongation of 2σ, 4σ, 6σ, and 8σ contours. The area of the stereonet outside the 2σ contour comprises a density of no preferred orientation (Kamb, 1959).

Finally, only six axial planes of minor folds at locality 92TT046 could be measured with any degree of certainty, three of which are vertical, and the remaining three dipping steeply (75 to 80°) to the northwest (fig. 84E). At first appearance, these data could be interpreted as a non-pervasive vergence of minor folds to the southeast. Such orientations of axial planes must be examined cautiously because of apparent strains superposed on the rocks here by post-Antler deformational events. By applying a 49° counterclockwise rotation to these six axial planes along a vector plunging 25° to S. 34° W.—the approximate bisectrix between geometric fold axes at locality 92TT046 and in the remaining rocks of the Valmy Formation below the Oyarbide fault—all six axial planes then assume dips to the west. The mean axial plane determined from the center of maximum of these six counterclockwise-rotated axial planes has a N. 14° E. strike and a 39° west dip, an attitude which suggests that the east vergence of these minor folds at locality 92TT046 is compatible with west-to-east transport along a N. 76° W. trend during emplacement of the Roberts Mountains allochthon.

Rocks of the Valmy Formation, mostly quartzarenite of unit 1, in the hanging wall block of the Oyarbide fault show attitudes of bedding that define a geometric fold axis that plunges at 3 to 5° to almost due south (fig. 83D, E). These data are defined by a pole-to-bedding girdle that is dominated by a strong maximum, showing contours as high as 16σ, that plunges steeply to the east. This maximum is the result of the large number of outcrops in this part of the Valmy Formation that dip at shallow angles to the west. The fact that the geometric fold axes for the bulk of the rocks of the Valmy Formation in both the hanging wall and footwall blocks of the Oyarbide fault both plunge at shallow angles to the south suggests that significant reorientation by post-Antler structures has not occurred throughout these two structural blocks. Nonetheless, the geometric fold axis in the hanging wall plunges at an angle shallower than that in the footwall which further suggests that the hanging wall block, either in part or in its entirety, may have been rotated somewhat during the Tertiary. This apparent Tertiary rotation probably occurred mostly in association with displacements along the Oyarbide fault. Primarily because of the scarcity of ribbon chert sequences

in unit 1 in this area, however, only a limited number of minor folds were found (fig. 83F). In addition, a number of hingelines of anticlines and synclines are present in the rocks of the Valmy Formation generally south of Mud Spring (pl. 2). These hingelines mostly have northeast trends.

Rocks of the Harmony Formation crop out widely in the Snow Gulch quadrangle, and they show attitudes of bedding that define a geometric fold axis that plunges at 10° to N. 15° E. (fig. 85A). This geometric fold axis was determined from a cylindrical-best-fit girdle that was applied to the data. A contoured plot of these data is dominated by a strong elongate maximum, showing contours as high as 16σ, whose center of gravity plunges almost vertically (fig. 85B). The geometric orientation of the maximum reflects the large number of outcrops of the Harmony Formation that dip at shallow angles. Unfortunately, only a small number of outcrop-scale folds are present in the area of outcrop of the Harmony Formation (fig. 85C). Three of five measured outcrop-scale fold axes show northeast trends. The rocks of the Harmony Formation, nonetheless, are characterized structurally by presence of a large number of broad open folds whose hinge lines show northerly trends (pl. 3). In addition, the steep dips to the east of the Dewitt thrust at the base of the Harmony Formation west of the Elder Creek porphyry system are probably a reflection of Antler orogeny-related eastward-directed movements and compression that caused the thrust fault between units 3 and 4 of the Valmy Formation to be overturned so that they now show dips to the west (pl. 3).

Orientation of minor folds and statistically determined geometric fold axes apparently related to deformation during the Antler orogeny in the quadrangles, therefore, have north-northeast orientations similar to other minor folds elsewhere in the Roberts Mountains allochthon. Structural data obtained from the Scott Canyon Formation—structurally the deepest part of the allochthon in the southeast part of the Battle Mountain Mining District—reveal that most minor folds trend northnortheast and plunge at shallow angles either to the north or to the south (Evans and Theodore, 1978). In addition, these relatively abundant minor folds in the Scott Canyon Formation have vertical or near-vertical axial planes, yielding an apparent horizontal component of structural shortening that is oriented N. 70° to 75° W., remarkably similar to the direction of shortening that could be inferred from data, though admittedly limited, in the successively higher Valmy and Dewitt plates of the Roberts Mountains allochthon. Because of absence of a significant number of axial planes that are overturned predominantly either to the east or west in rocks of the Scott Canyon Formation, however, a sense to the direction of tectonic transport could not be ascertained by Evans and Theodore (1978) for the structurally lowest part of the allochthon in the mining

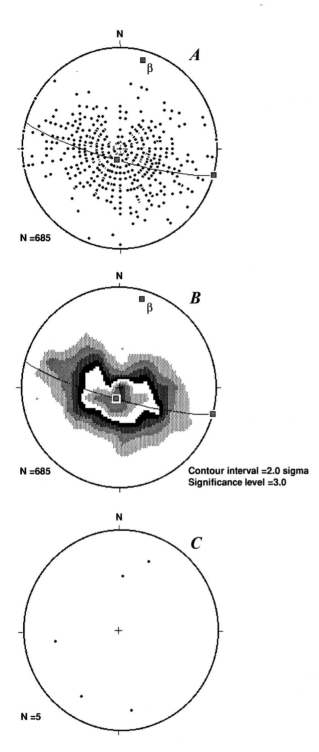

N =685

N =685

Contour interval =2.0 sigma
Significance level =3.0

N =5

Figure 85 Poles to bedding (*A*), contoured poles to bedding (*B*), and minor fold axes (*C*) in Upper Cambrian Harmony Formation in the Snow Gulch quadrangle. Contouring procedures same as figure 83 except 3σ is the expected number of data points within a counting area for a uniform distribution across the entire stereogram.

district. In the area of Carlin, Nev., general orientation of small-scale folds in upper plate rocks of the Roberts Mountains thrust, and presumably associated with the Antler orogeny, also is north-northeast (Evans and Theodore, 1978; Stephen G. Peters, oral commun., 1994).

Sonoma orogeny deformation

Rocks of the Havallah sequence in the quadrangles (pls. 1, 2) were emplaced along the Golconda thrust onto rocks of the overlap assemblage during the late Permian or early Triassic Sonoma orogeny (Silberling and Roberts, 1962; Roberts, 1964). Major structures in the Golconda allochthon associated with the Sonoma orogeny are made up of numerous imbricate thrust faults, including the regionally extensive Willow Creek thrust, and a number of relatively broad, open folds that are particularly well-defined in the Havallah Hills part of the area. Most folds show hinge lines that are north trending and that are parallel roughly to the trace of the nearby Golconda thrust which is covered by Tertiary and Quaternary gravels just to the east of the Havallah Hills (pl. 1). Regionally, autochthonous rocks below the Golconda thrust, that is those rocks belonging to the Antler sequence in the entire Battle Mountain area, are not folded at the outcrop scale comparable to the folds present in allochthons either of the Roberts Mountains thrust, or of the Golconda thrust (Evans and Theodore, 1978). Most strata in the Antler sequence show homoclinal sheet dips in the Battle Mountain Mining District either to the west in the western part of the mining district or to the east near Elephant Head in the eastern part of the mining district (Roberts, 1964; Theodore and Jones, 1992). Attitudes of these rocks belonging to the Antler sequence have been used to place a major antiformal hinge line through the mining district (Roberts, 1964; Madrid, 1987), the warping of which anticline apparently occurred before Late Cretaceous (Theodore and others, 1992). Thus, rocks of the Roberts Mountains allochthon that are stratigraphically beneath the Antler sequence apparently have no evidence that they were deformed by strains transmitted through the Antler sequence during emplacement of the Golconda allochthon.

Structural data, mostly attitudes of bedding, were gathered from throughout the Golconda allochthon in the study area, and these data then were divided for comparative purposes into those obtained from each of three domains of the allochthon, which are referred to as the Havallah Hills block, the Cottonwood block, and the south Oyarbide block (fig. 86). Limits of these three blocks are defined by the two major normal faults that bound them, the Cottonwood Creek fault and the Oyarbide fault. In the Havallah Hills block, a cylindrical-best-fit girdle through the pole-to-bedding data defines a geometric fold axis that plunges about 2° to N. 4° E. (fig. 87*A*), which confirms the almost horizontal attitudes of many of the fold hinges shown on the geologic map of this part of the allochthon (pl. 1). The overwhelming bulk of bedding attitudes measured in the Havallah Hills have roughly north-south strikes and shallow dips to either east or west that thus outline a strong single maximum, as high as 22σ that plunges almost 90° (fig. 87*D*). In the Cottonwood block, a cylindrical-best-fit girdle through pole-to-bedding data reveals a somewhat similar orientation for the geometric fold axis, which in this area plunges approximately 7° to N. 10° E., but with a strong preponderance of bedding attitudes that dip to the west-northwest (fig. 87*B*, *E*). Further analysis of the three maxima that make up pole-to-bedding data in the Cottonwood block reveals that the centers of these three maxima correspond to bedding attitudes with N. 18° E., N. 18° E., and N. 11° W. strikes, and 23°, 47°, and 56° dips, respectively, to the west. However, in the south Oyarbide block, or footwall block of the Oyarbide fault, a cylindrical-best-fit girdle of bedding attitudes in the Havallah sequence defines a geometric fold axis that plunges instead to the southwest, 12° to S. 12° W. (fig. 87*C*, *F*). The possibility that this difference in orientation of bedding attitudes on either side of the Oyarbide fault reflects some rotation that accompanied Miocene displacements along the Oyarbide fault is discussed in the subsection below entitled "Tertiary Deformation."

Only a sparse number of minor, outcrop-scale fold axes are present in the three tectonic blocks of the Havallah sequence, and the attitudes of some of these minor fold axes are somewhat similar in orientation with geometric axes or β–axes determined statistically from pole-to-bedding data in the three blocks. Most minor folds show shallow plunges to either the northeast or to the southwest, although attitudes of minor fold axes in the Cottonwood block show more scatter in their orientations than those from the other two blocks (fig. 88). All minor folds in the Cottonwood block were measured in rocks belonging to LU–1 of the Havallah sequence, and are thus in the footwall of the Willow Creek thrust (pl. 1). Some outcrops of highly deformed argillite in the footwall of the Willow Creek thrust also show development of a pervasive strain-slip cleavage whose bedding-plane intersections are coaxial with nearby minor folds.

The imbricate style of the numerous thrust faults associated with emplacement of the Golconda allochthon during the Sonoma orogeny is probably best displayed on the geologic maps (pls. 1, 2). Of particular importance is the Sonoma-orogeny-related Willow Creek thrust which juxtaposes basin deposits and turbidites in its upper plate (LU–2) onto mostly slope deposits (LU–1) of the Havallah sequence in its lower plate (Murchey, 1990; Doebrich, 1995). It is not possible from data currently available (1999) to make a reliable estimate on amount of

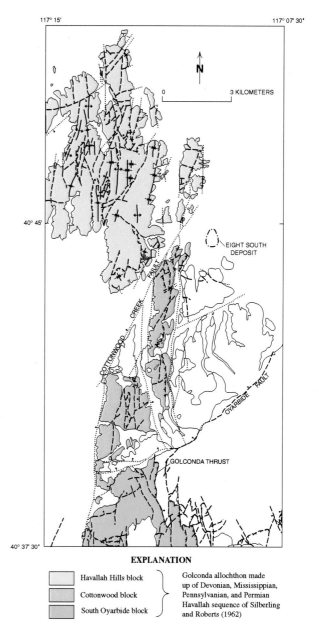

Figure 86 Areal extent of three areas in Golconda allochthon defined as Havallah Hills block, Cottonwood block, and south Oyarbide block from which structural data were gathered for fabric diagrams shown in figure 87.

supracrustal shortening involved in juxtaposition of these two packages of rock along the Willow Creek thrust. As noted by Murchey (1990), Mississippian basalt and chert at the base of LU–2 tectonically rests on Pennsylvanian and Permian sequences of mostly argillite in LU–1.

Approximately 1 km east of the stock at Trenton Canyon, in general area of sec. 19, T. 32 N., R. 43 E., the Willow Creek thrust apparently overrides the leading edge of the Golconda allochthon itself, and the Willow Creek thrust brings into direct tectonic juxtaposition metamorphosed chert and calcareous siltstone of subunit hcs in its upper plate with conglomerate of the Battle Formation in

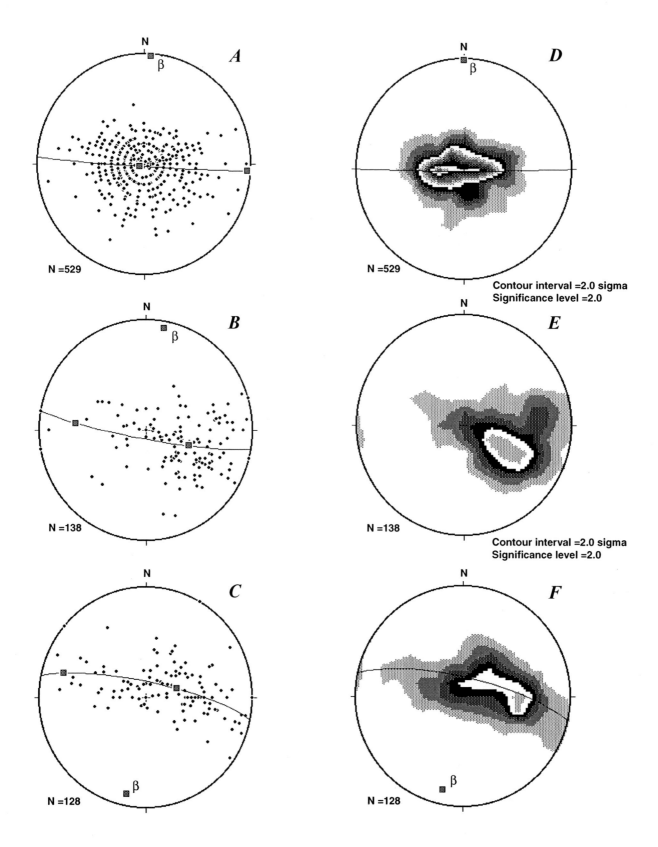

Figure 87 Lower hemisphere, equal-area fabric diagrams showing poles to bedding and contoured poles to bedding in Devonian, Mississippian, Pennsylvanian, and Permian Havallah sequence of Havallah Hills block (*A, D*), Cottonwood block (*B, E*), and south Oyarbide block (*C, F*). Contouring procedures same as figure 83.

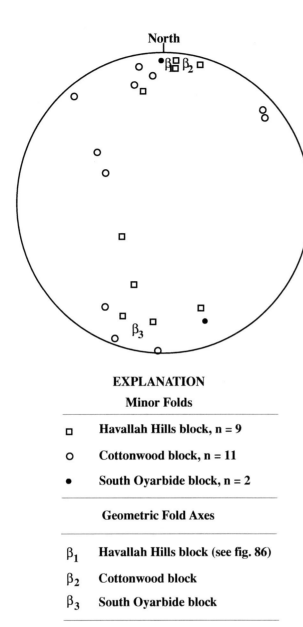

EXPLANATION

Minor Folds

□	**Havallah Hills block, n = 9**
○	**Cottonwood block, n = 11**
●	**South Oyarbide block, n = 2**

Geometric Fold Axes

β_1	**Havallah Hills block (see fig. 86)**
β_2	**Cottonwood block**
β_3	**South Oyarbide block**

Figure 88 Synoptic fabric diagram showing attitudes of minor fold axes in Havallah Hills block (squares), Cottonwood block (crosses), and south Oyarbide block (filled circles) of Devonian, Mississippian, Pennsylvanian, and Permian Havallah sequence, together with geometric fold axes (β_{1-3}, from fig. 87) determined respectively from each block.

its lower plate (pl. 2). Thus, the sole of the Golconda allochthon in this area locally can be considered to be the Willow Creek thrust. The gradual northward tectonic thinning over an approximately 10–km strike length of LU–1 in the lower plate of the Willow Creek thrust is well shown by overall map patterns of these rocks to the south (Doebrich, 1992, 1995). From the general area of the abandoned Oyarbide ranch, the Willow Creek thrust can be traced just to the west of the main drainage of north-south-trending Trout Creek fairly continuously along almost the entire length of the Cottonwood block of the Golconda allochthon. The footwall of the thrust here includes mostly

argillite of subunit ha (pl. 2). The lowermost subunit in the hanging wall of the thrust is almost everywhere Pennsylvanian and (or) Permian basalt, subunit hba, as described above.

Some high-angle fractures and faults in the Golconda allochthon also are related apparently to displacements during emplacement of the allochthon. Some of these fractures show development along their traces of prominent slickensides whose centers of gravity plunge, in the two areas examined in detail, at shallow angles to due east—the inferred trend of tectonic transport of the Golconda allochthon (fig. 89). No difference in orientation of slickensides with vertical position in the overall tectonic stacking pattern appears to be present within the allochthon. Data obtained from an area in tectonostratigraphic plate A and close to the trace of the Willow Creek thrust (fig. 89*A*), are similar in orientation to those obtained from tectonostratigraphic plate E (fig. 89*B*), which is much higher in the stacking pattern.

The apparent difference in the thickness of the Golconda allochthon below the Buffalo thrust at the base of tectonostratigraphic plate E suggests that there may have been some high-angle normal faulting in the Golconda allochthon prior to emplacement of the Buffalo thrust (pl. 2). The Golconda allochthon below plate E north of the Cottonwood Creek fault is much thicker than that in the Cottonwood block, which is south of the Cottonwood Creek fault. Thus, initial displacements along the general buried trace of the Cottonwood Creek fault may have been initiated sometime during latest Paleozoic time. The Golconda allochthon is much thicker in the Antler Peak 7–1/2 minute quadrangle part of the mining district than it is in the North Peak and Valmy quadrangles (Doebrich, 1995).

MESOZOIC DEFORMATION

No direct evidence is present in the area that a particular set of structures is associated unequivocally with deformation during the Mesozoic. Mesozoic sedimentary rocks do not crop put in the area. However, evidence is present elsewhere in north-central Nevada that northwest-trending shattered antiformal hingelines of a regional scale may have controlled emplacement of some Cretaceous and Tertiary magmas and their associated ore deposits of various types (Madrid and Bagby, 1986; Madrid, 1987; Madrid amd Roberts, 1991). As pointed out by Madrid and Roberts (1991), these hingelines probably developed sometime during a Jurassic compressional event because the oldest igneous rocks that cut the orientation of folds in their Cortez-Shoshone fold belt, which includes the Battle Mountain Mining District, are the Late Jurassic coarse-grained syenite and monzogranite of Buffalo Mountain

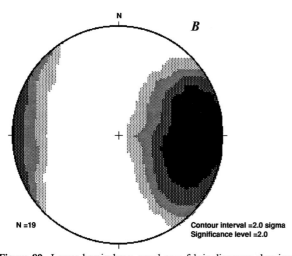

Figure 89 Lower hemisphere, equal area fabric diagrams showing contoured attitudes of slickensides formed along high-angle fractures in LU–2, subunit hpls, of Devonian, Mississippian, Pennsylvanian, and Permian Havallah sequence in Havallah Hills. *A*, Locality 90TT184-185 in tectonostratigraphic plate A near north end of Havallah Hills; *B*, locality 90TT295 in tectonostratigraphic plate E near southeast corner of Havallah Hills. Contouring procedures same as figure 83.

(fig. 1), and the youngest rocks involved in the fold belt are Late Triassic in age. Emplacement of the northwest-elongate Late Cretaceous pluton at Trenton Canyon may have been controlled, in part, by a zone of weakness that was inherited from an inferred Jurassic compressional event (Doebrich and others, 1995). Nonetheless, there is no suggestion in attitudes of bedding in the general area of the pluton for significant warping or forceful emplacement by the pluton at Trenton Canyon. In addition, a number of northwest-striking faults, some mineralized, and some northwest-trending drainages, which are shown on the geologic map without mapped faults along them, are present in that part of the hanging wall of the Oyarbide fault that is underlain by mostly quartzarenite of the Valmy Formation (pl. 2). These northwest-trending features may be related to Mesozoic tectonism. In addition, the range

front in the Snow Gulch quadrangle shows a linear northwest trend, as does the alignment of small Tertiary stocks in the general area of upper Paiute Canyon, approximately 1.5 km to the southwest of the range front (pl. 3; Ivosevic and Theodore, 1995; 1996; Doebrich and others, 1995). However, Late Cretaceous stocks, dikes, and quartz stockworks associated with the Late Cretaceous Buckingham stockwork Mo system in the east-central part of the mining district are strongly aligned in an east-west direction (Theodore and others, 1992). In addition, some northwest-striking faults in the general area of the Buffalo Valley Au mine, approximately 5 km south of the study area, are filled by northwest-striking, compositionally extended 31- to 36-Ma dikes that are associated with Au mineralization there (Seedorff and others, 1990; Doebrich, 1995; Doebrich and others, 1995).

Relations between Cretaceous magmatism and the highly varied orientation of associated structures suggest that Mesozoic-age structures are not confined to any particular orientation in the Battle Mountain Mining District. The fact that some dikes mapped as Cretaceous or Tertiary are obviously emplaced along north-striking, high-angle faults in the mining district in turn suggests that some faults may be Mesozoic in age (pls. 1, 2). Although it is possible that these dikes are Cretaceous, it is more likely that they are Tertiary—the dikes could not be dated radiometrically because of their alteration, and they are designated as being either Cretaceous or Tertiary. Nonetheless, the predominant structural fabric overall the mining district has a north-trend, including many north-striking mineralized faults in the Copper Canyon area of the mining district (Theodore and Blake, 1975; Myers, 1994; Doebrich and others, 1995). This north-south orientation undoubtedly must reflect mostly middle Tertiary extension, and it must also mirror a pre-existing structural grain inherited from both the Roberts Mountains and Golconda allochthons. In addition, throughout the mining district an even earlier and more fundamental north-south control than that exerted by the two allochthons may have been operative—that control is the north-trending feature inferred to be present at depth in this area between oceanic and Proterozoic cratonic crust (Bloomstein and others, 1993).

TERTIARY DEFORMATION

Deformation during the Tertiary, entirely of an extensional nature, has had a profound impact on the geology, geomorphology, and location of ore deposits in the quadrangles. Perhaps the most striking example of this is the Virgin fault which crops out in the Copper Canyon area of the Battle Mountain Mining District (Roberts, 1964; Roberts and Arnold, 1965; Theodore and Blake, 1978;

Myers, 1994; Doebrich and others, 1995). Many north-striking faults in the Valmy, North Peak, and Snow Gulch quadrangles are mineralized substantially by Au–bearing fluids, including the Wayne zone at the Lone Tree Mine (see subsection below entitled "Lone Tree Gold Deposit"), and the Red Rock and East Hill/UNR deposits, which are located approximately 1.5 km southwest of the Eight South deposit (see subsection below entitled "Geology of the Marigold Mine Area"). In addition, somewhat smaller concentrations of Au are known at the Five North and the Eight North deposits, both of which also are associated with north-striking mineralized faults (pl. 1). A number of other north- and approximately north-striking faults as well as major fracture zones elsewhere in the quadrangles also are altered and some mineralized by Au–bearing fluids. Alteration along some of these faults is described in the subsection below entitled "Geology of Selected Mineralized Areas in the Valmy and North Peak Quadrangles." Most north-striking mineralized faults are undoubtedly Tertiary in age, but more precise ages of mineralization along them have not been determined up to 1999. Dilles and Gans (1995) suggest that north-south striking, closely spaced normal faults across a broad region of Nevada—including the general area of the mining district—showed movements from about 30 Ma to about 15 Ma. Presumably, movement along many north-striking normal faults in the mining district occurred initially somewhat earlier than this in response to east-west extension roughly during late Eocene or early Oligocene, more or less contemporaneous with the well-documented emplacement of mineralizing dikes and small stocks of that age in the Copper Canyon area (Theodore and others, 1973; Myers, 1994; Doebrich and others, 1995). An orientation of the regional least principal stress along an east-west trend probably occurred some time during the late Eocene and early Oligocene.

The Battle Mountain Mining District includes three broad zones or belts where northwest-trending structures—including faults and some fold hinges—and Cretaceous and Tertiary intrusive rocks are concentrated, and five additional zones where north-south structures and Tertiary elongated plutons and dikes are also concentrated (Doebrich and others, 1995). Most major loci of Tertiary metallization in the mining district are present in the broad areas where zones with these two orientations intersect. Northwest-trending zones in places (for example, Trenton Canyon) are probably pre-Late Cretaceous in age, and elsewhere (for example, Buckingham) Late Cretaceous stocks are aligned east-west. Furthermore, the northwest-trending zones in the mining district may coincide spatially with shattered hingelines associated with a regional Jurassic compressional event (Madrid, 1987; Madrid and Roberts, 1991). However, the district-scale antiformal hingeline that Roberts (1964) shows through the mining district

(based on bedding orientations of rocks of the Pennsylvanian and Permian Antler sequence) has a more northerly trend than the northwest-trending zones that recent mapping of the mining district has delineated—approximately N. 20° W. versus approximately N. 45° W. The northwest-trending zones are not as well defined, however, as the north-south zones. The dominance of Tertiary faults and dikes with north-south strikes in the mining district probably results partly from these structures having inherited their orientations from the strong north-south structural grain throughout the lower Paleozoic Roberts Mountains and upper Paleozoic Golconda allochthons (Evans and Theodore, 1978; Theodore, 1991a, 1991b; Doebrich, 1992, 1995). Certainly, geologic evidence is present that some north-south faults predate emplacement of late Eocene and early Oligocene felsic magmatism in the Copper Canyon area (Theodore and Blake, 1975), and, in the Willow Creek area of the district, evidence exists that north-south faults predate outflow facies cooling units of the 34–Ma Oligocene Caetano Tuff (Doebrich, 1995). North-west striking dikes at the Buffalo Valley Gold Mine have ages that range from about 36 Ma to 31 Ma (Doebrich and others, 1995). In addition, many localities are present in the mining district where north-south-striking dikes, mostly Oligocene granodiorite porphyry, and faults cut late Eocene and early Oligocene intrusions that crop out in northwesterly trending zones and that are an integral component of the northwest-elongated plutons. One of these places is in the Bluff area, near the south edge of the Snow Gulch quadrangle (pl. 3), where late Eocene and early Oligocene porphyritic monzogranite is cut by north-south striking Oligocene granodiorite dikes (Theodore and others, 1992; Theodore, 1994; Ivosevic and Theodore, 1995; 1996). Moreover, both sets of north- and northwest-trending structural zones and belts of elongated plutons and dikes throughout the mining district may be considered as being active over a relatively short time span in the Tertiary from some time prior to 40 Ma to approximately 31 Ma.

Therefore, significant east-west extension and its accompanying initial breakup of the crust in this part Great Basin during the Tertiary may be considered to have begun at some time prior to 40 Ma on the basis of geologic relations in the mining district. If the least principal stress were oriented roughly east-west at approximately 40 Ma, then emplacement of felsic magmas of about this age along northwest-trending zones would have to be envisioned as being controlled entirely by pre-Late Cretaceous zones of weakness that are possibly associated with shattered hingelines of Jurassic age (Madrid, 1987; Madrid and Roberts, 1991). In other words, the northwest-trending structural zones in the mining district could be envisioned as having been reactivated tectonically and magmatically during the time when the bulk of the extension of the mining district was being accomodated by normal offsets along the north-south striking zones. A necessary corollary of this hypothesis would then be that there also should be present some components of lateral offsets along the northwest-trending zones.

Orientation of regional stress to accommodate synmineral faulting and igneous activity simultaneously along both these structural trends that were mostly active during the same period of time in the Tertiary may have involved a regional least principal stress regime that fluctuated between an east trend and a N. 45° E. trend approximately between 40 Ma and 31 Ma. At the time(s) that the least principal stress was oriented roughly N. 65° E., potential components of lateral offset along both north-south-striking and northwest-striking high-angle faults in the mining district should be maximized. However, data to verify such lateral offsets are extremely difficult to obtain because of the absence of piercing points along faults.

In addition, local geologic evidence is present generally east of the Elder Creek porphyry Cu system that the least principal stress sometime after 38 Ma may have been oriented slightly south of due east (pl. 3). In this area, a number of north-northeast-striking normal faults—they have strikes of approximately N. 5–10° E.—locally cut some of the late Eocene and early Oligocene plutons and their surrounding recrystallized rocks (Theodore, 1994).

The regional least principal stress apparently was oriented along a N. 65–70° E. trend during the 17– to 14–Ma time interval when bimodal volcanism along the Northern Nevada Rift, which trends N. 20–25° W., occurred (Zoback and others, 1994). As these authors point out, the rift is now known to be at least 500 km in length based on outcrops of volcanic rocks and basaltic dike swarms, and regional aeromagnetic surveys.

Finally, clockwise rotation of the regional least principal stress direction to an orientation of about S. 65° E. can be called upon to yield the Oyarbide fault, the preeminent northeast-striking post-mineral fault in the Battle Mountain Mining District. This fault parallels many other faults in the region related to development of the prominent Midas geomorphic trough which cuts and postdates the northern Nevada rift north of the mining district. Additional post-rift faults having roughly the same strike as the Midas trough (approximately N. 65° E.) are widespread throughout a 240–km-wide segment of the rift that extends from the Midas trough on the north to the Cortez fault on the south. The Battle Mountain Mining District is located approximately in the center of this segment of the crust showing southeast-northwest extension sometime after development of the 17– to 14–Ma rift. Eventually, the regional least principal stress direction again must have assumed an approximate east trend in the mining district to account for roughly north-striking range faults, primarily along the west side of the mining district, that cut unconsolidated gravels.

Some additional major faults in the immediate region of the Battle Mountain Mining District have strikes roughly similar to the Midas trough and are probably Miocene in age also. These faults include some with a slightly more northerly strike, including a presumed strand of the Buffalo Mountain fault, the West Poplar fault, that is present below Tertiary and Quaternary gravels in the general area of Lone Tree Hill (pls. 1, 2). Both of these faults have some components of lateral offsets as does the similarly oriented Cottonwood Creek fault. A small number of northwest- and northeast-striking faults also are mineralized in the area, some to concentrations as much as 30 g Au/t (H. Duerr, oral commun., 1993). However, presence of apparently young, mineralized northeast-striking faults here is not that surprising in light of significant Au mineralization, as much as 1 million oz, present in the Mule Canyon deposit, which is hosted by mineralized structures in Miocene basalt along the northern Nevada rift itself (Caldwell and others, 1992).

Latest movements along the Oyarbide fault, which dips approximately 60° NW (fig. 90) are inferred to have obliquely offset Tertiary gravels now exposed in the hanging wall of the fault. These gravels are interbedded with Oligocene olivine augite basaltic andesite that has an approximate 31–Ma age (see section above entitled "Potassium-argon Chronology of Cretaceous and Cenozoic Igneous Activity, Hydrothermal Alteration, and Mineralization"), and a large part of the gravels themselves also must be approximately the same age. In addition, presence of a small patch of Tertiary gravel perched on top of a ridge at a 7,940–ft elevation in the footwall of the fault provides a surface whose dip-slip component of offset along the Oyarbide fault is approximately 700 m when compared to the present trace of the Tertiary erosion surface in the hanging wall of the fault (pl. 2). The projected leading edge of rocks in the Golconda allochthon in the hanging wall of the Oyarbide fault shows an approximate right lateral separation of approximately 1 km when compared to location of these rocks in the footwall of the Oyarbide fault. This amount of horizontal offset cannot be accommodated entirely by a stepping to the right along the Oyarbide fault by the west-dipping sole of the Golconda allochthon.

Although latest displacements along the Oyarbide fault certainly appear to be later than Tertiary gravels interbedded with some basalt, many strike-and-dip attitudes of Oligocene or Miocene calc-alkaline rhyolite tuff in the hanging wall of the Oyarbide fault (maximum 10° dip) reveal that much of the calc-alkaline rhyolite tuff dips southeast toward the projected plane of the fault at depth. These relations suggest that attitudes of rocks older than calc-alkaline rhyolite tuff on either side of the fault should

Figure 90 Photograph of Oyarbide fault viewed to northeast across Cottonwood Creek.

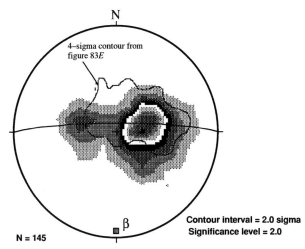

N

4–sigma contour from figure 83*E*

N = 145

β

Contour interval = 2.0 sigma
Significance level = 2.0

Figure 91 Fabric diagram showing contoured pole-to-bedding data of Ordovician Valmy Formation in hanging wall block of Oyarbide fault rotated 10° counterclockwise about a horizontal axis that trends N. 60 E., and 4σ contour of figure 83*B* showing pole-to-bedding data of the Valmy Formation in footwall block of the Oyarbide fault without any superposed rotation. Contouring procedures same as figure 83.

be examined in order to eliminate possibility of displacements along the Oyarbide fault that might predate the tuff. Rocks in the hanging wall of the Oyarbide fault appear to have been rotated a maximum 10° clockwise about a horizontal axis when viewed to N. 60 E. along the strike of the fault. Comparison of contoured pole-to-bedding attitudes of rocks of the Valmy Formation rotated counterclockwise 10° in the hanging wall of the Oyarbide fault with the 4σ contour of bedding attitudes in the footwall of the Oyarbide fault reveals a significant overlap of statistically significant concentrations of data points on either side of the fault (fig. 91). Therefore, no significant displacement probably occurred along the present trace of the Oyarbide fault prior to deposition of rhyolite tuff.

The west side of the range front in the North Peak quadrangle partly is bounded by a pair of parallel range-front faults, the Buffalo Valley fault and the West range-front fault, each of which has visible fault scarps along its trace generally south of the study area (Roberts, 1964; Doebrich, 1992, 1995). The Buffalo Valley fault, the westernmost of the two faults, is well exposed at the Buffalo Valley Au mine, approximately 2.5 km south of the study area, where it cuts Quaternary fanglomerate deposits along an unmineralized sharply defined planar surface. The Buffalo Valley fault can be traced by linear topographic depressions only approximately 1.5 km into the North Peak quadrangle near the southwest corner of the quadrangle (pl. 2). The West range-front fault is the more continuous of the two faults and has been traced approximately 7 km north from the south edge of the study area to a point where it appears to terminate against the projected trace of the northeast-striking Cottonwood Creek fault. The West range-front fault is now (1999) well exposed in the workings of the North Peak Au deposit. These faults that bound the mountain range on the west are the faults in the area that show the youngest displacements. Another, fault that appears to be quite young is the Elder Creek fault, an approximately N. 10° E.–striking normal fault, whose west

block apparently has been downdropped and which appears to offset young gravel deposits (pl. 3).

Additional discussion concerning relations between Tertiary faulting and mineralized rocks are included in the subsection below entitled "Lone Tree Gold Deposit."

ECONOMIC GEOLOGY OF THE VALMY, NORTH PEAK, AND SNOW GULCH QUADRANGLES

History of mining activity in the Battle Mountain mining district

Historically, the Battle Mountain Mining District has been one of the largest producers of Cu in Nevada—mining in the district spans a period of more than 130 years. Many economically important metal deposits of various types were discovered in the mining district during the 1860s and a significant number of them brought into production in the succeeding decades. The Copper Canyon Cu–Pb–Zn underground mine in the southern part of the mining district—one of the significant early sites for the production of metal—was operated sporadically between

1917 and 1955 (Roberts and Arnold, 1965) (fig. 92). How-ever, the first large-scale attempt by open-pit methods to mine base and precious metals, mostly Cu, did not begin until 1967 in the Copper Canyon and Copper Basin areas. Both of these large-scale mining operations were centered on porphyry-type mineralized rock that has widely divergent ages: 39 Ma at Copper Canyon (Theodore and others, 1973), and 86 Ma at Copper Basin (Theodore and McKee, 1983; McKee, 1992). The first large orebody mined by open-pit methods in the Copper Canyon area was the the East Orebody (Theodore and Blake, 1975), also known as the East Copper Pit (fig. 92; see also, Doebrich and others, 1995).

Exploration activity in the Copper Basin part of the mining district, however, substantially predates these activities at Copper Canyon as it apparently began during the earliest stages of prospecting in the 1860s, although exploration and development there were not formalized until 1897 when Glasgow Western Exploration Co., Ltd. consolidated the surrounding land holdings (Blake, 1992). In addition, mineralized rocks at Copper Basin include several secondarily enriched Cu ore bodies related to the Buckingham stockwork Mo deposit, which was not fully delineated until many years later in the 1980s (Blake, 1992; Loucks and Johnson, 1992).

In each of these centers of widespread mineralized rocks, at Copper Canyon and at Copper Basin, porphyry-type mineralized rock is surrounded in areas measuring as much as 20 km^2 by a halo of polymetallic veins whose metal contents systematically vary laterally with distance from the porphyry centers (Roberts and Arnold, 1965; Theodore and others, 1986; Kotlyar and others, 1998). Especially well-developed at Copper Canyon, a Cu–Au–Ag zone in the central part of the porphyry center is mantled by a Au–Ag zone, which is, in turn, surrounded by a Pb–Zn–Ag zone. The Cu–Au–Ag zone at Copper Canyon subsequently was determined to host the Fortitude Au skarn (Wotruba and others, 1986; Myers, 1994), which was discovered during the winter of 1980–1981 and ceased production in early 1993 (Doebrich and others, 1995). The Fortitude deposit also is mantled by elevated abundances of Hg which form a prominent hook-shaped pattern of anomalies around the northern part of the deposit (Kotlyar and others, 1995). In addition, deposit-scale metal zonation of Cu, Au, Ag, Pb, and Zn is well developed at Fortitude (Kotlyar and others, 1998). It is anticipated that the Copper Canyon area of the mining district will eventually achieve a global or geologic resource of approximately 374 t Au (12 million oz Au) when all near-surface ore bodies have been delineated by the early 2,000s (A.M. Campo, oral commun., 1998). Of this amount of metal, approximately 65.6 t Au (2.1 million oz Au) is credited to the Fortitude Au deposit. In middle 1998, mining operations were suspended in the Copper Canyon area until requisite

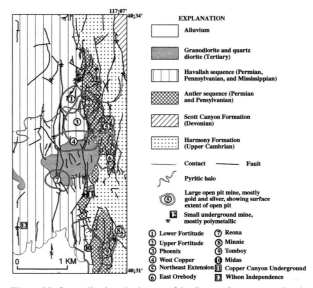

Figure 92 Generalized geologic map of the Copper Canyon area, Battle Mountain Mining District, Nev., showing distribution of various lithologic rock types without faults and location of major base- and precious-metal deposits. Coordinates are from Battle Mountain Gold Co. topographic grid. Modified from Doebrich and others (1995).

permits were obtained for the processing of additional reserves.

Gold-bearing rocks also were recognized during early stages of exploration in the northern part of the Battle Mountain Mining District (Roberts and Arnold, 1965). The Au–bearing rocks at the historic workings of the small Old Marigold Mine (fig. 93), approximately 1.6 km southwest of the Eight South deposit, are extremely important from an exploration standpoint, because of their impact upon subsequent discoveries that are currently (1999) in production nearby (fig. 94). The mineralized rocks at the Old Marigold Mine belong mostly to the Permian Edna Mountain Formation and provided one of the cornerstones upon which Ralph J. Roberts based his exploration concepts to site three drill holes for VEK/Andrus Associates on the pediment near the present open pit at the Eight South deposit (R.J. Roberts, oral commun., 1987; Graney and Wallace, 1988; Graney and McGibbon, 1991; see section below entitled "Geology of the Marigold Mine Area"). These concepts and presence of Au–bearing rock in one of the holes drilled by VEK/Andrus Associates eventually focused subsequent exploration efforts in nearby pediment areas that eventually culminated in discovery of the Eight South deposit. Although production records are somewhat incomplete for the Old Marigold Mine during the period when it was first placed into operation in the late 1930s, several thousand tons of Au ore that were shipped at that time apparently contained roughly $7.00 per ton Au (Roberts and Arnold, 1965). To provide some notion as to what such a value might reflect in terms of average content of Au in the deposit, Britain and South Africa had reinstated the price of Au between 1925 and 1933 to an internation-

Figure 93 Photograph showing historic workings at Old Marigold Mine, a small Au deposit in northern part of North Peak quadrangle. Note vehicle in central part of bench for scale.

ally accepted value of £4.248 per troy ounce Au, whereas in January 1934 the U.S. Government subsequently fixed the price at $35.00 per troy ounce Au (Shawe, 1988). Thus, ore mined from the Old Marigold Mine during the late 1930s probably had a grade of about 6.25 g Au/t (0.2 troy oz Au/ton). During 1981 and 1982, approximately 2,900 tonnes of rock, which contained an average content of 3.44 g Au/t (0.11 oz Au/ton), and which presumably belonged to debris flows in the lower parts of the Permian Edna Mountain Formation, were mined from the small open cut at the Old Marigold Mine and placed onto a heap-leach pad (Larie K. Richardson, oral commun., 1984). Approximately 79 percent of the contained Au subsequently was recovered from this pad in 1983.

Although drill holes sited by Roberts in 1985 did not penetrate the Eight South orebody itself, one of the drill holes encountered an intercept of 0.53 g Au/t (0.017 oz Au/ton) at a depth of 220 to 233 m (Graney and McGibbon, 1991). This latter drill hole subsequently was determined to have penetrated part of the uneconomic halo of mineralized rocks that surround the Eight South deposit (fig. 94). As described by Graney and McGibbon (1991), the alluvium-covered Eight South deposit included approximately 12.8 tonnes Au (410,000 troy oz Au) in millable ore and leach ore before startup of mining in 1989, but eventual production from the Marigold group of deposits apparently will exceed 28.1 tonnes Au (900,000 troy oz Au) (table 21).

Currently (1999), most productive areas for Au in the northern part of the Battle Mountain Mining District are from deposits in the general area of the Eight South deposit, near the north boundary of the North Peak quadrangle, as well as from the Lone Tree deposits (fig. 94). However, additional Au deposits are present in the central and southern parts of the North Peak quadrangle, including the North Peak, Valmy, and Trenton Canyon deposits, which are described in the section below entitled "Geol-

ogy, Mineralization, and Exploration History of the Trenton Canyon Project." Initial mining of ore from the North Peak deposit began in late 1996 with its placement onto heap-leach pads near the range front (pl. 2).

The most northern of the three quadrangles, the Valmy quadrangle, includes the Lone Tree and the Stonehouse Au deposits—the two major mineral deposits known to date (1999) in the quadrangle (fig. 94). However, these two deposits should be considered as belonging to one large, contiguous metallized system. In addition, drilling of the Stonehouse deposit subsequent to its discovery revealed that it is made up of two mineralized zones: the Wayne zone, which extends from the north, and the Chaotic zone (B.L. Braginton, written commun., 1996). Prior to onset of large-scale mining activities in 1990, the deposits were reported to contain an aggregate geologic resource of approximately 103 tonnes (t) Au (3.3 million troy oz Au), including as much as 65.6 t Au (2.1 million troy oz Au) that had been delineated initially in the Lone Tree deposit (The Northern Miner, Nov. 13, 1989, p. 13; Offering Circular, Santa Fe Pacific Corporation, July 31, 1990, p. 22; Santa Fe Gold Corporation, Lone Tree Mine Dedication Day Factsheet, May 20, 1992; Santa Fe Pacific Corporation, Press Release, September 2, 1992). In all, these Au reserves at Lone Tree appeared at one time to consist of 8.2 million t oxide ore at a grade of 3.1 g Au/t (0.1 troy oz Au/ton) and 12.7 million t sulfide ore at a grade of 2.7 g Au/t (0.085 troy oz Au/ton) (Bloomstein and others, 1992). Drilling in 1993, however, more than replaced the amount of Au that was mined at Lone Tree during the last half of 1993—proven and probable contained Au in the deposit subsequently was determined to amount to 125 t Au (4 million troy oz Au) as of December 31, 1993 (Skillings Mining Review, January 22, 1994, p. 11). A geologic resource of approximately 18.5 t Au (577,000 troy oz Au) was blocked out initially in the Stonehouse deposit prior to subsequent step out drilling (table 21). The Stonehouse deposit, which was discovered in 1988, somewhat earlier than the Lone Tree deposit, at one time was estimated to contain a minable resource of approximately 8.7 t Au (292,000 troy oz Au) in 3.27 million t of ore to a depth of 207 m—as evaluated in April 1992 (D.H. McGibbon, oral commun., 1992). In 1995, the deposits at Lone Tree—including the four mineralized zones that make up the entire system (see the section below entitled "Lone Tree Gold Deposit")—were estimated to contain proven and probable reserves of approximately 55 million t at a grade of 2.41 g Au/t or (B.L. Braginton, written commun., 1996).

The deposits at Lone Tree apparently are aligned along a north-south trend with the Chimney Creek (Osterberg and Guilbert, 1991) and Rabbit Creek (Bloomstein and others, 1991; Parratt and others, 1999) Au deposits that are located about 70 km to the north at

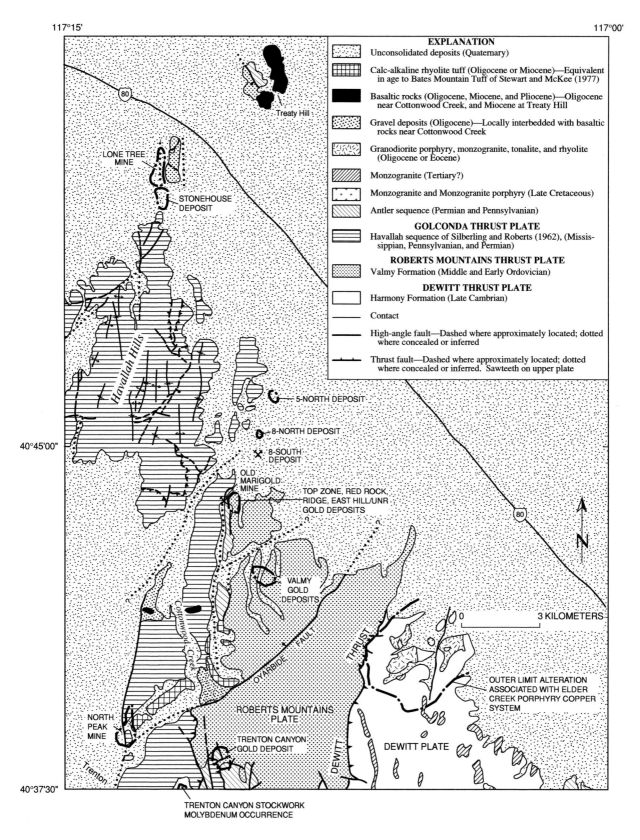

Figure 94 Sketch map showing mineralized and altered areas in the Valmy, North Peak, and Snow Gulch quadrangles recognized up to late 1995. Geology modified from plates 1–3.

Table 21—Gold grades and tonnages of precious-metal systems at Lone Tree, Marigold, and Trenton Canyon/Valmy in the Valmy and North Peak quadrangles, Humboldt and Lander Counties, Nevada.

	Gold (in g/t)	Tonnes (in thousands)
Lone Tree[1]	2.22	37,835.0
Marigold[2]	1.2	31,495.
Trenton Canyon/Valmy[3]	.88	16,763.

[1] Reserves at $350/oz Au as of December 31, 1997.

[2] Includes 1988–1994 production from Eight South deposit, production plus December 31, 1997 reserves at Top Zone, Red Rock, East Hill/UNR, and Ridge deposits, and December 31, 1997 reserves at Resort and Old Marigold/Ponds deposits.

[3] Reserves at $350/oz Au as of December 31, 1997 composited from three widely separate deposits, including the Trenton Canyon, Valmy, and North Peak deposits (see text).

the northeast end of the Getchell (Potosi) Mining District (fig. 1). After purchase of the Chimney Creek deposit by Santa Fe Pacific Mining Company in 1993, the Chimney Creek and Rabbit Creek deposits were consolidated into one operation that was then referred to as the Twin Creeks deposits, which were announced as containing 266 t contained Au (8.5 million troy oz Au) as of December 31, 1993 (Skillings Mining Review, January 22, 1994, p. 11). Santa Fe then established, by an exchange of stock on September 30, 1994, a wholly independent company, Santa Fe Pacific Gold Co., whose major assets included the Lone Tree and Twin Creeks deposits. In late 1996, Santa Fe Pacific Gold Co. was the subject of separate takeover bids from Homestake Mining and from Newmont Mining. In May, 1997, Santa Fe Pacific Gold Co. was purchased by Newmont Mining in a stock-for-stock exchange.

The fundamental north-south controlling structure that contributed significantly to concentration of the Twin Creeks, Lone Tree, and other deposits in the Battle Mountain Mining District appears to be a major, deep-seated structural discontinuity that has been referred to as a "suture" by Bloomstein and others (1991). This "suture" may be a continental-scale rift that affected Proterozoic craton and subsequently acted as a zone of weakness that may have channeled, in the general area of the mining district, magmas and their evolving fluids during the Mesozoic and Tertiary. In addition, the Valmy quadrangle near its southern boundary also includes the Five Northeast, Five North, and Eight North Au deposits (fig. 94). All of these relatively recently discovered Au deposits are buried under varying thicknesses of Quaternary gravel and alluvium,

and, for some of the deposits, Tertiary gravel and alluvium. This apparently deep Au–controlling structure or zone, which must be at least several km wide where displayed at the surface by large numbers of narrow mineralized faults, probably extends much farther to the south than the general area of the Old Marigold Mine—the structure probably is one of the fundamental geologic features that is responsible for the extent of Au–enriched pluton-related systems as far south as the McCoy Mining District in the northern part of the Fish Creek Mountains (fig. 1).

The North Peak quadrangle includes the Eight South deposit (fig. 94). The Eight South deposit was discovered in late 1985 by Cordex Exploration Company, which at the time was composed of Rayrock Mines Inc., Placer Dome Exploration (U.S.) Ltd., and Lacana Gold Inc. Further exploration and early development of the property was conducted as a joint venture among these companies—conducting business as Roby Exploration Company—and Santa Fe Pacific Mining Company, which in exchange for a 30 percent interest in the project had provided certain additional lands for exploration that were then under its control. As of March 31, 1992, ownership of the Marigold operation reverted entirely to the Roby (Cordex) partners as a part of an exchange of land holdings and interest in the Marigold Joint Venture between Santa Fe Pacific Mining Company and Roby (Rayrock Yellowknife Resources Inc., Press Release, March 31, 1992). Furthermore, as a result of this exchange, Santa Fe Pacific Mining Company obtained control of the Stonehouse deposit and some other tracts of land in the immediate vicinity of Lone Tree Hill. A subsequent press release by Rayrock

Yellowknife Resources Inc. provided details of an additional restructuring of ownership of the Marigold Joint Venture (Skillings' Mining Review, June 20, 1992, v. 81, no. 25, p. 7). As reported, Rayrock Yellowknife Resources Inc. agreed to purchase all of Placer Dome's remaining one-third interest in the Marigold Joint Venture retroactive to April 1, 1992, and thereby increased its share of the joint venture to 66 2/3 percent. The other 33 1/3 percent is held by Homestake Mining Company, which took over Lacana successor, Corona Gold. Rayrock, operating partner since 1996, was purchased by Glamis Gold Ltd. in 1999.

Although most other metal prospects in bedrock areas of the North Peak quadrangle historically were relatively unproductive compared to those in the southern half of the mining district (Roberts and Arnold, 1965; Seedorff, 1991; Theodore and others, 1992), several recent major precious-metal discoveries by various companies have dramatically changed these early preconceptions about metal endowment. In 1995, Santa Fe Pacific Gold Corp. announced that its Trenton Canyon Project contains proven and probable reserves that amount to 13.25 million tonnes (t) (14.6 million tons) ore at a grade of 1.09 g Au/t (0.035 oz Au/ton) (Santa Fe Pacific Gold Corp., Press Release, May 25, 1995). Combined reserve estimates in January, 1996 at the Trenton Canyon Project were revised upwards to 18.21 million t at a grade of 0.91 g Au/t (Santa Fe Pacific Gold Corp., Press Release, January 30, 1996). Initial production of Au is expected to take place in 2000. The deposits that make up the Trenton Canyon Project include the five Trenton Canyon deposits, the Valmy deposits, and the North Peak deposit, all of which are in the North Peak quadrangle (fig. 94; see also, section below entitled "Geology, Mineralization, and Exploration History of the Trenton Canyon Project").

As described above in this section, many ore deposits known in the mining district up until the late 1980s were clustered in its southern and eastern parts, and these ore deposits are closely related spatially and genetically to felsic intrusions (fig. 92). They are pluton related, mostly of the porphyry Cu type or the low fluorine (quartz monzonite) porphyry Mo type, as well as their fringing, genetically affiliated polymetallic veins, precious-metal skarns, and distal-disseminated Ag–Au deposits (Roberts and Arnold, 1965; Theodore and Blake, 1975,1978; Schmidt and others, 1988; Seedorff and others, 1990; Myers and Meinert, 1991; Theodore and others, 1992; Myers, 1994).

The area of the Snow Gulch quadrangle, which makes up the northeast quadrant of the Battle Mountain Mining District, includes the Elder Creek porphyry Cu system and its satellitic polymetallic veins and mineralized faults (pl. 3; fig. 94). In addition, a large number of mineralized plutons and associated skarns and polymetallic veins are also present in the southeast part of the quadrangle, including a weakly developed porphyry Cu–Mo system present near the south-central part of the quadrangle and associated with a N. 45° W alignment of late Eocene or early Oligocene felsic plutons (Ivosevic and Theodore, 1996). In the general area of the Surprise Mine (fig. 2), just south of the southern boundary of the Snow Gulch quadrangle near its southeast corner, Au–Ag mineralized rocks associated with these Tertiary plutons were apparently superposed onto the fringing Au–Ag mineralized rocks associated with the Late Cretaceous Buckingham stockwork system. The quadrangle also includes, near its southeast corner, the Bailey Day Au skarn deposit (Theodore and others, 1991).

Although the Elder Creek porphyry system was originally recognized as a porphyry Cu system in the middle 1960s, subsequent exploration efforts by a number of companies over the last several decades have failed to delineate an economic porphyry-type orebody there (Theodore, 1994, 1996). Nonetheless, some prominent concentrations of mineralized veins and mineralized faults in the area had already attracted significant exploration activity at a time when some of the earliest exploration targets in the mining district were first being evaluated in the late 1800s and early 1900s (Roberts and Arnold, 1965). The presence of secondary Cu minerals and the fact that the Elder Creek area is in the low foothills close to emigrant trails followed since the 1840s to Au fields in central California both contributed to early discovery of underlying mineralized rocks. Many polymetallic veins and mineralized faults near Elder Creek—such as at the Morning Star Mine, Big Pay Mine, Gracie Mine, and the Ridge Mine, pl. 3—show minor amounts of Au in their production records. For example, during the six years for which metal extraction occurred between 1905 and 1938, the Morning Star Mine is reported to have produced approximately 194 oz Au from a total shipment of approximately 209 tonnes ore (Roberts and Arnold, 1965). The average Cu content of the oxidized ore produced from this mine is about 4 weight percent. In addition, this particular mineralized occurrence is close to the inferred center of the Elder Creek porphyry system, whereas the Ridge and the Gracie Mines are located just inside the mapped outer limit of widespread pyritic epigenetic alteration associated with the porphyry system. These latter two mines are reported respectively to have yielded, in all, 469 and 1,584 g Au prior to 1939 during, at most, four years of production. Exploration activities contunued during late 1994 and early 1995 in the general area of the Elder Creek porphyry system. These activities were focused primarily on the pediment developed on the northeast margin of the system. Other minor amounts of Au in the Snow Gulch quadrangle were produced by Battle Mountain Gold Co. in early 1993 from the Au skarn deposit at the Bailey Day Mine (334 kg Au in 137,000 t ore; Doebrich and others,

1995), and sporadic Au placer operations were focused at various times over the past 80 years along some of the drainages that emanate from the northeastern flank of the mining district (Roberts and Arnold, 1965; Theodore and others, 1992).

Exploration and mining activities began to shift back to the southern part of the mining district in the early 1990s—these were again focused in the general area of Copper Canyon. Mining by Battle Mountain Gold Co. at the open-pit Iron Canyon Mine, approximately 1.6 km due east of the Fortitude deposit (fig. 94), as well as mining at the Surprise, Empire, and Labrador Mines—the latter three deposits are in the Copper Basin part of the mining district—continued during 1992, with mill-grade ore trucked from these sites to the ore-reduction facility at Copper Canyon and heap-leach ore placed on a pad near the mouth of Long Canyon. The small open-pit operation at the Iron Canyon Mine, 703,000 tonnes (t) (775,000 tons) of ore at a grade of approximately 3.13 g Au/t (0.1 oz Au/ton) (P.R. Wotruba, oral commun., 1992; Doebrich and others, 1995), was placed on standby during the third quarter of 1993.

In late 1992, Battle Mountain Gold Co. announced the presence at its Reona Project at Copper Canyon (fig. 92) of an additional geologic resource of 25 tonnes (t) Au (800,000 troy oz Au) in three deposits named the Reona, South Canyon, and Sunshine (Anonymous, 1992; Doebrich, 1995; Doebrich and others, 1995). Approximately 11.3 t of the 25 t Au is expected to be converted into heap-leachable minable reserves. Mining at the Reona Project was to continue at a rate of approximately 1.4 t Au (45,000 troy oz Au) per year from early 1994 through 1999. Production from the Reona Project also was expected to reach 0.5 t Au (16,000 troy oz Au) in the fourth quarter of 1994 (The Northern Miner, 1994, v. 80, no.33, p. 1). Finally, Battle Mountain Gold Co. during the last quarter of 1993 began the process of obtaining requisite permits to mine the Phoenix deposit, also located in the Copper Canyon area (fig. 92). The Phoenix deposit is a Au–skarn deposit contiguous with the Fortitude ore body on the north and the West ore body or West Pit on the south. The Phoenix deposit contains approximately 26.1 million t of millable ore at a grade of roughly 1.34 g Au/t (0.043 oz Au/t), and approximately 8 million t heap leach ore at a grade of 0.875 g Au/t (0.028 oz Au/t) (Battle Mountain Gold Co., press release, 1995; see also, Doebrich and others, 1995; Kotlyar and others, 1998). Production from the deposit is expected to last at least 10 years.

Pluton-related mineralized rocks in the mining district also are present in the general areas of Trenton Canyon near the south border of the North Peak quadrangle, west-southwest of Rocky Canyon, and at the Buffalo Valley Au deposit (Kizis and others, 1997)—the latter is considered by Seedorff and others (1990) to reflect a style of porphyry metallization that may be somewhat closer to the paleosurface than mineralized rocks at Copper Canyon described by Theodore and Blake (1975, 1978), Myers and Meinert (1991), and Myers (1994). Fairmile Gold was continuing to explore in the general area of the Buffalo Valley Au deposit in early 1996, and identified two alluvium-covered exploration targets west of the main pit at Buffalo Valley Au (Fairmile Gold, press release, January, 1996; see also, Kizis and others, 1997). A summary of exploration activities at Trenton Canyon is included in the subsection below entitled "Trenton Canyon Stockwork Molybdenum System."

Placer Au was discovered in the Copper Canyon area of the Battle Mountain Mining District in 1909, and intermittent small-scale placer operations were carried on into the early 1940s Roberts and Arnold, 1965; Doebrich and others, 1995). From 1944 to 1955, a dredge was operated on the alluvial fan at the mouth of Copper Canyon (fig. 92), and these dredge operations reportedly produced 3.3 t Au (107,000 oz Au) that averaged 860 fineness (Johnson, 1973; Doebrich and others, 1995). In the late 1980s, this fan was projected again to yield Au at a rate of approximately 0.56 t Au (18,000 oz Au) per year (Battle Mountain Gold Co., 1987, Annual Report to Stockholders, February 5, 1988), and some gravels from the fan eventually were mined during 1990 prior to suspension of placer operations there in 1991. In addition, many drainages that emanate from the northeast-facing flank of the mining district have been explored sporadically over the years for placer Au (Roberts and Arnold, 1965; Theodore and others, 1992; Theodore and Jones, 1992). One of these placers, the B&M Placers at the mouth of Snow Gulch, yielded about $47,000 in relatively coarse Au during 1940 and 1941 from a 10–m-wide channel that extended approximately 700 m along the fan from the mouth of the canyon (Willden, 1963). In 1983, Stellar International Inc. conducted an in-depth examination of the placer–Au potential of Au–bearing gravels near the mouth of Paiute Gulch, another small northeast-trending drainage that emerges from the northeast-facing flank of the mining district (fig. 2; Theodore and others, 1992).

Metal endowment of the Battle Mountain mining district

Many previous reports by a number of authors have addressed various aspects of the extraordinarily varied and copious metal endowment of the Battle Mountain Mining District as resulting from the location of the mining district in a shallow-seated geologic environment at the intersection of regional-scale metallotects of several ages. In addition, presence within the eastern and southern parts of the mining district of many exposed metal-bearing plutons of highly diverse ages contributed significantly to overall metal endowment of the mining district. This magmatism is calc-alkaline and it roughly coincides throughout broad areas of north-central Nevada with large amounts of extension during the Tertiary, although the bulk of the extension in north-central Nevada culminated about the time there was a shift from calc-alkaline to bimodal magmatism (Seedorff, 1991; Steve Ludington, written commun., 1995). Some crustal extension in the Copper Canyon area predates a small, late Eocene to early Oligocene, 39–Ma granodiorite body that was emplaced there astride all of the major tectonic blocks in the mining district (Theodore and Blake, 1975). In addition, several extremely important metallized trends within the mining district seem to characterize the known major ore bodies and their genetically related granitic, *sensu lato*, rocks (Blake and others, 1979; Doebrich, 1995; Doebrich and others, 1995) (fig. 95*A*, *B*). However, some important regional-scale metallotects also impact the mining district, and these include the following: (1) a Late Cretaceous magmatic arc resulting from trench-related magmatism in a continental margin mobile belt (for example, Westra and Keith, 1981), (2) a highly mineralized trend, the northwest-trending Eureka mineral belt of Shawe and Stewart (1976), previously termed the "Eureka-Battle Mountain mineral belt" by Roberts (1966) and also referred to as the "Battle Mountain gold belt " by Madrid and Roberts (1991), and (3) an apparently 5– to 8–km-wide, loosely constrained, north-south mineralized trend, termed the "Rabbit Creek-Marigold mineral belt," that essentially extends through roughly the middle of the Valmy and North Peak quadrangles. In the general area of the Twin Creeks deposits, the north-south Rabbit Creek-Marigold mineral belt is somewhat more tightly defined (Bloomstein and others,1991). If the hypothesis developed below concerning relations between distribution of the clearly porphyry Cu and (or) Mo styles of base- and precious-metal deposits and the enigmatic disseminated–Au deposits in the northwest part of the mining district are generally correct, then the east-west breadth of the mineral belt must be as wide as the entire mining district.

Interpretation of regional gravity data in Nevada, from which effects of shallow-seated responses have been removed by computer-aided geologic-based algorithms, suggests a major pre-Cenozoic crustal boundary correlates spatially with the Eureka-Battle Mountain mineral belt (Jachens and others, 1989). However, Grauch and others (1995) redefine slightly the inferred trace of the Eureka-Battle Mountain mineral belt and relate its geographic position to the edge of a deep Proterozoic terrane.

As pointed out by Barton and others (1988), Mesozoic Cu–mineralized rocks show a strong function with emplacement level in the Western United States, and it is viewed by them as having formed at depths less than four

km in the general area of the Battle Mountain Mining District. The intradistrict metallotects at Battle Mountain include the caprock effect of impervious chert, argillite, and siltstone of the Devonian, Mississippian, Pennsylvanian, and Permian Havallah sequence (Silberling and Roberts, 1962; Roberts and Thomasson, 1964; Stewart and others, 1986; Miller and others, 1992), which structurally overlies some chemically favorable replacement horizons in the Pennsylvanian and Permian Antler sequence (Roberts and Arnold, 1965; Nash and Theodore, 1971; Theodore and Blake, 1975, 1978; Graney and McGibbon, 1991). The latter metallotect was genetically important in localizing ores in the Copper Canyon area, as well as at a number of places in the northern part of the mining district, including the general areas of the Lone Tree and Eight South deposits.

In addition, most ores in the mining district are present in a number of broad areas of intersections between north- and northwest-trending structural zones (Doebrich and others, 1995) (fig. 95). Some concentrations of deposits along northwest trending zones may have been focused along the shattered hingelines of broad anticlines formed during the Mesozoic (Madrid, 1987; Madrid and Roberts, 1992; Doebrich and others, 1995). Recently, Mesozoic hingelines with similar trends also have been reported from as far west as the westernmost pendants in the Sierra Nevada batholith of central California (Sawlan, 1992).

Geology and geochemistry of selected mineralized areas in the Valmy, North Peak, and Snow Gulch quadrangles

In this section of the report, a number of areas in the Valmy, North Peak, and Snow Gulch quadrangles that have been mineralized during the Cretaceous and during the Tertiary are described. These include the Late Cretaceous Trenton Canyon stockwork Mo system, the Tertiary Lone Tree Au deposit, the apparently Tertiary Marigold Mine area, the Tertiary porphyry Cu system at Elder Creek, the relatively widespread deposits that make up the Trenton Canyon Project, and several other minor altered areas that in middle 1998 were considered as untested occurrences of epigenetic alteration. As shown, mineralized areas are quite widespread in the quadrangles, the largest number of the precious-metal mineralized areas are present north of the Oyarbide fault (fig. 94). No known occurrences of syngenetic Ordovician-age mineral deposits are known in the quadrangles, including bedded barite deposits, similar to the many extraordinarily economic barite deposits known elsewhere in the region (Peters and others, 1996). In addition, as described previously, no known occurrences

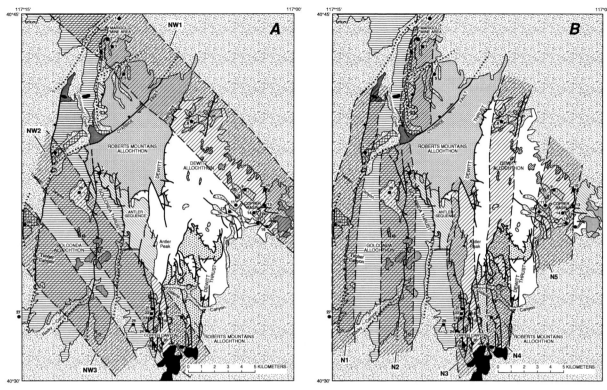

Figure 95 Major structural zones in the Battle Mountain Mining District, Nev., characterized by elongated plutons and granodiorite porphyry dikes, aeromagnetic lineaments, and alignment of mineralized areas. Structural zones modified from Doebrich and others (1995). Geology modified from Roberts (1964). *A*, Northwest-trending zones; *B*, north-south trending zones.

of massive sulfide-type mineralized rocks are present that are associated genetically with Mississippian or Pennsylvanian and (or) Permian basaltic rocks of the Havallah sequence, although minor occurrences of exhalative silica are fairly common in many areas of outcrop of BTT–suite rocks belonging to the Havallah sequence.

TRENTON CANYON STOCKWORK MOLYBDENUM SYSTEM

Exploration activities focused on the Late Cretaceous stockwork Mo system at Trenton Canyon apparently were conducted initially by Utah Mining and Construction Co. during 1966–69 at which time core drilling of three holes was completed in an introductory exploration phase entirely devoted to the porphyry system. In all, up until 1992, approximately 30 diamond and rotary drill holes had been put down in the immediate area of the pluton at Trenton Canyon into different targets delineated by various exploration groups since initial recognition of pluton-related mineralized rocks in the area. Most holes were drilled in those parts of the pluton where it crops out to the south of the North Peak quadrangle. Three short unpublished summary reports prepared by S.A. Taylor (Reno, Nev.) were provided to the U.S. Geological Survey by R.W. Schafer (at that time Regional Manager, BHP Minerals Interna-

tional Inc.), and another St. George Minerals internal report by R.W. Thomssen (written commun., 1992) was provided by Jim McGlassen (at that time Vice President, St. George Minerals). These four reports provide the basis for much of the following parts of this subsection that outlines the results of exploration activities through late 1992.

Three altered and mineralized zones at the surface, all within the pluton at Trenton Canyon, were identified in the mid 1960s by exploration geologists of Utah Mining and Construction Co. as the North, Central, and South zones, and they collectively underlie an area measuring approximately 1,200 m by 650 m (fig. 96). The South zone crops out approximately 0.2 km south of the southern border of the North Peak quadrangle. Each zone is characterized at the surface by moderate to strong white mica-clay alteration with locally superimposed silicification including quartz veinlets. Mineralized rocks are of both a disseminated type and a veinlet type, amounting to as much as 1 to 3 volume percent introduced minerals, and including visible pyrite, molybdenite, chalcopyrite, chrysocolla, malachite, ferrimolybdite, goethite, and jarosite. Geochemical surveys indicate that the North and Central zones of alteration in the pluton are anomalous in Mo, Cu, and As; no geochemical anomalies were discovered in the South zone, however. From interpretation of leached capping, concentration of primary sulfide minerals in the three main zones of alteration is estimated to vary from 1 to 3 volume percent, and no textural evidence is present in the associated gossans that significant enrichment of secondary Cu sulfide minerals may have occurred at depth in the pluton. In addition, variable signs of some mineralized rocks are present in several small areas of altered rock outside of the three zones outlined at the surface.

Two relatively large bodies of breccia adjacent to the main mass of the pluton show some signs of alteration and mineralization (fig. 96). These bodies are interpreted primarily to reflect contact brecciation during emplacement of the pluton during the Late Cretaceous.

Geochemical studies at Trenton Canyon entailed collection of 181 soil and 158 rock chip samples as part of a more extensive geochemical survey conducted previously (Utah Mining and Construction Co., unpub. data, 1969). The soil samples were analyzed only for Cu and Mo, and the rock-chip samples were analyzed for Cu, Mo, Pb, Zn, Mn, and As. In the North zone, anomalous Mo concentrations in soil define a northeast-southwest trend, whereas rocks show as high as 0.101 weight percent Mo contents in granodiorite on the fringes of some of the most intense concentrations of vein quartz. Some high concentrations of As coincide with the highest concentrations of Mo. In the Central zone, Cu in rock has spotty high concentrations that coincide spatially with those rocks including white mica and clay mineral assemblages, and quartz-veined-white mica-clay alteration assemblages. Molybdenum has elevated abundances, to as much as 0.032 weight percent, that also coincide spatially with mapped extents of the latter type of alteration assemblage. Arsenic concentrations in rock vary quite closely with concentrations of Mo. The South zone is not characterized by significant and areally widespread concentrations of any of the metals analyzed, although overall intensity of hydrothermal alteration that has affected the rocks in this zone seems comparable to outcrops in the North zone that contain strong geochemical anomalies. Finally, all three of the zones are characterized by relatively low concentrations of Mn.

Diamond core-drilling during the first stage of exploration efforts to depths of 230 to 350 m in three holes confirmed presence of a pervasive, veinlet-disseminated mineralized system containing molybdenite, pyrite, and minor Cu sulfide minerals apparently associated with porphyry phases of the Late Cretaceous pluton (S.A. Taylor, written commun., 1969; R.W. Schafer, written commun., 1992). At depth in the drill holes, widespread tongues and dikelets of porphyritic igneous phases are present that may be comparable to some of the large-K–feldspar quartz monzonite porphyry emplaced into the core of the Buckingham molybdenum system during its latter stages of evolution (Loucks and Johnson, 1992). Grades of Mo-mineralized rocks at Trenton Canyon, however, are eco-

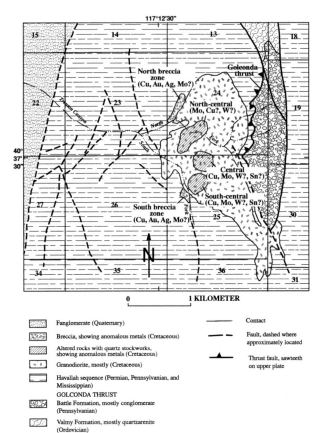

Figure 96 Pluton at Trenton Canyon and its surrounding area including locations of the North, Central, and South alteration zones within pluton. Modified from S.A. Taylor (written commun., 1969).

nomically submarginal to depths of 230 to 350 m vertically below drill collars. The widespread presence of tongues and dikelets of various types of porphyry-textured rock encountered in the drill holes, moreover, suggests that they may be offshoots from a deep central stock associated with the stockwork Mo system, and subsolidus alteration in the drill holes is most strongly concentrated close to porphyry-textured tongues and dikes. Strength of Mo–mineralized rock, furthermore, is generally proportional to intensity of alteration encountered in the drill holes.

The first three holes drilled by Utah Mining and Construction Co. in the North and Central zones of the Trenton Canyon pluton all revealed pervasive but overall economically low concentrations of Mo (S.A. Taylor, written commun., 1969). It should be recalled that unmineralized low-calcium and high-calcium granites are reported to contain approximately 1.3 ppm Mo and 1.0 ppm Mo, respectively (Parker, 1967). Grades in the holes at Trenton Canyon vary in the range 0.002 to 0.08 weight percent Mo, with the highest assayed intercept having 0.31 weight percent Mo along an approximately 1–m-long interval. As a comparison, grades in the Buckingham system, much of which is exposed at the surface, average approximately 0.06 weight percent Mo for a tonnage well in excess of 1,000 million tonnes (Loucks and Johnson, 1992). In addition to molybdenite and pyrite in the Trenton Canyon system, other minerals encountered in the sulfide zone include chalcopyrite, bornite, and very minor chalcocite, as well as traces of sphalerite, hematite, magnetite, rutile, arsenopyrite, and possibly cassiterite. Several intervals assayed as much as 0.2 ounces per ton Ag, although no mineral hosts for the Ag were identified. In the Buckingham system, tetrahedrite contains 3 to 30 weight percent Ag (Loucks and Johnson, 1992). On the fringes of the porphyry dikes in the drill holes at Trenton Canyon and in some of the adjoining equigranular granodiorite, weak-to-strong concentrations of quartz-K–feldspar-calcite veinlets and cavity linings that contain fine-grained to dusty-appearing molybdenite are present together with fine- to coarse-grained pyrite. The wallrocks of the porphyry dikes are altered in their entirety to white mica-clay assemblages, and they include disseminated pyrite, generally 1 to 3 volume percent, and both disseminated molybdenite and molybdenite in veinlets without quartz. The most intensely mineralized rocks are present in the North zone and along the west and north sides of the Central zone (fig. 96).

Molybdenum also is present along narrow breccia zones as much as 0.6 m wide that course through the system at Trenton Canyon (S.A. Taylor, written commun., 1969). Brecciation is present along nearly every contact between porphyry-textured dikes and equigranular granodiorite cut by the drill holes. The matrix of these breccias is filled by quartz, pyrite, and fine-grained molybdenite.

Intensity of Mo introduction corresponds with overall intensity of alteration that has affected the rocks (S.A. Taylor, written commun., 1969). The interior portions of some porphyry dikes have sparse concentrations of quartz-molybdenite-pyrite veinlets, as well 1 to 3 volume percent disseminated pyrite and some pyrite in them that is concentrated in patches. These interior portions of the dikes are altered partially to totally by white mica-clay assemblages. Equigranular granodiorite in some areas relatively far away from porphyry dikes has a gradation from white-mica-bearing assemblages to assemblages dominated by clay minerals. Concentrations of disseminated pyrite and quartz-molybdenite-pyrite veinlets are extremely sparse in clay mineral-altered areas of equigranular granodiorite. Zones of equigranular granodiorite characterized by propylitic alteration assemblages show sparse concentrations of quartz-molybdenite-pyrite veinlets and approximately 0.5 volume percent disseminated pyrite. Finally, in the Trenton Canyon system, an inverse relationship is present between overall concentrations of Cu and Mo. A similar inverse relationship between Cu and Mo is present in the Buckingham stockwork Mo system where increased concentrations of Cu are found, laterally and at depth, on the fringes of the most

intensely Mo–mineralized rocks (Loucks and Johnson, 1992; Theodore and others, 1992).

Subsequent to the initial exploration activities detailed above, the Trenton Canyon stockwork Mo system was explored again by an additional two holes put down by Utah Mining and Construction Co. during a second stage of drilling that lasted approximately two years (D.W. Blake, oral commun., 1992). The deeper of these two holes was approximately 800 m and also tested the deep porphyry target in the general area of the North alteration zone outlined in the stock (fig. 96). Utah Mining and Construction Co. apparently retained an interest in its claims in the general area of the Trenton Canyon system until 1983. Much of the yearly assessment work during this period of time was accomplished by small-scale exploration for turquoise, mostly in SW 1/4 sec. 24, T. 32 N., R. 42 E. (R.W. Thomssen, written commun., 1992).

Following these exploration activities, a shift occurred from a search for pluton-related Cu–Mo targets to a search for precious metal targets. Auritech Joint Venture obtained in 1983 a lease from Santa Fe Pacific Mining Inc. of sec. 25, T. 32 N., R. 42 E., just south of the south border of the North Peak quadrangle in the Trenton Canyon area, and proceeded to outline geochemically some areas in that section that are highly anomalous in Au, As, and Hg (R.W. Thomssen, written commun., 1992). As further pointed out by Thomssen, five rotary percussion drill holes, aggregating approximately 300 m in all, then were put down to test these geochemical anomalies. Only low concentraions of Au were found in each of the five holes drilled.

In 1984, Santa Fe Pacific Mining Inc. began its own exploration activities for precious metals in the area of Trenton Canyon by drilling seven rotary percussion holes in secs. 23 and 25, T. 32 N., R. 42 E. Only one of the seven holes was collared in sec. 23, and two of the holes collared in sec. 25 attempted to test the rocks of the Battle Formation close to the east border of the pluton of Trenton Canyon (R.W. Thomssen, written commun., 1992). Unfortunately, both of these holes failed to penetrate the lower sequences of the Battle Formation, and those rocks of the formation encountered at depth did not contain any economic concentrations of Au; one of the holes, however, did penetrate a 23–m interval that contains 0.5 oz/ton Ag, 0.5 weight percent Pb, and 0.1 weight percent Cu, in addition to a 15–m interval near the surface that contains 0.11 weight percent Mo.

In late 1988, another round of exploration was conducted in the area of the pluton at Trenton Canyon by Billiton Minerals U.S.A., Inc. which included the drilling of 16 reverse circulation rotary holes totaling approximately 2,200 m (R.W. Thomssen, written commun., 1992). Four of these holes were collared in NE 1/4 sec. 25, T. 32 N., and one of the holes was collared in SW 1/4 sec. 9, T.

32 N., R. 43 E., near the southwest corner of sec. 9. Two of the holes in NE 1/4 sec. 25 finally penetrated successfully the entire sequence of the Battle Formation and bottomed in quartzarenite of the Valmy Formation. In one of these two holes, an 8-m interval near a depth of 100 m was found to contain as much as 75 ppb Au and as much as 15.2 ppm Ag; all other reported analyses of the rocks of the Battle Formation penetrated by this hole are less than 10 ppb Au. In the other hole that penetrated the rocks of the Battle Formation entirely, a concentration of 1,140 ppb Au was detected apparently in a narrow intercept in the lower part of the formation.

During the interval 1992 to 1995, Santa Fe Pacific Mining Inc. conducted a relatively widespread exploration program centered approximately 1 km northeast of the pluton at Trenton Canyon that culminated in discovery of the Trenton Canyon Au deposits described below (fig. 94).

ELDER CREEK PORPHYRY COPPER SYSTEM

The Elder Creek porphyry Cu system is in the central part of the Snow Gulch quadrangle, and it was first tested as a porphyry Cu target in 1965 by Duval Corp. (pl. 3). No commercial concentration of metals was known in the system as of middle 1998, although there was some minor production in the early part of the 20th century of base- and precious-metals from some fringing polymetallic veins associated with the system. Many polymetallic veins and mineralized faults in the general area of Elder Creek (such as at the Morning Star Mine, Big Pay Mine, Gracie Mine, and the Ridge Mine, pl. 3) show minor amounts of Au in their production records. In addition, the Ridge and Gracie Mines are each located close to the centers of two aeromagnetic anomalies, each about 100 to 150 gammas magnetic field strength above local background, that are part of a donut-shaped pattern of anomalies that encompass the Elder Creek system (Battle Mountain Gold Company, unpub. data, 1994). These anomalies largely appear to be just interior to the outer limit of the introduction of pyrite±pyrrhotite that surrounds the system. The anomalies, in places, result partly from the presence of as much as several volume percent disseminated pyrrhotite in rocks of the Harmony Formation that are recrystallized to biotite hornfels (D.W. Blake, oral commun., 1994).

Approximately 0.3 km south-southwest of the Ridge Mine, the position of the outer limit of Fe sulfide mineralized rock that surrounds the Elder Creek porphyry Cu system seems to be a function also of some intensely phyllic-altered and pyrite-veined Oligocene granodiorite porphyry (pl. 3). This occurrence (in NW 1/4 sec. 12, T. 32 N., R. 43 E.) is along a mineralized fault that is parallel to the mineralized structure that controls mineralized rocks

at the Ridge Mine. A few rock samples examined petrographically from the occurrence also show presence of traces of disseminated chalcopyrite in the groundmass of phyllic-altered granodiorite porphyry. However, the occurrence had been prospected previously, most likely for precious metals, by a number of shallow workings. Abundant quartz-pyrite veins cut granodiorite porphyry at the locality, and the fluid-inclusion signature of the vein quartz includes an absence of halite-bearing fluid inclusions and wide-ranging liquid-vapor proportions in two-phase fluid inclusions suggestive of boiling (Gostyayeva and others, 1996).

Ratios of Ag/Au from mineralized occurrences in the general area of the Elder Creek porphyry Cu system appear to vary systematically across the system (Theodore, 1996). The overall ratio of Ag/Au in production data from the four above-listed mines ranges from a low value of approximately 2.5 at the Morning Star Mine to a high value of approximately 129 at the Big Pay Mine (Roberts and Arnold, 1965). These changes in the values of the Ag/Au ratio are consistent with the geographic positions of the four mineralized occurrences relative to the system because they show dramatic increases of the Ag/Au ratio toward the margins of the system. Actually, the Big Pay Mine is located along a mineralized north-striking fault roughly 0.4 km outside the outer limit of pyrite±pyrrhotite alteration surrounding the porphyry system. However, the relatively low Ag/Au ratios at the Morning Star Mine, which is situated near the center of the system, also might be related to the fact that the Morning Star Mine appears to be in the down-faulted block of the post-mineral Elder Creek fault (pl. 3). In the Copper Canyon area of the mining district, a notable decrease in the Ag/Au ratio is present in production data in a Au–Ag zone that mantles a central Cu–Au–Ag zone near the core of the porphyry system (Theodore and others, 1986). In addition, nine of the ten large Au–Ag orebodies are present within an area wherein Ag/Au weight ratios in ore are less than 10 (Kotlyar and others, 1998; see also, Einaudi, 1990).

Duval Corporation was the first to test the Elder Creek porphyry system as a porphyry target by drilling three core holes in 1965 near the west border of sec. 6, T. 32 N., R. 44 E., which consequently were placed on the eastern margins of the system because of the land situation at the time of drilling (Joe Lamanna, written commun., 1994). These three holes were part of a regional Cu exploration program that was being conducted at that time.

Rocky Mountain Energy Company, formerly a subsidiary company of Union Pacific, apparently was the next company to test the porphyry Cu target in the Elder Creek area by drilling eight holes during 1968 (P.R. Wotruba, written commun., 1994). Six of the holes were located in W 1/2 sec. 1, T. 32 N., R. 43 E., another hole was drilled in the south-central part of sec. 36, T. 33 N., R. 43 E., and the last hole was drilled in NW 1/4 sec. 6, T. 32 N., R. 44 E. Subsequently, in the middle 1970s, AMOCO tested the pediment areas of the system geophysically (S.W. Ivosevic, written commun., 1995).

Battle Mountain Gold Co. then drilled an additional 40 holes in the Elder Creek area since they staked some additional claims there in 1985 (Joe Lamanna, written commun., 1994). However, all of this drilling by Battle Mountain Gold Co.was targeted toward discovery of economic Au–mineralized rocks peripheral to the porphyritic monzogranite of Elder Creek. The 40 drill holes were spread throughout secs. 1, 2, 11, and 12 of T. 32 N., R. 43 E.; and sec. 36 of T. 33 N., R. 43 E. Of the 40 holes, seven were drilled in sec. 36, five in sec. 2, six in sec. 11, 13 in sec. 1, and nine in sec. 12.

Late in 1994 and early in 1995, Western Mining Corporation (U.S.A.) drilled an additional 19 holes in the northeastern part of the Elder Creek porphyry Cu system (Tom Gray, written commun., 1995). The deepest hole is approximately 500 m. Most holes were collared in unconsolidated gravels on the pediment areas of the system, and they were targeted to explore the precious-metal potential on the margins of the quartz stockwork-enriched core of the system. Approximately 30 intercepts of roughly 3–m thicknesses were penetrated that have in excess of 100 ppb Au in 11 of the holes. Most elevated concentrations of Au appear to be associated with arsenopyrite, which yields abnormally high temperatures of formation on the basis of its composition (Gostyayeva and others, 1996). Small grains of argentian tetrahedrite are present in some polished thin sections of arsenopyrite-rich rock. However, significant concentrations of chalcopyrite also were encountered sporadically in various holes, in places amounting to approximately 10 to 20 volume percent across short intercepts of drill core. The most impressive visually consists of hornfels of the Harmony Formation that has been bleached by widespread, fine-grained white mica and impregnated with abundant chalcopyrite at a depth of approximately 220 m in one of the drill holes. Chalcopyrite is associated with pyrrhotite, vein quartz, and carbonate minerals, as well as minor amounts of chlorite and traces of sphene—all suggestive of a propylitic assemblage. In addition, along this particular approximately 0.3–m-long chalcopyrite-rich intercept, there were only 26 ppb Au in an approximately 2–m-long drill core that brackets the intercept. There are, as well, 4 ppm Pb, 12 ppm Zn, and 23 ppm As in the 2–m-long drill core.

Lone Tree gold deposit

by Edward I. Bloomstein, Bruce L. Braginton, Roy W. Owen, Ronald L. Parratt, Kenneth C. Raabe and Warren F. Thompson

Introduction

LOCATION, OWNERSHIP, AND DEPOSIT SIZE

The Lone Tree Au mine is located about 45 km east-southeast of the city of Winnemucca on the west and south flank of Lone Tree Hill in Humboldt County, Nev. The deposit lies at the north end of the low, rolling informally named Havallah Hills about 1 km south of Interstate Highway 80, in secs. 11, 13, and 14, T. 34 N., R. 42 E., Mount Diablo Base and Meridian (fig. 94). The Lone Tree deposit is herein considered to be in the northernmost part of the Battle Mountain Mining District. All lands in the area of Lone Tree Hill are owned or controlled by Newmont Gold Co. Much in the present report that follows is modified from Bloomstein and others (1993) and from B.L. Braginton (written communs., 1996, 1998). Proven and probable ore reserves at Lone Tree as of December, 1995, based on a price of $350/oz Au, are approximately 37.8 million tonnes (t) averaging 2.22 g Au/t for an aggregate economic reserve of approximately 92.2 t Au or 2,965,000 oz Au. This total may be divided further into oxide and refractory categories. Oxide mill reserves are 1,790,000 t averaging 4.25 g Au/t. Oxide run-of-mine heap leach ore reserves are 8,703,000 t averaging 0.66 g Au/t. Refractory ore is further classified into two sub-categories, sulfide mill ore and run-of-mine heap leach ore. Refractory sulfide mill ore reserves are 22,418,000 t averaging 3.0 g Au/t. Refactory run-of-mine ore reserves are 4,923,000 t averaging 0.75 g Au/t.

ACKNOWLEDGMENTS

The authors would like to express their appreciation and gratitude to all workers involved in the Lone Tree project, at the time part of Santa Fe Pacific Mining, Inc. Their hardwork and attention to detail have provided the basis for the understanding and insights contained in this paper. The authors wish to especially recognize Gary L. Massingill for the insightful technical review of this report. Any errors or omissions, however, remain solely the responsibility of the authors. Appreciation is also expressed to management of Santa Fe Pacific Mining, Inc.

for permission to release the data contained herein. Much of the following represents the state of geologic knowledge regarding Lone Tree in late 1993. However, key elements of subsequent interpretations are included such as recognition of the Antler High, Chaotic, and Sumac ore zones. The increase in detailed knowledge and recognition of additional major structures has not substantially changed overall interpretation of Lone Tree geology and controls of mineralization.

Exploration and discovery history

Prospecting in the general area of Lone Tree Hill started in the middle 1860s about the time of construction of the Central Pacific part of the transcontinental railroad whose track is located about 3 km northeast of Lone Tree Hill. Secondary Cu–mineralized rocks were exposed on the top and sides of the hill and were explored subsequently by a number of adits and shafts. These mineralized rocks contained erratic concentrations of Au, but exploration failed to develop any significant Cu or Au. Additional sporadic exploration occurred during the next several decades with the most serious exploration conducted for porphyry Cu in the 1960s and 1970s by Duval Corp. and Bear Creek Mining Co., but again with no success, although some low-grade Au–mineralized rocks were found. In the early and middle 1960s, the area of Lone Tree Hill was explored for Au by Nerco Inc., Freeport Inc., and several other Canadian companies, which yielded narrow drill intercepts of apparently insignificant low grade, fracture-controlled Au–mineralized rocks. Cordex Exploration began exploration on and near Lone Tree Hill in 1988 under the Marigold Joint Venture with Santa Fe Pacific Mining Inc., and in early 1989 discovered Au–mineralized rocks about 1 km south of Lone Tree Hill. Subsequently in July 1989, Santa Fe Pacific Mining Inc. encountered ore-grade intercepts in the first drill hole of a pediment exploration program in sec. 11 west of Lone Tree Hill and outside of the property under control of the Marigold Joint Venture. During the remainder of 1989, 12 additional holes were drilled in sec. 11. Many of those holes encountered ore-grade mineralization along a suspected north-south fault (figs. 97, 98). That part of the deposit in sec. 11 was drilled out during the first six months of 1990, and the first Au was poured on August 23, 1991, 25 months after drilling the discovery hole. In May of 1992, Santa Fe Pacific Mining Inc. exchanged its interest in the Marigold joint Venture for the southern part of the deposit in secs. 13 and 14 and contiguous land, thereby consolidating ownership of the property. The Lone Tree deposit now (1999) includes four mineralized zones—Wayne, Chaotic, Antler High, and Sequoia zones (fig. 97).

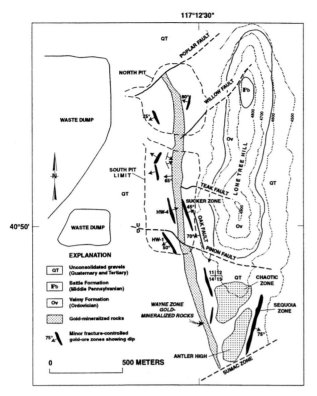

Figure 97 Geology of Lone Tree Mine including outlines of North and South pits as of September 1992. Major concentrations of mineralized rock are within Wayne zone (dotted pattern), and minor fracture-controlled ore zones also are shown by elongate areas in black.

Regional geology

STRATIGRAPHY

Parts of three major Paleozoic packages of rock, which are present widely throughout the Battle Mountain Mining District, also are present in the Lone Tree area: the allochthon of the Roberts Mountains thrust, the autochthonous Antler sequence, and the allochthonous Havallah sequence, which comprises the upper plate of the Golconda thrust (fig. 99). The oldest rocks penetrated by drilling are quartzarenite, chert, minor basalt and argillite of the Valmy Formation, which is part of the Roberts Mountain allochthon (Roberts and others, 1958; Roberts, 1964; Madrid, 1987). Rocks of the Valmy Formation crop out on Lone Tree Hill and have a general north strike, and they dip 30 to 70 degrees to the west. Tight, east-verging folds in these rocks are probably the result of strong east-directed compression during the Late Devonian to Early Mississippian Antler orogeny (Roberts and others, 1958).

Resting on rocks of the Valmy Formation are shallow-water, clastic rocks of the Pennsylvanian and Permian Antler sequence, which is made up of three formations of which the basal one rests with angular unconformity upon lower Paleozoic rocks of the Roberts Mountain allochthon (Roberts, 1964). All of these rocks are tectonically below the Golconda thrust (fig. 99). The autochthonous Antler sequence was deposited within the Antler orogenic belt during postorogenic submergence. Two formations of the

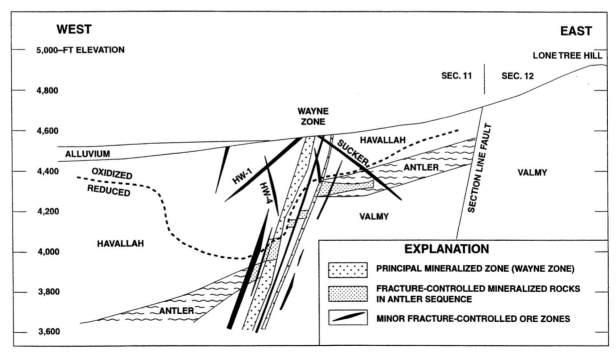

Figure 98 Schematic east-west cross section through Lone Tree Mine.

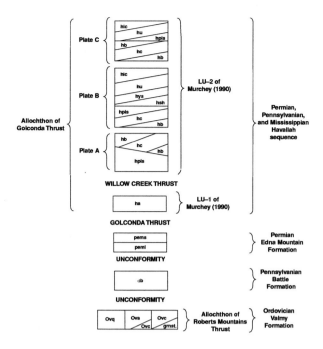

Figure 99 Correlation chart for stratigraphic units in area of Lone Tree Mine (see text for details).

Antler sequence are present at Lone Tree in the northern parts of the mineralized system, the Battle Formation at the base and the overlying Edna Mountain Formation. The Antler Peak Limestone, normally above the Battle Formation (Roberts, 1964), is present in the southern part of the Lone Tree system, specifically in the open pit in sec. 13. The Battle Formation (unit ꓷ b, figs. 97, 99) is present as a thin 0– to 6–m veneer of cobble conglomerate overlying the erosional surface developed on quartzarenite of the Valmy Formation homocline in the north-central part of Lone Tree Hill. The Edna Mountain Formation does not crop out in the Lone Tree area, but rocks widely encountered at depth in drill holes are correlated on the basis of their lithologic similarity and geologic position with rocks assigned to the Edna Mountain Formation by Erickson and Marsh (1974) in the vicinity of Iron Point, SW 1/4 sec. 14, T. 35 N., R. 41 E. A generalized stratigraphic column showing lithology of strata of the Antler sequence and its relations to unconformably underlying rocks of the Valmy Formation is shown in figure 100.

Overlying rocks of the Antler sequence are rocks belonging to the Havallah sequence, which extensively crop out to the south of Lone Tree Hill. The Havallah sequence is composed of at least six tectono-stratigraphic plates or packages that have been assigned to two lithotectonic units (LU–1 and LU–2, fig. 99) separated by the Willow Creek thrust (Murchey, 1990). The lowest rocks in LU–2 are a Mississippian, deep-water, oceanic assemblage of green chert and basalt (fig. 99), which are overlain by several tectonic plates of a clastic-dominated rocks composed of calcareous sandstone, shale, mudstone,

sandy limestone and conglomerate. The origin of the Havallah sequence is controversial (Miller and others, 1992). According to Brueckner and Snyder (1985), the Havallah sequence is an accretionary sedimentary wedge scraped off the deep ocean floor ahead of an advancing thrust that overrode the continental margin and developed numerous imbricate thrusts within the allochthon. An alternative interpretation favored by Miller and others (1984, 1992) and Murchey (1990) and the authors requires extrabasinal clastic sediments in the Havallah basin to be derived primarily from erosion of the Antler highlands. The basin in which these sediments are interpreted to be deposited became shallow with time and generated an upwards shallowing sequence of radiolarian chert and basalt at the base that progressively grade into shallow-water clastic rocks near the tops of individual packages. The resulting rock pile was subsequently disrupted tectonically by varying intensities of thrusting during emplacement of the pile along the late Permian and (or) Early Triassic Golconda thrust associated with the Sonoma orogeny (Silberling and Roberts, 1962). The Golconda thrust is exposed in the Lone Tree pit, to the southeast at the open-pit of the Eight South deposit, and on Golconda summit to the west of Lone Tree Hill. Megascopic folding of Havallah sequence rocks is evident to the southwest at Buffalo Mountain and to the south in the Havallah Hills (pl. 1), and outcrop-scale folding of the Havallah rocks also is present in rocks exposed in the open pit at the Lone Tree Mine. In the Havallah Hills, strata generally are gently dipping, and are folded into a series of broad, open north-trending anticlines and synclines (pl. 1). The precise age of folding is unknown, and is most likely of multiple ages, as well as being related to some of the last sub-horizontal, east-oriented compressional stresses developed during tectonic emplacement of the Golconda allochthon during the Sonoma orogeny.

STRUCTURE

North-south normal faults are the principal faults of the Lone Tree Mine area, and also are some of the most highly mineralized structures. Such faults also are predominant in the Havallah Hills to the south (fig. 94), producing pronounced topographic lineaments strongly evident in LANDSAT imagery. Throughout the area, these faults have relatively recent motion, and have produced a series of north-trending horsts and grabens. Lone Tree Hill is a prominent north-trending horst. The boundary fault on the east side of Lone Tree Hill is a major normal down-to-the-east fault of unknown displacement (fig. 97). The west side of the hill also has been disturbed by a number of stair-stepping, down-to-the-west faults. The most easterly of these faults, the Section Line fault, has been penetrated by drill holes, and possesses an apparent 100 m of offset. Numerous other faults parallel to these are present, but have relatively minor offset. North-south or perhaps north-northeast normal down-to-the-west faults with minor displacements have been identified in the pit. It is possible that these young normal faults are listric at depth and diverge from the nearly vertical mineralized structural zone. The west-dipping homoclinal strata at Lone Tree Mine have been interpreted to lie near the axis of a broad asymmetrical Mesozoic anticlinal warp that plunges gently to the south (Radu Conelea, written commun., 1990). Open extensional fractures striking north-south may have developed along shattered hinge lines of fold axes of such anticlinal structures, allowing upward movement of mineralizing fluids. As discussed in the section entitled "Tertiary Deformation" above, north-south faults and fault zones are prominent throughout the Battle Mountain Mining District (fig. 94).

Northwest-striking faults also are prominent in the Lone Tree Mine and on Lone Tree Hill itself. These extensional structures locally have several meters of normal post-mineral displacement (fig. 97).

Faults and fracture sets of other orientations also are prominent in the general area of Lone Tree Hill. Many orientations can be related to structural features in the surrounding region that have similar orientations and are extensive over broad areas. Northeast-striking faults are some of the youngest structures—they apparently are related to regional, northwest-southeast, Miocene extension in this part of the Great Basin along the Midas trough, which is part of the Humboldt structural zone. This 200–km-wide, northeasterly oriented zone is believed by some to have been a zone of left lateral strike-slip, as well as dip-slip movement (Rowan and Wetlaufer, 1981). Displacement of at least 1 km has been documented in the Midas area (Wallace, 1991). Buffalo Mountain also is bounded on its southeast by a range-front fault that has a similar strike. Tertiary gravels along the northeast-striking, 13–km-long Oyarbide fault on the northwest side of Battle Mountain are offset with the down dropped block on the northwest (fig. 94). This suggests a dip-slip component of displacement of approximately 700 m (see section above entitled "Tertiary Deformation"). Normal offsets along these structures generally appear to have been among the youngest in the region, possibly occurring during the last 17 to 14 m.y. Two northeast-striking normal faults in the North pit at Lone Tree, the Willow and the Poplar faults (fig. 97), drop bedrock down to the northwest and offset on them appears to be on the order of few meters and to be post-mineral. Also on Lone Tree Hill, several underground workings follow an irregular near-vertical fault-breccia zone striking N. 40° E. This breccia contains abundant coatings of Fe and Cu oxide. All of these northeast-striking faults are brittle-type structures. A northeast-striking south-dipping set of faults known as the Sumac zone truncates mineralized rocks at the south end of the deposit in the general areas of secs. 13 and 14.

The north end of the Havallah Hills, about 2.4 km south of Lone Tree, also is disrupted by a series of northeast-striking, high-angle faults whose associated joint patterns on aerial photographs give the appearance of a northeast strike to the rocks. Northeast-striking extensional structures offset all other structural features, even the relatively recent north-south fractures, which may be related to the prominent north-south fault zones throughout the Battle Mountain Mining District (fig. 95B; see also, Doebrich and others, 1995).

Many faults in the area are filled with intrusive dikes. A string of small Tertiary intrusive dikes extends from the area of the Eight South deposit in a N. 10° W. alignment toward the east side of Lone Tree Hill. Graney and McGibbon (1991) describe these as Tertiary "feldspar porphyry" dikes that are present along steep north- and northwest-striking faults at the Eight South deposit. In the Lone Tree Mine area, an altered and sparsely mineralized north-trending granodiorite dike has been penetrated by drilling to the west of the main zone of mineralization. Sparsely occurring altered rhyolite dikes strike roughly east in the pit at the Lone Tree Mine. Potassium-argon dating of biotite and hornblende mineral separates from intrusive rocks obtained from the south end of Lone Tree Hill indicate a 36– to 39–Ma age for this magmatism (see section above entitled "Potassium-argon Chronology of Cretaceous and Cenozoic Igneous Activity, Hydrothermal Alteration, and Mineralization"). Similar age, 35– to 40–Ma intrusive activity also is widespread to the southeast in the mining district, and it is associated with formation of the Fortitude Au skarn deposit as well as other base and precious metal deposits (Theodore and others, 1973; Theodore and Hammarstrom, 1991; Theodore and others, 1992; Myers, 1994).

East-west structures, generally subtly expressed as joints, are widely present. These structures probably are of early derivation and have been reactivated locally. A major east-west feature, suggested by aeromagnetic as well as topographic lineaments, corresponds to a bend in the Humboldt River north of Lone Tree Hill. A strong east-northeast joint set is evident on the south half of Lone Tree Hill, while those on the north half strike east. At Buffalo Mountain, east-west structures slightly offset northwest striking structures. A distinctive set of conjugate east-northeast and almost east-west vertical joints are present in the Lone Tree pits. These fractures are commonly coated with pyrite and barite, and a locally heavy stain of Cu oxide.

West-northwest and east-northeast striking faults in the Lone Tree pit may be related to the east-west structures. These faults are steeply dipping, and they show apparent normal offsets. However, an east-west structure with oblique slickensides, the Teak fault, has been mapped near the midpoint of the South pit (fig. 97), and has down-to-the-south offset probably on the order of ten meters. The Piñon fault zone, a major zone of west-northwest- to west-striking normal faults, crosses the south end of the South pit (fig. 97). This structure has an apparent normal offset of at least 100 m with a possible oblique or scissors component. To the east, it truncates the south end of Lone Tree Hill. Offset also is present along a number of minor faults. One prominent west-northwest-striking fault offsets the bedrock surface about eight meters down to the south in the southeast corner of the South pit.

Additional structural complexities were revealed by detailed studies in the open pit at Lone Tree. Structures with a north strike are less common than those striking north-northwest or north-northeast. The north-striking structures appear to have developed early. However, multiple episodes of movement are evident, and the north-striking faults have been deflected and intersected by north-northeast and north-northwest striking ones. The overall map-scale feature is a complex zig-zag fault network with a north-trending line of bearing. Local structural complexities include dilational jogs and fault bends. The bulk of the mineralized rocks are contained within the north-striking segments, but several subsidiary ore zones are controlled by north-northwest and north-northeast segments. Similar structural relations are present in some other nearby Au orebodies.

Structural relations at the Twin Creeks Mine, 32 km to the north of the Lone Tree Mine, are consistent with the generally north trend of mineralized rocks present at Lone Tree. The Twin Creeks orebody is hosted by Ordovician rocks which are folded into north-northwest trending anticlines and synclines of presumed Antler-orogeny age. These folds are, in turn, cut by northeast-striking, 50° northwest-dipping extensional faults with dextral, strike-slip offset (Bloomtein and others, 1990). The north-northeast striking DZ fault at the Twin Creeks Mine is interpreted to be a Riedel shear associated with a regionally significant north-south basement structure showing right-lateral offset called the "Rabbit Suture." This structure may constitute a part of an even larger regional suture zone inferred by Soloman and Taylor (1989) that may have been repeatedly reactivated by Ordovician, late Paleozoic, and middle-Tertiary tectonics. The suture zone is presently expressed as a north-south belt of Au–As–Hg mineralized rocks, extending from the Twin Creeks deposit to the area of Lone Tree, Old Marigold Mine, and beyond. Its position once was considered to coincide with the western margin of the Proterozoic craton as defined by Elison and others (1990), who used an initial $^{87}Sr/^{86}Sr$ ratio of 0.706 to mark its boundary, although, as described above, a 0.705 ratio now (1999) is considered to better reflect the present-day geo-

graphic position of the Proterozoic craton. The Lone Tree Mine is located squarely within this belt of deep-seated crustal dislocation.

MINERAL DEPOSITS

Structurally controlled Au–, Cu–, and Mo–mineralized rocks are present farther to the south in the Battle Mountain Mining District (Roberts and Arnold, 1965). Brittle deformation zones along faults and at structural intersections are especially important sites for mineralization. Large Cu orebodies in the Copper Canyon area are associated with major north-south faults as well as northwest and northeast fracture zones. The Fortitude Au skarn orebody also is present in the Copper Canyon area, approximately 1 km north of the 38– to 39–Ma pluton that lies astride major accreted terranes there (Wotruba and others, 1986; Myers, 1994). Brittle deformation zones along faults and at structural intersections are especially important sites for mineralization. Blake and others (1979) outlined various trends of fractures and mineralized intrusions within the mining district. At the Copper Basin Mine, a N. 70° W.– to N. 90° W.–striking system of dikes, faults, and fractures host quartz–Mo veins and secondarily enriched Cu ore bodies. This trend developed in conjunction with emplacement of a composite calc-alkaline intrusive complex centered on the Buckingham stockwork Mo system during the Late Cretaceous (Theodore and others, 1992). Faults striking N. 45° W. and N. 45° E. developed as the Buckingham stockwork system waned, and they are associated with several Pb–Zn–Ag vein deposits. The last mineralizing event is peripheral to the porphyry Mo system, and is related to emplacement of north-trending granodiorite porphyry dikes. This event produced fracture-controlled Cu–Pb–Zn–Ag veins that also contain minor Au.

Some other nearby areas document well the importance of structural controls on distribution of mineralized rocks in the region. Twenty-five km to the south of the Lone Tree Mine at the Buffalo Valley Mine, mineralized rocks, classified as Au skarn and (or) distal-disseminated Ag–Au, were localized within a northwest-striking, normal fault zone filled with a quartz latite dike (Doebrich, 1995). In addition, a north-striking normal fault controls distribution of Au–mineralized rocks at the Kramer Hill deposit, 27 km west-northwest of Lone Tree (Kretschmer, 1991). High-grade mineralized rocks in this deposit are confined to quartz veins within the fault zone. In addition, hanging-wall rocks host disseminated Au. A granodiorite dike is present along the vein, and east-striking, post-mineral faults slightly offset ore (Kretschmer, 1991).

As many as 15 Au deposits have recently been discovered in the northwest parts of the Battle Mountain Mining District in the general area of the historic workings at the Old Marigold underground mine (fig. 94). The Eight South deposit, was discovered under alluvial cover in 1985 by the Cordex Syndicate (see section below entitled "Geology of the Marigold Mine Area"). The Eight South deposit formed as a replacement of Pennsylvanian and Permian Antler sequence rocks below the Golconda thrust, and the deposit is localized at the intersection of northwest- and north-striking faults and fractures (Graney and McGibbon, 1991). Northwest-, northeast-, and east-striking faults also are present. Reactivation of north-trending horst and graben styles of normal faulting has offset the ore zone during Tertiary extension. Mineralized rocks are disseminated primarily within previously calcareous rocks of the Antler sequence.

Two other structurally-controlled Au orebodies are close to the Eight South deposit, and another area that also contains structurally-controlled, non-economic Au–mineralized rocks is present somewhat farther to the southwest. The Top Zone orebody, approximately 2 km south-southwest of the Eight South deposit (fig. 94), generally trends northwest and dips gently to the west (see also, section below entitled "Geology of the Marigold Mine Area"). Ore is contained within several brecciated, flat-lying, possibly thrust-fault bounded, zones in interbedded quartzarenite and shale of the Valmy Formation. North- to northwest-striking, near vertical fractures may have served as feeders. Gold mineralized rocks at the historic workings of the Old Marigold Mine, also about 2 km south-southwest of the Eight South deposit (fig. 94), are present as irregular silicified bodies within a thin sequence of the Antler sequence, and are best developed in debris flows of the Edna Mountain Formation that unconformably overlie the irregular paleotopographic surface of the Valmy Formation. At Buffalo Mountain, 15 km to the southwest of the Lone Tree Mine, rocks of the Havallah sequence are broken by weakly mineralized, north- and northwest-striking, jasperoid-bearing structures with little or no displacement. Pods of high-grade Au–mineralized rocks are formed where these structures intersect. Mineralization is strongly fracture-controlled and forms stockwork bodies of Au–bearing quartz veinlets. Mineralized rocks and faults post-date silification that caused the rocks to fracture in a brittle manner. Although adjacent rocks are permeable, they are barren.

Mine geology

Stratigraphy

The Havallah sequence at the Lone Tree Mine area is a series of different lithologic units distributed in four stacked thrust-bounded plates (fig. 99). The four plates are the lowermost of six plates identified to the south in the Havallah Hills, and including parts of LU–1 and LU–2 (pl. 1). A plan view of the Lone Tree Mine area at the 4,000–ft elevation (fig. 101) illustrates the relationship among the various plates of the Havallah sequence in the Golconda allochthon and steep-dipping major faults. The steep-dipping faults strike northeasterly and northwesterly and have normal dip-slip displacements. The mapped subunits of the Havallah sequence are lensoidal in plan view, partly due to their initial depositional character, and partly due to imbrications during thrusting. Rocks of the Havallah sequence include: (1) argillite (subunit ha, fig. 99); (2) pelagic chert, argillite and submarine basalt flows (subunit hc); (3) sandy limestone and pebble conglomerate (subunit hpls); and (4) siliciclastic and calcareous shallow-marine sediments (subunits hsh, hys, hu and hlc).

The thrust plates interpreted by drilling and outcrop mapping are shown on cross-sections (figs. 102, 103) and are designated plates A, B and C from bottom to top (fig. 99). However, stratigraphic relations of rocks exposed in the open pits currently (1999) are being revised substantially from the nomenclature used herein (K. Kunkle, oral commun., 1998). Nonetheless, bedding in rocks of the Havallah sequence in the Lone Tree pit strikes generally north, and dips 20° to 50° west Radical lithologic changes across short distances laterally in drill holes have been interpreted as faults, and, in places, these faults show approximately 100 m of offset. Individual thrust plates are composed of similar lithostratigraphic units and are identified by repetitions in the sequences of rock. Plate A is the lowermost plate, directly, above the Golconda thrust and contains sandy limestone and pebble conglomerates (subunit hpls) overlain by chert, argillite and submarine basalt (subunit hc). The contact between the Havallah sequence and the underlying rocks of the Antler sequence is interpreted to be the Golconda thrust; however, evidence of shearing and brecciation along the contact is generally lacking. Plate B contains a thick sequence of subunit hc overlain by varying thickness of subunits hpls, hys, hsh and hu. Plate C is only present approximately 2.5 km to the southwest in the vicinity of the process plant and it has a not been identified in the area of the open pits at the Lone Tree Mine.

The most characteristic feature of rocks of the Havallah sequence is boudin-like structures that were produced either by sediment-loading or shearing during thrusting. The boudin-like features are less than 1 meter wide in the pit walls. They also are present in drill core and thin sections. Phacoids of coarse-grained sandy limestone and sandstone are suspended in a foliated mudstone matrix. Commonly silicified, in places converted to skarn, the augens also contain disseminations and veinlets of base-metal sulfide minerals and quartz. The phacoids generally dip gently to the west, but in places are parallel to minor shear zones. The structural fabric as seen in thin sections is defined by mineral segregation, cataclasis, and illite-clay stringers. Calc-silicate minerals in siltstone or skarn define highly contorted, ductile-deformed lamination. Cherty silica forms stringers in mudstone, and also is present as hairline quartz microveins oriented normal to bedding surfaces, probably as the product of dissolution of silica during diagenesis or during development of the foliation. Interleaved slivers of altered basalt also show effects of intense ductile deformation, with abundant calcite veinlets parallel to foliation. Mineralized basalt of the Havallah sequence generally is unsilicified, argillized, and contains numerous Fe–oxide veinlets. Mineralized intervals normally contain less than 1–cm-wide pyrite veinlets, which generally parallel foliation, and wrap around and cut across phacoids. Disseminated pyrite is quite abundant locally. In places, fine-grained, disseminated pyrite has been dismembered from veinlets which have been disaggregated due to shearing. Some pyrite veinlets terminate at augen margins indicating post-augen shearing. Post-mineral seams of gouge commonly parallel foliation, and are produced either by rotation into high-angle shear planes, or by low-angle thrust-parallel shearing. Havallah sequence-hosted high-grade mineralized rocks are contained commonly in clayey gouge zones where silica has been removed from the gouge zones by leaching.

The Antler sequence consists of siltstone, lithic sandstone, and basal sandy conglomerate, and it is generally about 50 m thick; however, its composition and thickness are extremely variable. Evidence of thrusting within units has not been observed, and bedding is generally undisturbed. The Antler sequence consists of the Edna Mountain Formation present as siltstone and sandstone (Pemu)

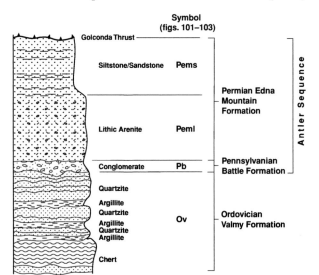

Figure 100 Lithostratigraphic column of Pennsylvanian and Permian Antler sequence and Ordovician Valmy Formation in area of Lone Tree Mine.

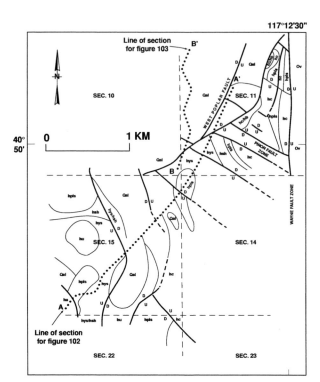

Figure 101 Projection to surface of geology encountered by drill holes at 4,000-ft elevation in general area of Lone Tree Mine, showing traces of cross sections AA' and BB', figures 102 and 103, respectively. Symbols same as figure 99.

underlain by a coarse lithic arenite (Peml) and further underlain by the Battle Formation (℗ b), which is present as a thin basal conglomerate near the contact with the Valmy Formation (fig. 100). The Antler sequence has been strongly silicified, especially toward its base, and the sequence contains both fracture-controlled and disseminated mineralized rocks. Mineralized rocks in the upper siltstone contain argillic alteration, with and without local silicification, and very fine disseminated pyrite. Hairline stockwork pyrite veinlets are present along weakly preserved bedding planes. The lithic arenite is, in many places, intensely silicified and contains near-vertical open fractures filled with drusy pyrite, arsenopyrite, barite, and quartz along with minor pyrrhotite and stibnite. The Au grades are moderate to high. The basal conglomerate of the Battle Formation locally contains a truly disseminated type of mineralization, apparently developed where premineral silicification was incomplete. This unit generally contains the lowest Au grades. The rocks of the Antler sequence were deposited uncomformably on an irregular, channeled surface developed on rocks of the Valmy Formation.

The Valmy Formation at the Lone Tree Mine includes massive quartzarenite, along with interbedded to interlaminated argillite and chert. The massive quartzarenite is usually featureless; however, interbedded lithologies generally show effects of some differential slip

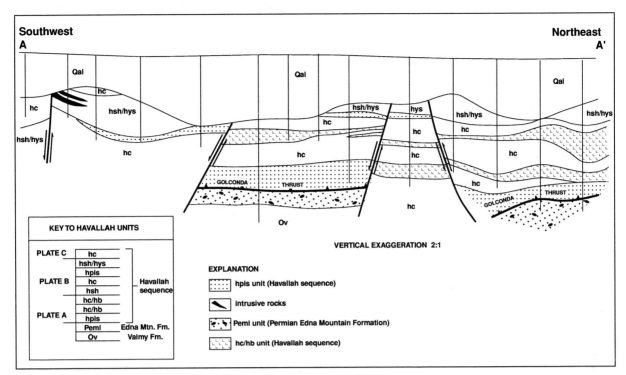

Figure 102 Normal faults that displace trace of Golconda thrust along AA' (fig. 101).

along bedding surfaces. Cataclastic shearing and an oblique foliation of mineral fabric is common. Argillite-quartzarenite contacts and interbeds show cataclastic quartzarenite and trains of disaggregated quartz grains in a well-indurated clay matrix, giving the appearance of primary sandy argillite interbeds. Milled seams of gouge also are common along quartzarenite-argillite contacts. Thin pyrite-quartz "gash" type veinlets are, in places, at right angles to bedding, which is commonly rotated or dragged into later shear planes. These shear zones possess argillite gouge seams produced by slippage parallel to bedding and concordant veins of quartz. Ptygmatic folds locally have been developed in thin chert beds. Mineralized quartzarenite is present predominantly in hydrothermal breccias that have been completely or partially cemented by quartz, adularia, and pyrite. Pyrite and blue covellite have coated east-west open fractures in quartzarenite in many places. Mineralized argillite is commonly represented by crackle-breccia networks of illite, adularia, and pyrite. Repeated brecciation and cementation suggest that faulting and alteration were contemporaneous.

STRUCTURAL CONTROL OF MINERALIZATION

The distribution of Au–mineralized rocks at Lone Tree is primarily controlled by three principal fault systems known as the Wayne zone, the Sequoia zone, and the Antler High zone, as well as a structurally complex area referred to as the Chaotic zone (fig. 98). The Wayne zone includes more than 50 percent of the Au in the deposit, and the zone is restricted primarily to a 50– to 120–m-wide tract of intense fracturing and oblique-slip faulting (fig. 98). It is named in memory of Wayne R. Bruce, a Santa Fe Pacific Mining, Inc. geologist, whose untimely death occurred shortly after discovery of the Lone Tree deposit.

The Wayne zone is a mostly normal fault zone that strikes north and dips 65 to 75° to the west and that has a significant strike-slip component (figs. 97, 98). This fault has displaced rocks of the Havallah sequence, Antler sequence, and Valmy Formation, as well as Tertiary lake sediments present in the gravel-covered areas. Paleozoic rocks have been displaced vertically about 130 m across the zone. Horizontal offset is indicated by slickensides and by the large number of bends of the fault along their traces. The exact horizontal component of displacement is unknown. Another normal fault, the Section Line fault, is parallel to the Wayne zone, but to the east of the Wayne zone, and shows about 100 m of vertical offset (fig. 98).

The Wayne zone has a strike length of nearly 2.5 km at the Lone Tree Mine. Located at the south end of Lone Tree Hill is a zone of west-northwest to west-trending normal faults, which are termed the Piñon fault zone (figs. 97, 101). The Wayne zone apparently offsets the Piñon fault zone. Offset is present on a number of minor structures, including a prominent west-northwest-striking fault, which displaces the bedrock surface about 8 m down to the south in the southeast corner of the South pit. South of the Piñon fault zone, offset is present on a number of

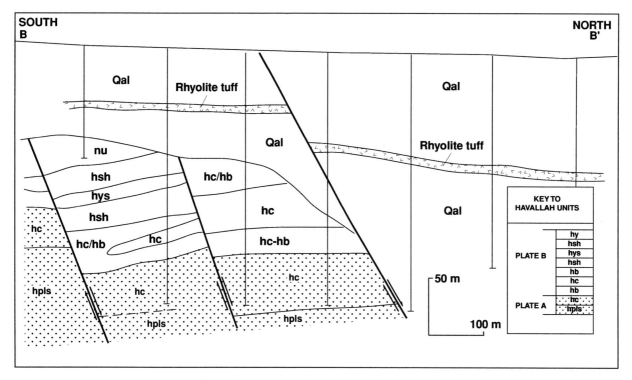

Figure 103 Significant normal offset along BB' (fig. 101) in Quaternary sands and gravels, north block down, along Poplar fault.

minor structures that deflect the main trace of the Wayne zone to the southeast (fig. 97). The north end of the Wayne zone has been truncated by a number of faults that make up the Poplar fault zone (fig. 97), which is parallel to and temporally associated with the northeasterly-trending Midas trough (see above). Erosion, combined with movement along the Poplar fault zone, has dropped bedrock on the north-end of the Lone Tree Hill as well as the area of the mine.

The internal parts of the Wayne zone comprise a system of relatively narrow, north-northwest- and north-northeast-striking faults that form in section an anastomosing complex of brittle shear zones enveloping rhomboid-shaped blocks of relatively competent, but highly fractured domains, of lesser strain. Multiple episodes of motion are indicated along the major structures within the zone. The highest grade oreshoots tend to be localized at intersections or deflections where zones of dilation were produced. Individual mineralized structures dip steeply both to the west and to the east. Drilling has encountered mineralized rocks along the Wayne zone approximately 300 m down dip from the original ground surface; subeconomic mineralized rocks remain open below this depth. The principal characteristic of the Lone Tree deposit is the apparent spatial coincidence of several structurally-controlled episodes of mineralization.

A major strand within the Wayne zone is the Powerline fault which contains the majority of high grade Au in the deposit. The Powerline fault is composed of a

set of three or more faults within the Wayne zone that vary in width from 3 to 30 m. The strike of the Powerline fault averages north-south, but varies from N. 10° W. to N.10° E. In the South pit, the Powerline fault strikes north-northwest. Premineral structural preparation along eventual trace of the Powerline fault appears to be most extensive at intersections of a number of faults. Highest Au grades within the Powerline fault are concentrated in the central part, which is a highly shattered, fault-hosted, hydrothermal breccia. Silicified, multiple cementation breccias also are present at distal margins of the Powerline fault, but, in general, these are low in contents of Au. Early silica-rich, Au–poor fluids may have circulated in the general environment of the Powerline fault and effectively formed an impervious seal. Continued motion and cataclasis along the fault then formed a zone of secondary permeability, which created effective pathways for major fluid flow of late Au–rich fluids, and subsequent supergene leaching.

The Sequoia zone is located southeast of Lone Tree Hill, and essentially marks the southeast margin of the deposit (fig. 97). The structural fabric of the Sequoia zone is quite similar to that of the Wayne zone, although there are some important differences. The known strike length of the Sequoia zone is approximately 700 m, which is significantly less than that of the Wayne zone. Overall dip of the Sequoia zone is 75°, as opposed to the 65° dip of the Wayne zone. Post-mineral faults have displaced or cut out mineralized rocks within the Sequoia zone, and they have had significant effects on the continuity of mineralized rocks, both down dip and along strike.

The Antler High zone is located within a horst block of rocks belonging to the Antler sequence between the Wayne zone and the Sequoia zone, and it is limited to approximately the southern one-third of the deposit (fig. 97). Mineralized rocks in the Antler High are primarily developed within rocks of the Edna Mountain Formation, Battle Formation, and the Valmy Formation, and are concentrated commonly parallel or sub-parallel to bedding. Some evidence suggests that mineralized rocks in the Antler High are hosted within a dense network of narrow fractures similar to a stockwork. A number of steeply dipping mineralized faults are known to cut through the Antler High, and these mineralized faults may have served as feeder structures. The strike and dip angles of the latter structures are sub-parallel to those of the Wayne and Sequoia zones. Along the northern edge of the Antler High, several of these structures trend upward through the Golconda thrust and into the overlying highly siliceous rocks of the Havallah sequence. The combination of these specific high-angle structures and local, low-angle mineralized faults is referred to as the Chaotic zone (fig. 97). In addition to the high-angle structural control in the Antler High, a low angle, 45° east, west emergent compressional feature termed the Redwood fault appears to control a substantial part of the distribution of mineralized rocks in that zone. The Redwood fault effectively doubles the thickness of the Edna Mountain Formation within the Antler High. Mineralized rocks are present within the fault plane and within highly shattered rocks of the adjacent hangingwall block. The age of the Redwood fault is not known, but some evidence suggests that it predates the Sonoma orogeny.

Mineralized rocks also are present within narrow north-northwest and north-northeast striking faults and fractures, both internally and externally to the Wayne zone. Dilatant zones at intersections of these structures host high grade ore shoots. These zones are narrow and have discontinuous strike lengths away from the Wayne zone. The dips of these faults vary from vertical to 50° east or west. An example of this type of subsidiary zone is the HW–4 zone in the South pit at Lone Tree (figs. 97, 98). The HW–4 zone dips steeply, more than 75°, to the east into the Powerline fault. This structure, and others like it, may continue at depth beneath the Powerline fault into the footwall.

Zones of shallow-dipping, discontinuous mineralized rocks also have been mapped at the Lone Tree Mine. Examples of these are the HW–1 zone and the Sucker zone (figs. 97, 98). The HW–1 zone is interpreted to be a bedding plane discontinuity or intra-formational thrust fault within the Havallah sequence. The zone is relatively thin, less than 5 m wide, but contained as much as 8.5 g Au/t.

As a point of interest, mineralized rocks along the HW–1 zone were encountered in the original intercept in the discovery hole at the Lone Tree Mine. The Sucker zone is exposed in the South pit, and strikes northwest, intersecting the Powerline fault near the center of the pit. However, in contrast with the Powerline fault, the Sucker zone dips to the east at approximately 45°. Grades of up to 34 g Au/t have been obtained from this thin zone that is generally less than 2 m wide (fig. 98).

A principal characteristic of the Lone Tree deposit is the spatial coincidence of several structurally-controlled episodes of mineralization. Hydrothermal breccias, containing as much as 25 volume percent matrix expansion, host a significant part of the Au. High grade ore is present at fault or fracture intersections, or at jogs in the trace of the faults, which form dilatant zones. Silicified, multiple-phase breccias are present along the margins of the major mineralized zones. These appear to be relatively early paragenetically, and, in general, they are of low grade. Late tectonic breccias have been superposed on the hydrothermal breccias. The most recently active faults tend to be milled-breccia, post-mineral faults and shears, which commonly contain greater than 50 volume percent clay gouge, and display a crude lamination produced by streaks of Fe–oxide minerals, pyrite, or clasts. Reactivation of high-angle faults is demonstrated by barren, vuggy silica-cemented structures that overprint similarly oriented mineralized zones. Crackle breccias developed within brittle rocks of the Edna Mountain and Valmy Formations also are mineralized, and the crackle breccias are crosscut by the Wayne zone. Areas of intense micro-fracturing in highly silicified rocks of the Edna Mountain Formation are the closest approximation to "classical" disseminated styles of mineralization yet noted at Lone Tree.

Although not conclusively documented, especially on the ore shoot scale, many Tertiary structural elements currently known at Lone Tree appear to fit into a general scheme of extensional tectonics, notwithstanding the fact that the Lone Tree deposit is present at the margin of a bedrock block essentially surrounded by alluvium, which precludes complete understanding of the district-scale setting of the deposit. The deposit also may have formed in response to a combination of strike-slip and normal faulting related to regional wrench faults.

HYDROTHERMAL ALTERATION

The main alteration types in the area of the Lone Tree Mine are argillic, silicic, potassic, propylitic, and skarn. However, some pre-mineral crystallization of secondary quartz is characteristic for sandy limestone of the hpls subunit in the Havallah sequence and weakly calcareous sandstone and lithic arenite of the Edna Mountain

Table 22—Chemical compositions of major rock types in Devonian, Mississippian, Pennsylvanian, Permian Havallah sequence showing contrasts between unaltered and altered rocks at Lone Tree gold deposit, Humboldt County, Nevada.

Alter-ation[1]	Lithology	Unit[2]	No. of Analyses	SiO_2	Al_2O_3	Fe_2O_3[3]	MgO	CaO	Na_2O	K_2O	TiO_2	P_2O_5	Loss on ignition
				Chemical analyses (weight percent)									
unalt.	chert	hc	8	87.65	5.27	2.5	0.52	0.16	0.06	1.03	0.24	0.05	2.24
unalt.	argillite	hc	52	77.63	8.14	4.42	1.27	.91	.21	1.9	.51	.14	3.33
unalt.	siltstone	hc	20	76.77	7.5	5.17	.91	1.44	.23	1.61	.47	.18	5.15
alt.	mineralized siltstone,sandstone, avg. Au = 0.158 oz/t	hc	21	78.56	7.59	5.16	.99	.79	.12	1.83	.47	.27	3.05
alt.	mineralized breccia avg. Au = 0.105 oz/st	hc	5	77.91	9.18	4.44	.57	.66	.07	2.1	.51	.44	3.9
unalt.	basalt	hb	56	47.38	15.82	9.82	4.9	5.35	1.38	1.54	2.36	.53	9.7
unalt.	basalt	hpls	5	50.11	16.75	11.78	4.28	4.24	.71	2.71	1.32	.21	7.85
alt.	mineralized basalt avg. Au = 0.307 oz/st	hb	13	46.93	16.77	10.72	2.34	2.85	.27	3.23	3.84	.77	10.6
unalt.	sandy limestone	hpls	2	53.8	2.51	1.35	.75	21.59	.3	.32	.17	.36	18.75
unalt.	prebble conglomerate	hpls	4	80.19	6.79	4.47	1.3	.99	1.45	.74	.36	.33	3.15
alt.	mudstone/siltstone avg. Au = 0.182 oz/st	hpls	14	68.72	11.95	6.94	1.02	.95	.09	2.83	.71	.45	5.73
alt.	sandstone avg. Au = 0.05 oz/st	hpls	41	73.33	10.33	5.47	1.1	.98	.26	2.18	.56	.36	5.61
unalt.	siltstone/mudstone	hys/hsh	6	80.38	7.76	2.86	.96	1.66	.44	1.55	.43	.24	3.4
unalt.	siltstone	hy	5	72.49	8.37	3.65	2.52	3.74	.43	2.12	.61	.2	5.14
alt.	mineralized breccia avg. Au = 0.318 oz/st	hys/hsh	5	83.05	5.51	4.16	.36	.39	.08	1.26	.36	.2	2.18
	skarn	various, mainly hpls	6	56.66	14.58	6.26	3.52	9.4	1.05	2.57	.59	.1	4.42

[1]Alteration; unalt., unaltered; alt., altered.
[2]Units same as fig. 98.
[3]Total Fe as Fe_2O_3.

Formation throughout the Lone Tree Mine. The chemistry of this process is discussed further in this section. Pre-mineral recrystallization was required for repeated brittle fracturing and development of competency contrast. The overall type of alteration changes downward from an oxidized argillic alteration in rocks of the Havallah sequence into unoxidized argillic, silicic, and potassic alteration in rocks belonging to the Antler sequence and Valmy Formation. Propylitic alteration (calcite, epidote, chlorite, nontronite, and smectite clay) has affected basaltic rocks of the Havallah sequence and also sandstones of the Antler sequence. Minor development of pre-mineral skarn is associated with sparse calcareous sandstone and sandy limestone of the Havallah seqience. The strong structural control of mineralized rocks in the deposit resulted in these various alteration assemblages being intermixed in the mineralized fault zones.

The upper part of the Lone Tree Mine has been intensely oxidized by supergene fluids. A distinctive type of oxidized argillic alteration is confined to a network of fractures and faults within the Wayne zone. The host rocks in the zone are mudstone, siltstone, and basalt of the Havallah sequence. The predominant alteration assemblage consists of quartz, kaolinite, dickite, jarosite, and alunite. In high-grade oxidized rocks of orange-colored gouge in the Powerline fault, abundant coarsely crystalline kaolinite and dickite are present. Jarosite replaces kaolinite outward from the Powerline fault and is present in bright yellow-colored, low-grade oxide ore. Alunite apparently is cogenetic with kaolinite, and also is present in geothite-cemented, high-grade oxide ore. It is a soft, pink, fracture-filling material which has an excellent alunite pattern on an X–ray diffractogram. Unit-cell dimensions for the alunite are a = 6.987 Å and c = 17.289 Å (Stroffregen and Alpers, 1992). The "c" value indicates 93 mol percent of K component. Sulphur isotope data ($\delta^{34}S$ = +2.4 per mil, Geochron Laboratory, Cambridge, MA, written commun., 1992) provide evidence against a "magmatic-hydrothermal" origin for the sulfate in this alunite at the Lone Tree Mine. Rye and others (1992), consider magmatic-hydrothermal alunite to invariably have $\delta^{34}S$ in excess of 10 per mil and, generally, in excess of 20 per mil. The small positive values for $\delta^{34}S$ indicate that the main control of its sulfur isotopic signature is the surficial (atmospheric) oxidation of primary or magmatic sulfide sulfur. This apparently confirms that alunite at the Lone Tree Mine, as well as its accompanying kaolinite, must post-date mineralization and it must have formed in the weathering environment. The kaolinite-alunite-goethite oxidation descends along the faults of the Wayne zone well below the water table to a depth of about 200 m below the surface.

Hydrothermal phyllosilicate minerals generally are confined to structurally-controlled fluid conduits of the Wayne zone. Only diagenetic, detrital, mixed layer illite-smectite appears to be present in mudstone and siltstone of the Havallah sequence in the immediate area of the Lone Tree Mine.

The main alteration type associated with Au–mineralized rocks at the Lone Tree Mine is potassic alteration, which is represented by the assemblage adularia, sericite, calcite, pyrite, and rutile. At the Lone Tree Mine, this alteration type is present in rocks of the Havallah and Antler sequences, but is especially intense in quartzarenite and argillite of the Valmy Formation. Mostly fine-grained, 5- to 10-μm-wide adularia is intimately intergrown with quartz and is present where megascopic silicic alteration is strong. Adularia typically is present as small euhedral rhombs and aggregates, which, locally, make up as much as 35 volume percent of the rocks. Secondary alunite commonly replaces adularia rhombs. Illite (sericite) is coarse grained, about 40 μm wide, and contains approximately 10 volume percent smecite inter-layers. The illite has a well-ordered crystal structure and represents a 2M polytype that appears to have formed at high temperature (Deer and others, 1962b). It is spatially associated with adularia.

Late-stage vein fillings and matrix alteration in breccias and vein stockworks consist of quartz, adularia, calcite and sericite (potassic alteration), as well as Au–bearing pyrite and arsenopyrite accompanied by minor amounts of other sulfide minerals. The overall volume increase during this process is estimated at about 25 volume percent suggesting that matrix expansion is due to the movement of a hydrothermal fluid. Most quartz in the breccias is well-crystallized, and there is no widespread occurrence of metastable low-temperature varities of silica such as opal and crystobalite typically encountered in near-surface epithermal environments elsewhere.

GEOCHEMISTRY

Whole-rock chemical analyses further define chemical changes that occurred during the process of alteration and mineralization at the Lone Tree Mine. Chemical composition of units in three of the plates of the Havallah sequence (fig. 99) are given in table 22. In subunits hpls and hu, unaltered sandy limestones contain about 21.6 weight percent CaO and as much as 6.9 weight percent MgO (table 22). These dolomitic limestones are present outside of an approximately 700 m wide aureole of decalcification surrounding the Lone Tree deposit. Close to the Wayne zone, the CaO content in sandy limestone is depleted down to 0.79 weight percent in ore-hosting sandstone (table 22). Also near the Wayne zone, silicified rocks apparently increase to about 20 weight percent added SiO_2. The apparent addition also includes about 8 weight percent Al_2O_3, 4 weight percent total Fe, and 2 weight percent K_2O. How-

Table 23—Chemical compositions of major rock types in Pennsylvanian and Permian Antler sequence and Ordovician Valmy Formation showing contrasts between unaltered and altered rocks at Lone Tree gold deposit, Humboldt County, Nevada.

[1]Alteration	Lithology	[2]Unit	No. of Analyses	SiO_2	Al_2O_3	[3]Fe_2O_3	MgO	CaO	Na_2O	K_2O	TiO_2	P_2O_5	Loss on ignition
				Chemical analyses (weight percent)									
				Pennsylvanian and Permian Antler sequence									
unalt.	lithic arenite	Peml	12	91.3	2.63	1.76	0.01	0.62	0.05	0.31	0.15	0.47	2.1
unalt.	sandstone	Pems	8	83.61	6.41	2.26	.66	.84	.08	1.14	.36	1.01	3.5
alt.	lithic arenite	Peml	27	88.43	1.49	4.58	.13	.46	.09	.35	.08	.32	3.9
	avg. Au = 0.104 oz/st												
alt.	sandstone	Pems	35	83.5	5.32	3.68	.31	.74	.1	1.16	.36	.56	3.3
	avg. Au = 0.116 oz/st												
				Ordovician Valmy Formation									
unalt.	quartzite	Ova	4	90.81	1.15	4.59	0.07	0.09	0.05	0.27	0.05	0.07	2.58
alt.	quartzite	Ova	54	89.87	1.54	4.39	.09	.14	.1	.48	.08	.1	2.7
	avg. Au = 0.23 oz/st												
unalt.	argillite	Ova	9	71.65	13.53	3.76	1.22	.54	.06	3.75	.61	.43	3.4
alt.	argillite	Ova	13	71.11	12.59	4.61	.92	.63	.11	3.77	.55	.53	5.
	avg. Au = 0.255 oz/st												

[1]Alteration; unalt., unaltered; alt., altered.
[2]Units same as fig. 99.
[3]Total Fe as Fe_2O_3.

Table 24—Analytical data summarizing geochemistry of ore by host unit and lithology at Lone Tree gold deposit, Humboldt County, Nevada.†

Host Unit	Host Lithology	No. of Samples	Au grade, oz/st	As	Sb	Hg	Cu	Mo	Pb	Zn	Ag	Ba	Sr	Au/Ag
				Chemical analyses (parts per million)										
			Strata of Devonian, Mississippian, Pennsylvanian, and Permian Havallah sequence											
Pha	siltstone	19	0.121	1388	53	N.d.	327	10	25	95	1	1197	191	4.2
Phc	mudstone	9	.226	7131	37	N.d.	74	3	18	144	0.2	3046	66	6.4
Phc	siltstone	13	.146	5076	66	N.d.	203	4	44	944	2	1597	66	6.3
Phpls	weakly calcareous sandstone	17	.130	2129	32	0.3	41	2	13	61	1	398	77	5.66
			Permian Edna Mountain Formation											
Peml	lithic arenite	26	0.075	1639	131	10	170	98	14	72	2	2264	59	1.6
Pems	sandstone	30	.115	1406	69	2.7	107	31	24	52	3	5282	110	2.1
			Ordovician Valmy Formation											
Ova	quartzite	51	0.141	1637	118	N.d.	191	61	18	229	3	2958	23	3.6
Ova	argillite	10	.135	1352	89	N.d.	74	15	21	83	1	1056	88	9.5

†[Analyses reported are based on 5–ft channel samples; N.d., not determined]

Table 25—Analytical data showing local geochemical background of various map units near Lone Tree gold deposit, Humboldt County, Nevada.†

Host Unit[1]	Host Lithology	No. of Samples	As	Sb	Hg	Cu	Mo	Pb	Zn	Ag	Ba	Au/Ag
					Chemical analyses (parts per million)							
			Devonian, Mississippian, Pennsylvanian, and Permian Havallah sequence									
Pha	siltstone	11	51	7	0.1	86	3	31	73	0.4	2385	64
Phc	mudstone	45	130	6	.1	105	5	21	91	.2	4505	68
Phc	siltstone	15	78	6	.06	138	4	15	77	.3	3532	75
Phpls	calcareous sandstone	26	184	25	.8	127	4	32	166	.1	801	87
			Permian Edna Mountain Formation									
Peml	lithic arenite	12	202	52	3.6	63	22	15	42	1.3	3219	58
Pems	sandstone	8	55	41	1.4	99	8	38	65	.8	1531	204
			Ordovician Valmy Formation									
Ova	quartzite	6	248	15	N.d.	19	6	6	15	0.3	218	16
Ova	argillite	10	108	13	N.d.	110	3	9	36	.6	1307	99

[1]Units same as figure 99.

†[N.d., not determined]

ever, the amount of truly added components may be less than these values because of the loss of CO_2 during dissolution and alteration (see also, Stanley and Madeisky, 1994; Madeisky, 1995). A few remnants of incompletely decalcified rocks of the Havallah sequence are present within the Wayne zone, which rocks contain from 6 to 10 weight percent CaO. The apparent intense pre-mineral silicification of rocks of subunit hpls also is accompanied by early Pb–Zn sulfide minerals.

Near Lone Tree Hill, a galena-sphalerite zone evolves into a sphalerite-chalcopyrite skarn zone. The skarn protoliths also are sandy limestone and calcareous siltstone of subunits hpls and hu. Calc-silicate alteration is not extensive and it is present in bedded siltstone and, in places, phacoidal pods. Microprobe data show that the skarn contains clinopyroxene, intermediate in composition between Fe–rich hedenbergite and Mg–rich diopside— it also contains a small percentage of Mn–rich johannsenite (J.M. Hammarstrom, oral commun., 1992). K–feldspar, actinolite, epidote, sphene, and clinozoisite also are present. No garnet or vesuvianite have been found. The sulfide minerals include pyrite, arsenopyrite, and Cd–Co–Ni-rich pyrite (bravoite), as well as sphalerite containing blebs of chalcopyrite. Whole-rock chemistry of skarn at the Lone Tree Mine also is shown in table 22. These skarns are somewhat similar to other 39–Ma-old Antler sequence-hosted Pb–Zn skarns in the Battle Mountain Mining District studied by Theodore and Hammarstrom (1991). However, skarn at Lone Tree predates Au mineralization, and is cross-cut by subsequent fractures filled with silica-potassic alteration assemblages. The overall alteration pattern described above, that is, Pb, Zn (outer zone) —> Cu, Zn (near porphyry zone) —> 36– to 40–Ma small porphyry cropping out at the south end of Lone Tree Hill, resembles the metallogenic zones around Cu and Mo porphyry systems described in the mining district (Blake and others, 1979; Theodore and others, 1992).

Apparent additions and subtractions of cations in some major-element oxides accompanied mineralization at the Lone Tree Mine. The data show that chemical changes related to mineralization in siliciclastic rocks of subunit hc consist of additions of only Al_2O_3 and K_2O (as much as 7.2 weight percent in some of the individual analyses averaged in table 22), reflecting thereby the intensity of potassic alteration. Conversely, basaltic rocks in subunits hb and hpls have considerable chemical changes in many oxides upon alteration. The altered and mineralized samples of basalt contain less SiO_2, MgO, CaO and Na_2O and more total Fe, K_2O, TiO_2 and P_2O_5 than the unaltered ones (table 22). Addition of Fe and K is, again, related to potassic alteration, while residual enrichment in TiO_2 and P_2O_5 is caused by depletion of silica, Mg and Ca. Potassic alteration in this and other units becomes strong and intensive when K_2O content varies from 3 to 7.5 weight percent.

The chemical composition of the main lithological units in the Edna Mountain Formation and Valmy Formation is given in table 23. The most notable chemical characteristic of apparently unaltered sandstone of the Edna Mountain (subunit Pems, table 23) is its phosphatic character. The average P_2O_5 is 1.01 weight percent reaching as much as 3.46 weight percent in some individual analyzed samples. Phosphate is present as amorphous ooids and nodules among framework quartz grains. The sandstone also contains 3 to 5 volume percent apatite which may represent recrystallized conodonts. Phosphatic enrichment in the sedimentary environment is generally interpreted as an indicator of shallow-water or moderately deep water sedimentation (Nichols and Silberling, 1990). Altered sandstone of subunit Pems is thoroughly decalcified and silicified at the Lone Tree Mine. Because all rocks of the Edna Mountain Formation from which these chemical data have been obtained are proximal to the Wayne zone, their original pre-alteration CaO content is not known, although it is thought to be small—possibly from 3 to 5 weight percent CaO. As in subunit hpls of the Havallah sequence, silicification here is pre-mineral. Organic matter (3 to 5 weight percent) survived the hydrothermal activity and was converted to pyrobitumen.

Chemical changes in mineralized versus unaltered quartzarenite and argillite of the Valmy Formation are minimal given the non-reactive nature of the rock (table 23). In contrast with quartzarenite, unaltered argillite of the Valmy Formation has higher K_2O average contents (3.75 weight percent). The amount of diagenetic, exhalative(?), bedding-parallel white mica and clay in argillite varies from 55 to 68 volume percent. Mineralized argillite contains more total Fe and some increase in K.

Trace element data for 175 approximately 2–m composite samples of high-grade ore—obtained from rotary drill cuttings and core splits—are summarized in table 24. The same samples have been analyzed for their major-element chemical composition and mineral chemistry— their opaque minerals, as well, have been studied petrographically. Geochemical data from 133 approximately 2–m composite samples of relatively unaltered rocks from outside the Wayne zone (table 25) provide local trace-element background concentrations necessary for comparison with trace-element composition of ore. Comparisons of the two datasets allow estimates of gains and losses during hydrothermal activity and recognition of the dispersion halo accompanying gold mineralization. The emphasis of this section of the report is on geochemistry of the Au system centered at the Lone Tree Mine. Pre–Au mineralization, such as Pb–Zn disseminations in limestone of subunit hpls and Cu–Zn skarn in calcareous siltstone of subunit hc, are not discussed here.

The geochemistry of ore zones differs in each lithologic host (table 23). Comparison of the Havallah sequence-, Antler sequence-, and Valmy Formation-hosted ore zones shows that relatively few of the trace elements introduced together with Au have anomalously high concentrations. Arsenic correlates best with Au and varies from approximately 1,350 to 7,130 ppm. Antimony is moderately anomalous in the 37- to 131-ppm range. Mercury ranges from 0.3 to 10 ppm, and Se varies from 15 to 75 ppm. However, the data set for Se is still incomplete.

The addition of As to sulfide mineral-bearing, mostly pyritic ore is significant, amounting to 10 to 30 times background concentrations, and shows that As is one of the main metals associated with Au–bearing fluids. Arsenic in oxide ore is two to four times that of the sulfide ore due to As mobility during oxidation. The main carrier of As in oxidized rocks is Au–bearing goethite. The Sb gain in sulfide ore is rather small, from two to three times background, and only correlates weakly with Au and As. The oxidation process leads to a 50 percent decrease in Sb concentration in oxide ore.

The mineralized system at the Lone Tree Mine is not rich in Hg, and Hg does not play a significant role in the uppermost parts of the geochemical halo. The majority of high Hg concentrations are present in samples that (1) display an open-space texture that is filled with pyrite, and (2) in sulphosalt-bearing hydrothermal breccias that are hosted by rocks of the Antler sequence and Valmy Formation. These high Hg concentrations are not associated with the upper parts of the Havallah sequence, as might be expected if Hg were to penetrate these rocks and be concentrated high in the overall mineralized column of rocks. Oxidation causes depletion of Hg.

In addition to the above-mentioned trace elements, some other metals have variable relations to ore. Molybdenum locally is enriched in Valmy Formation-hosted hydrothermal breccias, to as much as 800 ppm Mo in a few individual samples. The Havallah sequence-hosted Au ore apparently is depleted in Ba, because unaltered rocks contain higher Ba background concentrations than ore. However, the part of the system at the Lone Tree Mine sampled for this study is low overall with respect to Ag, Cu, Pb, and Zn. These elements, however, are anomalously high in the precursor base metal-rich mineralized rocks possibly related to a nearby 34- to 39-Ma porphyry Cu system. However, base metals generally are sparse in the main stage of sulfide deposition and even in the late-stage sulfide minerals associated with the Au event.

Spatial distributions of Au, As, Sb, Hg, and other metals at the Lone Tree Mine show a north-trending multimetal geochemical anomaly to be associated spatially with the Au–mineralized rocks almost precisely coincident with the center of the system. Local highs of As within the overall anomaly coincide with structural intersections that also formed loci of high-grade pods of Au–mineralized rocks.

MINERALIZATION

The total amount of sulfide ore at the Lone Tree Mine constitutes about 55 percent of in-place ounces found up to 1993. Slightly more than half of this amount of ore is hosted by the rocks of the Havallah sequence; 15 percent is contained in rocks of the Antler sequence, and about 4 percent is hosted by rocks of the Valmy Formation. Most Au–sulfide mineralized rocks at Lone Tree are contained within mosaic clasts—supported by hydrothermal breccias with pyrite, arsenopyrite and silica cement. These breccia zones are overprinted by narrow mineralized shear veins containing silica and pyrite. The shear veins are commonly graditional into a mosaic breccia that include open-space filling textures containing drusy and botryoidal pyrite. Hydrothermal breccias and shear veins both host sinuous fractures which are associated with late barite-marcasite-pyrite-regular-orpiment veinlets. Mineral paragenesis in different ore samples were studied in polished sections. Subsequently, key mineral grains were analyzed by Dr. Graham Wilson at Isotrace Laboratory, University of Tronto (Canada) using an electron microprobe and, for Au only, Accelerator Mass Spectrometry (AMS), an ultrasensitive ion microprobe technique with detection limit of 1 ppb gold. Four-millimeter diameter cores were drilled from polished rock slices and target grains were optically aligned and sputtered with a 0.5 mm Cs primary beam. Also used were electron microprobe and ion probe (SIMS) data on ore composites from the Lone Tree Mine that were studied for metallurgical purposes by Drs. Stephen Kesler (University of Michigan) and Stephen Cryssoulis (Surface Science Laboratory, University of Western Ontario, Canada).

The sulfide ore at the Lone Tree Mine contains a relatively low amount of sulfide sulphur—approximately 2.2 weight percent in Havallah sequence- and Valmy Formation-hosted ore and 2.9 weight percent in Antler sequence-hosted ore. The most common sulfide minerals are pyrite and marcasite (approximately 75 volume percent of total sulfide mineral population)—arsenopyrite is the second most abundant sulfide mineral (approximately 15 volume percent). Associated with pyrite and arsenpyrite is a complex sulfide suite consisting of pyrrhotite, loellingite ($FeAs_2$), sphalerite, galena, covellite, chalcopyrite, regular, and orpiment, as well as barite and a sulphosalt assemblage consisting of tetrahedrite $(CuFeAg)_{12}(SbAs_4S)_{13}$, polybasite ($Ag_{16}Sb_2S_{11}$), wolfsbergite ($Cu_2Sb_2S_4$) and a Pb–Ag selenide of an appropriate PbAgSe composition.

Arsenopyrite was deposited in two stages: (1) in high-grade sulfide veins hosted by rocks of the Havallah and Antler sequences, arsenopyrite, tetrahedrite, and wolfsbergite are the earliest minerals in the ore paragenesis. This arsenopyrite contains rounded corroded blebs of Pb–Ag selenide and loellingite as well as inclusions of sphalerite and galena, and the arsenopyrite is intergrown with chalcopyrite. The arsenopyrite is overgrown by euhedral pyrite, that is in turn coated by fine-grained colloform pyrite. However, later (2) arsenopyrite, which is in a another paragenesis in some high-grade quartzarenite-hosted breccia of the Valmy Formation, formed later than pyrite, and contains inclusions of pyrite. Rims of arsenopyrite around pyrite crystals are common in these rocks. Mineral chemistry of eight arsenopyrite grains using SIMS Microprobe from various ore samples representing both parageneses has been determined (G. Wilson, written commun., 1992). Although further work is required to confirm validity of arsenopyrite geothermometry, the results suggest that sulfide mineral assemblages at the Lone Tree Mine initially formed at relatively high temperatures. The chemistry of arsenopyrite associated with loellingite and pyrite is relatively constant and suggests that these sulfide minerals in the system were deposited in a mesothermal environment, roughly from 400 to 450 °C, although the applicability of arsenopyrite compositions to thermometry in a shallow-seated hydrothermal environment is questionable because of the possibility of a failure to attain equilibrium (Sharp and others, 1985). Most arsenopyrite, arsenical pyrite, and other associated minor sulfide minerals were formed during the main stage of Au mineralization defined by potassic alteration (adularia, sericite, quartz, pyrite and arsenopyrite). Banded pyrite and marcasite, however, show distinct variations in minor-element contents (including As, Sb, Co, and Ni) and suggest deposition from hydrothermal fluids of rapidly changing composition and declining temperature.

Pyrite at the Lone Tree Mine varies in habit and mineral chemistry. Two principal morphological types of pyrite can be distinguished primarily on the basis of their size. The first, most abundant type is present as small, anhedral grains cemented by fine-grained spongy matte of pyrite, chalcopyrite, or quartz and forms aggregates, vuggy clots, and fine stockworks, and this type of pyrite shows a variety of parageneses with its surrounding minerals. This fine-grained pyrite, in places, contains zones of fine-grained, strikingly banded, colloform material. The presence of crystalline cores, dark intermediate zones, and bright outer zones all suggest major fluctuations in fluid chemistry during crystallization. Most Au, according to electron microprobe and Accelerator Mass Spectrometry (AMS) probe data, is present as small, less than 1 μm in size, native Au inclusions residing in fine grained pyrite, specifically in the pyrite grains contained in arsenopyrite.

The proportion of Au present in pyrite and arsenopyrite has not been established conclusively. Pyrite, however, is most likely the principal carrier of μm–sized Au because it has a relatively high Au content and it is the most abundant sulfide mineral present. Conversely, higher average content of Au in arsenopyrite may suggest that most of the Au is located in volumetrically minor amounts of arsenopyrite. A much smaller amount of Au resides in the thin bands of colloform-type pyrite. Regardless of microscopic location of Au in pyrite and arsenopyrite, the sulfide minerals must be physically decomposed in order to obtain a high recovery of Au from this refractory ore. Low-temperature oxidation has been demonstrated as the most economic process of gold recovery (Simmons, 1992).

Several AMS probe traverses across visibly zoned pyrite grains of the fine-grained variety indicate that euhedral cores are rich in As and poor in Sb. A dark intermediate zone in pyrite contains an average of 0.2 weight percent Co and 0.4 weight percent Ni. Arsenic concentrations decline from 3.30 weight percent in the core to 1.35 weight percent As at the rim. Antimony has a reverse tendency, increasing from almost zero in the core to 2.5 weight percent Sb at the margin. In some pyrite, contents in the outer zone reach 12 weight percent Sb. Correlation between Au, As, and Sb in pyrite is not strong at the Lone Tree Mine. The bands of colloform pyrite contain a great deal of As and Sb and only a small amount of Au. Conversely, the submicron-sized inclusions of pure Au in pyrite and arsenopyrite may contain no As or Sb.

The second type of pyrite is relatively coarse-grained single cubic grains that show no internal fabric and are present in vein fillings, breccia cement, and as disseminated euhedral grains. Microprobe analysis shows little gold is present in this coarse-grained pyrite. In breccia, the coarse pyrite cement locally contains growth zones with euhedral crystals forming a core surrounded by massive pyrite. This pyrite cement is mostly medium-grained and non-porous. It is partially replaced by marcasite and mixed with paragenetically late, fine-grained quartz. Marcasite is present in veinlets as porous anhedral grains, pseudomorphs after pyrite, and also forms bladed rosettes and intergrowths with pyrite.

Minor amounts of additional sulfide minerals and sulphosalts are present in different parageneses, some of which appear to predate the introduction of the bulk of the Au. Pyrrhotite and chalcopyrite are present with Au in early pyrobitumen-quartz veinlets. An apparently early assemblage of sulphosalts is the main carrier of Ag in the hydrothermal system at the Lone Tree Mine. Tetrahedrite has a silver content varying from 0.6 to 3.8 weight percent Ag, while the Pb–Ag selenide mineral contains as much as 18 weight percent Ag and polybasite contains 10 weight percent Ag. Low-iron sphalerite and galena are common in lithic arenite and quartzarenite. Intergrown chalcopyrite and sphalerite are widespread in skarn developed in siltstone of the Havallah sequence. Covellite commonly rims pyrite and is present together with chalcopyrite in lithic arenite.

The main deposition of sulfide minerals and Au was followed by shearing, additional local brecciation, and development of open space. At this time, a low-temperature supergene alteration and oxidation became intensive in the upper part of the Lone Tree deposit. This resulted in replacement of pyrite by marcasite and in development of porous marcasite. The presence of marcasite indicates acidic conditions and an epithermal temperature of about 200 °C at the time of its crystallization (Murowchik and Barnes, 1986). Late-stage sulfide mineralization consisting of pyrite, marcasite, quartz, barite, minor realgar, and orpiment formed along discontinuous fractures. Open cavities contain euhedral cubic pyrite, realgar, orpiment and drusty quartz. The timing of this late-stage mineralization is most likely post-oxidation. Deposition of marcasite in acidic hydrothermal conditions also is evidenced by kaolinite inclusions in marcasite. Some pyrite also is present as inclusions in late barite. Pyrite-filled fractures crosscut kaolinite-altered clasts in lithic arenite. Cinnabar, orpiment, and realgar are present as thin veinlets overprinted by the marcasite and pyrite. Locally, chalcopyrite and chalcocite partially filled late fractures cutting pyrite veins and rimmed open cavities. These late stage mineral assemblages have an epithermal character, and appear to have the same structural control as the early stages of mineralization in the deposit, as well as probably being associated with warm geothermal waters present in the southern parts of the Lone Tree system.

Oxide ore has been developed by acid leaching and oxidative processes in the upper part of the Lone Tree deposit, predominantly in rocks of the Havallah sequence. Free Au is present in some unsilicified fractures and vugs, and no Au was found in breccia fragments. This indicates remobilization of Au in the oxide environment on a local scale. Gold grains of 30- to 60–μm-size are common. Gold recovery by heap leaching can be as high as 90 percent in these rocks.

FLUID-INCLUSION STUDY
by
D.M. DeR. Channer

A preliminary fluid-inclusion petrographic and microthermometric study was undertaken at the University of Toronto, Department of Geology, using three ore-grade samples from the Lone Tree Mine that were selected and previously studied by E.I. Bloomstein. This study adds to the fulid-inclusion database on the LoneTree Mine previously published by Kamali and Norman (1996) and Norman and others (1996). The three selected samples are as follows:

Sample number	Type of ore	In this study designated as
CST9–784.5– 789.5	Sulfide	No. 17
CST 12-786	Sulfide	No. 26
DST 745–75D–755	Sulfide	No. 38

DEFINITIONS OF TERMS AND ABBREVIATIONS

In the discussion below, a number of commonly accepted fluid-inclusion terms are used in the following manner:

Fluid-inclusion types I, II, and III: In this study, the designation of fluid-inclusion types is based on relative ages of fluid inclusions, which were established mostly by cross-cutting relations among trails of fluid inclusions, and relations of fluid-inclusion trails to quartz crystal boundaries. This classification scheme is somewhat different from that used by others (Nash, 1976; Roedder, 1984; and many others), but was adopted in an attempt to establish which fluid inclusions, all of which are essentially free of daughter minerals, are associated with introduction of Au in the three samples studied by microthermometric methods. Type I fluid inclusions are judged to be the oldest, and type III fluid inclusions the youngest.

Bubbles: Fluid inclusions less than 1 or 2 μm wide are too small to be measured thermometrically on the fluid-inclusion stage used, and they are referred to as bubbles. Commonly, optics associated with these fluid inclusions are so poor that relative proportions of liquid and vapor cannot be estimated adequately, and, in many cases, even the presence of two phases cannot be determined. These bubble trails are designated as type I fluid inclusions and they were not studied by heating and freezing methods because of their small size. They are considered to pre-date mineralization in the deposit.

Fluid inclusions: Inclusions that trapped a fluid, and, in this study, found to be mostly greater than 2 μm in size where unequivocally associated with Au mineralization. The optical constraints of the heating and freezing system used during the study make it difficult to obtain good heating and freezing measurements on fluid inclusions less than 4 to5 μm wide.

Te: Eutectic temperature indicated by the temperature of first observed melting of the ice phase accompanying increase in temperature after the fluid inclusion has been frozen.

Th: Temperature of total homogenisation to either liquid (L) or or vapor (V), or critical point behavior (CR) e.g., Th–L.

METHODOLOGY

Standard heating and freezing techniques for fluid-inclusion microthermometry were followed (Roedder, 1984). Fluid inclusions first were mapped and described on the basis of their room temperature appearance and their textural relations with other minerals and microstructural features. Approximately 100–μm-thick, doubly polished fluid-inclusion sections were used for fluid-inclusion petrography. Fluid-inclusion maps were made of selected areas which were then cut out and prepared for microthermometry, which was carried out on a Linkam TH600 programmable heating/cooling stage (Shepherd, 1981) in the F.G. Smith Fluid Inclusion Laboratory at the University of Toronto. Calibration in the range –56.8 °C to +28.7 °C was carried out using a standard H_2O–CO_2–NaCl fluid inclusion itself calibrated against organic compounds of known melting points. Most phase changes in this temperature range, with the exception of initial melting temperatures (Te), can be determined to ±0.1 °C. For some fluid inclusions, final ice melting temperature was determined ±0.5 °C owing to lack of clear indicators of the end point of melting, even when the freezing technique was used. Above 28.7 °C, the stage was calibrated using the method of Macdonald and Spooner (1981)—the usual precision is ±1 °C in this range. The uncertainty on values of Th–V can , however, be much greater than this value owing to difficulty of observing when the vapor has totally filled the fluid-inclusion cavity. In this study, values of Th–V were taken as the lower observed temperature for possible homogenisation; therefore, these values are minima. Although fluid inclusions were cooled to as low as –180 °C, no phase changes indicative of the presence of separate CO_2–, CH_4–, or N_2–bearing phases were observed. Many fluid inclusions in one of the samples stud-

ied (sample 38) did not fill completely during heating tests because the upper limit of the stage was reached.

FLUID-INCLUSION PETROGRAPHY

GENERAL FEATURES OF SAMPLE *17*

This sample, which was obtained from drill hole CST–9 at a depth of approximately 260 m, consists of quartz clasts, mostly less than 1 mm in size, in a fine-grained, altered matrix (fig. 104*A*). Detrital quartz grains have rounded shapes except where affected by subsequent recrystallization. Sulfide grains with regular to irregular shapes are present in altered matrix and in interstices among quartz grains. The quartz clasts and altered matrix are cut by microveins of quartz, which also cut some sul-

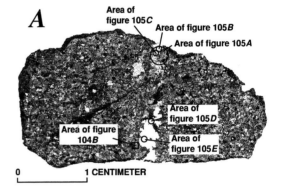

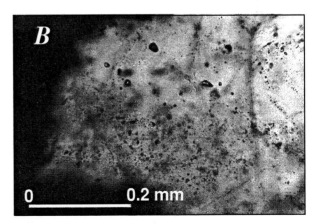

Figure 104 Photomicrographs of quartz-veined sample 17 selected for fluid-inclusion study from mineralized grit belonging to Permian Edna Mountain Formation in Lone Tree Mine. From drill hole CST–9, approximately 260–m depth. *A*, Photomicrograph of doubly-polished plate showing locations of areas of fluid-inclusions studied by heating and freezing stage (fig. 105) and area of *B*. *B*, Photomicrograph showing secondary healed microfractures that cross-cut a detrital quartz clast. Type II fluid inclusions and sulfide minerals are present along the microfractures.

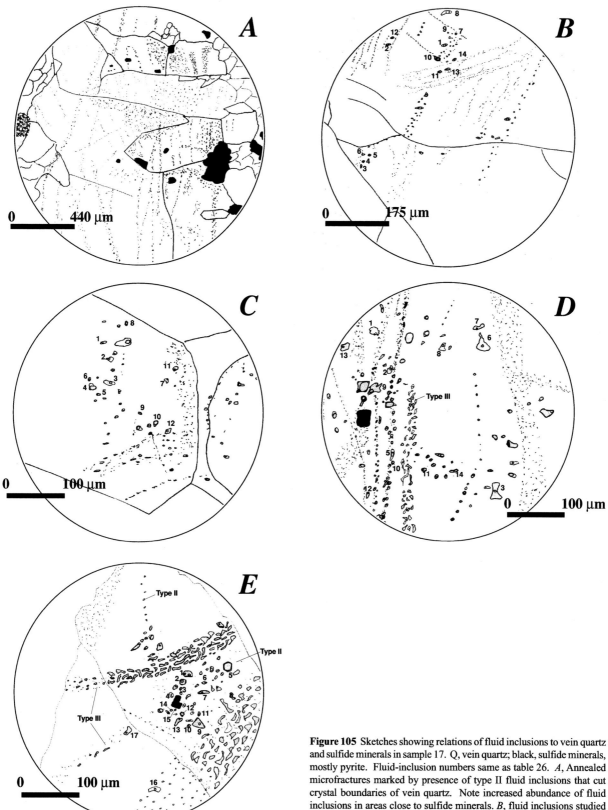

Figure 105 Sketches showing relations of fluid inclusions to vein quartz and sulfide minerals in sample 17. Q, vein quartz; black, sulfide minerals, mostly pyrite. Fluid-inclusion numbers same as table 26. *A*, Annealed microfractures marked by presence of type II fluid inclusions that cut crystal boundaries of vein quartz. Note increased abundance of fluid inclusions in areas close to sulfide minerals. *B*, fluid inclusions studied by microthermometric methods at location *A*-1. *C*, fluid inclusions studied by microthermometric methods at location *A*-2. *D*, fluid inclusions studied by microthermometric methods at location *B*-1. *E*, fluid inclusions studied by microthermometric methods at location *C*-1.

fide grains. Sulfide minerals are associated with quartz veins and are present as small grains within and among the quartz crystals. Vein quartz commonly forms euhedral crystals with little evidence of internal strain. Some crystals have discontinous undulose extinction, but no evidence of advanced sub-grain formation or dynamic recrystallization was noted. Quartz clasts have, in places, recrystallized during mineralization. Quartz-quartz grain boundaries commonly are marked by fine-grained alteration minerals, irregularly-shaped fluid inclusions, and sulfide minerals, all of which suggest significant fluid movement along grain boundaries, as well as the obvious fracture-related fluid flow described above.

Type I fluid inclusions are present as wispy bubble trails, commonly contained entirely within quartz clasts. These trails may have formed during the sedimentary history of the quartz clasts. However, type II fluid inclusions are rare, dark, and generally vapor-rich wherever they are present in the clasts of detrital quartz. They have variable shapes and are present along annealed microfractures that cut the quartz clasts. Hence, fluids associated with the type II fluid inclusions formed after the quartz clasts. In some cases, sulfide grains of a size similar to the fluid inclusions are present along the annealed microfractures (fig. 104*B*). Type III fluid inclusions form bubble trails which also cut detrital quartz clasts and the type II trails. The type III fluid inclusions are therefore late in the fluid evolution of the system. Quartz-quartz grain boundaries commonly also show nearby concentrations of dense and complex swarms of fluid inclusions of types II and III.

The single quartz vein present in sample 17 provided excellent paragenetic relations among populations of fluid inclusions. The trains of fluid inclusions cut across crystal boundaries (fig. 105*A*), and the type II fluid inclusions along these trains are therefore secondary with respect to this vein (evidence from sample 26 described below suggests that elsewhere in the deposit type II fluid inclusions and quartz veins are related). No unequivocally primary fluid inclusions were found in the quartz vein of sample 17. Type II fluid inclusions are generally vapor dominated. Coexistence of penecontemporaneous liquid- and vapor-rich fluid inclusions indicates that two immiscible fluids may have been trapped. In some small domains, liquid-rich fluid inclusions apparently are the result of a superposition and cross-cutting by late-stage type III bubble trails. Type II fluid inclusions generally contain one phase (vapor only) or two phases (liquid + vapor); no daughter minerals were observed. Some fluid inclusions contain a rounded, moderate birefringence phase which, because it is not present in all type II fluid inclusions, is interpreted as an accidentally trapped phase. Type III fluid inclusions are present mostly as bubbles on annealed microfractures or in clouds and swarms. Commonly, their relations with type II fluid inclusions are complex, but they

generally postdate them. In the rare cases where type III fluid inclusions are somewhat larger and phase proportions in them discernible, they are two phase and liquid dominated. Some type III fluid inclusions contain a low birefringence, trapped solid within them.

GENERAL FEATURES OF SAMPLE 26

The overall premineralization lithogy of sample 26 is similar to sample 17; that is, a pebbly grit belonging to the Permian Edna Mountain Formation. However, sample 26, obtained from drill hole CST–12 at a depth of approximately 239 m, shows evidence of much more extensive mineralization by introduced vein quartz and sulfide minerals than sample 17 (fig. 106*A*). Although alteration and recrystallization are pervasive in this sample, detrital quartz clasts are present within altered matrix and are cut by numerous secondary trails of fluid inclusions (fig. 106*B*). Coarse patches of sulfide minerals are intergrown with euhedral, undeformed vein quartz throughout the sample (fig. 106*C*). Fluid inclusions were studied in detrital quartz clasts and in hydrothermally introduced quartz adjacent to, and within, sulfide minerals.

Fluid-inclusion relations among detrital quartz clasts, type II fluid inclusions, and type III fluid inclusions in sample 26 are essentially the same as in sample 17 (fig. 107).

Type II fluid inclusions in hydrothermal quartz associated with sulfide minerals in sample 26 are present as randomly distributed, isolated fluid inclusions with highly variable liquid-to-vapor ratios (fig. 106*D*). Hence, type II fluid inclusions are both primary and secondary relative to the deposition of quartz and sulfide minerals, suggesting that circulation of some fluids continued after their crystallization.

GENERAL FEATURES OF SAMPLE 38

Sample 38 is intermediate between samples 17 and 26 in terms of the amount of sulfide minerals present, and its lithology includes detrital, rounded quartz and lithic clasts set in an altered fine-grained matrix. Presumably, this sample also is part of the Permian Edna Mountain Formation. The lithic clasts are altered and in some cases are traversed by a fine network of quartz- rich veinlets. The sulfide minerals form irregular grains, which in some areas appear to be coalescing into larger accumulations and enveloping some detrital quartz clasts. Sulfide minerals also are present as fine grains along quartz-quartz grain boundaries or dispersed through altered matrix and lithic clasts (fig. 108*A*). The main control on sulfide distribution may be fractures which cut quartz and lithic clasts. Recognizable clasts of quartz are relatively rare because silicification that accompanied mineralization has been

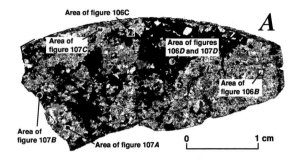

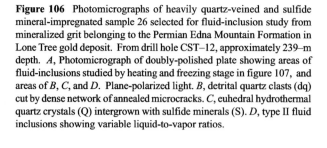

Figure 106 Photomicrographs of heavily quartz-veined and sulfide mineral-impregnated sample 26 selected for fluid-inclusion study from mineralized grit belonging to the Permian Edna Mountain Formation in Lone Tree gold deposit. From drill hole CST–12, approximately 239–m depth. *A*, Photomicrograph of doubly-polished plate showing areas of fluid-inclusions studied by heating and freezing stage in figure 107, and areas of *B*, *C*, and *D*. Plane-polarized light. *B*, detrital quartz clasts (dq) cut by dense network of annealed microcracks. *C*, euhedral hydrothermal quartz crystals (Q) intergrown with sulfide minerals (S). *D*, type II fluid inclusions showing variable liquid-to-vapor ratios.

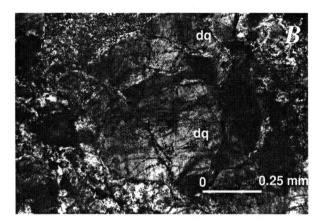

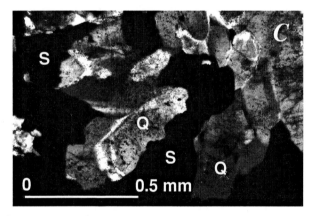

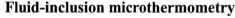

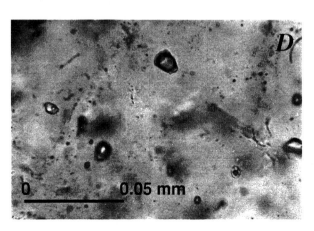

quite extensive throughout the sample. However, some quartz clasts are present and are distinguished by their rounded shapes and their surrounding fine-grained matrix. Recrystallized quartz or newly introduced quartz grains form quartz-rich patches comprised of crystals with euhedral shapes.

Quartz clasts in sample 38 contain fluid-inclusion types similar to those in the other two samples examined. The cogenetic association between type II fluid inclusions and fine-grained sulfide minerals is illustrated by sub-parallel annealed microfractures that show sulfide grains along one microfracture and type II inclusions on the other (fig. 108*B*).

In addition to the typical type II fluid inclusions along annealed microfractures, other type II fluid inclusions are present as randomly distributed and clustered networks of liquid- and vapor-rich fluid inclusions in quartz that texturally is associated closely with sulfide minerals (fig. 108*C*). This textural association among these fluid-inclusion types (fig. 109) is remarkably similar to the fluid-inclusions in many porphyry Cu systems, with the exception of the lack of NaCl daughter minerals (Nash, 1976).

Fluid-inclusion microthermometry

Sample 17

Type II fluid inclusions at location A–1 in the examined quartz vein of sample 17 are present along subparallel, annealed microfractures which crosscut and also parallel quartz crystal boundaries (fig. 105*B*). Clouds of late-stage type III fluid inclusions, mostly of bubble size, cut the trains of type II fluid inclusions. Salinities range from approximately 4 to 7 weight percent NaCl equivalent in the type II fluid inclusions, and Th values range from 240 to 389 °C to liquid and vapor (table 26). The range in degree of fill, from 10 to 90 volume percent at 40 °C, combined with the total homogenization to both liquid and vapor, indicate that immiscible or boiling fluids were trapped, and that the Th values are therefore actual trapping temperatures (Roedder, 1984).

Fluid-inclusion location A–2 is actually contiguous with location A–1 along the same vein and comprises a continuation of the prominent type II annealed microfracture studied at location A–1 (fig. 104*A*). Location A–2 also contains some fluid inclusions that are localized along grain boundaries (fig. 105*C*). The measured filling temperature and salinities are similar to those at location A–1; that is, salinities 4.8 to 5.1 weight percent NaCl equivalent, and Th from 212 to 374 °C to both liquid and vapor, and degrees of fill that are in the range 2 to 80 volume percent at 40 °C (table 26).

Location B–1, also in hydrothermal quartz along the same vein (figs. 104*A*, 105*D*), shows closely spaced subparallel annealed microfractures that contain large numbers of type II fluid inclusions at 40 °C of both the liquid-rich and vapor-rich varieties (table 26). Type II fluid inclusions also are present in isolated clusters as exemplified by fluid-inclusions numbered 8, 11, and 14 on figure 105*D*. These three fluid inclusions may be primary. A shallow dipping plane of irregularly shaped type III fluid inclusions (those numbered 1, 2, 6, 7, 13 on figure 105*D*) cuts the plane defined by the type II fluid inclusions. Type II fluid inclusions show widely varying salinities, from 1.2 to 24 weight percent NaCl+CaCl$_2$ equivalent (table 26). Temperatures of homogenization range from 249 to 434 °C to both liquid and vapor; degrees of fill range from 5 to 7 volume percent. Type III fluid inclusions, however, have restricted salinities—4.0 to 4.7 weight percent NaCl equivalent—and Th values (152 to 190 °C), and they are liquid dominated (table 26). These low temperatures of homogenization confirm that the fluids associated with the type III fluid inclusions are part of the the late paragenetic stages of the mineralized system.

At location C–1 in sample 17, an area also along the same vein as locations A–1, A–2, and B–1, type II fluid inclusions are present both in clusters and along annealed microfractures (fig. 105*E*). These are cut by type III annealed microfractures. Type II fluid inclusions in clusters have salinities of 4 to 6 weight percent NaCl equivalent, variable phase ratios, and Th values ranging from 249 to 374 °C to both liquid and vapor (table 26). Type III fluid inclusions examined by heating and freezing tests have properties similar to type III fluid inclusions at location B–1 (that is, approximately 4 weight percent NaCl equivalent, and Th of about 150 °C).

SAMPLE 26

Type II fluid inclusions in sample 26 at location 1, which is an area of recrystallized quartz close to a patch of sulfide minerals, are present in clusters and on annealed microfractures (fig. 107*A*). There are a few small clots of sulfide minerals intergrown with quartz close to the fluid inclusions on which the heating and freezing tests were done. The type II fluid inclusions have salinities in the range 6.7 to approximately 27 weight percent NaCl+CaCl$_2$ equivalent (table 27). They also have variable proportions of liquid and vapor at 40 °C, and Th is in the range 214 to 399 °C to both liquid and vapor. In fluid inclusions numbered 1, 2, and 3 (fig. 107*A*), melting of a hydrate phase was observed at temperatures from –45 to –40 °C which, when combined with final melting of ice values of approximately –34 °C, suggest a X_{Na} (where X_{Na} = Na/(Na+Ca)) of approximately 0.1 (Oakes and others, 1990). An unknown solid phase present in fluid inclusions 1 and 2 melted at approximately 150 °C during heating tests.

Location 5 in sample 26 lies adjacent to some relatively large sulfide grains, and contains type II fluid inclusions on annealed microfractures (fig. 107*C*). Some fluid inclusions also are present along grain boundaries. Type II fluid inclusions have variable salinities (3.2 to 29 weight percent NaCl equivalent), and all of these fluid inclusions homogenized to liquid in the range 185 to 264 °C (table 27).

Location 9 of sample 26 is partially enclosed by sulfide minerals and contains type II fluid inclusions on annealed microfractures and in clusters (fig. 107*D*). Type II fluid inclusions show salinities from 3 to 14 weight percent NaCl equivalent, variable phase ratios (5 to 80 volume percent at 40 °C), and Th values from 160 to 396 °C to both liquid and vapor (table 27). The type III fluid inclusions show salinitites near the low end of the range of salinities found in the type II fluid inclusions (3.4 to 4.5 weight percent NaCl equivalent), and the type III fluid inclusions have Th values in the range 105 to 160 °C (table 27).

SAMPLE 38

The quartz grain which hosts the fluid inclusions examined on the heating and freezing stage at location 5 of sample 38 is partially enclosed by sulfide minerals (fig. 109*A*). In addition, the quartz grain contains prominent annealed microfractures containing a variety of type II fluid inclusions along them. One microfracture, defined by fluid inclusions numbered 1 to 8, has fluid inclusions of extremely restricted salinities (approximately 5.7 weight percent NaCl equivalent), degree of fill that is in the range 30 to 60 volume percent at 40 °C, and Th in the range 360 to 380 °C (table 28). Another microfracture in the quartz grain contains liquid-rich fluid inclusions whose degrees of fill are 20 to 30 volume percent, and whose salinities are 22 to 25 weight percent NaCl+CaCl$_2$ equivalent with homogenization to liquid in the range 314 to 465 °C (table

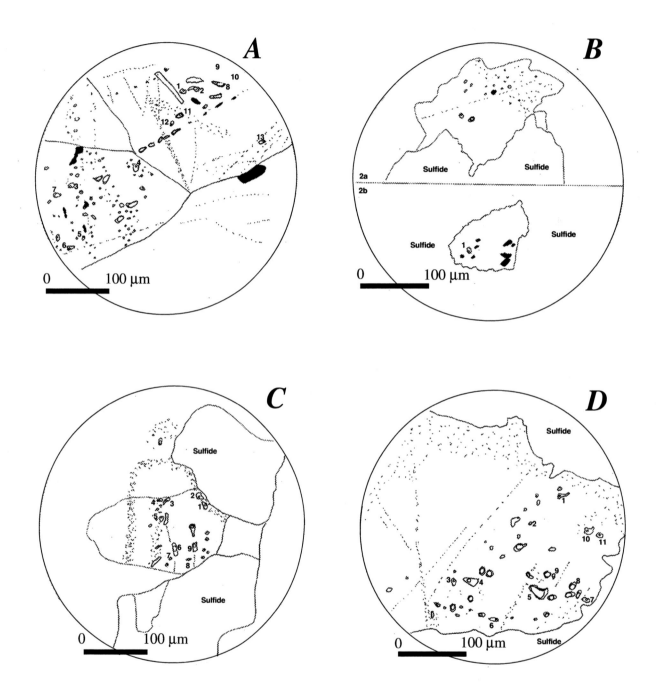

Figure 107 Sketches showing relations of fluid inclusions to vein quartz and sulfide minerals in sample 26. Q, vein quartz; black or S, sulfide minerals, mostly pyrite. Fluid-inclusion numbers same as table 27. *A*, Fluid inclusions studied at location 1. *B*, fluid inclusions studied at location 2. *C*, fluid inclusions studied at location 5. *D*, fluid inclusions studied at location 9.

28). A third microfracture in the quartz grain contains vapor dominated type II fluid inclusions which were not measured by heating and freezing tests. These three microfractures are interpreted as having trapped coexisting vapor-dominant and liquid-dominant fluids from a boiling environment.

Location 9 of sample 38 is within a recrystallized or newly grown part of a quartz crystal and contains a group of scattered, well-formed type II fluid inclusions (fig. 109B). Paragenetically later type III fluid inclusions define annealed cross-cutting microfractures. Scattered type II fluid inclusions have high salinities of approximately 23 weight percent NaCl equivalent with degrees of fill from 10 to 40 volume percent, and homogenization to liquid in the range 296 to at least 474 °C (table 28). A large number

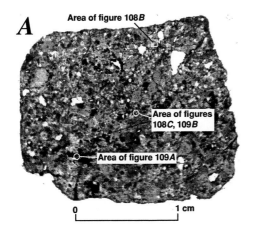
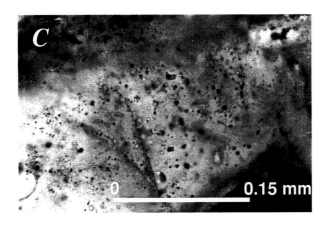
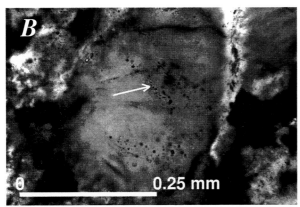

Figure 108 Photomicrographs of moderately quartz-veined and moderately sulfide mineral-impregnated sample 38 selected for fluid-inclusion study from mineralized grit belonging to Permian Edna Mountain Formation in Lone Tree Mine. From drill hole DST–745, approximately 230–m depth. A, Photomicrograph of doubly-polished plate showing locations of fluid-inclusions studied by heating and freezing stage in figure 109, as well as areas of B, and C. Plane-polarized light. B, detrital quartz clasts cut by two prominent subparallel annealed microcracks, one of which contains sulfide minerals (at head of arrow) and the other type II fluid inclusions; crossed nichols. C, coexisting primary liquid- and vapor-rich type II fluid inclusions at location 9 (table 27); crossed nichols.

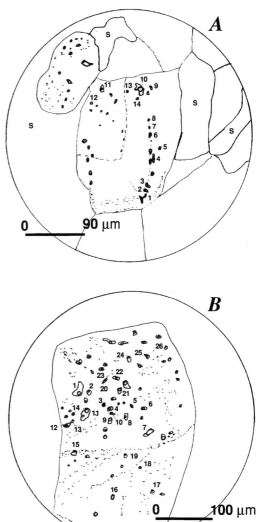

Figure 109 Sketches showing relations of fluid inclusions to recrystallized or introduced quartz and sulfide minerals in sample 38. Q, quartz; S, sulfide minerals. Fluid-inclusion numbers same as table 28. A, Fluid inclusions studied at location 5; B, fluid inclusions studied at location 9.

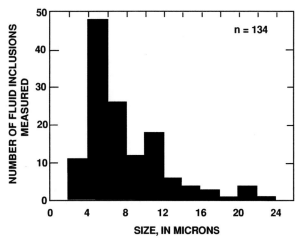

Figure 110 Histogram showing size distribution for 134 fluid inclusions in quartz from samples 17, 26 and 38 in Lone Tree Mine.

of these type II fluid inclusions did not completely homogenize at 474 °C, the upper limit of the stage. In addition, a large number of vapor-dominated fluid inclusions are also present at this location, but many of them were not measured during the heating and freezing experiments. However, one type II fluid inclusion (no. 16, fig. 109*B*) on an annealed microfracture which contains mainly vapor-rich fluid inclusions had a salinity of approximately 6 weight percent NaCl equivalentt, and a Th value (400 °C) to vapor that is similar to the high salinity, nearby fluid inclusions that fill to liquid.

Summary of fluid inclusions

A number of physical characteristics of the fluid inclusions are amenable to graphical representation. Of a total of 134 measured fluid inclusions, the minimum size

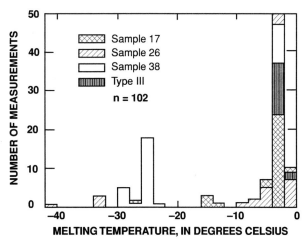

Figure 111 Histogram showing distribution of values of Tmice for 102 fluid inclusions in quartz from samples 17, 26 and 38 in Lone Tree Mine.

measured was 2 μm, the maximum was 22 μm, the mean was 7 μm, and the median was 6 μm (fig. 110). Certainly, the smallest diameter type I and type III fluid inclusions that constitute the bubble trains are much smaller than 2 μm. However, overall size of fluid inclusions in the three samples apparently is somewhat smaller than the fluid inclusions found to be associated with mineralized rocks in some porphyry systems farther to the south in the mining district (Nash and Theodore, 1971; Theodore and others, 1992).

A bimodal distribution is present for 102 values of the melting temperature of ice in the three samples, including a main peak at –4 to –2 °C and a secondary peak at –24 to –26 °C (fig. 111). Although sample 38 contributed most points to the lower peak, not enough data are available to know if this reflects a real difference with the other samples in terms of relative amounts of trapped fluid. All samples contribute approximately equally to the main

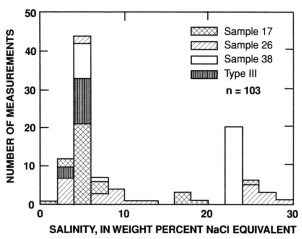

Figure 112 Distribution of salinity, in weight percent NaCl equivalent, for 103 fluid inclusions in quartz from samples 17, 26 and 38 in Lone Tree Mine.

peak. In addition, the peak of Tmice values for type III fluid inclusions, which are representative of late-stage fluids that circulated through the deposit, coincides with the main peak for the type II fluid inclusions. An almost continuous range of salinities exists from ~1 to 29 weight percent NaCl equivalent in type II fluid inclusions (fig. 112). In contrast, the type III fluid inclusions show a relatively restricted range of salinities from ~2 to 6 weight percent NaCl equivalent. Type II fluid inclusions in all three samples have degrees of fill from ~5 to ~90 volume percent vapor at 40 °C (fig. 113). The bias in numbers towards liquid-rich fluid inclusions reflects the fact that these types of fluid inclusions can be measured more easily than the vapor-rich varieties. For vapor-dominated fluid inclusions where, as in this case, the vapor is predominantly H_2O it is commonly impossible to determine either the salinity of the liquid rim or the value of Th. There is no doubt, however, from visual inspection of the polished

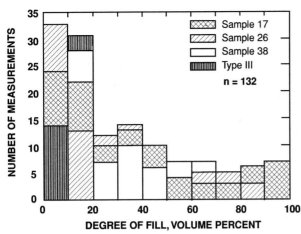

Figure 113 Histogram showing degree of fill, in volume percent at 40 °C, for 132 fluid inclusions in quartz from samples 17, 26, and 38 in Lone Tree Mine.

plates that, in terms of actual numbers of trapped type II fluid inclusions, the vapor-dominated fluid inclusions are the most abundant. Type III fluid inclusions are all liquid-rich with degrees of fill up to 20 volume percent.

Values of Th are shown in figure 114 with the type of homogenization shown for each sample. Type II fluid inclusions in all three samples homogenize to both liquid and vapor over a similar temperature range of ~125 °C to ~480 °C and maybe even higher because some fluid inclusions had not completely homogenized at the upper temperature limit of the stage (~480 °C). The degree-of-fill data described above combined with the Th data are consistent with the interpretation that immiscible fluids have been trapped. Although the Th and degree-of-fill data present an apparently simple phase separation scenario, the wide range of salinitites determined for fluid inclusions of the same type is somewhat confusing, especially because some trapped fluids are NaCl–dominated while others are $CaCl_2$–rich.

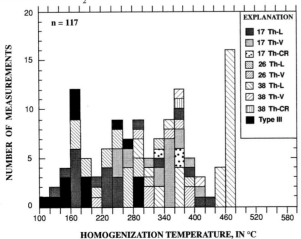

Figure 114 Histogram showing temperatures of total homogenization for 117 fluid inclusions in quartz from samples 17, 26, and 38 in Lone Tree Mine.

Whereas low- to moderate-salinity fluids are present in fluid inclusions that homogenize to either liquid or vapor, high salinity liquids only are present in fluid inclusions that homogenize to liquid (fig. 115). Because high salinity fluids apparently are compositionally distinct from low salinity NaCl–H_2O fluids, it is unlikely that they were formed by phase separation. However, some unknown amounts of Ca also may be present in low salinity fluids associated with the type II fluid inclusions, and the resolution of the chemical composition of these low salinity fluids must await further in-depth fluid-inclusion investigations. On the one hand, the high salinity fluids may represent residual fluids from a pre-existing porphyry Cu system which were caught up in an epithermal system but did not mix well, owing to their density contrast with the more dulite epithermal fluids. On the other hand, the low salinity fluids simply may represent the entry of meteoric-

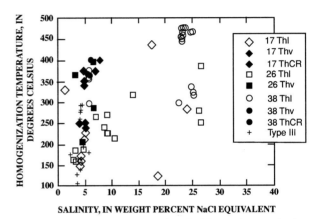

Figure 115 Plot showing salinity versus liquid-vapor homogenization temperature for 77 fluid inclusions in quartz from samples 17, 26, and 38 in Lone Tree Mine.

dominated fluids into a porphyry environment during its final stages as the fluid shells were collapsing into system as it cooled. The overall configuration of the data plotted on the salinity-versus-homogenization-temperature diagram (fig. 115) is somewhat similar to analogous diagrams for the Buckingham stockwork Mo system (Theodore and others, 1992).

Fluid-inclusion and petrographic observations have established that pyrite-arsenopyrite crystallization in the studied samples is associated with dominantly vapor-rich fluid inclusions which were immiscible with a liquid phase of variable salinity at the time of trapping. Because immiscible fluids were trapped, the Th values then represent the trapping temperatures (Roedder, 1984). No pressure correction is required. Th values for the type II fluid inclusions range from 125 to 480 °C with a mean of approximately 325 °C. Salinities of type II fluid inclusions are variable but most determinations, for both Th–L and Th–V fluid inclusions, range from 2 to 8 weight percent NaCl equivalent. This suggests that the salinity of the origi-

Table 26—Temperature and salinity data from fluid inclusions hosted by quartz in the Lone Tree gold deposit.†

Location number (fig. 105)	[1]Type	[2]Textural setting	Size (μm)	[3]Wt% NaCl equiv.	[4]Vol. %	Filling temp. (°C)	[5]Type of fill	[6]Comments
NaCl–H₂O fluid inclusions								
A - 1 - 1	II	HF	10	7.02	40	374	CR	
A - 1 - 2	II	HF	9	4.65	10	243	L	1
A - 1 - 3	II	HF	4	4.8	50	350	V	1
A - 1 - 4	II	HF	6	4.96	90	350	V	1
A - 1 - 5	II	HF	4	4.8	90	250	V	1
A - 1 - 6	II	HF	4	5.11	90	240	V	1
A - 1 - 7	II	HF	4	4.03	40	250	V	1
A - 1 - 8	II	HF	16	N.d.	30	389	V	
A - 1 - 9	II	HF	4	N.d.	60	330	V	1
A - 1 - 10	II	HF	8	N.d.	70	379	CR	
A - 1 - 11	II	HF	4	N.d.	30	340	V	1
A - 1 - 12	II	HF	5	N.d.	40	334	CR	1
A - 1 - 13	II	HF	6	N.d.	40	340	V	1
A - 1 - 14	II	HF	3	N.d.	70	320	V	1
A - 2 - 1	II	HF	7	4.96	2	225	L	
A - 2 - 2	II	HF	7	5.11	15	226	L	
A - 2 - 3	II	HF	20	4.8	5	212	L	
A - 2 - 4	II	HF	10	N.d.	80	359	V	
A - 2 - 5	II	HF	4	4.96	20	374	V	
A - 2 - 6	II	HF	3	4.8	30	340	V	1
A - 2 - 7	II	HGB	6	4.96	5	225	L	
A - 2 - 8	II	HF	5	N.d.	50	369	V	
A - 2 - 9	II	HF	4	N.d.	60	N.d.	N.d.	
A - 2 - 10	II	HF	5	N.d.	80	304	V	
A - 2 - 11	II	HGB	8	N.d.	10	259	L	
A - 2 - 12	II	HGB	12	N.d.	10	237	L	
B - 1 - 1	III	HF	10	4.65	5	190	L	2
B - 1 - 2	III	HF	8	4.03	5	N.d.	L	
B - 1 - 3	II	HF	14	1.22	5	329	L	
B - 1 - 6	III	HF	16	4.18	2	165	L	
B - 1 - 7	III	HF	10	4.18	2	181	L	
B - 1 - 8	II	CLU	10	3.87	10	N.d.	L	
B - 1 - 9	II	HF	20	3.87	10	249	N.d.	
B - 1 - 11	II	CLU	4	7.86	70	400	V	1
B - 1 - 13	III	HF	4	N.d.	5	152	L	
B - 1 - 14	II	CLU	8	N.d.	20	404	L	
C - 1 - 1	II	CLU	10	4.34	10	163	L	
C - 1 - 2	II	CLU	10	6.01	50	383	L	
C - 1 - 3	II	CLU	4	4.18	5	161	L	
C - 1 - 4	II	CLU	8	5.56	60	374	L	3

Table 26—Temperature and salinity data from fluid inclusions hosted by quartz in the Lone Tree gold deposit.†

Location number (fig. 105)	[1]Type	[2]Textural setting	Size (μm)	[3]Wt% NaCl equiv.	[4]Vol. %	Filling temp. (°C)	[5]Type of fill	[6]Comments
C - 1 - 5	II	CLU	9	N.d.	85	351	V	
C - 1 - 6	II	CLU	6	4.03	5	162	L	
C - 1 - 7	II	CLU	17	4.18	2	163	L	
C - 1 - 9	II	CLU	20	4.03	5	160	L	
C - 1 - 10	II	CLU	6	4.03	5	150	L	
C - 1 - 11	II	CLU	4	N.d.	90	260	V	1, 3
C - 1 - 12	II	CLU	4	N.d.	90	260	V	1, 3
C - 1 - 13	II	CLU	10	4.18	10	171	L	
C - 1 - 14	II	CLU	2	4.18	90	N.d.	V	1, 3
C - 1 - 15	II	CLU	2	N.d.	90	N.d.	V	1, 3
C - 1 - 16	III	HF	12	N.d.	5	162	L	
C - 1 - 17	III	HF	10	4.18	5	140	L	
$NaCl$–$CaCl_2$–H_2O Fluid inclusions								
B - 1 - 4	II	HF	12	17.73	N.d.	N.d.	N.d.	
B - 1 - 5	II	HF	18	17.42	20	N.d.	N.d.	
B - 1 - 10	II	HF	5	17.58	50	434	L	
B - 1 - 12	II	HF	4	24.	15	281	L	
C - 1 - 8	II	CLU	3	18.53	1	125	L	

[1]At room temperature: II, rare, dark, and generally vapor-rich fluid inclusions. They have variable shapes and are present on healed fractures cutting quartz clasts. Hence, type II fluids formed after the quartz clasts and are secondary in origin. In some cases, sulfide grains of a size similar to the fluid inclusions are also present along the healed fractures; III, bubble trails which cut quartz clasts and the type II trails. Grain boundaries frequently contain dense and complex swarms of fluid inclusions of types II and III.

[2]CLU, cluster of primary fluid inclusions; HF, healed fracture containing secondary or pseudo-secondary fluid inclusions; HGB, high angle grain boundary fluid inclusions of secondary origin.

[3]Salinities expressed in equivalent weight percent NaCl equivalent were calculated using the equation of Hall and others (1988), whereas salinities of more saline fluid inclusions are expressed in weight percent NaCl + $CaCl_2$ equivalent and were calculated using the equation of Oakes and others (1990).

[4]Degree of filling of a fluid inclusion by the vapor phase; this is a qualitative observation made at 40 °C.

[5]Phase present upon heating to total homogenization: L, liquid; V, vapor; CR, critical point behavior.

[6]1, Th value is a minimum owing to the difficulty of observing final Th in fluid inclusions homogenizing to vapor; 2, fluid inclusion contains a trapped solid which in several cases dissolved at ~150 °C; 3, fluid inclusion formed a clathrate, as suggested by Tm_{clath} in the range 7.0 to 8.3 °C for five fluid inclusions at loc. C–1 (see fig. 105E).

†[Sample 17 (see text); N.d., not determined]

Table 27—Temperature and salinity data from fluid inclusions hosted by quartz in the Lone Tree gold deposit.†

Location number (fig. 107)	[1]Type	[2]Textural setting	Size (μm)	[3]Wt% NaCl equiv.	[4]Vol. %	Filling temp. (°C)	[5]Type of fill	[6]Comments
				NaCl–H_2O fluid inclusions				
1•4	II	HGB	4	10.49	5	214	L	
1•5	II	HF	4	9.21	1	225	L	
1•6	II	HF	6	9.21	5	269	L	
1•7	II	ISO	8	6.74	85	284	V	2
1•8	II	CLU	10	N.d.	60	339	V	
1•9	II	CLU	3	N.d.	20	300	V	1
1•10	II	CLU	4	N.d.	10	259	L	
1•11	II	CLU	6	N.d.	5	300	V	1
1•12	II	CLU	4	N.d.	10	289	L	
1•13	II	HF	4	N.d.	10	299	L	
2a•1	II	ISO	4	N.d.	10	332	L	
5•1	II	HGB	7	5.71	5	180	L	
5•2	II	HGB	10	2.24	5	175	L	
5•3	II	HF	6	7.17	5	264	L	
5•4	II	HF	5	9.73	30	N.d.	N.d.	
5•5	II	HF	4	3.39	10	N.d.	N.d.	
5•7	II	HF	3	3.23	2	N.d.	N.d.	
5•8	II	HF	4	3.23	2	185	L	
5•9	II	HF	10	8.68	20	239	L	
9•1	III	HGB	14	4.49	5	159	L	
9•2	II	CLU	7	13.94	10	315	L	
9•3	II	CLU	5	2.9	5	163	L	
9•4	II	CLU	14	4.65	10	187	L	
9•5	II	CLU	22	6.45	70	396	V	2
9•6	II	HF	8	3.23	10	160	L	
9•7	II	CLU	6	2.9	10	161	L	
9•8	II	CLU	9	3.23	80	365	V	
9•9	II	CLU	4	4.49	70	204	V	1
9•10	III	HGB	7	3.39	5	130	L	
9•11	III	HGB	5	3.55	10	105	L	
9•12	II	CLU	10	N.d.	80	N.d.	N.d.	
9•13	II	CLU	13	N.d.	65	348	V	
9•14	II	CLU	6	N.d.	10	279	L	
				NaCl–$CaCl_2$–H_2O fluid inclusions				
1•1	II	CLU	6	26.57	5	249	L	
1•2	II	CLU	10	26.57	15	383	L	
1•3	II	HF	5	26.22	10	278	L	
5•6	II	HF	12	29.	15	N.d	N.d.	2

[1]At room temperature: II, rare, dark, and generally vapor-rich fluid inclusions. They have variable shapes and are present on healed fractures cutting quartz clasts. Hence, type II fluids formed after the quartz clasts and are secondary in origin. In some cases, sulfide grains of a size similar to the fluid inclusions are also present along the healed fractures (fig. 107A); III, bubble trails which cut quartz clasts and the type II trails. Grain boundaries frequently contain dense and complex swarms of fluid inclusions of types II and III.

[2]ISO, isolated primary fluid inclusion; CLU, cluster of primary fluid inclusions; HF, healed fracture containing secondary or pseudo-secondary fluid inclusions; HGB, high angle grain boundary fluid inclusions of secondary origin.

[3]Salinities expressed in equivalent weight percent NaCl equivalent were calculated using the equation of Hall and others (1988), whereas salinities of more saline fluid inclusions are expressed in weight percent NaCl + $CaCl_2$ equivalent and were calculated using the equation of Oakes and others (1990).

[4]Degree of filling of a fluid inclusion by the vapor phase; this is a qualitative observation made at 40 °C.

[5]Phase present upon heating to total homogenization: L, liquid; V, vapor.

[6]1, Th value is a minimum owing to the difficulty of observing final Th in fluid inclusions homogenizing to vapor; 2, determination of Tm_{ice} to ±0.1 °C was difficult so that a precision of ±0.5 °C is more likely.

†[Sample 26 (see text); N.d., not determined]

Table 28—Temperature and salinity data from fluid inclusions hosted by quartz in the Lone Tree gold deposit.†

Location number (fig. 108A)	[1]Type	[2]Textural setting	Size (μm)	[3]Wt% NaCl equiv.	[4]Vol. %	Filling temp. (°C)	[5]Type of fill	[6]Comments
				NaCl–H₂O fluid inclusions				
5•1	II	HF	10	5.71	40	365	L	2
5•2	II	HF	6	5.71	50	366	L	2
5•3	II	HF	10	5.41	30	368	V	
5•4	II	HF	20	5.71	30	359	L	2
5•5	II	HF	6	5.71	60	375	L	2
5•6	II	HF	7	5.71	40	361	L	2
5•7	II	HF	7	5.71	50	N.d.	N.d.	2
5•8	II	HF	6	5.71	50	364	L-CR	2
9•11	II	CLU	8	5.71	10	296	L	
9•15	III	HF	6	4.18	10	289	L	
9•16	II	HF	5	6.16	60	400	V	1
9•17	III	HF	4	4.18	10	282	L	
9•18	III	HF	4	4.18	5	291	L	
9•19	III	HF	4	4.18	5	277	L	
9•20	III	HF	5	4.03	5	259	L	
				NaCl–CaCl₂–H₂O fluid inclusions				
5•9	II	HF	5	25.14	20	314	L	
5•10	II	HF	14	24.76	20	465	L	3
5•11	II	HF	10	24.83	30	334	L	
5•12	II	HF	6	22.76	N.d.	N.d.	N.d.	
5•13	II	HF	3	25.14	20	320	L	
5•14	II	HF	3	25.14	20	465	L	3
9•1	II	CLU	10	23.2	10	474	L	3
9•2	II	CLU	4	23.38	15	474	L	3
9•3	II	CLU	3	22.58	20	296	L	
9•4	II	CLU	8	23.2	15	457	L	
9•5	II	CLU	4	23.16	15	444	L	
9•6	II	CLU	6	23.16	30	474	L	3
9•7	II	CLU	12	23.63	30	474	N.d.	3
9•8	II	CLU	6	23.16	20	449	L	
9•9	II	CLU	5	23.07	30	474		3
9•10	II	CLU	6	22.98	20	454	L	
9•12	II	CLU	5	23.46	40	474	L	3
9•13	II	CLU	3	23.5	30	474	L	3
9•14	II	CLU	4	23.42	40	474	L	3
9•21	II	CLU	5	23.2	15	465	L	
9•22	II	CLU	5	23.25	30	474	N.d.	3
9•23	II	CLU	4	23.03	30	474	N.d.	3
9•24	II	CLU	5	23.42	30	474	N.d.	3
9•25	II	CLU	4	23.46	40	474	N.d.	3
9•26	II	CLU	4	23.29	40	474	N.d.	3

[1]At room temperature: II, rare, dark, and generally vapor-rich fluid inclusions. They have variable shapes and are present on healed fractures cutting quartz clasts. Hence, type II fluids formed after the quartz clasts and are secondary in origin. In some cases, sulfide grains of a size similar to the fluid inclusions are also present along the healed fractures (fig. 108B); III, bubble trails which cut quartz clasts and the type II trails. Grain boundaries frequently contain dense and complex swarms of fluid inclusions of types II and III.

[2]CLU, cluster of primary fluid inclusions; HF, annealed microfracture containing secondary or pseudo-secondary fluid inclusions.

[3]Salinities expressed in equivalent weight percent NaCl equivalent were calculated using the equation of Hall and others (1988), whereas salinities of more saline fluid inclusions are expressed in weight percent NaCl + CaCl₂ equivalent and were calculated using the equation of Oakes and others (1990).

[4]Degree of filling of a fluid inclusion by the vapor phase; this is a qualitative observation made at 40 °C.

[5]Phase present upon heating to total homogenization: L, liquid; V, vapor.

[6]1, Th value is a minimum owing to the difficulty of observing final Th in fluid inclusions homogenizing to vapor; 2, determination of Tm$_{ice}$ to ±0.1 °C was difficult so that a precision of ±0.5 °C is more likely; 3, value of Th is a minimum because the upper heating limit of the stage was reached.

†[Sample 38 (see text); N.d., not determined]

nal fluid prior to phase separation also may have been in the range of 2 to 8 weight percent NaCl equivalent.

The high salinity $CaCl_2$–NaCl fluids in the Lone Tree deposit, which are part of the type II fluid-inclusion assemblage, are suggestive of the presence of a porphyry Cu system at depth somewhere in the general area of the deposit. These high salinity fluids may have been entrained into the evolving Au–rich hydrothermal system, reheated, and then trapped in crystallizing phases. High $CaCl_2$–fluids are inferred to be associated with fluid inclusions in Tertiary skarns approximately 1 km north of the Buckingham stockwork Mo system (Theodore and Hammarstrom, 1991). The slight spread of salinities toward intermediate values between the main low salinity domain to the high salinity domain at Lone Tree may be interpreted to reflect some mixing. Clathrate development in some type II fluid inclusions during freezing tests suggests that some minor amounts of CO_2 may be present in the fluids but not in such quantities that a separate liquid–CO_2 phase forms at room temperatures. Since this fluid-inclusion study was completed, halite-bearing fluid inclusions have been discovered in some mineralized quartz veins from the Chaotic zone (D.A. John, oral commun., 1996). In addition, many fluid inclusions from these veins contain as many as seven translucent daughter minerals, and the liquid-vapor proportions in the fluid inclusions suggest that the fluids were trapped at high temperatures in a boiling environment. These observations support the presence of a porphyry system somewhere below the Lone Tree Mine.

Fluids associated with the type III fluid inclusions postdate the type II fluids and have salinities of 2 to 6 weight percent NaCl equivalent, and Th to the liquid is in the range from 100 to 300 °C. These fluids apparently are associated with mineralized rocks that are rich in Hg and low in Au. They may represent liquid-rich components of a separated fluid phase which continued circulating after the main pulse of Au deposition. Alternatively, and more probably, they may be completely unrelated to Au–bearing fluids associated with the type II fluid inclusions.

Bulk gas chemistry of fluid inclusions

A preliminary analysis of volatile composition of bulk fluid inclusions of the same samples studied above has been completed by D.C.J. Bray (written commun., 1993) using a gas-chromatography technique. This section of the report is modified slightly from Bray's written communication describing his results. The gas chemistry of fluid inclusions can be used for two major purposes: (1) as a comparison with similar ore deposits elsewhere and modern geothermal systems, and (2) to study of the effects of phase separation and water-rock interaction on concentrations of major and trace elements.

Most gas chemistry data available in 1993 are related to Archean mineralized systems with high CO_2 contents (Bray and others, 1991). A recent review of gases from geothermal systems is given by Sasada and others (1992). Kesler and others (1986) presented some very useful data on the gas chemistry of fluid inclusions associated with ore deposits.

The gas chromatography technique applied to the samples from the Lone Tree deposit—that is, bulk crushing of about 2 grams of sample—results in a composited analysis of all fluid inclusions present, including both liquid- and vapor-rich varieties of all ages of fluid inclusions. By definition, the technique is not applicable to sulfur species hosted by the fluid-inclusion waters because the sulfur species may interact during the heating and crushing of the samples (table 29).

The fluid-inclusion microthermometric study above has shown that boiling fluids were trapped and that vapor-rich fluid inclusions are contained in the samples in an abundance greater than liquid-rich varieties. Ratios of CO_2 to gas obtained in this study, can be used as indication of the degree of phase separation which has occurred. During phase separation, gases such as CH_4 and N_2 partition into the vapor phase preferentially relative to CO_2. Therefore, the liquid-rich end member will record progressively higher CO_2-to-CH_4 or CO_2-to-N_2 ratios as phase separation proceeds. In contrast, such ratios in the vapor-rich phase will decrease gradually.

Ratios of CO_2-to-CH_4 for the Lone Tree samples range from 0.4 to 9.6, and ratios of CO_2-to-N_2 vary from 7.9 to 20.4 (table 28). In comparison with the value of these ratios determined by Sacada and others (1992) and Thomas and Spooner (1992), the ratios at the Lone Tree Au deposit are low which may reflect a case of closed system phase separation. In this case, vapor- and liquid-rich fluid inclusions are trapped together in the samples and yield bulk analyses which are close to that of the original fluid. The analysis is preliminary because the samples studied do not represent either vertical or lateral sections through the entire Lone Tree mineralized system. In addition, these analyses also include the late stage, type III fluid inclusions which may not be related to mineralization in the deposit. Nonetheless, such sample sets can provide useful information about the relative position in a column of mineralized rock where phase separation begins and dies. The data from the Lone Tree deposit also show that fluids associated with mineralization apparently became relatively depleted in gases, possibly as a reflection of their position high in the system without a suitable cap rock to retard fluid flow. The presence of some hydrothermal breccias at Lone Tree further suggests that gas

Table 29—Analytical data from gas chromatographic analyses by bulk crushing of samples from the Lone Tree gold deposit.†

	Sample No.: CST12-786' Crush #1 3200 psi	Sample No.: CST-802 Crush #1 3800 psi	Sample No.: DST-745 750-755 Crush #1 3200 psi	Sample No.: CST-9 784.5-789.5 Crush #1 2900 psi
N_2^+	0.023	0.067	0.279	0.15
CH_4	.871	.357	.229	.611
CO_2	.357	1.48	2.205	3.056
H_2O	98.743	98.011	97.265	96.163
C_2H_4	—	19 ppm	2 ppm	8 ppm
$C_2H_6^+$	53 ppm	—	39 ppm	50 ppm
COS	—	.083	.018	.013
C_3H_4	—	—	—	—
C_3H_6	—	—	—	—
C_3H_8	—	—	3 ppm	6 ppm
C_4H_8	—	—	—	—
C_4H_{10}	11 ppm	—	—	—
SO_2	—	—	—	—
Weight (g)	2.3	2.0238	2.1748	2.2331
Total Moles	4.034E-05	5.508E-06	9.498E-06	2.962E-06
Moles/g	1.749E-05	2.722E-06	4.367E-06	1.327E-06

†[in molecular percent, unless stated otherwise; ppm, parts per million; —, not detected]

loss was, in part, due to locally intense boiling that accompanied a rapid release of pressure.

***

SUMMARY AND CONCLUSIONS

The Lone Tree Au deposit is hosted by Paleozoic siliciclastic rocks composed of interleaving thrust slices. The rocks are folded but predominantly strike north and dip west. The rocks are displaced predominantly by the Wayne zone, a system of north-striking faults with unknown strike-slip and some dip-slip displacement. The latest recorded displacements were primarily a cumulative dip-slip offset of approximately 130 m.

At the mine scale, mineralized rocks are structurally-controlled by north-striking faults. High-grade ore is localized mainly at fault intersections in brittle rocks. The ore is hosted by hydrothermal breccias, containing as much as 25 volume percent matrix expansion, veins, and mineralized shear zones.

Hydrothermal alteration and mineral chemistry suggest that the Lone Tree deposit is dominated by adularia-sericite alteration, including anomalous amounts of Se as well as generally low sulphidation. The average sulfide sulfur amount is 2.5 weight percent and Se concentrations reach 150 ppm. The post-skarn ore mineralogy shows two overprinted assemblages reflecting compositions of fluids from two hydrothermal episodes at the Lone Tree Mine. During the main Au–bearing episode that is associated with hydrothermal breccias, microinclusions of Au in arsenical pyrite and arsenopyrite were deposited from fluids at temperatures possibly ranging from mesothermal (450 °C) to epithermal (200 °C), although the former temperature needs to be verified. The second episode, associated with open-space filling textures, is related to dilute epithermal fluids which deposited a small amount of weakly auriferous, Hg–rich pyrite, marcasite, orpiment, and realgar. The combination of hydrothermal breccia and open-space textures suggests significant fluctuation in chemistry and pressure between the main stage of ore deposition and the late fluids.

Comparison of the Lone Tree Mine with other Au deposits in the region shows it to be rather unique. Almost without exception, all other Au deposits or mining districts of adularia-sericite-(Se) type in the western United States are volcanic-hosted (such as Sleeper, Republic, Bodie, Comstock, Tonopah, and many others). The Lone Tree deposit is not closely related to volcanism, nor does it have the layered vein fillings and banded quartz-chalcedony veins typical of volcanic-hosted deposits. Such veins require a period of quiescence for their formation, which the multistage brecciation at the Lone Tree Mine does not indicate. The Lone Tree system also is quite different from typical Carlin-type deposits. For example, the Lone Tree Au system does not have jasperoid resulting from pervasive ore-stage silicification, nor does it contain significant organic carbon. In addition, at Lone Tree abundant indications of potassic metasomatism are present, a type of alteration that is notably absent from most Carlin-type systems. Also, distribution of mineralized rocks strictly is structurally controlled even at the scale of a thin section and is not disseminated. The Lone Tree deposit has more sulfide minerals than most Carlin-type deposits, including arsenopyrite, in particular, which is generally absent in the latter. The deposits at Lone Tree have been classified by Peters and others (1996), as well as Doebrich and Theodore (1995) as distal-disseminated Ag–Au—that is, associated genetically with a pluton-related environment. However, some geochemical aspects such as As enrichment and low base-metal content make Lone Tree similar to Carlin-type deposits.

At the Lone Tree Mine, a 2.25–km-long zone of high-grade Au–mineralized rocks containing more than 150 tonnes of Au—as of this writing in middle 1998—is associated with a north-striking fault system that includes some strike-slip displacements. Large volumes of mineralized fluid were channeled along this conduit and mixed with a large amount of near-surface groundwater, producing approximately 60 million tonnes of ore. This Au resource combined with a high intensity of oxidation to a 200–m depth, as well as many stages of brecciation, suggest that the fault system may be a part of a regional-scale deep-seated structural feature. This structural feature may represent part of a long-lived structural zone (Soloman and

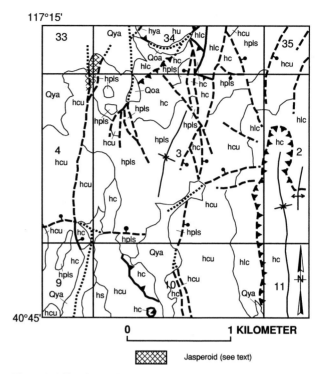

Figure 116 Sketch map showing location of jasperoid described in west-central part of Havallah Hills. Geology modified from plate 1. Geologic units same as plate 1.

Taylor, 1989; Bloomstein and others, 1990), and it may serve as a regional exploration guide.

JASPEROID ALONG HIGH-ANGLE FAULTS IN WEST-CENTRAL PART OF HAVALLAH HILLS

Jasperoid is present along the northernmost approximately 0.4–km-long segment of an at least 1.6–km-long system of high-angle, north-striking faults that crop out in the west-central part of the Havallah Hills, and that project farther north into some gravel-covered areas (fig. 116). The original definitions of jasperoid used by Spurr (1898) and followed by Lovering (1972) are followed in this report; that is, "a rock consisting essentially of cryptocrystalline, chalcedonic, or phenocrystalline silica, which has formed by the replacement of some other material, ordinarily calcite or dolomite." As mapped, the area of jasperoid alteration along the faults straddles the common section corner among secs. 33 and 34, T. 34 N., and secs. 3 and 4, T. 33 N., R. 42 E. At the time of geologic mapping in 1990, and for several years thereafter, no prospects were present on the jasperoid, nor did the jasperoid appear to have been sampled. In addition, the jasperoid has no striking color anomaly to attract prospectors and exploration geologists. The bulk of the jasperoid formed in pebbly limestone, subunit hpls, of LU–2 of the Havallah sequence (pl. 1). In addition, the rocks of the Havallah

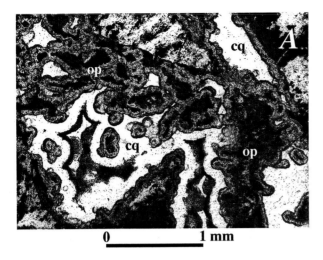

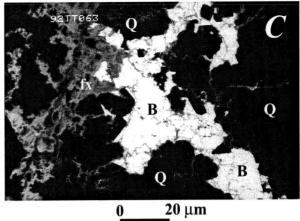

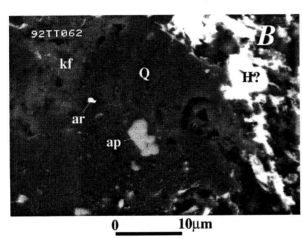

Figure 117 Photomicrographs and scanning electron micrographs of jasperoid along high-angle faults in west-central part of Havallah Hills. Q, quartz. *A*, Photomicrograph showing multiple generations of silica, including colloform opaline silica (op) and crossfiber-textured chalcedonic quartz (cq). Sample 92TT062. Plane polarized light. *B*, Backscattered electron micrograph showing 1–μm-wide grain of argentite (ar) hosted by quartz in general area of K–feldspar (kf), barium-Fe manganate, possibly hollandite (H?), and apatite (ap). Sample 92TT062. *C*, Barite (B) associated with Fe oxide (fx). Sample 92TT063.

sequence in this general area probably belong to tectonostratigraphic plate B as the various plates have been defined above. As pointed out in the subsection immediately above, subunit hpls hosts a significant part of the Au ore in the Lone Tree Mine.

Initial reconnaissance petrographic studies of five polished thin sections of rock variably altered to jasperoid and jasperoid-associated alteration assemblages along the fault reveal that these rocks record a complex and protracted history of fluid flow in their fabrics. These petrographic studies were followed by SEM investigation of three of the polished thin sections of jasperoid. Multiple generations of various types of silica in jasperoid include early-stage mammillary- or botryoidal-textured opaline silica, followed by granular-textured jasperoid, and even-

tually crossfiber-textured chalcedonic quartz—all filling what were obviously abundant open spaces at one time (fig. 117*A*). These open spaces probably resulted from initial decalcification reactions of the host rocks belonging to subunit hpls. The decalcification reactions also must have preceded deposition of the various kinds of silica now found along the fault. Included among the different types of silica in the rocks are a motley concentration of barite, sparse K–feldspar and pyrite, abundant Fe oxide replacing pyrite, Fe–Mn oxide, and apatite, together with detrital monazite, zircon, and rutile. Small blebs of argentite (Ag_2S) are present in two of the samples examined by SEM (fig. 117*B*). Iron oxide typically is associated with barite (fig. 117*C*), and some Fe oxide contains small amounts of As and Zn as revealed in 100–second counts of small domains of Fe oxide using the X–ray analyzer of the SEM. The As may be relict from the oxidation of As–bearing pyrite. Barium also seems to be present in jasperoid as Ba–Fe manganate, possibly the mineral hollandite (fig 117*B*). Thus, overall textural and paragenetic associations in jasperoid are similar to jasperoid in the Elephant Head part of the mining district (Theodore and Jones, 1992), and the mineral assemblages also bear notable similarities to those present in the Lone Tree Au deposit, particularly the presence of K–feldspar (see section above entitled "Lone Tree Gold Deposit").

Thin sequences of sandy biosparite nearby the main trace of the jasperoid-altered fault attest to lateral extent of recrystallized rocks adjacent to the fault, in as much as these sequences show 1–to 3–mm-wide seams of highly deformed calcite impregnated abundantly by Fe oxide that probably is replacing pyrite. Chemical analysis of a single composite sample of jasperoid, mostly from outcrops just to the south of the above-described section corner, reveals a concentration of 512 ppb Au, roughly 50 times what many exploration geologists consider to be local background in these types of rock (Theodore and Jones, 1992). Other metals contained in the analyzed sample of jasperoid include 7.5 ppm Ag, 145 ppm As, 10 ppm Sb, 414 ppb Hg, 64 ppm Cu, 24 ppm Pb, and 248 ppm Zn—all anomalous metal concentrations. By way of comparison, an outer halo of approximately 75 ppm As encompasses all Au–mineralized rocks present at the Lone Tree Au deposits (D.H. McGibbon, oral commun., 1992). Therefore, minor-element signature of jasperoid along high-angle faults mapped in the west-central part of the Havallah Hills appears to be remarkably similar to that at the large Au–mineralized system at the Lone Tree Mine.

Geology of the Marigold Mine area

by Douglas H. McGibbon and Andy B. Wallace

INTRODUCTION

The Marigold Mine area is in north-central Nevada about 25 km northwest of the community of Battle Mountain. Oxide-disseminated Au ore is mined from several different open pits, and is present in Paleozoic rocks of the Ordovician Valmy Formation and the Pennsylvanian and Permian Antler sequence. The deposition of Au ore at Marigold was influenced by both structural and stratigraphic controls. Mineralized rocks are inferred to have formed between approximately 39 Ma and 23 Ma and the ore fluids may have had a magmatic source. Geochemically anomalous elements in the mine area include Au, As, Sb, Hg, and Tl.

Marigold Mining Company mines disseminated Au ore from several ore zones located in southeastern Humboldt County, Nevada, about 25 km northwest of the town of Battle Mountain (fig. 94). The main gate to the minesite is about 6 km south by gravel road from the settlement of Valmy, which straddles Interstate U.S. Highway 80. The ore is treated on-site, near the base of the northern slopes of the Battle Mountain Range, in a combined carbon-in-leach mill and run-of-mine heap leach facility.

ACKNOWLEDGMENTS

The following descriptions of the geology of the Marigold Mine area are the result of the efforts of many people and would not have been possible without them. The authors extend our sincere appreciation to consultants Wayne Kemp and Peter Chapman; former Cordex geologists Joe Graney, Don Avery, Lindsay Craig, and Valerie Bilbo; former Marigold geologists David Eastwood and Roy McKinstry; Cordex geologists Tom Putney and Jim Greybeck; and Marigold geologist James Balagna. Discussions with Ted G. Theodore, Benita L. Murchey, Jeff L. Doebrich of the U.S. Geological Survey and Stephen S. Howe of the University of Vermont have been informative and thought provoking. We thank Marigold Mining Company and Rayrock Mines, Inc., for allowing us to spend time in preparation of this contribution and for granting us their permission to publish it.

HISTORY

The earliest known mining claims-related activity in the Marigold Mine area took place on July 1, 1927, when Walter A. Browne located three claims that covered part of a low ridge situated just north of Trout Creek (fig. 118). These claims, the Emma B., and Emma B. #1 and #2, covered the area that would later come to be called the Marigold Mine. In 1930, Robert Cannon and E. Douglas staked the Lillian, Opal, and Opal 1 claims nearby. In 1936, the claims were purchased by The Marigold Mines Incorporated, who operated a mine by that name in the Unionville Mining District south of Winnemucca, Nev. Frank E. Horton was named president of The Marigold Mines Inc. in 1940, and additional claims were located adjacent to the original ones.

Mining from the underground workings soon began and approximately 9,100 tonnes (t) of ore averaging 6.25 g Au/t (0.20 oz Au/ ton) were treated at nearby smelters until World War II put a stop to operations. Following Mr. Horton's death in 1943, the claims passed through a number of owners. Sporadic exploration activities were conducted around the old mine by various companies over several decades, and more claims were staked.

In 1979, Donald and Suzanne Decker purchased the claims and drilled two holes in what would later be known as the Red Rock zone. Later in 1979, Placer Amex leased the claims. During that year and 1980, Placer drilled 38 holes totalling approximately 3,100 m. This drilling tested regional structural trends and associated geochemical anomalies—some of these holes were located in what would later be called the East Hill/UNR zone.

Several small companies explored the property for the next few years. Marigold Development Company and successor companies headed by Larie Richardson crushed and heap leached about 2,800 t of ore mined from a small open pit dug above the old underground workings in 1983 and 1984, with a recorded production of 8.5 kg Au (271 troy oz Au).

Also during 1983 and 1984, VEK Associates staked several claims in the area south of Valmy, including some that covered sec. 8 on the pediment about 1.6 km north of the Old Marigold Mine (fig. 118). They conducted geophysical surveys and drilled three mud-rotary holes (VEK 8–1 through VEK 8–3) during the winter of 1984–1985 in the claim block that covered sec. 8 (fig. 118).

Gold-mineralized rocks at the Old Marigold underground mine had been mined from strata lying immediately beneath the Golconda thrust fault, which is a major regional structure (Silberling and Roberts, 1962; Roberts, 1964). Drilling of the three holes in sec. 8 was guided by the idea that the thrust fault might be the major control over Au deposition. One of these holes, VEK 8–2, report-

edly intersected approximately 15 m of well-oxidized, leach-grade, oxide Au-mineralized rocks that averaged 0.84 g Au/t (0.027 oz Au/ton) . These Au concentrations, however, did not begin until a depth of approximately 240 m from the surface.

Andy Wallace, of Cordex Exploration Company—at the time a partnership of Placer Dome, Rayrock, and Corona—visited sec. 8 with VEK's Ralph J. Roberts and other Cordex geologists in the Spring of 1985, when assay results had been received for only the first two VEK drill holes. Wallace was encouraged by the thick section of favorable Antler sequence rocks visible in the drill cuttings, the deep level of oxidation, as well as the reported presence of Au-mineralized rocks, and the strong alteration in the third hole (which had not yet been assayed, but later turned out to be barren), and he made an offer to lease the property. A few months later, in September of 1985, Cordex acquired a lease on the claims in exchange for performing approximately 3,300 m of drilling.

The exploration program of Cordex began with the drilling of two cross sections of two holes each that were designed to offset VEK holes 8–2 and 8–3 in the up-dip direction based on regional structure. Holes NM–1 through NM–4 (fig. 118) were located in alluvium-covered areas, approximately 130 and 270 m east of the respective VEK holes. The Cordex holes were nearly 1.6 km from the nearest exposures of bedrock.

Assay results of the first two Cordex holes showed only low Au contents, but holes NM–3 and NM–4 intersected approximately 25 m of 2.2 g Au/t (0.07 oz Au/ton), and approximately 30 m of 6.9 g Au/t (0.22 oz Au/ton) respectively. As luck would have it, the site of NM–4 was located almost at the center of what would become the 8 South pit. Interestingly, by the time Cordex's involvement began, approximately 200 holes had been drilled in the general Marigold area. Most holes were located adjacent to the old underground mine.

Santa Fe Pacific Mining Inc. was invited in 1986 to pool certain lands with those of the Cordex partners because it appeared that the mineralized rocks in sec. 8 would extend into sec. 17, which is a "railroad checkerboard" section. A joint venture was formed for the continued exploration of the property. In return for providing these lands to the joint venture for exploration, Santa Fe Pacific Mining Inc. received a thirty percent working interest in the project.

Later that year, additional lands that included the Old Marigold Mine and today's Top Zone, East Hill/UNR, and Red Rock ore bodies were acquired by joint venturing with Welcome North and Nevada North, Inc., two small Canadian companies that also were operated by Larie Richardson.

Following additional drilling and completion of a feasibility study, the decision was made in March of 1988

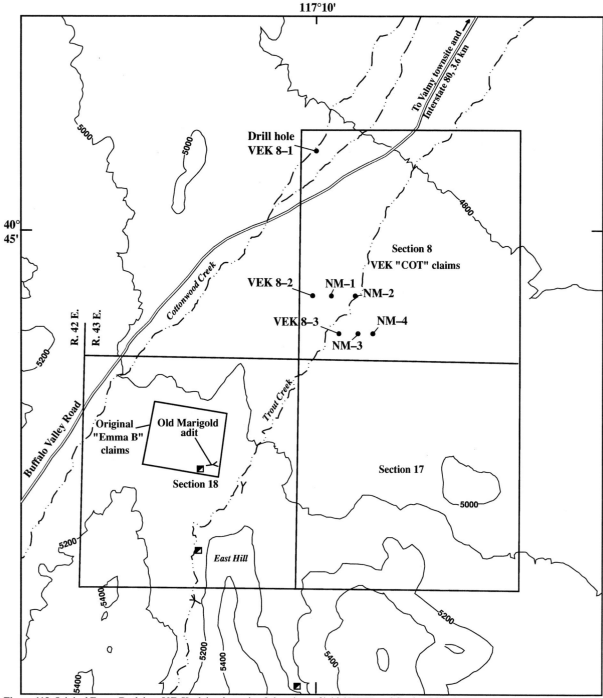

Figure 118 Original Emma B. claims, V.E. K. claims in section 8, locations of initial V.E.K. and Cordex drillholes, as well as locations of early mine workings described in text.

to develop a mine and mill-heap leach operation, with Rayrock Mines Inc. named as the operating partner. Stripping of the main "8 South" deposit—referred to in other parts of this report as the Eight South deposit—began in September of that year, utilizing a 15 cubic yard electric shovel and five 120–ton trucks. The first doré bar was poured in August of 1989.

The various joint ventures were rationalized by 1992 through purchasing the Welcome North-Nevada North interests and exchanging the then recently discovered

Table 30—Descriptions of Paleozoic units in the general area of Marigold Mine complex.

Age	Unit	Formation	Lithology	Maximum Thickness
Devonian, Mississippian, Pennsylvanian, Permian[1]	Havallah sequence	"Havallah" of Roberts (1964)	Tectonically stacked thrust plates dominantly composed of chert plus siltstone (with minor metabasalt), sandstone, shale plus dirty limestone, calcareous sandstone, or chert plus limestone	Thousands of feet
		"Pumpernickel" of Roberts (1964)	Siliceous siltstone and chert with local metabasalt. Should be combined with Havallah "formation" in Havallah sequence	>600 ft
Permian[2]		Golconda thrust Edna Mountain Formation	Spiculitic brown siltstone overlying debris flow sedimentary breccia and siltstone	>600 ft
Late Pennsylvanian and Early Permian[3]	Antler sequence	Disconformity Antler Peak	Locally fossiliferous gray limestone or micrite, some siltstone and sandstone	200 ft
Middle Pennsylvanian[3]		Disconformity Battle Formation	Chert-quartzite cobble conglomerate with some sandstone	400 ft
Ordovician[3]	Allochthon of Roberts Mountains thrust	Angular unconformity Valmy Formation	Contorted quartzite and siltstone, shale, minor chert, and greenstone	Thousands of feet

[1] Stewart and others (1986); see also, this report above.
[2] B.L. Murchey (written commun., 1994).
[3] Roberts (1964).

Stonehouse orebody plus additional land to Santa Fe Pacific Mining Inc. for their 30 percent interest and certain other lands. A short while later, Rayrock purchased the Placer Dome one-third interest. The 1998 ownership of the project was Rayrock two-thirds, and Homestake Mining Company one-third. The interest of Homestake came about as a result of their having acquired Corona Gold Inc., the successor to Cordex VI partner Lacana Gold.

From the start of pre-production stripping in September, 1988, through December, 31, 1997, about 91.7 million tonnes (t) of ore plus waste have been mined. This figure includes about 17.6 million t of combined leach-plus-mill ore that collectively contained 819,567 oz Au. Approximately 4.2 million t of mill ore averaging 3.4 g (0.109 oz) Au/t has been processed in a conventional cyanide-in-leach mill with the remaining 13.4 million t of 0.69 g (0.021 oz) Au/t material treated by run-of-mine heap leaching. Milling recovery has been better than 90 per-

cent since startup and leach recoveries are near 70 percent. December 31, 1997 minable reserves at $325/oz Au of approximately 13.9 million t at an average grade of 1.0 g Au/ t (0.033 oz per ton) gold provide a mine life through 2004, but the potential for extending this by further discoveries is quite good.

MINE AREA GEOLOGY

STRATIGRAPHY

The regional geology will not be treated in detail here as it has been documented elsewhere in this report (see sections above entitled "Regional Structural and Stratigraphic Features of Lone Tree Hill and the Battle Mountain Mining District," and "Geology of the Valmy, North

Peak and Snow Gulch Quadrangles"). The stratigraphic units of interest in the mine area (table 30) include rocks belonging to three structurally and lithologically distinct packages. These are, in ascending order; the Valmy Formation, the Antler sequence, and the Havallah sequence. Figure 119 is a simplified geologic map of the mine area showing the locations of the most significant areas of known mineralized rock. The descriptions and comments below refer to what has been observed by us in the mine area, and may differ in some respects from other published reports with more regional emphases (see also, Roberts, 1964).

Valmy formation

The oldest rocks exposed in the mine area are Early to Middle Ordovician (Roberts, 1964) and are assigned to the Valmy Formation. The Valmy Formation is part of the allochthon emplaced during Devonian-to-Mississippian time in the Antler orogeny. The entire formation is severely deformed, and its stratigraphic base is not exposed in the mine area. The Valmy Formation is composed of metaquartzarenite, silty shale, and argillite, as well as minor quartzarenite and minor bedded chert. Our understanding of the stratigraphy of the Valmy Formation is hindered by lack of distinctive marker beds in the formation and by its high degree of deformation, but it can be subdivided roughly by lithologies.

The rocks of the Valmy Formation have undergone at least three major orogenic events, including the Antler orogeny, Sonoma orogeny, and extensional tectonics associated with development of the Basin and Range. Thus, the rocks of the Valmy Formation show markedly more deformation than the overlying Antler sequence. Among the types of deformation found in the Valmy Formation are tightly to isoclinally folded beds, low-angle faults that truncate packages of folded beds, and high-angle faults that have various strikes. Some of these high-angle faults are older than the low-angle structures, while others are younger.

Thickness of the Valmy Formation is unknown, but is in the range of a few thousand meters. The upper contact is a regional angular unconformity, and we have not seen or drilled through the basal contact on the property.

Antler sequence

The Late Devonian and (or) Early Mississippian Antler orogeny created a regional, uplifted belt composed of "western facies" or siliceous-volcanic assemblage rock units such as the Valmy Formation (Roberts and others,

1958; Roberts, 1964). Erosion of this highland produced coarse sediment that was redeposited during Pennsylvanian and Permian time in the general area of the Battle Mountain Mining District to form the "overlap", or Antler sequence, clastic rocks.

The Antler sequence includes clastic rocks composed of recycled material derived from the eroded Valmy Formation of the Antler orogenic belt, along with interbedded impure carbonate rocks of shallow-marine origin (Saller and Dickinson, 1982). These upper Paleozoic rocks present in the mine area include: the Battle Formation, immediately overlying the Valmy Formation; the Antler Peak Limestone, which lies above the Battle Formation; and, capping the sequence, the Edna Mountain Formation. Subunits of each of these formations are present locally. Elsewhere in the region, several other formations also are present in the overlap assemblage.

Formations comprising the Antler sequence generally tend to have lens-like geometries and they also have marked facies changes over short lateral distances, suggesting the presence of numerous subsidiary depositional basins during sedimentation. The Antler sequence generally has not been folded. The units of the Antler sequence are much less deformed than both the underlying Valmy Formation and overlying rocks of the Havallah sequence, both of which are believed to have been emplaced along thrust faults.

Battle Formation

The Battle Formation typically contains conglomerate beds resting on eroded Valmy Formation, along with some beds of arenite and shale. The age of the Battle Formation is reported as Middle Pennsylvanian by Roberts (1964), as well as in this report above. It locally may pinch out against highs developed on eroded Valmy Formation. The basal part of the Battle Formation in the mine area is typically composed of coarse, chert and metaquartzarenite cobble conglomerate, which contains a variable amount of coarse sand matrix with the same grain compositions as the cobbles. A distinctive shale unit with ripple marks that are enhanced by fine Fe–oxide staining is present just above the conglomerate in some places. Elsewhere, rocks of the Battle Formation may be a coarse siliceous arenite with interbedded conglomerate. Thickness of the Battle Formation in the general Marigold Mine area varies from zero to approximately 130 m.

Antler Peak Limestone

The unit overlying the Battle Formation is the Late Pennsylvanian and Early Permian Antler Peak Limestone (Roberts, 1964). Where fresh, the Antler Peak Limestone typically appears as a massive to well bedded, gray, micritic limestone, and commonly contains abundant whole or fragmentary fossils including corals, brachiopods, bryozoans, and crinoids. The Antler Peak Limestone locally contains cherty lenses resulting from diagenetic(?) silicification. The thickness of the Antler Peak Limestone ranges from zero to more than 65 m in the mine area and varies abruptly over fairly short distances.

Edna Mountain Formation

The Antler Peak Limestone is overlain in the mine area by a distinctive unit that we have correlated with the Permian Edna Mountain Formation (Roberts, 1964) on the basis of its lithology, stratigraphic position, and the presence of fossils in equivalent strata south of the mine area. Murchey and others (1994) recently have reported the presence of rhax spicules of possible Leonardian to Wordian age in samples of the spiculitic siltstone upper member of this unit that were collected from locations just south of the Red Rock ore body. However, according to Murchey and others (1994), only Wordian megafossils have been reported from units assigned to the Edna Mountain Formation. The Golconda thrust fault forms the upper contact of the formation. The lower contact, on the basis of the data in Murchey and others (1994), is a disconformity. The maximum thickness of the Edna Mountain Formation in the mine area is at least approximately 200 m.

The lower unit of the Edna Mountain Formation, as described in the mine area, consists mostly of coarse, poorly sorted debris flows or sedimentary breccias with intercalated beds of punky siltstone and lesser "grit," or coarse arenite, composed of chert and metaquartzarenite fragments. The sedimentary breccia or debris-flow beds contain mainly matrix-supported clasts of Valmy Formation-derived metaquartzarenite and chert, as well as subordinate clasts of siltstone and limestone that resemble the Antler Peak Limestone and contain similar fossils, and, rarely, clasts that appear to be recycled basal conglomerate belonging to the Battle Formation. Clast diameters in the debris-flow beds reach a maximum of approximately 2 m, with individual beds as much as 3.5 m thick. The debris-flow beds interfinger with calcareous siltstone and local distinctive "grit" beds of pebbly arenite that is composed of angular chert and metaquartzarenite grains. These grit beds have a speckled salt-and-pepper appearance.

The upper unit of the Edna Mountain Formation consists of a thick brown, or where less oxidized, gray, spicule-bearing siltstone that contains minor interbedded grit and debris flows or sedimentary breccia beds. Good exposures may be seen in the pit wall above the Old Marigold underground mine and other places. Thin sections reveal that the spicules commonly are filled by collophane, and B.L. Murchey (oral commun., 1993) reports that they are identical to rhax spicules found in the Permian Phosphoria Formation in Idaho.

Havallah Sequence

The package of late Paleozoic rocks overlying the Golconda thrust fault is referred to, in aggregate, as the Havallah sequence. The Havallah units are strongly deformed, and include a variety of lithologies. The Havallah sequence previously was considered by Roberts (1964) to include rocks referred to by him as the Pennsylvanian(?) Pumpernickel and Pennsylvanian and Permian Havallah Formations. Recent studies, including that of Stewart and others (1986) and Murchey (1990), have shown that the Havallah sequence actually is a tectonically interleaved collection of similar-appearing rocks that range in age from Early Mississippian to Early Permian in the general area of the Battle Mountain Mining District and in the Tobin Range. Among rock types commonly present in the Havallah sequence at Marigold are bedded ribbon chert, siliceous siltstone, metabasalt, dirty limestone, pebbly and limy arenite, and fine-grained calcareous arenite. Owing to the high degree of internal folding and faulting that affects rocks of the Havallah sequence, its thickness is not known. In places, however, the Havallah sequence, in total, is clearly a few thousand meters thick. The upper contact here is an angular unconformity overlain by Tertiary or Quaternary alluvium, while the sole of the Golconda thrust defines the base of the Havallah sequence.

GEOLOGY OF THE ORE ZONES

The present Marigold property includes several identified ore zones. Figure 119 shows the locations of these zones and simplified geology in the general area of the mine. Gold in all of the known ore zones is sub-microscopic in size, and all known ore is thoroughly oxidized. Regardless of these similarities among the ore bodies, one of the most interesting aspects of the Marigold property is the highly variable character of orebody geometry and hosting stratigraphy. When one considers how closely the ore zones are located with respect to one another, the degree of variation in form they display is remarkable.

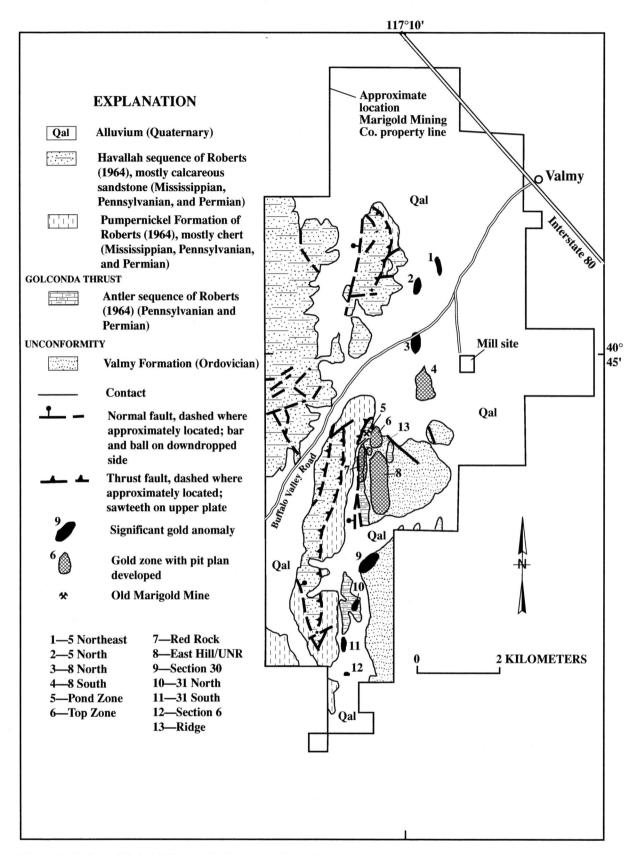

Figure 119 Geology of Marigold Mine area showing property line and location of mineralized zones.

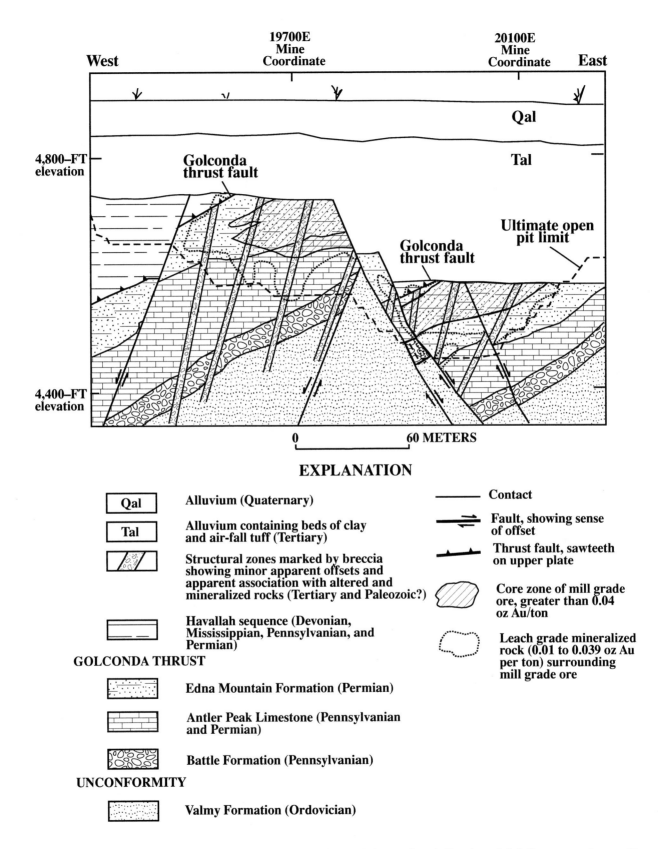

Figure 120 Vertical cross section across 8 South ore zone at 20,200-ft North (mine coordinates). Breccia symbols indicate structural zones with little apparent offset that appear to have been important controls for alteration and mineralization.

The most important ore zone known to date, and the one that made development of the property economically feasible, is called the 8 South deposit, after the section of land in which it is located. The 8 South ore zone had been the major source of mill feed to the end of 1994 at which time it was mined out. Other mineralized zones at Marigold for which at least preliminary ore reserves have been calculated are the Top, Ridge, East Hill/UNR, Red Rock, Old Marigold/Ponds, Resort, 8 North, and 5 North zones. Mineralized rocks in most of these deposits are primarily leach grade, with local pods of mill-grade ore. Mining is presently (1999) underway at the East Hill/UNR and Top zones. Much of the northern end of the Red Rock zone has been mined, but mining had not begun on the southern part as of June, 1998. The 5 North Au reserves have been determined, and the deposit is scheduled for mining later in the life of the mining complex. The Au reserves present in the 8 North zone require a higher Au price than was current in the first half of 1998 to be economic. Other occurrences of Au are known on the property, but they require further exploration development in order to be evaluated properly.

8 SOUTH ZONE

The 8 South deposit was covered by Quaternary and Tertiary alluvium ranging in thickness from approximately 50 to 100 meters (fig. 120). Original reserves for 8 South were 4.5 million tonnes (t) of ore, all oxide, at an average grade of 2.7 g (0.085 oz) Au/t. Mining of 8 South essentially has been completed, with production totalling 5.81 million t of 2.19 g (0.07 oz) Au/t ore. As can be seen on the cross section (fig. 120), part of the eastern, "down" block of the ore zone has been removed by erosion.

Gold concentrations at 8 South were restricted almost entirely to debris flows and certain siltstone beds of the lower Edna Mountain Formation and upper Antler Peak Limestone that has been leached of its carbonate minerals, leaving a punky, porous siltstone. Continuity of grades in the ore zone was excellent both along strike and laterally. Typically, a fringing envelope of leach-grade Au contents in the range 0.31 to 1.22 g Au/t (0.010 to 0.039 oz Au/t) surrounded a core of mill grade mineralized rock, which is defined as ore with a grade of 1.25 g Au/t (0.040 oz Au/t) or higher.

Types of alteration and (or) physical changes present in the 8 South deposit include: (1) leaching of carbonate cement and matrix grains in some areas, mainly leaving clay; (2) strong and varicolored pervasive Fe–oxide staining that shows variable relations with ore grade-bearing rocks; (3) several forms of barite introduction; (4) late deposition of calcite and some botryoidal silica; (5) subtle microscopic silica overgrowths on siltstone grains; and (6) minor, local development of jasperoid in pods and along feeder structures.

Alteration at Marigold is manifested by a number of different mineral assemblages throughout much of the Antler sequence, which is believed to have been calcareous. Alteration, most of which was probably related to Au mineralization, has removed some of the carbonate minerals from wall rock, although some ore, while not showing late calcite when viewed by hand lens, still reacts vigorously with dilute HCl, and may include calcite relict from the sedimentary environment. Strong goethite or limonite staining is common in the ore zone. Vivid red hematite staining is most commonly associated with high grades of Au, particularly when other colors of oxide minerals also are present. Yellow limonite in siltstone or debris-flow beds is associated with high Au grades in some places. Purple hematite(?) commonly is developed in the

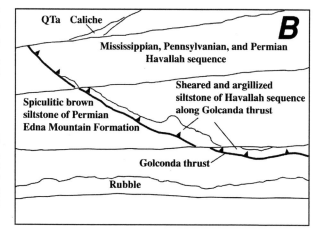

Figure 121 South towards Golconda thrust fault exposed in south wall of 8 South pit. Chert and siltstone of Devonian, Mississippian, Pennsylvanian, and Permian Havallah sequence are thrust over Permian Edna Mountain Formation brown siltstone containing sparse sedimentary breccia beds. Thrust fault has been well exposed during much of mining of 8 South pit, and must be one of the clearest exposures of thrust fault relations in Nevada. Height of lowermost bench is approximately 7 m.

"brown siltstone" upper unit of the Edna Mountain Formation, which is generally not mineralized in the deposit. Some Mn oxide also is present, but it appears to have no relation to Au–mineralized rocks. Instead, it seems to be present preferentially in certain beds.

Barite is abundant in 8 South and is present in many forms. Some of these are: massive to fine-grained replacements of certain silty beds; isolated, semi-transparent, multiply-zoned euhedral crystals as much as 15 cm long, floating in a clay and (or) silt matrix that results from leaching of carbonate minerals from originally calcareous siltstone; veins composed of aggregates of small euhedral crystals; white, subhedral aggregates forming steeply dipping veins as much as approximately 2 m thick; veins a few cm thick, transecting bedding, that appear to originate from zones of bedding replacement; and disseminated, poorly crystalline barite that partially replaces variably altered siltstone. Howe and Theodore (1993) report that the sulfur isotope signature of the barite at 8 South displays a strongly magmatic character, without much evidence of a great deal of wallrock interaction with the fluid (uniform, generally light δ^{34}S values that cluster tightly in the +10 to +12 per mil range). This is less the case for the Top Zone and East Hill/UNR barite samples analyzed to

date; however, some barite samples from Top Zone host fluid inclusions whose oxygen and deuterium isotopic ratios are nearly magmatic (Howe and others, 1995).

Silicification in 8 South was, for the most part, quite subtle. It was best seen in thin sections of siltstone host rock, appearing as very fine, doubly terminated quartz overgrowths on silt-sized detrital quartz grains that were probably exposed by removal of carbonate matrix. Late, fine, white botryoidal silica also was present locally along open fractures. Local poddy zones of jasperoid also were present, chiefly toward the north end of the pit, where they comprised a small fraction of the mineralized rocks. A late stage of introduced calcite was present in some places, and some late calcite could be seen to coat barite crystals.

Structures that apparently controlled ore deposition in 8 South were not well understood during rotary reverse-circulation delineation drilling, but have since been mapped during pit development and show up rather well in plots of blast-hole assays. These structures are north- to north-northwest striking, steeply westward-dipping zones as much as 7 m wide, that yielded large boulders during mining because of introduced silica, barite, and calcite. The rest of the ore was typically much less consolidated than the material from these structures. In many places, these

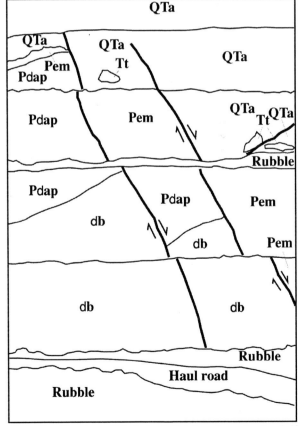

Figure 122 North showing north-trending normal fault system with distinct footwall and hanging wall fault strands that downdropped eastern half of 8 South ore zone and protected it from erosion during much of Tertiary. Light colored tuff layers (Tt) at north end of pit, offset by minor cross faults at base of alluvium, have been dated at 22.9±0.7 Ma. Tuff layers rest directly on eroded mineralized rocks of Permian Edna Mountain Formation just off right edge of photograph. Height of lowermost bench is approximately 7 m.

structural zones appeared to have been repeatedly brecciated and healed. They were commonly more silicified, baritized, and calcified than other rocks from the 8 South zone. Increased alteration and high trace-element content were typical in the structures and are present less frequently outside them. In addition, the structures were concentric with respect to local areas of high-grade Au–mineralized rocks. The maximum offset across the mineralizing structures is perhaps 7 m, and the Golconda thrust fault truncates these metal-controlling structures. These observations explain why the structures were not recognized during rotary exploration drilling. Because of the regular spacing of the drill holes and the fairly regular intersections of the structures, they were interpreted during drill-hole logging to form breccia horizons conformable with the Antler sequence, and thus were interpreted at one time to be stratigraphic in nature.

The Golconda thrust fault is spectacularly displayed in the south wall of the pit, where chert and siltstone of the Havallah sequence have been thrust over brown-to-gray, locally calcareous siltstone of the upper part of the Edna Mountain Formation (fig. 121). The thrust fault served as another control over ore deposition. Highly sheared and somewhat argillized siltstone immediately above the sole of the thrust fault, in places, capped the permeable and reactive debris flows and siltstone of the Edna Mountain Formation. Elsewhere, the upper, brown siltstone unit of the Edna Mountain Formation served as a cap or barrier to

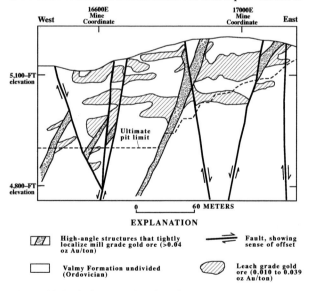

Figure 123 Vertical cross section through Top Zone deposit at 15,480–ft North mine coordinates. Stippled areas coincident with shear symbols indicate high-angle controlling structures that tightly localize mill grade ore (greater than 0.040 oz/ton Au). Gold grades drop abruptly to leach values (0.010 to 0.039 oz/ton Au) where structures abut shattered metaquartzarenite beds or thinly interbedded metaquartzarenite and shale. Locally, mineralized rocks may extend for as much as approximately 30 m along bedding. Thick sequences of shale generally do not contain Au mineralized rocks. Because of extreme structural complexity, individual lithologies are not shown on cross section.

fluid flow because of its massive, tight, non-reactive character. As seen during pit mapping, the sheared zone along the thrust fault also beheaded the mineralizing structures. The "damming" effect beneath the thrust fault probably caused the ascending hydrothermal fluids to pool beneath the thrust contact and helped focus their movement laterally, away from the source structures and along and through the more chemically and physically favorable lithologies of the rocks of the Edna Mountain Formation, in particular, and other rocks of the Antler sequence in general. This passage of Au–bearing hydrothermal fluids then also contributes to dissolution of carbonate minerals and alteration of the bedrock; Au–mineralized rock results from these and probably other changes in the physical and chemical environment.

The pediment surface, Golconda thrust fault, and the eastern half of the 8 South ore zone (fig. 120) have been downfaulted as much as 60 m by a normal fault system (fig. 122) that strikes northward and dips steeply eastward. These faults preserved part of the downthrown eastern portion of the ore zone from erosion. The fault zone extends beyond the pit walls both to the north and south—footwall- and hanging-wall fault strands have been mapped over most of the fault zone's extent. The contact of the fault with alluvium at the north end of the pit is very abrupt and quite steep; this abrupt contact may be due, in part, to recurrent movement of the fault. Several faults offset parts of the alluvial cover to within a few meters of the present surface, but none of the faults actually reached the premining surface.

Fresh biotite-bearing tuffaceous rock immediately overlies the eroded ore zone in the "down" block (fig. 122), and has been dated at 22.9±0.7 Ma based on potassium-argon analysis of biotite (see section above entitled "Potassium-argon Chronology of Cretaceous and Cenozoic Igneous Activity, Hydrothermal Alteration, and Mineralization"). This date is less than the minimum age of the mineralized rocks, because the top of that part of the ore zone that lies immediately beneath the tuff was eroded before deposition of the tuff, and presently lies at the bedrock-alluvium contact at the base of more than 70 m of alluvial cover.

Dikes and small intrusive bodies cut by mineralization are present at Stonehouse, and in sec. 31, T. 33 N., R. 43 E. (about 6 km south of the 8 South deposit), where Garside and others (1993) have reported potassium-argon ages of 39.4 Ma (biotite) and 36.4 Ma (hornblende) from unaltered granodiorite porphyry from Lone Tree Hill, although for some unknown reasons their respective error limits do not overlap (see also section above entitled "Potassium-argon Chronology of Cretaceous and Cenozoic Igneous Activity, Hydrothermal Alteration, and Mineralization"). This porphyry is probably the same rock that, where altered, hosts part of the Au at the adjacent

Table 31—Summary of trace-element geochemistry of selected samples from the 8 South deposit.†

Unit	Oz Au/Ton	As (ppm)	Hg (ppm)	Sb (ppm)	Tl (ppm)	Remarks
Shale of "Pumpernick-el" Formation	—	215.0	4.1	1.7	<0.98	Waste; above Golconda thrust and ore zone
Brown siltstone of Edna Mountain Formation	Tr.	80.	7.	1.	.2	Waste; above ore zone, below Golconda thrust
"Grit" of Edna Mountain Formation	0.016	186.	6.	1.	.6	Leach grade
Sedimentary breccia of Edna Mountain Formation	.079	342.	1.6	17.	6.1	Low grade mill ore
Yellow silt-stone of Edna Mountain Formation	.367	3060.	10.9	145.	170.	"Super" high grade mill ore
Limey siltstone of Antler Peak Limestone	.006	461.	5.8	29.	4.1	Waste; below ore zone
Chert-quartzite cobble conglo-merate of Battle Formation	—	139.	<.48	3.2	<.965	Waste; below ore zone

†[All analyses using inductively-coupled plasmas methods by Geochemical Services Inc. except Au; Au, fire assay gravimetric by Marigold Mining Company laboratory; —, not detected; Tr., trace]

Stonehouse deposit. Garside and others (1993) considered alteration and Au mineralization at Stonehouse to be late-stage events related to the igneous activity. Therefore, the probable age of mineralization at Marigold lies between about 39 and 22.9 Ma.

The geochemistry of the 8 South deposit is summarized in table 31. Trace elements that are considered anomalous include a typical suite of epithermal indicators such as As, Sb, Hg, Tl, and Ba. While some barite is present in many of the other sediment-hosted epithermal disseminated Au deposits in Nevada, it is particularly abundant at 8 South, as discussed above. Since economically significant bedded barite is present elsewhere in strata of the Valmy Formation, it is likely that the abundant, obviously epigenetic barite present here might have been remobilized from sedimentary or diagenetic barite horizons in the underlying Valmy Formation. However, ongoing sulfur isotope studies by Howe and Theodore (1993) and Howe and others (1995) indicate a strongly magmatic signature for the barite sulfur, with uniformly light $\delta^{34}S$ values.

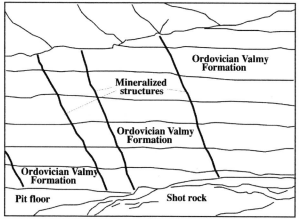

Figure 124 Southeast showing high-angle, generally steeply west-dipping structures, including those here indicated by symbols, that controlled Au deposition in Top Zone deposit. Such structures very tightly localize highest grades of Au; lateral dissemination of Au in shattered metaquartzarenite and shale shows abrupt drops in Au grades adjacent to walls of structures. Widths of mineralized bedding-conformable zones are then dependent upon geometries of individual packages of deformed beds of Ordovician Valmy Formation that are juxtaposed against structures. Bench height, approximately 13 m.

Top Zone

The Top Zone deposit is located near the northern end of East Hill (fig. 119). Mineralized rock was partially exposed at the surface, and an adit extending about 60 m along a structure there had been opened by earlier operators in the late 1930s and early 1940s. A few drill holes by previous operators also cut the deposit. Minable reserves at the start of production were estimated at approximately 2.96 million t of mainly leach-grade ore at an average grade of 1.1 g Au/t (0.035 oz Au/t). Minable reserves for Top Zone, as of December 31, 1997, including its recently discovered southward extension called Ridge Zone, based on a $325/oz Au price, are 4.76 million t at an average grade of 1.19 g (0.038 oz) Au/t. Production through December, 1997 was 5.62 million t mill ore plus leach material, at an average grade of 0.91 g Au/t (0.028 oz Au/t).

All mineralized rock in the Top Zone deposit is oxidized and is hosted by the Valmy Formation (fig. 123). Ferruginous calcite or ankerite veinlets, in places, form weak stockworks in the metaquartzarenite of the Valmy Formation at Top Zone, and barite is locally common in the form of fracture coatings and very sparse veinlets. More rarely, isolated prismatic replacement barite crystals as much as about 1.5 cm wide are present in the metaquartzarenite. Some beds of metaquartzarenite in the mine area contain black intergranular specks that are believed to be graphite. In places, some metaquartzarenite at Top Zone is silicified by microstockwork veins; other metaquartzarenite appears to have been flooded by creamy chalcedonic silica. Some distinctive, friable quartzarenite beds above or peripheral to ore have been altered by "sanding," but this texture is more likely due to the removal of originally calcareous or argillaceous matrix or cement and may be either supergene or hydrothermal alteration. These beds are now composed almost entirely of unlithified, well-rounded, and well-sorted quartz grains. Shale, at many places in the pit, is pervasively Fe stained in shades of maroon or purple, but this staining is unrelated to Au–mineralized rocks, and the relatively thick sequences of shale in the deposit are barren of economic concentrations of Au. Clay alteration of shale beds is common, however.

The areas of high (milling grade) Au concentrations are present along mappable high-angle structures that dip steeply to the west and, on average, strike northward (fig. 124). The structures clearly are delineated on assay maps of the blast-hole samples from the pit. These structures range from several centimeters to 1- to 7-m in thickness,

Figure 125 West showing typical intensely deformed beds of Ordovician Valmy Formation exposed in west wall of East Hill/UNR open pit, immediately south of Top Zone deposit. Dark beds are metaquartzarenite, and light ones are shale. While the beds themselves are intensely folded, individual packages of beds (predominantly metaquartzarenite, interbedded metaquartzarenite and shale, or thick sequences of shale) are present in structural contact with one another. Where they have been exposed by mining, packages of beds are bounded by both high-angle and low-angle faults that have various relative ages. Exposure height, approximately 13 m.

and Au–assay grades decline abruptly into the adjoining mineralized, shattered metaquartzarenite or interbedded metaquartzarenite and shale. Away from the structures, favorable strata host somewhat flatter and broader, gently westward-dipping, tabular zones of leach-grade Au concentrations contained mainly in limonite or goethite on fractures. Micro-brecciated metaquartzarenite with heavily limonite-stained fractures or free limonite in vugs is usually Au bearing. Where thick sequences of shale abut the structures, the shale is not mineralized and the structures may be "dead" where they pass through the shale beds. Where the structures encounter more favorable metaquartzarenite or interbedded metaquartzarenite and shale along strike, minable Au grades again may be present.

The rocks of the Valmy Formation everywhere in the Top Zone deposit are severely deformed internally, and they show lensoid packages of favorable strata yielding lensoid packages of strata favorable for mineralization (fig. 125). The resulting irregularly-shaped mining blocks require close attention to grade control during mining.

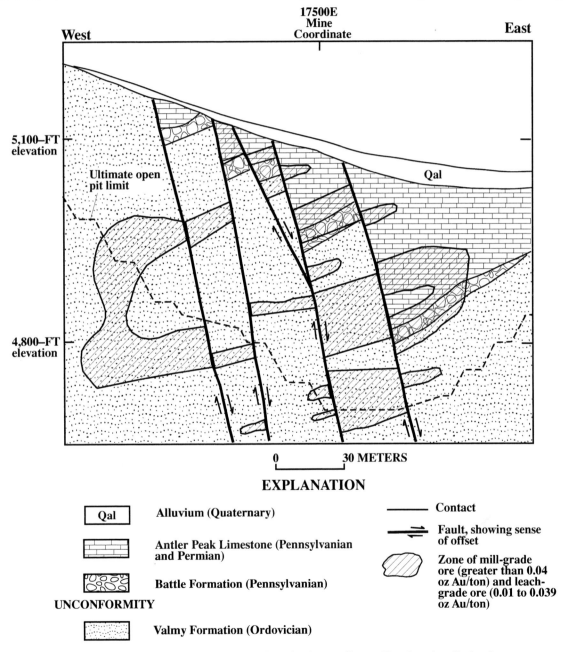

Figure 126 Northern part of East Hill/UNR ore zone at 14,400–ft North mine coordinates. Note that mineralized rocks are present at several stratigraphic levels. Best Au grades are usually found in rocks of the Pennsylvanian and Permian Antler sequence where present, but most ore is hosted by rocks of the Ordovician Valmy Formation. Antler sequence has been removed by erosion from most of southern end of deposit. Hatched pattern indicates extent of Au–mineralized rocks, including mill and leach grades.

EAST HILL/UNR ZONES

The East Hill/UNR Zones are located immediately southeast of the Top Zone pit, on the east flank of "East Hill" and along the minor headless drainage to the east (fig. 119). The mineralized rocks at East Hill/UNR have a known strike length of approximately 1,500 m and are still open to the south, where they narrow. The East Hill Zone was discovered in holes drilled by Placer Amex, and the southern part of the mineralized rocks was discovered in a shaft sunk by the Hortons. The July, 1997, design pit has two deep areas with an intervening "hump." The northern area is referred to as the East Hill zone, while the southern area is known as the UNR zone, since it is on a section of land deeded to the University of Nevada, Reno. Preliminary minable reserves for the original (East Hill) pit were estimated during the project feasibility study as 2.1 million t at an average grade of 1.1 g Au/t (0.036 oz Au/t). Mining through the end of 1996 totals 5.58 million t at an average grade of 1.0 g Au/t (0.032 oz Au/t). Recent drilling has increased this figure substantially. The published reserves on December 31, 1997, were 7.2 million t at an average grade of 0.94 g Au/t (0.030 oz Au/t). The total geologic resource is substantially larger than this minable reserve tonnage. All mineralized rocks known to date are oxidized.

Gold-mineralized rocks at East Hill are hosted by both the Antler sequence and the Valmy Formation (fig. 126). Morphology of the ore zones is influenced strongly by a number of north-striking faults which have post-ore movement, dividing the ore zones into a series of stair step-like bodies. Individual fault blocks have good continuity of Au grades for distances of as much as 100 m along strike, but ore lenses can abruptly terminate against the faults. High-angle structures, such as those mapped in the Top Zone pit, likely controlled Au mineralization at East Hill/UNR.

Mineralized rocks belonging to the Antler sequence at East Hill are present immediately above or overlapping the contact of strongly Fe–stained Battle Formation with the Valmy Formation, as well as higher in the section in very locally silicified or leached siltstone and limestone belonging to the Antler Peak Limestone. Mineralized rocks also are present in local areas of debris flows belonging to the Edna Mountain Formation, which only is present sparsely in the north end of the ore zone. Minor barite is present mainly in silty limestone of the Antler Peak Limestone. The silty limestone commonly alters to punky, pale yellow, red, or tan silty rock owing to decalcification and oxidation.

Mineralized rocks of the Valmy Formation are made up of shattered and oxidized metaquartzarenite similar to that seen in the Top Zone deposit. The lateral continuity of zones of high Au grades is somewhat better at East Hill/

UNR than is the case at Top Zone. Intervals in which micro-breccia textures were noted during the logging of drill holes almost invariably contain Au. Some high-grade mineralized rocks hosted by the Valmy Formation are black metaquartzarenite that contains intergranular specks of probable graphite. Some metaquartzarenite ore has undergone addition of silica, appearing as micro-stockwork veinlets or chalcedonic flooding.

Several north-striking, east-dipping normal faults that downdrop the ore zones to the east have been recognized in the northern (East Hill) end of the deposit, based upon obvious offset of lithologies belonging to the Antler sequence. These faults have been inferred to extend southward, towards the UNR end of the deposit, based upon apparent offsets of mineralized zones encountered in drilling. As mentioned above, rocks of the Valmy Formation are complexly deformed and lack distinctive marker beds. Thus, it is difficult to work out its internal structure and stratigraphy. However, good evidence is available for the presence of isoclinal and recumbent folds (figs. 125, 126).

As shown in figure 119, the East Hill/UNR, Red Rock, and Top Zone ore bodies are located on the flanks and nose of a prominent ridge that early on was informally named "East Hill" because of its location with respect to the hill that hosts the old underground workings. Chemical analyses of gridded float-chip samples collected along and across the entire ridge show anomalous As, Sb, and Hg over most of the ridge. The most obvious geochemical anomalies are those of As and Au; most samples analyzed showed more than 250 ppm As contents, and Au was as high as 2.3 ppm. Antimony, Hg, and Tl contents also are significantly anomalous, in places.

RED ROCK ZONE

The Red Rock Zone lies southwest of the Top Zone, and extends southward along the original Trout Creek drainage (fig. 119). Minable grades and thicknesses extend to the south for over 800 m along strike. Some mineralized rock was exposed at the surface and had been explored by a short adit, a shallow vertical shaft, and several drill holes prior to Cordex's involvement. Minable reserves at the end of July, 1996 were calculated at 1.09 million t at an average grade of 0.94 g Au/t (0.030 oz Au/t). Production through the end of 1996 was 603,583 t at 1.19 g Au/t (0.038 oz Au/t). The original reserves were approximately 2.17 million t at 0.036 oz Au/t. Approximately two-thirds of the ore zone remained unmined in July, 1998.

The Red Rock deposit is hosted mainly by the Antler Peak Limestone (fig. 127), but Au also is present in the Edna Mountain debris-flow unit. In places, significant zones of elevated Au concentrations are present in the basal conglomerate of the Battle Formation as well, where it and the Antler Peak Limestone are juxtaposed by faults. Locally, the Antler Peak Limestone has been converted to a very dense jasperoid by silicification. Some jasperoid appears to have been repeatedly brecciated and rehealed.

The deep red color displayed by the jasperoid is responsible for the name of the ore zone.

A generalized pattern of progressive alteration of the limestone is present southward along the west flank of East Hill, passing from fresh micritic limestone exposed near

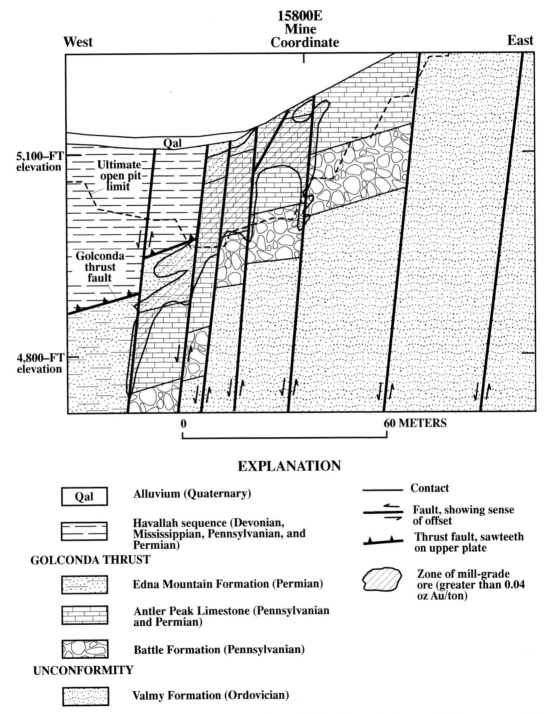

EXPLANATION

Qal	Alluvium (Quaternary)	——— Contact
	Havallah sequence (Devonian, Mississippian, Pennsylvanian, and Permian)	Fault, showing sense of offset
GOLCONDA THRUST		Thrust fault, sawteeth on upper plate
	Edna Mountain Formation (Permian)	Zone of mill-grade ore (greater than 0.04 oz Au/ton)
	Antler Peak Limestone (Pennsylvanian and Permian)	
	Battle Formation (Pennsylvanian)	
UNCONFORMITY		
	Valmy Formation (Ordovician)	

Figure 127 Red Rock ore zone at 13,700–ft North mine coordinates. Certain faults of west-dipping normal fault system appear to have been important in controlling Au deposition and spotty silicification throughout deposit. Ore zone is further complicated by a number of cross faults. Mill grade ore (greater than 0.040 oz/ton Au) is commonly hosted by Permian Edna Mountain Formation where present, but all rocks of Pennsylvanian and Permian Antler sequence may be mineralized. Pennsylvanian and Permian Antler Peak Limestone locally is converted to jasperoid.

the "narrows" of Trout Creek, through weakly recrystallized limestone, and finally into jasperoid. These changes are accompanied first by an increase in white crystalline calcite veinlets in the limestone, then by fine-grained tan dolomite(?) veinlets, and finally by fine-grained, Fe–stained silica veinlets as jasperoid is approached. Corals and other fossils present in the Antler Peak Limestone are preferentially silicified. Where it has not been converted to jasperoid, limestone in the ore zone commonly is strongly stained by Fe–oxide minerals and limestone has undergone some leaching of calcite. Rocks of the Edna Mountain and Battle Formations also show strong and, commonly, multicolored staining by Fe oxides. These units may be brecciated where mineralized, and they may, as well, contain abundant orange, brown, and red clay gouge along minor mineralized fractures. A small amount of barite also is present locally in the ore zone.

Structural control of Au deposition at Red Rock was by one or more steeply westward-dipping, north-striking faults (fig. 127). The mineralized zones, when viewed in cross section, appear to have roughly the shape of ice cream cones. Locally, tops of two adjacent cones may extend laterally from the structures along favorable stratigraphy to join, but the width of the deposit still is typically less than 35 m. The ore zone has a distinct, fairly shallow bottom. Several cross faults have been mapped along the length of the mineralized zone, primarily on the basis of offsets in the Antler sequence and abrupt thickness changes in lithology from drill cuttings. Additional minor cross fractures with little apparent offset can be found intersecting the main north-striking ones, particularly in areas where large pods of jasperoid are present.

8 North and 5 North zones

Two additional mineralized zones are known as the 8 North and 5 North deposits. The 8 North zone lies in the northern part of sec. 8, north of the main pit at 8 South (fig. 119). The deposit is not exposed, and is known only from drill cuttings. Mineralized rock extends as much as 350 m in a north-south direction and 230 m east-west. It is geologically quite similar to the 8 South deposit, except that about two-thirds of the mineralized zone has been eroded away. As with 8 South, the deposit is hosted by debris flows and siltstone of the Edna Mountain Formation. The 8 North zone was discovered in the fifth hole drilled in the original drilling program by Cordex. The most striking alteration at 8 North involves introduction of barite and Fe staining, but the Fe staining is less intense here than in 8 South. Barite is very abundant. Some leaching of calcite is present in 8 North, but much of the Au–bearing rock still has a strong reaction to dilute HCl. Silicification appears to be minor.

The 5 North deposit lies in the northern part of sec. 5, T. 33 N., R. 43 E. (fig. 119). Preliminary reserves are 666,062 tonnes at an average grade of 1.34 g Au/t (0.043 oz Au/t). The zone is covered by alluvium and is known from drill holes only. Structurally, this zone rather resembles the East Hill zone, except that most mineralized rock at 5 North is hosted by debris flows and siltstone of the Edna Mountain Formation, although a few thin drill intercepts of mineralized Antler Peak Limestone were encountered at 5 North. In contrast to East Hill, significant mineralized rock has not been encountered in the rocks of the Battle or Valmy Formations at the 5 North deposit. Iron oxide staining is pronounced in shades of red and yellow. However, some of the most strongly stained rock is present in holes drilled outside of the area that contains significant concentrations of Au. Minor leaching of calcite has occurred at 5 North, and introduction of barite is minor, although much of the barite lies outside Au–mineralized zones. Both the upper and lower plates of the Golconda thrust fault, as well as the mineralized zone, have been cut at 5 North by at least five, throughgoing, normal faults that dip steeply eastward and strike nearly due north. Discontinuous splays branch off the main faults along strike in several places and may connect adjacent faults. At 5 North, the dips of beds in the Antler sequence are interpreted to be steeper and more variable than we see in most places at Marigold. This may be due to rotation of beds caused by differential movements of fault blocks.

Summary

Gold deposits at Marigold are hosted by the Valmy Formation and by all formations of the Antler sequence. In excess of 36.9 minable tonnes Au (1.3 million oz Au) are known to be present in these deposits. The Au deposits show many characteristics typical of the "Carlin type" and are inferred to have formed between 38 and 36 Ma, on the basis of geologic relations at nearby pluton-related Au deposits in the Copper Canyon area of the Battle Mountain Mining District. The relative importance of structural and stratigraphic control over Au deposition varies in all of the deposits at Marigold. Both structural and stratigraphic controls are present to some degree in all of the mineralized zones, and the geometry of the ore bodies has been modified profoundly by post-ore faults and erosion. Preliminary sulfur isotopic work indicates a magmatic source for the fluids, but no magma body of any size is known at the depths drilled to date. All of the Marigold ore zones are thoroughly oxidized. The depth of oxidation exceeds the maximum depth of holes drilled to date, approximately 330 m. Alteration includes full or partial decalcification, silicification in some zones, argillization,

Table 32—Grades and tonnages for distal-disseminated Ag–Au deposits in Nevada and Idaho.†

	State	Tonnes (in thousands)	Au (g/t)	Ag (g/t)	Comments	Source
Eight South	Nevada	5,810.0	2.19	—	Includes entire production completed through 1994	This report
Top Zone	Do.	9,400.	1.	—	Includes production plus December 1996 reserves	—Do.—
East Hill/UNR	Do.	14,100.	.92	—	——Do.——	—Do.—
Red Rock	Do.	1,640.	.97	—	——Do.——	—Do.—
Five North	Do.	666.1	1.34	—		—Do.—
Lone Tree	Do.	55,180.	2.41	—	Proven and probable reserves as of December, 1995	B.L. Braginton, written commun., 1996
Trenton	Do.	11,490.	1.09	—	Proven and probable reserves as of February, 1997	This report
Valmy	Do.	9,493.6	.72	—	——Do.——	—Do.—
North Peak	Do.	2,674.2	.66	—		—Do.—
Empire	Do.	1,541.5	1.8	5.9	Production 1982, and 1991–1994	Doebrich and others (1995)
Northern Light	Do.	391.1	1.6	22.1	Production only during 1993	—Do.—
Sunshine	Do.	1,335.	.6	5.1	Proven and probable reserves as of January, 1994	—Do.—
Reona	Do.	7,659.1	1.01	6.6	Possible reserves as of January, 1994	Doebrich and others (1995)
NE Extension Leach	Do.	4,029.6	.85	4.1	——Do.——	—Do.—
Iron Canyon	Do.	764.7	1.25	16.1	Proven, probable, and possible reserves, January, 1994	—Do.—
Cove	Do.	83,147.	1.25	56.4	—	Emmons and Eng (1994)
Starpointer	Do.	1,360.	4.8	10.3	—	Cox and Singer (1994)
Hilltop	Do.	10,350.	2.5	2.	—	—Do.—
Taylor	Do.	7,000.	—	103.	—	—Do.—
Candelaria	Do.	27,000.	.19	50.	—	—Do.—
Buffalo Valley	Do.	1,470.	1.37	—	Classified as gold skarn by Theodore and others (1991)	Seedorff and others (1990)
Nos. 1–5	Do.	5,082.	1.97	—	Bald Mountain Mining District	A.D. Hitchborn and others, written commun., 1995
RBM	Do.	1,543.	2.03	—	——Do.——	—Do.—
Rat	Do.	2,787.	2.56	—	——Do.——	—Do.—
Top	Do.	12,250.	1.97	—		—Do.—
West End	Idaho	4,202.9	1.68	.62	1,307,186 t Ag ore	A.A. Bookstrom, written commun., 1996
Homestake	Do.	997.9	2.4	.52		—Do.—
Yellow Pine	Do.	13,683.6	1.65	14.52	Also referred to as Meadow Creek. 3,509,442 t Ag ore	—Do.—
Elk City district	Do.	8,346.1	.99	—		—Do.—
Beartrack	Do.	33,515.2	1.89	—		—Do.—
Humbug	Do.	14,500.	1.28	—	—	—Do.—
Atlanta East	Do.	8,074.	2.81	8.16	—	—Do.—
Atlanta West	Do.	3,175.2	1.78	5.6	—	—Do.—
Yellowjacket	Do.	524.1	3.63	—	—	—Do.—
Moscow	Do.	3,810.2	1.71	—	—	—Do.—
Red Mountain	Do.	9,071.9	.86	—	—	—Do.—

and widespread veining and replacement by barite. Arsenic, Sb, Tl, Hg, and Au are geochemically enriched.

Geology, mineralization, and exploration history of the Trenton Canyon project

By Robert P. Felder

INTRODUCTION

The Trenton Canyon Project arose from three separate exploration projects, North Peak, Valmy, and Trenton Canyon (fig. 94). The North Peak deposit was discovered in 1988, then part of a joint venture with Bow Valley Mining. A regional stream-sediment sampling program conducted in 1987 identified a strong Au, Hg, and As anomaly in sec. 11, T. 32 N., R. 42 E., along the west range front of Battle Mountain (pl. 2). Follow-up sampling and mapping were completed in 1988 and 1989, and reverse circulation drilling identified the area of Au–mineralized rocks in 1989. The deposit was drilled out during the period 1990 to 1994, and currently (1998) contains 176 holes. Santa Fe Pacific Gold Corporation bought out Bow Valley's interest in the property in 1994. The Valmy deposit was found in 1989, just north of and contiguous with Hecla's Trout Creek deposit, which was discovered in 1988—these were consolidated in 1992 when Santa Fe Pacific Gold bought out Hecla's interest in the property. Geologic mapping, as well as soil and rock-chip sampling were initiated in Santa Fe's sec. 29, T. 33 N., R. 43 E., during 1988 (pl. 2). Two areas of anomalous Au, As, and Sb were identified by soil sampling, and were first drilled in 1989, identifying the Au deposits subsequently planned for mining. Development drilling continued during the period 1990 to 1996, with more than 500 holes completed by 1998 on the property.

The Trenton Canyon deposit was discovered through grass roots exploration in 1991, and contains the majority of the reserves on the project (table 32). Surface geologic mapping and sampling were initiated on the Trenton Canyon property in 1989, in an area identified to have good potential for the presence of Au–mineralized rocks. Results from preliminary surface rock-chip sampling encouraged expansion of the scope of the program, and grid soil sampling was conducted over selected areas. During 1990, the first two drill holes were completed, and hole DNT–02 encountered approximately 15 m of 1.56 g Au/t (0.050 oz Au/t). In 1991, efforts were intensified to include grid soil sampling and 1:4,800–scale geologic mapping over the whole property. Several areas of highly

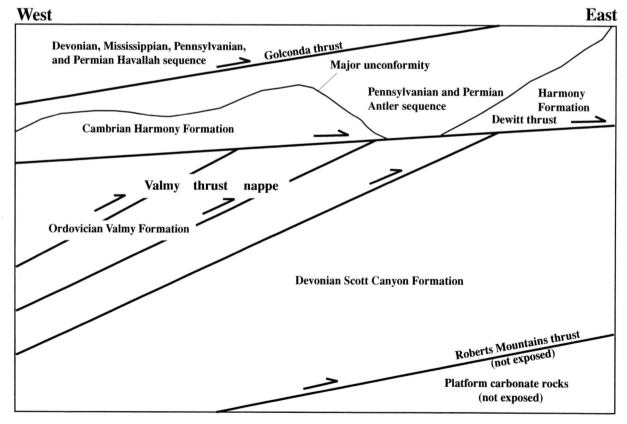

Figure 128 Schematic cross section showing structural relations among Paleozoic rocks in Battle Mountain area. Modified from Madrid (1987).

anomalous to ore-grade Au contents were identified with this program, and drilling in these areas identified the main Trenton Canyon deposit in secs. 18 and 19, T. 32 N., R. 43 E. (fig. 94; pl. 2). The discovery hole, DNT–09, encountered 19 m of 1.84 g Au/t (0.059 oz Au/t) from 0 to 19 m in depth. Development drilling continued through 1996, and more than 600 holes have been completed on the property. Santa Fe Pacific Gold Corporation's Board of Directors approved the Trenton Canyon Project for mining in May, 1995. Santa Fe Pacific Gold merged with Newmont Gold Co. in 1997.

REGIONAL GEOLOGY

The geology of the Battle Mountain Mining District reflects a complex history of sedimentation, volcanism, tectonism and intrusive activity. Sedimentary and volcanic rocks ranging in age from Cambrian to Permian are present in two major allochthons (Roberts, 1964). The Roberts Mountains allochthon (made up of the Cambrian Harmony Formation, Ordovician Valmy Formation, Devonian Scott Canyon Formation) includes a complex sequence of siliciclastic sedimentary rocks and associated volcanic rocks, which were complexly deformed during emplacement in the Late Devonian to Mississippian Antler orogeny (Roberts, 1964; Madrid, 1987). These rocks are unconformably overlain by shallow water carbonate and clastic rocks of the Antler overlap assemblage (Middle Pennsylvanian Battle Formation, Pennsylvanian and Permian Antler Peak Limestone, and Permian Edna Mountain Formation). Coeval basinal rocks of the Golconda allochthon (Devonian, Mississippian, Pennsylvanian, and Permian Havallah sequence) were emplaced over the Antler Sequence and underlying Roberts Mountains allochthon during the Late Permian to Triassic Sonoma orogeny (Silberling and Roberts, 1962; Roberts, 1964). Mesozoic and Tertiary intrusive activity is widespread in the Battle Mountain region, with spatially associated zones of contact metamorphism and metasomatism (see section above entitled "Potassium-argon Chronology of Cretaceous and Cenozoic Igneous Activity, Hydrothermal Alteration, and Mineralization"). Oligocene and younger unconsolidated gravel deposits and ash-flow tuff, and Pliocene basalt unconformably overlie all older rock types (Roberts, 1964). A schematic cross-section illustrating structural and stratigraphic relations among Paleozoic rocks in the Battle Mountain area is shown in figure 128.

PROJECT AREA GEOLOGY

STRATIGRAPHY

The stratigraphic section in the project area is summarized below:

AGE	FORMATION	ABBREVIATIONS (on fig. 129)
Tertiary and Quaternary	Alluvium and (or) colluvium	QT
Tertiary	Bates MountainTuff	Tt
Cretaceous	Granodiorite and monzogranite	Ki
Devonian, Mississippian, Pennsylvanian, and Permian	Havallah Sequence	DPh
Pennsylvanian and Permian	Antler sequence	d Pa
Ordovician	Valmy Formation	Ov

The Valmy Formation is a highly variable sequence of interbedded and structurally interleaved quartzarenite, siltstone, chert and greenstone. The units are highly fractured and isoclinally folded. Thrust faults form many contacts between large blocks of competent quartzarenite and the less competent units. These rocks are unconformably overlain by the Battle Formation, the basal formation of the Antler sequence, which formation comprises a relatively undeformed sequence of interbedded conglomerate and sandstone, as well as minor siltstone and limestone. The Antler Peak Limestone, the middle formation of the Antler sequence, consisting of gray, sandy limestone, is present only at the south end of sec. 19, T. 32 N., R. 43 E., at Trenton Canyon, where it overlies the Battle Formation (pl. 2). The Edna Mountain Formation, the uppermost formation of the Antler sequence, consists of angular pebble conglomerate and siltstone, and is present in one locality at the northwest end of the Trenton Canyon property in sec. 12, T. 32 N., R. 42 E. (pl. 2). The Havallah sequence, in thrust and (or) high-angle fault contact with underlying units, is a variably metamorphosed sequence of chert, siltstone, and minor sandstone. The Oligocene Caetano Tuff, a Tertiary crystal rich, quartz latite ash-flow tuff is present as an erosional remnant sitting on Havallah sequence in sec. 30, T. 32 N., R. 42 E., at the south end of the Trenton Canyon project area (Doebrich, 1995). Tertiary, rhyolitic tuffs, dated at approximately 25 Ma, are present in a northeast-trending band of outcrops at the North Peak property, and are considered to be Oligocene

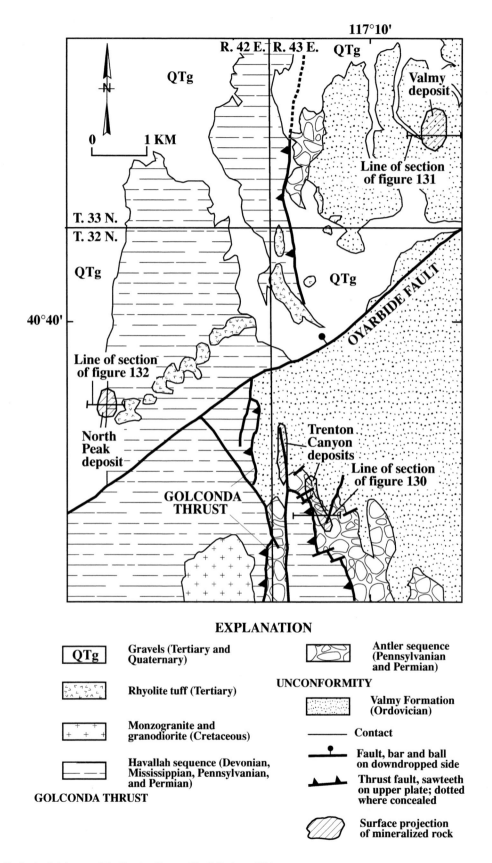

Figure 129 Geologic sketch map of the Trenton Canyon, North Peak, and Valmy areas.

or Miocene Bates Mountain Tuff equivalent (see subsection above entitled "Oligocene or Miocene Calc-alkaline Rhyolite Tuff").

INTRUSIVE ROCKS

At least three suites of intrusive rocks are present in the project area. The largest is the 89–Ma granodiorite and monzogranite of the stock at Trenton Canyon, which crops out for as much as 3 km in a northwest-southeast direction on the southwest side of the Trenton Canyon project area (fig. 129; Theodore and others, 1973). The most widespread phases at the surface are medium-grained granodiorite porphyry. A younger suite of intrusive rock (approximately 34.9 ± 1.0 Ma.; K–Ar, sericite—see section above entitled "Potassium-argon Chronology of Cretaceous and Cenozoic Igneous Activity, Hydrothermal Alteration, and Mineralization") of intermediate composition dikes and sills is present along northerly striking structures in the area of the Trenton Canyon deposit. Similar appearing dikes have been mapped and drilled at the Valmy and North Peak properties, but have not been dated. All of these intrusive rocks have been intensely sericitized. The sericite date stated above is interpreted as an alteration age, and a best estimate of the age of Au mineralization at Trenton Canyon, based on the presence of Au in sheared, sericitic fault zones adjacent to altered intrusive rocks. The true age of the dikes is still not known, but based on field relations, is considered to be Tertiary. In addition, a suite of strongly propylitized dioritic dikes have been drilled at the North Peak property—these rocks have not been dated.

STRUCTURE

The structural geology of the Trenton Canyon area is complex, having been subjected to multiple igneous and deformational events. For a detailed treatment of this subject, the reader is referred to Roberts (1964), Blake and others (1979), and Madrid (1987), as well as the section above entitled "Structural Geology". The main structural fabric in the project area is represented by north-striking structures, ranging generally from N. 30° W. to N. 20° E. Gold-mineralized rocks and Tertiary dikes and sills are controlled by these structures, which have been cut by younger, Midas-trough associated, post-mineral faults striking N. 45° E. to N. 80° E. Noteworthy of these is the Oyarbide fault, which strikes N. 50° E. across the project area, and has in excess of 700 m of down-to-the-north displacement (fig. 129). These faults are superimposed on pre-mineral east-west-, west-northwest-, and northeast-striking faults, and on thrust faults of the Antler and Sonoma orogenies. The Golconda thrust strikes north-south across the project area, superimposing rocks of the

Havallah sequence over those of the Antler sequence and Valmy Formation.

ALTERATION AND GEOCHEMISTRY

Three types and ages of alteration are evident in the three areas that comprise the Trenton Canyon Project. From oldest to youngest, they are: (1) calc-silicate alteration associated with the 89–Ma stock of Trenton Canyon (present only in the Trenton Canyon area), (2) hydrothermal alteration associated with Au mineralization (present in all three deposit areas), and (3) supergene oxidation (present in all three deposit areas).

Extensive thermal metamorphism is present in the aureole of the stock of Trenton Canyon. The effects of this event are seen as widespread recrystallization to hornfels, with abundant calc-silicate alteration. Local garnet-diopside skarn is evident in calcareous units of the Battle Formation, as well as some associated Cu–mineralized rock. This metamorphism has affected rocks of the Havallah sequence and the westernmost fault block of Battle Formation most profoundly due to their closest proximity to the stock at Trenton Canyon (fig. 129). The pre-mineral calc-silicate alteration is characterized by extensive recrystallization, accompanied by the following mineral assemblages:

Havallah sequence:tremolite±calcite±muscovite
Battle Formation:
   tremolite, calcite±chlorite±muscovite±diopside±garnet±pyrrhotite± pyrite±magnetite (garnet skarn associated with Cu in sec. 24, T. 32 N., R. 42 E., at Trenton Canyon)
Valmy Formation: tremolite±muscovite± diopside±pyrite.

Metamorphism of more distal rocks of the Battle and Valmy Formations is overall less intense, due to a combination of lesser reactivity and increasing distance from the intrusion. The Au–mineralized rocks of the Trenton Canyon deposits are present almost completely outside the metamorphic halo associated with the Trenton Canyon intrusion. Local contact metamorphic effects are seen in sedimentary rocks on the Valmy and North Peak properties, but are of little significance.

A summary of the alteration mineralogy and geochemistry associated with Au mineralization in the three deposit areas is given below.

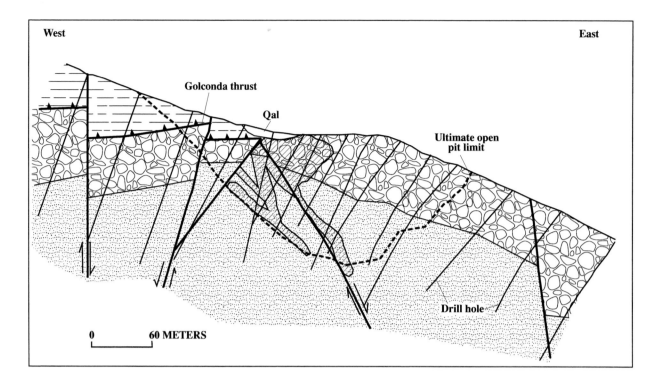

Figure 130 Trenton Canyon deposit at 759,000–ft North mine coordinates-profile AA' (fig. 129).

DEPOSIT AREA	ALTERATION SIGNATURE		OTHER
	Mineralogy	Geochemistry	
North Peak	qtz-dol-bar-py (Feox)	+ As, Sb, Hg, Ba	
	clays	+ SiO$_2$ CaO	
Valmy	qtz-py (Feox)-clays	+ As, Sb, Hg	bleaching/clays
	(barite)	+ K$_2$O, FeO, Al$_2$O$_3$	"sanding"
		CaO (?)	rare sulfosalt
Trenton Canyon	qtz-ser-py (Feox	+ As, Sb, Hg, (Ag)	rare Sb$_2$S$_3$
		+ SiO$_2$, TiO$_2$	
		CaO, MgO	

At Trenton Canyon, hydrothermal, Au–stage alteration is mineralogically characterized by a quartz-sericite-pyrite (Fe oxide) assemblage. Pyrite is generally oxidized. One stibnite vein was encountered in a core hole at the Trenton Canyon deposit; more commonly stibiconite (Sb$_2$O$_3$ (OH)$_2$) has been observed. Gold-stage alteration is characterized by silica enhancement, most likely due to the removal of CaO and MgO, rather than appreciable addition of SiO$_2$. The removal of CaO and MgO reflects the destruction of the tremolite-calcite mineral assemblage in the contact metamorphosed protolith. This is somewhat different from typical Carlin-type system alteration, where decalcification of carbonate rocks with associated decrease in density and increase in porosity is observed. In the case of Trenton Canyon, the term modified decalcification may be more appropriate, so as not to be confused with classic, Carlin-type deposit decalcification.

At Valmy, hydrothermal alteration and Au mineralization are associated with bleaching and argillization, certain quartz veinlets, and Fe–oxide minerals. "Sanding" of quartzarenite is present in some mineralized zones, probably reflecting destruction of matrix feldspar in what were originally feldspathic sandstones, leaving incompetent friable sand in a white powdery (clay) matrix.

At North Peak, alteration associated with Au mineralization is characterized by weak decalcification and silicification of sandstone of the Havallah sequence, ac-

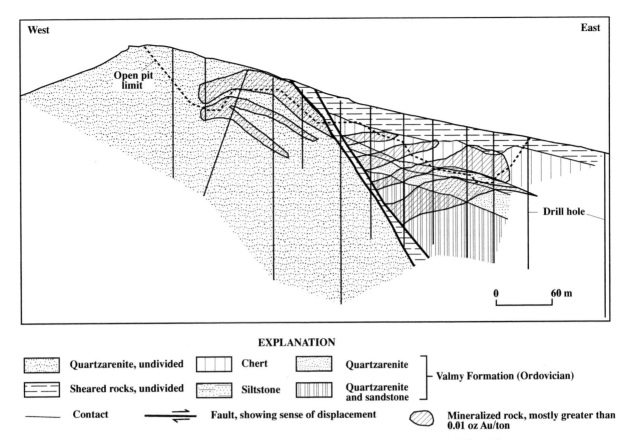

West **East**

Open pit limit

Drill hole

0   60 m

EXPLANATION

Quartzarenite, undivided   Chert   Quartzarenite

Sheared rocks, undivided   Siltstone   Quartzarenite and sandstone

⎤ Valmy Formation (Ordovician)

Contact   Fault, showing sense of displacement   Mineralized rock, mostly greater than 0.01 oz Au/ton

Figure 131 Vertical cross sections through Valmy deposit at 781,000–ft North mine coordinates-profile CC' (fig. 129).

companied by Fe–oxide minerals, clay minerals, and minor barite.

Supergene oxidation has resulted in the destruction of most of the pyrite and other sulfide minerals originally present in the deposits, forming Fe–oxide minerals, and rare stibiconite (the latter mineral is present at Trenton Canyon).

MINERALIZATION

Mineralization in the Trenton Canyon deposits primarily is controlled by structure. Geologic features influencing the distribution of mineralized rocks are high angle faults and their intersections with (1) thrust faults and (or) low angle lithotectonic contacts within the Valmy Formation and Havallah sequence; (2) favorable stratigraphic horizons (especially in the Battle Formation atTrenton Canyon); and (3) the unconformity between the Battle Formation and the Valmy Formation (at Trenton Canyon). Cross-sections show representative geology and ore controls for each of the three deposit areas (figs. 130–132). Mineralized rocks commonly are characterized by strongly limonitic fault gouge or breccia containing a finely-milled mixture of Fe–oxide minerals and silica±sericite±clay minerals. Favorable stratigraphic ho-

rizons (that is, the Battle Formation) display variable alteration products of sandy matrix material—these products include bleached rocks, argillized rocks(?), and Fe–oxide minerals±sericite, giving a white to light gray to orange color, and a crumbly to incompetent texture depending on the degree of alteration of the matrix. Fault breccia and (or) gouge zones in the Valmy Formation will commonly contain fragments of quartz vein material, probably indicating renewed movement (possibly hydrothermal stage) on older, quartz vein-filled fractures. These quartz vein fragments may be present within limonitic gouge, as described above, or with increasing depth may be cemented with pyrite. Quartz, which is associated in time and space with Au–mineralized rocks, is present as vuggy, open space fillings associated with limonite (pyrite) and (or) sericite with minor clays, and occasionally as veinlets.

A schematic stratigraphic column showing the distribution and style of Au–mineralized rocks at each of the deposits, as well as the rock units hosting mineralized rocks, emphasizes the strong contribution that structure plays in the generation of these deposits (fig. 133).

RESERVES

Proven and probable reserves at $350/oz Au in the three deposits that constitute the Trenton Canyon Project as of December 31, 1997 are 16,763,000 tonnes (t) (18,473,000 tons) at an average grade of 0.88 g Au/t (0.028 oz Au/ton, for a total of 517,000 oz Au. All reserve ounces are oxide. Reserves are distributed among the three deposit areas as follows: Trenton Canyon (5,558,000 t at 1.34 g Au/t), Valmy (10,298,000 t at 0.66 g Au/t), and North Peak (907,000 t at 0.53 g Au/t) (table 32).

DISTRICT-SCALE IMPLICATIONS OF LARGE DISTAL-DISSEMINATED PRECIOUS-METAL DEPOSITS

The geology and geochemistry of many precious-metal deposits in the North Peak and Valmy quadrangles provide insight into district-scale patterns of their distribution, as well as their ages, even though many deposits apparently do not have clear-cut spatial and genetic ties to felsic plutons and have not been dated radiometrically. The dominant theme, however, that is repeated through all of the deposits is the importance of faults and fractures as contributing elements to the overall genesis of the deposits. Nonetheless, large volumes of rock in many of the Au deposits contain truly disseminated mineralized rock which results from mineralized faults and fractures acting as feeders to deliver Au–bearing fluids to chemically and (or) physically receptive sites.

The Eight South gold deposit (fig. 94; Graney and McGibbon, 1991; see section above entitled "Geology of the Marigold Mine Area") was the first of relatively recent major discoveries in the immediate area of the historic workings at the Old Marigold Mine. This deposit has a relatively high Au/Ag ratio in its ore, compared to most pluton-related systems, as well as anomalous concentrations of As, Sb, Hg, Tl, and especially Ba. These geochemical signatures suggest, on the one hand, that it may be a Carlin-type system (Graney and McGibbon, 1991), or carbonate-hosted Au–Ag deposit as these deposits are referred to by Berger (1986). Albino (1993) points out, however, that common enrichment of As, Sb, and Hg in both Carlin-type systems and distal-disseminated Ag–Au deposits is partly the result of these elements having the ability to be transported as bisulfide complexes. In addition, Albino (1993) points out that many distal-disseminated Ag–Au deposits are enriched in Mn, whereas some Carlin-type deposits, in fact, may be leached of Mn.

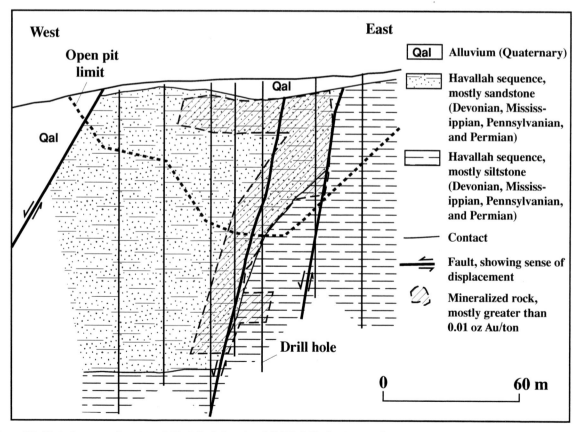

Figure 132 Vertical cross sections through North Peak deposit at 765,270–ft North mine coordinates-profile BB' (fig. 129).

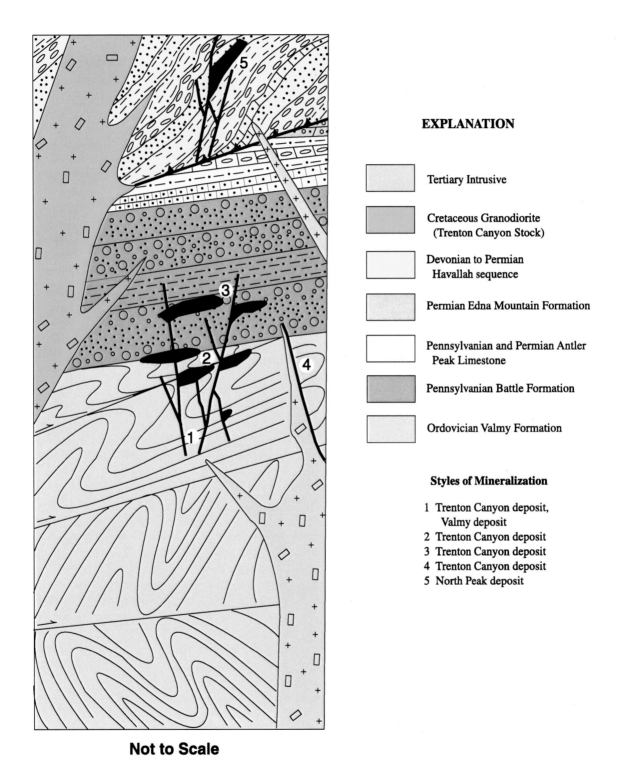

EXPLANATION

Tertiary Intrusive

Cretaceous Granodiorite
(Trenton Canyon Stock)

Devonian to Permian
Havallah sequence

Permian Edna Mountain Formation

Pennsylvanian and Permian Antler
Peak Limestone

Pennsylvanian Battle Formation

Ordovician Valmy Formation

Styles of Mineralization

1 Trenton Canyon deposit,
   Valmy deposit
2 Trenton Canyon deposit
3 Trenton Canyon deposit
4 Trenton Canyon deposit
5 North Peak deposit

Not to Scale

Figure 133 Schematic cross section showing tectonostratigraphy and ore controls for the Trenton Canyon, Valmy, and North Peak deposits.

From analyses reported in Graney and McGibbon (1991), one may infer that Au/Ag ratios in the oxidized ores from the Eight South deposit are approximately equal to 1. However, Au/Ag ratios prior to oxidation may have been different in the sulfidic rocks from which the oxidized ores have been derived. As described above, it appears likely that oxidation in the general area of the Eight South deposit may have been taking place for as long as approximately 23 m.y—possibly enhanced by the fact that the site of the deposit, at one time, was approximately 700 m higher topographically (D.H. McGibbon, oral commun., 1996). Moreover, Saunders (1993) suggests that chlorargyrite (AgCl), a readily soluble mineral resulting from oxidation of Ag–bearing sulfide ore by chloride-bearing groundwaters, probably controls the liberation of Ag into the surrounding groundwaters as $AgCl_2^-$ complexes. Thus, Au/Ag ratios in initial source rocks may be elevated significantly because of removal of Ag during oxidation, and percolation through the rocks of chloride-rich groundwaters. Recent additional mineral-deposit discoveries, which also are in the North Peak quadrangle, include at least 15 Au deposits in the hanging wall of the Oyarbide fault (fig. 94). All of these deposits may owe their origins to relatively far-traveled, hypogene Au–bearing fluids emanating from buried porphyry Cu systems somewhat akin to the evolution of immiscible fluids described for porphyry systems (Hedenquist and Lowenstern, 1994; Albino, 1994).

These 15 Au deposits, many without associated intrusions, are presumably distal-disseminated Ag–Au deposits, if we apply the definition of Cox and Singer (1990) to them, and broaden it to include genetically related mineral occurrences in the vertical dimension as described by Albino (1994). They are concentrated north of the Oyarbide fault because the northern block of this fault has been downdropped approximately 700 m, thereby placing tops of porphyry systems close to the present level of erosion. These relatively young displacements may have taken place at roughly the same time as the less-than-9–Ma displacements recorded along the similarly trending Midas trough to the north (Wallace and Hruska, 1991). The Oyarbide fault is classified by Dohrenwend and Moring (1991) as a young, post-17–Ma, deeply penetrating block fault. Exposed porphyry Cu and stockwork Mo systems are widespread south of the Oyarbide fault in the southern part of the mining district (Roberts, 1964; Roberts and Arnold, 1965; Theodore and Blake, 1975, 1978; Theodore and others, 1992; Howe and Theodore, 1993), as opposed to the general absence of near-surface plutons north of the Oyarbide fault.

In fact, these 15 deposits may be envisioned as providing one of the bridges across the conceptual gap between the pluton-related environment and the classic Carlin-type systems, if recent discoveries along the Carlin trend can be shown conclusively to have temporal and genetic links between base- and precious-metals and a magmatic environment (Teal and Jackson, 1997). However, one of the major differences between the pluton-related Au deposits in the Battle Mountain Mining District and the classic Carlin-type systems is the involvement of deep-sourced fluids in the genesis of the latter (Hofstra and Rye, 1998). Many classically defined Carlin-type systems, *sensu stricto*, elsewhere in Nevada are characteristically not closely associated genetically with a nearby intrusive rock (Hofstra and others, 1988, 1989; Seedorff, 1991; Arehart, 1996; and many others). They also appear to be confined to a part of the Great Basin that is characterized by tectonically thickened crust resulting from the accretion of a number of allochthonous terranes, and the Carlin-type deposits may owe their origins, in part, to metamorphism-derived fluids emanating from reactions at depth near the brittle-ductile transition zone (Seedorff, 1991; see also, Hofstra and Rye, 1998). However, such conclusions have been questioned as they might relate to metamorphic reactions across the ductile-brittle transition in the Ruby Mountains, Nev. (Berger and Oscarson, 1998). Carlin-type systems have a number of other characteristics, including presence of submicron-sized Au in arsenical rims of pyrite, as well as the widespread presence of solid carbon in many deposits (Peters and others, 1996) and even liquid oil in some others (Hulen and others, 1994). Indeed, if classic Carlin-type systems at some future time come to be considered as resulting primarily from magmatic processes (see also, Ressel, 1998; Hofstra and Rye, 1998), then the two types of occurrences—Carlin type and distal disseminated Ag–Au–might be variants of each other. Sillitoe (1994) suggests that these two types Au deposit should be considered as two broad, genetic varieties of a single category of deposits. Nonetheless, the classic Carlin-type deposits appear to have formed in a geologic environment significantly deeper and cooler than the pluton-related environment at Battle Mountain (Peters and others, 1996; Hofstra and Rye, 1998).

An original underestimation of base-metal contents of some mineralized systems in the hanging wall of the Oyarbide fault probably contributed to classification difficulties. For example, the Lone Tree deposit appears to show an increase in its base-metals with depth so that it is now (1999) beginning to display some chemical and physical attributes—as well as fluid-inclusion signatures (see above)—characteristic of distal-disseminated Ag–Au deposits. Another reason for the difficulty of classifying many Au deposits in the Lone Tree and Marigold areas includes the apparent presence of a protracted period of oxidation, exemplified by relations at Eight South and several other nearby deposits, which may have altered significantly base- and precious-metal ratios from those present in unoxidized mineralized rocks.

However, the overall geologic setting of the deposits in a pluton-related geologic environment at Battle Mountain, occurrence of sericitically altered dikes in direct association with sericitically altered and Au–mineralized rock in some deposits, and magmatic isotopic signatures of hydrothermal barite associated with Au–mineralized rock remain perhaps among the best evidence to assign deposits north of the Oyarbide fault to a pluton-related environment. The geologic setting of the deposits has been discussed thoroughly above. Sulfur isotopic ratios of hydrothermal barite—values of $\delta^{34}S$ clustering tightly at approximately +10 per mil—that is associated genetically with several of these Au deposits clearly indicate a significant magmatic component to ore-forming fluids (Howe and others, 1995). In addition, isotopic signatures of oxygen and deuterium in fluid inclusions in barite at Lone Tree indicate formation from magmatic waters (Hofstra and Rye, 1998). This is in marked contrast to well-studied Carlin-type systems elsewhere (Hofstra and others, 1988, 1989; Arehart, 1996).

Vertically stacked, large porphyry Cu systems thus may include relatively Au–enriched peripheral deposits, the distal-disseminated Ag–Au deposits—these peripheral deposits have been referred to as distal epithermal Au deposits by others (Jones, 1992). They are envisioned as representing the predominantly structurally controlled, high level parts of an idealized porphyry Cu system. Certainly, many distal-disseminated Ag–Au occurrences are akin genetically to polymetallic vein deposits, and may be considered to be variants of them (Cox and Singer, 1986). However, distal-disseminated Ag–Au deposits differ from the polymetallic veins in the commonly occurring disseminated aspects of many of them, which, in turn, have generated deposits as large as approximately 83 million tonnes at the Cove, Nev., deposit (table 32). They differ, as well, in absence of concentrations of Pb and Zn comparable to that found in most polymetallic veins.

SUGGESTIONS FOR EXPLORATORY PROGRAMS

The Battle Mountain Mining District seems to be positioned in the central part of a broad geologic province that is well-endowed in precious metals relative to surrounding provinces. Such enrichment may be an indication of an extensive precious-metal-enriched lower crust or mantle that continued periodically to provide precious metals to its upper continental crust as various geologic processes and igneous events became operative over long periods of geologic time. These geologic processes undoubtedly included widespread circulation of evolved meteoric and metamorphic fluids close in time to the onset of Tertiary extension in the Great Basin (Hofstra and others, 1988, 1989; Seedorff, 1991). A number of regional as well

as local exploration implications result from the present study, especially when it is considered in conjunction with some other recently completed investigations. The importance of regional tectonic controls to concentration of major Au–bearing systems at intersections of a number of broad north- and northwest-trending structural zones in the mining district is well recognized (Doebrich and Theodore, 1995; Doebrich and others, 1995). Previous studies also noted other major intradistrict metallized trends, including east-trending ones that may be Cretaceous in age and that may have been instrumental in the location of the major Mo systems at Buckingham and at Trenton Canyon (Blake and others, 1979). Each north-south and northwest structural trend in the mining district must be premineral or synmineral relative to introduction of late Eocene and (or) early Oligocene Au–mineralizing fluids. In addition, their surface projections apparently have been modified minimally by post-mineral displacements along the Miocene Oyarbide fault, as described above, because of the steeply dipping character of the mineralized zones. Some north-trending zones, such as the one along the west range front of the mining district, must have had continued post-mineral movements after cessation of circulation of Tertiary mineralizing fluids (Doebrich and others, 1995).

Data gathered from the Valmy, North Peak, and Snow Gulch quadrangles, as well as from the major ore deposits in the quadrangles, allow well-informed interpretations to be made of the economic geology beneath the gravel-covered region to the north and northeast (Peters and others, 1996). Further, the wide distribution of chemically reactive carbonate strata in LU–2 of the Havallah sequence in the Valmy and North Peak quadrangles was not anticipated at the beginning of this investigation, and, therefore, all areas underlain by such rocks must be evaluated carefully from an exploration standpoint to determine if any mineralizing fluids and (or) systems were able to penetrate through the largely non-reactive, underlying rocks belonging to LU–1 (see also, Doebrich and others, 1995; Kizis and others, 1997). This evaluation is especially warranted because more than 50 percent of the ore in the Lone Tree Mine is contained in rocks of the Havallah sequence (see section above entitled "Lone Tree Gold Deposit"). Furthermore, recent exploration activities in Buffalo Valley, just west of the mining district, have determined widespread presence of Au–mineralized calc-silicate assemblages in rocks of the Havallah sequence. However, the number of mineralized occurrences hosted by chemically non-reactive quartzarenite of the Valmy Formation also emphasizes the importance that mineralized structures or structural zones contribute to formation of Au deposits in the area. Gold deposits hosted mostly in Valmy Formation contain slightly lower overall Au grades than those hosted predominantly in chemically reactive calcareous rocks, as described below. In addition, all surface indications of mineralized rocks, including some prominently altered structures, in bedrock areas of the Valmy Formation have as yet (1999) not been tested by drilling. In addition, all gravel-covered pediment areas close to mineralized bedrock should be evaluated carefully, especially if mineralized structures are known to continue under gravels into adjoining pediment areas, such as the jasperoid described above in the west-central part of the Havallah Hills. Thoughtful deliberation of all exposed indications of mineralized rocks becomes especially important when one considers the number and economic importance of the ore deposits that have been discovered under alluvium in the three quadrangles during the last ten years.

Interpretation of the geology at reasonable depths on the basis of surface geologic relations in the study area provides additional specific targets that could be comprised, in places, of unexposed, chemically favorable strata. For example, the fault-bounded sliver of the Valmy Formation in contact with rocks of the Havallah sequence near the southeast corner of the Havallah Hills allows speculation that favorable strata of the Antler sequence might be present below the Golconda thrust at reasonable depths (pl. 2). Although rocks of the Valmy Formation along the fault sliver primarily have been uplifted structurally, a relatively large body of Cretaceous or Tertiary monzogranite partly adjacent to their outcrops also may have contributed to an overall arching of the surrounding sedimentary strata, including the sole of the thrust. Thus, chemically reactive strata of the Antler sequence below the thrust might be relatively close to the surface. As described above, this monzogranite, however, does not seem to have any signs of significant alteration in its adjoining wallrocks. Nonetheless, this area is present on trend with a number of broadly defined mineralized zones in the district, and is within 2.5 km of the large mineralized system centered at the Eight South deposit. A small number of approximately 200–m angle drill holes collared at right angles to the trend of the fault sliver tested this target during 1995. Results of this drilling program are unknown at the time of this writing (1999).

Exploration implications derived from metal ratios at the Elder Creek porphyry Cu system have been described above in the section entitled "Elder Creek Porphyry Copper System." Gravel-covered quartz stockworks near the core of the Elder Creek porphyry system could comprise a shallow—albeit high temperature (Gostyayeva and others, 1996)—Au target worthy of additional exploration. Such a target would be somewhat analogous to Au–bearing quartz stockworks present at the Reona deposit of the Copper Canyon area of the mining district.

A vast gravel-covered region immediately north of the Battle Mountain Mining District could conceal a high level porphyry system as well as other pluton-related types

of deposits (Peters and others, 1996). However, a concealed porphyry system should have a magnetic signature similar to the porphyry Cu system at Elder Creek, and thus should be readily recognizable from regional magnetic surveys. As described above, the system at Elder Creek includes a prominent donut-shaped ring made up of a number of aeromagnetic highs that surround the central stockwork-veined and K–silicate-altered core of the system. Some magnetically high areas in the pediment near the Elder Creek system have been evaluated by drilling (T. Gray, oral commun., 1995), and some arsenopyrite encountered at depth during this evaluation contains traces of Au (Gostyayeva and others, 1996). However, there are some other economically important Au–bearing pluton-related systems in the region that do not have a regionally prominent magnetic expression. For example, the combined McCoy-Cove, Nev., mineralized system in the northern part of the Fish Creek Mountains does not have a strong magnetic expression. Therefore, systems similar to those at McCoy-Cove could be obscured completely by a cover of gravel, and they may not be detected by aeromagnetic surveys, particularly those surveys that employ widely-spaced flight lines. Moreover, any Au deposits similar to the Eight South and Top Zone deposits also would be extremely difficult to detect from remotely sensed geophysical data because these deposits are oxidized entirely.

As noted above, approximately 15 separate occurrences of Au resources are present in a triangular-shaped area between Lone Tree Hill, on the north, and the trace of the Oyarbide fault, on the south. This triangular-shaped area comprises about 98 km². Some of these occurrences of Au resources are covered by Tertiary and Quaternary alluvial deposits in the pediment areas along the east flank of the Havallah Hills. A similarly-sized area to a depth of 1 km in the gravel-covered region north of Lone Tree Hill might contain, at a minimum, approximately the same number of occurrences of Au because roughly 30 to 40 percent of the surrounding large region to a depth of 1 km probably is made up of unconsolidated gravel (Peters and others, 1996). The grades and tonnages of these undiscovered deposits also should approximate those of known distal disseminated Ag–Au deposits in the general area of the Battle Mountain Mining District. Certainly, some, or perhaps many, of these Au deposits will never be found because of their presence under Tertiary and Quaternary gravels, as well as the absence of any diagnostic geophysical signature for the deposits. The tract considered highly prospective for the presence of distal-disseminated Ag–Au deposits probably extends to as far north as the approximate mid-point between the Twin Creeks deposit and the Battle Mountain Mining District (fig. 134; see also, Peters and others, 1996).

Grade and tonnage distributions for 36 distal-disseminated Ag–Au deposits in Nevada and Idaho, listed in

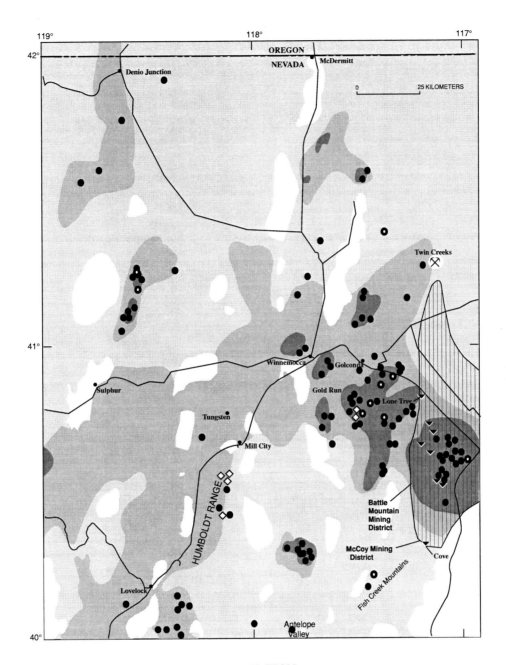

EXPLANATION

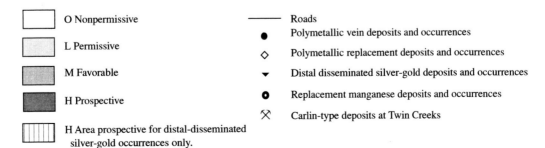

Figure 134 Sketch map showing extent of area, generally north of Lone Tree Hill, highly prospective for distal-disseminated Ag–Au deposits. Modified from Peters and others (1996).

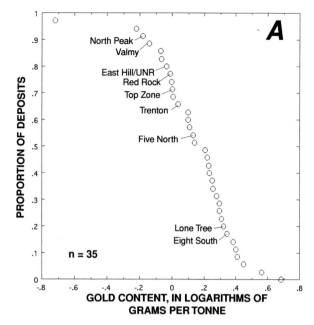

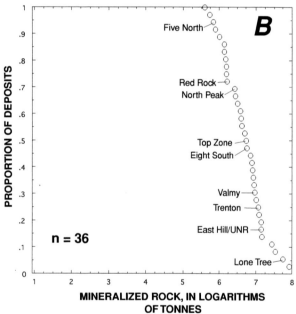

Figure 135 Cumulative distribution diagrams showing Au grades (*A*) and tonnage (*B*) for distal-disseminated Ag–Au deposits in Nevada and Idaho. Data from table 32.

table 32 and classified roughly employing the same geologic criteria used by Cox and Singer (1990), have approximately the same median tonnages as in the original model, but a somewhat higher median grade (approximately 1.62 g Au/tonne (t) versus 1.25 g Au/t) (fig. 135*A*). The locally barite-rich Eight South deposit probably is analogous to the barite-rich Tambo Au deposit in Chile (Siddeley and Araneda, 1986; Albino, 1994). Sixteen of the 36 deposits are in the Battle Mountain Mining District, and the deposits in the mining district north of the Oyarbide fault show no values for their Ag content prima-

rily because either no significant concentration of Ag is present because of leaching during oxidation, or Ag was typically not analyzed during exploration. Many deposits are either in the pre-mining stages or are in the early stages of mining at the time of this writing in 1999, and economically mineralized tonnages are quite likely to expand significantly as mining and exploration progress. In addition, the four deposits from the Bald Mountain Mining District, Nev., (Nos.1–5, RBM, Rat. and Top, table 32) are included tentatively in this classification scheme from data provided by A.D. Hitchborn and others (written commun., 1995). In the gravel-covered region north of the Battle Mountain Mining District, one half of the undiscovered deposits should be above the median Au values and ore tonnages, and the remaining one half should be below these values, provided the remaining undiscovered Au deposits north of Battle Mountain are similar geologically to those used to construct the grade-tonnage model.

The grade-tonnage data (table 32) further emphasize that there are a large number of mostly fracture-controlled, distal disseminated Ag–Au deposits that were ecomomically exploitable in 1997—spot price of Au was approximately $350/oz Au in early 1997—even though hosted by highly siliceous rocks. Six of the deposits (East Hill/UNR, Trenton, Valmy, North Peak, Red Rock, and Top Zone) have Au grades of approximately 1.0 g Au/t, which is well below the median 1.62 g Au/t value for the entire grade model (fig. 135*A*). The first and the last of these four, as well as the Valmy deposit, all are hosted by mostly siliceous rocks of the Valmy Formation. Median tonnage for the entire model is approximately 5.2 million t, and, in contrast to a pronounced concentration of Au grades below the median, tonnages for the nine major deposits in the Valmy and the North Peak quadrangles are disbursed along almost the entire tonnage-distribution curve (fig. 135*B*). Any distal-disseminated Ag–Au deposit formed in calcareous rocks of LU–2 in the upper plate of the thrust, or in calcareous strata of the Antler sequence below the thrust in the alluvium-covered areas north of Lone Tree Hill, should have somewhat higher Au grades than the approximately 1.0 g Au/t values for the deposits hosted by siliceous rocks. Furthermore, the intensely faulted and fractured character of rocks throughout the region make it likely that a strong structural control might be present in any buried deposit.

It is extremely difficult, however, to predict overall size and grade of any undiscovered mineralized system—especially those deposits which are concealed under post-mineral rocks and unconsolidated deposits (Peters and others, 1996).

The type of deposit inferred to be present north of the Battle Mountain Mining District (Peters and others, 1996) is critical for Au–endowment estimates simulated using Mark3 methodologies (see Root and others(1992) for an explanation of this technique). For example, the mean simulated Au endowment for 15 undiscovered Au deposits using the grade-tonnage distributions shown in figure 135 is approximately 336 t Au (10.5 million troy oz Au), or 280 t Au (8.75 million troy ozAu) if one were to use only 18 deposits from Nevada—the latter excluding also the ones from the Bald Mountain Mining District (G.T. Spanski, written commun., 1995). If one were to use grade-tonnage distribution curves for Carlin-type deposits and do a similar Au endowment simulation for 15 deposits, the simulated mean Au endowment would be 960 t Au (30 million troy oz Au). I believe that the former value (336 t Au) is more realistic than the latter for this 98–km^2 part of Nevada. By way of comparison, the approximately 50–km-long segment of the Carlin, Nev., Au trend between the Dee Mine and the Gold Quarry Mine has been estimated by McFarlane (1991) to contain identified resources of approximately 1,700 t Au (53 million troy oz Au) as of 1991. Furthermore, it is not unreasonable to assume that this premier Au belt or mining district will eventually be credited with a total endowment of minable Au resources measuring approximately 3,200 t Au (100 million troy oz Au) when one considers the additional discoveries made since 1991 along the belt (S.G. Peters, oral commun., 1995). Thus, simulated Au endowment estimates for the mostly alluvium-covered region north of Battle Mountain and south of the Twin Creeks deposit measure approximately 10 percent of estimated endowments for the similarly-sized, most productive segment of the Carlin trend. Such relative proportions of simulated Au endowment, furthermore, do not seem unreasonable geologically, provided the covered areas north of Battle Mountain do not contain some lower Paleozoic platform rocks close to the surface. If the latter relation were true, then Au endowment estimates would have to be revised significantly upwards. Finally, the overall simulated-plus-known Au endowment for the entire 75–km-long Twin Creeks-Marigold trend—including the area north of the Battle Mountain Mining District, the 98 km2 area comprising the original area with the 15 occurrences of Au resources in the mining district, and the Twin Creeks deposits themselves—appears to total some 944 t Au (approximately 30 million troy oz Au). This value, which measures approximately 30 percent of the total Au endowment estimated above for the most productive part of the Carlin trend, may be a better number to compare this part of Nevada with inferred Au endowments along the Carlin trend.

Regional dip of rocks of the Valmy Formation, where they crop out in the Battle Mountain Mining District, suggests that there is some likelihood that the tectonostratigraphic thickness of the formation may be relatively thin where the formation crops out near Lone Tree Hill, although deep drilling recently completed at the

Eight South deposit suggests that a considerable thickness of siliceous rocks may be present (D.H. McGibbon, oral commun., 1997). Nonetheless, lower Paleozoic carbonate platform rocks below the Roberts Mountains thrust might be at reasonable depths farther to the north, and they thus could provide an attractive host for the lower parts of a vertically stacked Au-mineralized system, the upper levels of which comprise the Lone Tree Au deposit. This hypothesis would be contingent upon all or a major part of the rocks of the Scott Canyon Formation, which is known to underlie the Valmy Formation (Roberts, 1964), being cut out structurally by displacements along the Roberts Mountains thrust. However, gravity data from the general area of Lone Tree do not suggest a near-surface presence of carbonate rocks (D.M. Cole, oral commun., 1998).

The relations among mapped extents of areas prospective for Tertiary porphyry Cu systems, distal-disseminated Ag–Au deposits, and the Twin Creeks Carlin-type deposits north of the Battle Mountain Mining District (fig. 134; see also, Peters and others, 1996) may have regional exploration implications far to the south of the mining district. The prospective area of mineralized rock north of the mining district appears to be elongated in a north-south direction partly because of the inferred presence of a deep-seated structural zone that controlled Tertiary emplacement of An–enriched magmas to yield the various porphyry Cu–related deposits, as well as their associated high-level distal-disseminated Ag–Au deposits. As shown, Carlin-type deposits, exemplified by the Twin Creeks deposits, appear to be present in a zone that mantles the distal-disseminated Ag–Au zone on the north (fig. 134). If one envisions the core of this structural zone being comprised of the Battle Mountain and McCoy Mining Districts and their predominantly porphyry Cu or pluton-related centers of metallization, then it is conceivable that there might be a mirror image relationship between zones of distal-disseminated Ag-Au and Carlin-type deposits in the Fish Creek Mountains south of the McCoy Mining District. However, geologic complications are present south of the McCoy Mining District—namely, a Miocene caldera in the southern part of the Fish Creek Mountains that is younger than late Eocene and (or) early Oligocene pluton-related mineralized rocks in the Battle Mountain and McCoy Mining Districts (McKee, 1970). The southern ring faults of the caldera appear to have formed near the position of the inferred outer limit of the distal-disseminated Ag–Au part of deposit zonation. Therefore, if geographic distributions among the three deposit-type zones north of the Battle Mountain Mining District (see also, Peters and others, 1996) roughly are developed symmetrically also to the south, then the broad pediment south of the Fish Creek Mountains in the northern part of Antelope Valley might be prospective for Carlin-type deposits.

However, recently completed estimates of depths to basement in this general area south of the Fish Creek Mountains suggest that pre-Cenozoic bedrock probably is at depths of approximately 1 km (David Ponce, written commun., 1997).

A corollary of the above evaluation of regional-scale distributions of the major types of Au deposits in this part of Nevada is that a close temporal relation may exist between porphyry-related metallization and Carlin-type metallization. If Carlin-type deposits in this part of Nevada—including Preble, Pinson, Getchell, and the Twin Creeks deposits—are related to deep-seated evolved metamorphic or magmatic fluids (Hofstra and others, 1988, 1989; Seedorff, 1991; Cline and others, 1996), then metamorphism- and (or) magma-derived fluids responsible for these deposits may have been channelled into a north-south structural zone below the brittle-ductile transition, which presumably is present at a depth of approximately 10 to 15 km in this part of Nevada (Parsons, 1995). A broad area of metamorphic rocks is present west of the Battle Mountain Mining District in the East Range and in the Humboldt Range (Peters and others, 1996). From geologic relations present in the Humboldt Range, it appears (1) that the age of the metamorphism must be younger than Late Cretaceous, which is the emplacement age of ductilely deformed igneous rocks in the Rochester Mining District (Vikre, 1981; Peters and others, 1996), and (2) that this metamorphism and deformation is probably associated with accretion from the west of the adjoining late Late Triassic basinal strata of the Jungo terrane (Silberling and others, 1987). On the basis of the assembled regional metallogenic relations, the general area of the Battle Mountain and McCoy Mining Districts probably represents an area where relatively shallow-seated magmas culminated in development of a number of precious-metal-enriched pluton-related mineral systems.

Finally, difficulties encountered during field discrimination of calcareous turbidite of LU–2 in the upper plate of the Golconda thrust from calacreous sandstone and siltstone in the tectonically underlying Edna Mountain Formation suggest that a regional re-evaluation of these formations using abundant microfossil collections is warranted. Such a re-evaluation is currently (1999) underway in the southern part of the Shoshone Range, Nev., where enigmatic geologic relations involving the Edna Mountain Formation are well exposed (Moore and Murchey, 1998). Such evaluations are especially appropriate because rocks of the Edna Mountain Formation host much of the ore at the Eight South Au deposit. Some areas previously considered to represent only strata belonging to the Havallah sequence in the upper plate of the Golconda thrust may in fact also contain sequences of rock belonging to the tectonically underlying Antler sequence.

GENERAL GEOLOGIC SUMMARY

A large part of the bedrock area at Battle Mountain is made up of allochthonous rocks in the upper plate of the Roberts Mountains thrust which was accreted eastwards onto the continental margin during the Devonian and Mississippian Antler orogeny. These allochthonous rocks include: mostly feldspathic arenite of the Upper Cambrian Harmony Formation; quartzarenite, chert, and altered basalt of the Ordovician Valmy Formation; and chert and argillite of the Devonian Scott Canyon Formation, all of which show fault contacts with one another dating from either the Antler orogeny or subsequent tectonism as young as middle Tertiary. The stratigraphic bases of these units are not present in this area. Of the three, the Harmony and Ordovician Formations are the most widely exposed in the quadrangles, whereas the Scott Canyon Formation is now recognized to crop out only in a small area in the Snow Gulch quadrangle that was previously considered to be part of the Valmy Formation. In this small area, rocks of the Scott Canyon Formation apparently are exposed in a window through the Dewitt thrust, which is a regional-scale thrust at the base of the Harmony Formation and which is related to the Antler orogeny.

The region was uplifted and covered unconformably by the Pennsylvanian and Permian Antler sequence of Roberts (1964) during the Antler orogeny. The Middle Pennsylvanian Battle Formation, in general poorly sorted calcareous and (or) siliceous chert-pebble conglomerate, makes up the lowest unit of the Antler sequence, and it crops out in a relatively small number of places. The middle unit of the Antler sequence, the Pennsylvanian and Permian Antler Peak Limestone, also has generally limited exposures in the area. The uppermost strata in the Antler sequence, made up of the rocks belonging to the Permian Edna Mountain Formation and present near the Old Marigold Mine as debris flows and fissile sandstone, probably represent phosphatic-bearing near-shore equivalents of the Permian Phosphoria Formation of Idaho. Although rocks of the Antler sequence make up only a small part of total rock exposures, they are extremely important from the standpoint of epigenetic metallization because they host some of the most important ore bodies formed in the mining district.

The Battle Mountain region also has been overridden by several structurally higher late Paleozoic and early Mesozoic thrust faults which are associated with the Sonoma orogeny and which resulted in eastward accretion of the Devonian, Mississippian, Pennsylvanian, and Permian Havallah sequence. The base of the Havallah sequence is the Golconda thrust, a post-late Early Permian to Early Triassic thrust, that crops out prominently in the area. Its upper plate strata comprise the Havallah sequence and include two major lithotectonic packages of rock (LU–1 and LU–2) that were originally defined by Murchey (1990). In the Valmy and North Peak quadrangles, these two packages include an upper plate (LU–2) of Mississippian, Pennsylvanian, and Permian basinal radiolarian chert, calcareous and phosphatic sand and silt turbidites, as well as basalt-basaltic andesite (BBA) and basaltic trachyandesite-trachyandesite (BTT) suites of volcanic rock. The base of LU–2 is the Willow Creek thrust, and rocks in the lower plate of this thrust fault are designated LU–1. Rocks of LU–1 are Pennsylvanian and Permian slope deposits that are made up mostly of siliceous argillite, cherty argillite and shale, and spicular chert. Rocks of both LU–2 and LU–1 crop out widely in the North Peak quadrangle, and are, in places, separated by extensive strike lengths of the Willow Creek thrust, which is an important imbricate thrust within the Havallah sequence and is related to the Sonoma orogeny. Locally east of Trenton Canyon, the Willow Creek thrust appears to override the Golconda thrust. Rocks of LU–2 in the Valmy and the North Peak quadrangles have been subdivided further into five lithotectonic plates, each of which typically includes a complex assortment of lithologies. These lithologies generally extend only for relatively short distances.

Rocks in the Havallah sequence at Battle Mountain show broad similarities and some sharp differences with rocks of the sequence that crop out elsewhere. Only one of the most monocrystalline-quartz-rich samples analyzed from calcareous clastic subunits plots within the field for Lower Permian monocrystalline quartz-rich petrofacies rocks determined from the Havallah sequence in nearby mountain ranges. Major differences in these ranges when compared with samples from Battle Mountain primarily involve a uniformly much higher concentration of monocrystalline quartz—approximately 95 volume. The total range in monocrystalline quartz at framework sites at Battle Mountain is about 50 volume percent. When compared to quartz-chert petrofacies of LU–2 in the Mount Tobin area, Nev., similarities in overall proportions of framework monocrystalline quartz suggest similar maturities, although samples from one of the calcareous sandstone subunits at Battle Mountain show elevated abundances of feldspar, particularly monocrystalline K–feldspar. However, samples from the quartz-chert petrofacies in the Mount Tobin area also differ significantly from samples of LU–2 at Battle Mountain in overall proportions of other framework grains, primarily because of elevated abundances of detrital chert fragments and depleted abundances of extrabasinal carbonate grains in the Mount Tobin area. Nonetheless, all these provenance data closely tie depositional environments of rocks in the Golconda allochthon to the continental margin of upper Paleozoic times as proposed by many others. As suggested previously, some debris shed into the Havallah basin probably was derived

from the Antler highlands. Presence of relatively widespread sponge-spicule-bearing sequences in the Edna Mountain Formation are interpreted to be near-shore equivalents of slope deposits made up of spicular-chert facies rocks that are so prominent in LU–1 of the Havallah sequence. Rocks belongong to the Havallah sequence are metamorphosed widely in the general area of Trenton Canyon by emplacement of a composite Late Cretaceous pluton, which also hosts an areally confined stockwork Mo system in its core.

Igneous rocks range in age from Ordovician to Tertiary. The oldest known igneous rocks are basalt and gabbro in the Ordovician Valmy Formation and the youngest igneous rock is Miocene basalt that crops out at Treaty Hill. The Upper Cambrian Harmony Formation is intruded by a large number of basaltic dikes and sills, some diabasic, that are locally folded along with surrounding sedimentary rocks of the Harmony Formation; the dikes and sills are assigned to the Ordovician and (or) Devonian periods. Thin flows of mostly Pennsylvanian and Permian basaltic trachyandesite-trachyandesite (BTT), as well, are present at the base of LU–2 of the Havallah sequence, and some areally limited occurrences of basalt-basaltic andesite (BBA) are present in Mississippian radiolarian ribbon chert in the northern part of the Havallah Hills. From various plots of BBA– and BTT–suite rocks on petrotectonic discrimination diagrams, volcanic rocks of the Havallah sequence are inferred to have been emplaced as submarine flows and (or) sills and dikes that reflect an evolution of petrotectonic environments with time. These environments appear to be: (1) initially during the Mississippian, an enriched mid-ocean ridge basalt (MORB) with relatively depleted abundances of high field strength elements (HFSE) and light rare earth elements (LREE) as the primary magmatic source—here exemplified by BBA–suite rocks—and (2) subsequently during Pennsylvanian and Permian time, either a continental margin regime reflecting effects of some subduction processes or a within-plate regime (exemplified by BTT–suite rocks). The latter rocks, in places, have a strong enrichment of HFSE mostly through increases in Zr/Nb and Zr/(Ti+Y) ratios, and they also have an enhancement of their heavy rare earth element (HREE) abundances (for example, ytterbium) possibly through fractionation processes at depth that may involve garnet. In addition, it appears unlikely that these elemental ratios have been modified by interactions with continental crust because the inferred pre-Sonoma-orogeny position of the Havallah basin was west of its present position and, therefore, probably west of the westernmost edge of the Proterozoic craton.

The next igneous activity in the area is represented by small bodies of presumably Paleozoic or Mesozoic gabbro and diabase that crop out close to the entrance to Trenton Canyon, and which are intensely contact metamorphosed by the Late Cretaceous pluton of Trenton Canyon. The age of these small bodies of gabbro and diabase cannot be established any more precisely than by their relationship to the Late Cretaceous contact metamorphism. Subsequent to intrusion of the pluton at Trenton Canyon, there was a hiatus of approximately 50 m.y. followed by emplacement of a small number of granodiorite dikes at roughly 36 to 39 Ma, as in the general area of Lone Tree Hill. During this same time interval, a large number of generally north-trending dikes in the Snow Gulch quadrangle and a number of small, similarly aged, northwest-elongated stocks near its southern border also were emplaced. These late Eocene and early Oligocene stocks and dikes are concentrated in one of several north-trending and northwest-trending structural zones in the Battle Mountain Mining District whose trends respectively reflect a structural grain inherited both from rocks in the Roberts Mountains and Golconda allochthons, and from pre-Late Cretaceous, probably Jurassic, zones of weakness. In addition, during late Eocene and early Oligocene time, the regional least principal stress axis may have periodically fluctuated between east and northeast trends contemporaneous with magmatic and tectonic activity in the mining district along the north- and northwest-trending structural zones.

Subsequent to emplacement of the small late Eocene or early Oligocene felsic intrusions in the area of Battle Mountain, the next youngest magmatic events are represented by outflow facies of the Oligocene (approximately 34–Ma) Caetano Tuff and Oligocene augite-olivine basalt, which, in the central part of the North Peak quadrangle, yields a 31.3±0.8–Ma potassium-argon age. The Oligocene basaltic rocks are interbedded with a relatively thick sequence of unconsolidated Tertiary gravel along the east margin of Bufflo Valley, and were followed chronologically by an outpouring of 25.5±0.8–Ma calc-alkaline rhyolite tuff and 22.9±0.7–Ma biotite-lithic tuff. Although these latter two rock types are present in limited areas in the quadrangles, they are probably correlative with different cooling units of the regionally widespread, late Oligocene and (or) early Miocene Bates Mountain Tuff (Stewart and McKee, 1977; McKee, 1994). The biotite-lithic tuff rests unconformably on mineralized debris flows of the Edna Mountain Formation in the Eight South Au deposit.

The Battle Mountain Mining District subsequently during the Middle Miocene was subjected to additional regional extensional stresses associated with outpouring of chemically bimodal lavas along the N. 20 to 25° W.–trending, 500–km-long northern Nevada rift. This rift crosses the region about 40 km to the northeast. The youngest igneous rocks in the area of this study, the approximately 12–Ma Miocene basaltic rocks at Treaty Hill, are related genetically to the rift, although most magmatism along the rift culminated during a 17–to 14–Ma interval

of time. The basaltic rocks at Treaty Hill are part of the widespread rift-related basaltic event which provided widespread basalt lavas in the central part of the Sheep Creek Range, and at many other places in north-central Nevada. Contemporaneous with the outpouring of basaltic lavas from the rift, an abundant outpouring of rhyolitic lavas occurred elsewhere in the region. By the time of rift development, the regional least principal stress axis apparently had a trend of about N. 65 to 70° E. Further clockwise rotation of the regional least principal stress to an orientation of about S. 65° E. is suggested by the northeast-striking, post-mineral Oyarbide fault, which crops out prominently in the North Peak quadrangle, and which has a strike similar to a large number of normal faults that cut the northern Nevada rift along a 240–km segment between Midas, Nev., on the north, and Cortez, Nev., on the south.

REFERENCES CITED

Albino, G.V., 1993, Application of metal zoning to gold exploration in porphyry copper systems by B.K. Jones: comments: Journal of Geochemical Exploration, v. 48, no 3, p. 359–366.

—1994, Time-pH-fO2 paths of hydrothermal fluids and the origin of quartz-alunite-gold deposits, *in* Berger, B.R., ed., Advances in research on mineral resources, 1994: U.S. Geological Survey Bulletin 2081, p. 33–42.

Anderton, R., 1985, Clastic facies models and facies analysis, *in* Brenchley, P.J., and Williams, B.P.J., eds., Sedimentology: Recent developments and applied aspects: Oxford, Blackwell Scientific Publications, Geological Society Special Publication, p. 31–47.

Anonymous, 1992, Battle Mountain Gold Co.: California Mining Journal, v. 62, no. 4, p. 35.

Arehart, G.B., 1996, Characteristics and origin of sediment-hosted disseminated gold deposits: a review: Ore Geology Reviews, v. 11, no. 6, p. 383–403.

Arehart, G.B., Foland, K.A., Naeser, C.W., and Kesler, S.E., 1993, ^{40}Ar/^{39}Ar, K/Ar, and fission track geochronology of sediment-hosted disseminated gold deposits at Post-Betze, Carlin trend, northeastern Nevada: Economic geology, v. 88, no. 3, p. 622–646.

Armstrong, R.L., 1975, Precambrian (1500 m.y. old) rocks of central Idaho—The Salmon River Arch and its role in Cordilleran sedimentation and tectonics: American Journal of Science, v. 275–A, p. 437–467.

Armstrong, R.L., and Ward, P.L., 1993, Late Triassic to earliest Eocene magmatism in the North American Cordillera: Implications for the Western Interior Basin, *in* Caldwell,W.G.E., and Kauffman, E.G., eds., Evolution of the Western Interior Basin: Geological Association of Canada, Special paper 39, p. 49–72.

Arribas, José, and Arribas, Eugenia, 1991, Petrographic evidence of different provenance in two alluvial fan systems (Palaeogene of the northern Tajo Basin, Spain), *in* Morton, A.C., Todd, S.P., and Haughton, P.D.W., eds., Developments in sedimentary provenance studies: Geological Society London Special Publication 57, p. 263–271.

Aruscavage, P.J., and Crock, J.G., 1987, Atomic absorption methods, in Baedecker, P.A., ed., Methods for geochemical analysis: U.S. Geological Survey Bulletin 1770, p. C1–C6.

Aurisicchio, C., Fererico, M., and Gianfagna, A., 1988, Clinopyroxene chemistry of the high-potassium suite from the Alban Hills, Italy: Mineralogy and Petrology, v. 39, p. 1–19.

Avé Lallement, H.G., and Oldow, J.S., 1988, Early Mesozoic southward migration of Cordilleran transpressional terranes: Tectonics, v. 7, no. 5, p. 1,057–1,075.

Bagby, W.C., and Berger, B.R., 1985, Geologic characteristics of sediment-hosted, disseminated precious-metal deposits in the western United States, *in* Berger, B.R., and Bethke, P.M., eds., Geology and geochemistry of epithermal systems: Society of Economic Geologists Reviews in Economic Geology, v. 2, p. 169–202.

Barton, M.D., Battles, D.A., Bebout, G.E., Capo, R.C., Christensen, J.N., Davis, S.R., Hanson, R.B., Michelson, C.J., and Trim, H.E., 1988, Mesozoic contact metamorphism in the western United States, *in* Ernst, W.G., ed., Metamorphism and crustal evolution of the western United States: Englewood Cliffs, N.J., Prentice-Hall, p. 110–178.

Batchelder, J.N., 1973, A study of stable isotopes and fluid inclusions at Copper Canyon, Lander County, Nevada: San Jose, California State University, M.S. thesis, 92 p.

—1977, Light-stable-isotope and fluid-inclusion study of the porphyry copper deposit at Copper Canyon, Nevada: Economic Geology, v. 72, no. 1, p. 60–70.

Batchelder, J.N., Theodore, T.G., and Blake, D.W., 1976, Stable isotopes and geology of the Copper Canyon porphyry copper deposits, Lander County, Nevada: Society of Mining Engineers Transactions, v. 260, no. 3, p. 232–236.

Bendix-Almgreen, S.E., 1966, New investigations on *Helicoprion* from the Phosphoria Formation of south-east Idaho, U.S.A.: Kongelige Danske Videnskabernes Selskab, Biolgiske Skrifter [Copenhagen], v. 14, no. 5, 54 p.

Berger, B.R., 1986, Descriptive model of carbonate-hosted Au-Ag, *in* Cox, D.P., and Singer, D.A., eds., Mineral deposit models: U.S. Geological Survey Bulletin 1693, p. 175.

Berger, V.I, and Oscarson, R.L., 1998, Tungsten-polymetallic- and barite-mineralized rocks in the Ruby Mountains, Nevada, *in* Tosdal, R.M., ed., Contributions to the gold metallogeny of northern Nevada: U.S. Geological Survey Open-File Report 98–338, p. 151–175.

Best, M.G., Christiansen, E.H., Deino, A.L., Grommé, Sherman, McKee, E.H., and Noble, D.C., 1989, Eocene through Miocene volcanism in the Great Basin of the western United States: New Mexico Bureau of Mines and Mineral Resources Memoir 47, p. 91–131.

Bhatia, M.R., 1983, Plate tectonics and geochemical composition of sandstones: Journal of geology, v. 91, no. 6, p. 611–627.

—1985, Composition and classification of Paleozoic flysch mudrocks of eastern Australia: Implications in provenance and tectonic setting interpretation: Sedimentary geology, v. 41, p. 249–268.

Blake, D.W., 1992, Supergene copper deposits at Copper Basin, *in* Theodore, T.G., Blake, D.W., Loucks, T.A., and Johnson, C.A., Geology of the Buckingham stockwork molybdenum deposit and surrounding area, Lander County, Nevada: U.S. Geological Survey Professional Paper 798–D, p. D154–D167.

Blake, D.W., and Kretschmer, E.L., 1983, Gold deposits at Copper Canyon, Lander County, Nevada: Nevada Bureau of Mines and Geology Report 36, p. 3–10.

Blake, D.W., Theodore, T.G., Batchelder, J.N., and Kretschmer, E.L., 1979, Structural relations of igneous rocks and mineralization in the Battle Mountain Mining District, Lander County, Nevada, *in* Ridge, J.D., ed., Papers on mineral deposits of western North America: International Association on the Genesis of Ore Deposits Symposium, 5th, Snowbird-Alta, Utah, 1978, Proceedings: Nevada Bureau of Mines and Geology Report 33, p. 87–99.

Blake, D.W., Theodore, T.G., and Kretschmer, E.L., 1978, Alteration and distribution of sulfide mineralization at Copper Canyon, Lander County, Nevada: Arizona Geological Society Digest, v. 11, p. 67–78.

Blake, D.W., Wotruba, P.R., and Theodore, T.G., 1984, Zonation in the skarn environment at the Tomboy-Minnie gold deposits, Lander County, Nevada, *in* Wilkins, Joe, Jr., ed., Gold and silver deposits of the Basin and Range province, western U.S.A.: Arizona Geological Society Digest, v. 15, p. 67–72.

Blatt, Harvey, Middleton, G.V., Murray, R.C., 1980, Origin of sedimentary rocks, Second Edition: Englewood cliffs, New Jersey, Prentice-Hall, Inc., 782 p.

Bloomstein, E.I., Braginton, B., Owen, R., Parratt, R, Raabe, K., and Thompson, W., 1992, Geology and geochemistry of the Lone Tree gold deposit, Humboldt County, Nevada [abs.]: Abstracts with Program, 93rd Annual Meeting, Reno Nevada, Society for Mining, Metallurgy, and Exploration, Inc. and Society of Economic Geology, February 15-18, 1993, p. 149.

— 1993, Geology and geochemistry of the Lone Tree gold deposit, Humboldt County, Nevada: 93rd Annual Meeting Reno Nevada, Society for Mining, Metallurgy, and Exploration, Inc. and Society of Economic Geology, February 15-18, 1993, Preprint 93–205, 23 p.

Bloomstein, E.I., Massingill, G.L., Parratt, R.L., and Peltonen, D.R., 1991, Discovery, geology, and mineralization of the Rabbitt Creek gold deposit, Humboldt County, Nevada, *in* Raines, G.L., Lisle, R.E., Schafer, R.W., and Wilkinson, W.H., eds., Geology and ore deposits of the Great Basin, Symposium Proceedings: Reno, Nevada, Geological Society of Nevada, p. 821–843.

Bouma, A.H., 1962, Sedimentology of some flysch deposits: Amsterdam, Elsevier, 168 p.

Bray, D.C.J., Spooner, E.T.C., and Thomas, A., 1991, Fluid inclusions volatile analysis by heated crushing, on-line gas chromatography; applications to Archean fluids: Journal of Geochemical Exploration, v. 42, p. 167–193.

Brooks, J.W., Meinert, L.D., Kuyper, B.A., and Lane, M.L, 1991, Petrology and geochemistry of the McCoy gold skarn, Lander County, Nevada, *in* Raines, G.L., Lisle, R.E., Schafer, R.W., and Wilkinson, W.H., eds., Geology and ore deposits of the Great Basin, Symposium Proceedings: Reno, Nevada, Geological Society of Nevada, p. 419–442.

Brooks, W.E., Snee, L.W., and Scott, R.B., 1994, Timing and effect of detachment-related potassium metasomatism on $^{40}Ar/^{39}Ar$ dates from the Windous Butte ash-flow tuff, Grant Range, east-central Nevada [abs.]: Geological Society of America Abstracts with Programs, v. 26, no. 2, p. 41.

Brooks, W.E., Thorman, C.H., and Snee, L.W., 1995, $^{40}Ar/^{39}Ar$ ages and tectonic setting of the middle Eocene northeast Nevada volcanic field: Journal of Geophysical Research, v. 100, no. B7, p. 10,403–10,416.

Brueckner, H.K., and Snyder, W.S., 1985, Structure of the Havallah sequence, Golconda allochthon, Nevada: Evidence for prolonged evolution in an accretionary prism: Geological Society of American Bulletin, v. 96, p. 1,113–1,130.

Bryan, W.B., 1972, Morphology of quench crystals in submarine basalts: Journal of Geophysical Research, v. 77, no. 29. p. 5,812–5,819.

—1983, Systematics of modal phenocryst assemblages in submarine basalts: Petrologic implications: Contributions to Mineralogy and Petrology, v. 83, p. 62–74.

Burchfiel, B.C., Cowan, D.S., and Davis, G.A., 1992, Tectonic overview of the Cordilleran orogen in the western United States, *in* Burchfiel, B.C., Lipman, P.W., and Zoback, M.L., eds., The Cordilleran orogen: Conterminous U.S. : Geological Society of America, The Geology of North America, v. G–3, p. 407–479.

Burchfiel, B.C., and Davis, G.A., 1972, Structural framework and evolution of the southern part of the Cordilleran orogen, western United States: American Journal of Science, v. 272, p. 97–118.

—1975, Nature and controls of Cordilleran orogenesis, western United States: Extensions of an earlier synthesis: American Journal of Science, v. 275A, p. 363–396.

Burchfiel, B.C., and Royden, L.H., 1991, Antler orogeny: A Mediterranean-type orogeny: Geology, v. 19, p. 66–69.

Caldwell, D.A., Brummer, J.E., McLachlan, C.D., Schumacher, A.L., and Thompson, K., 1992, Geology and geochemistry of the Mule Canyon gold deposits, Lander County, Nevada [abs.]: Abstracts with Program, 93rd Annual Meeting, Reno Nevada, Society for Mining, Metallurgy, and Exploration, Inc. and Society of Economic Geology, February 15–18, 1993, p. 150.

Carlisle, Donald, 1963, Pillow breccias and their aquagene tuffs, Quadra Island, British Columbia: Journal of Geology, v. 71, no. 1, p. 48–71.

Cashman, K.V., 1990, Textural constraints on the kinetics of crystallization of igneous rocks, *in* Nicholls, J., and Russell, J.K., eds., Modern methods of igneous petrology: Understanding magmatic processes: Mineralogical Society of America, Reviews in Mineralogy, v. 24, p. 259–314.

Castle, R.O., and Lindsley, D.H., 1993, An exsolution silica-pump model for the origin of myrmekite: Contributions to Mineralogy and Petrology, v. 115, p. 58–65.

Christiansen, R.L., and Lipman, P.W., 1972, Cenozoic volcanics and plate-tectonic evolution on the western United States. II. Late Cenozoic: Philosophical Transactions Royal Society of London, Series A, v. 271, p. 249–284.

Churkin, Michael, Jr., 1974, Paleozoic marginal ocean basin-volcanic arc systems in the Cordilleran foldbelt, *in* Dott, R.H., Jr., and Shaver, R.H., eds., Modern and ancient geosynclinal sedimentation: Tulsa, Oklahoma, Society of Economic Paleontologists and Mineralogists Special Publication 19, p. 174–192.

Cline, J.S., Hofstra, A.H., Rye, R.O., and Landis, G.P., 1996, Stable isotope and fluid inclusion evidence for a deep sourced fluid at the Getchell, Carlin-type, gold deposit, Nevada [Extended abs.]: Madison, Wisconsin, PACROFI VI, Program and Abstracts, p. 33–35.

Coats, R.R., 1969, Upper Paleozoic formations of the Mountain City area, Elko County, Nevada, *in* Cohee, G.V., Bates, R.G., and Wright, W.B., eds., Changes in stratigraphic nomenclature by the United States Geological Survey, 1967: U.S. Geological Survey Bulletin 1274–A, p. A22–A27.

Cox, D.P., and Singer, D.A., 1990, Descriptive and grade-tonnage models for distal disseminated Ag–Au deposits: A supplement to U.S. Geological Survey Bulletin 1693: U.S. Geological Survey Open-File Report 90–282, 7 p.

Cox, K.G., Bell, J.D., and Pankhurst, R.J., 1979, The interpretation of igneous rocks: London, George Allen and Unwin Ltd., 450 p.

Cressman, E.R., and Swanson, R.W., 1964, Stratigraphy and petrology of the Permian rocks of southwestern Montana: U.S. Geological Survey Professional Paper 313–C, p. 275–569.

Crock, J.G., and Lichte, F.E., 1982, Determination of rare earth elements in geological materials by inductively coupled argon plasma/atomic emission spectrometry: Analytical Chemistry, v.54, no. 8, p. 1,329–1,332.

Debrenne, Francoise, Gandin, Anna, and Gangloff, R.A., 1990, Analyse sédimentologique et paléontologie de calcaires organogènes du Cambrien Inférieur de Battle Mountain (Nevada, U.S.A.): Annales de Paléontologie (Vertebrate-Invertebrate), v. 76, no. 2, p. 73–119.

Deer, W.A., Howie, R.A., and Zussman, J., 1962a, Rock-forming minerals, Nonsilicates, vol. 5: London, Longmans, Green and Co. Ltd., 371 p.

—1962b, Rock-forming minerals, Sheet silicates, vol. 3: London, Longmans, Green and Co. Ltd., 270 p.

—1978, Rock-forming minerals, Single-chain silicates, vol. 2A: London, Longman Group Ltd., 668 p.

Deino, A.L., 1985, Stratigraphy, chemistry, K–Ar dating, and paleomagnetism of the Nine Hill Tuff, California-Nevada, Part I: Berkeley, University of California, Ph. D. Dissertation, 338 p.

De La Roche, H., Leterrier, J., Grandclaude. P., and Marchal, M., 1980, A classification of volcanic and plutonic rocks using R1R2-diagram and major element analyses; its relationships with current nomenclature: Chemical Geology, v. 29, no. 3–4, p. 183–210.

Dickinson, W.R., 1970, Interpreting detrital modes of graywacke and arkose: Journal of Sedimentary Petrology, v. 40, p. 695–707.

—1985, Interpreting provenance relations from detrital modes of sandstones, in Zuffa, G.G., ed., Provenance of arenites: Dordrecht, D. Reidel Publishing Co., p. 333–361.

Dickinson. W.R., Beard, L.S., Brackenridge, G.R., Erjavec, J.L., Ferguson, R.C., Inman, K.F., Knepp, R.A., Lindberg, F.A., and Ryberg, P.T., 1983, Provenance of North American Phanerozoic sandstones in relation to tectonic setting: Geological Society of America Bulletin, v. 94, no. 1, p. 222–235.

Dickinson, W.R., and Suczek, C.A., 1979, Plate tectonics and sandstone compositions: American Association of Petroleum Geologists Bulletin, v. 63, no. 12, p. 2,164–2,182.

Dilles, J.H., and Gans, P.B., 1995, The chronology of Cenozoic volcanism and deformation in the Yerington area, western Basin and Range and Walker Lane: Geological Society of America Bulletin, v.107, no. 4, p. 474–486.

Dilles, J.H., and Wright, J.E., 1988, The chronology of early Mesozoic arc magmatism in the Yerington district of western Nevada and its regional implications: Geological Society of America Bulletin, v. 100, no. 5, p. 644–652.

Doebrich, J.L., 1992, Preliminary geologic map of the Antler Peak 7-1/2 minute quadrangle, Lander County, Nevada: U.S. Geological Survey Open-File Report 92–398, 12 p.

—1994, Preliminary geologic map of the Galena Canyon 7–1/2 minute quadrangle, Lander County, Nevada: U.S. Geological Survey Open-File Report 94–664, 14 p.

—1995, Geology and mineral deposits of the Antler Peak 7.5–minute quadrangle, Lander County, Nevada: Nevada Bureau of Mines and Geology Bulletin 109, 44 p.

Doebrich, J.L., and Theodore, T.G., 1995, Geology and ore deposits of the Battle Mountain area, Nevada: New mapping, new findings, new ideas [abs.]: Reno, Nevada, Geological Society of Nevada Monthly Newsletter, January, 1995, [p. 5–6].

Doebrich, J.L., Wotruba, P.R., Theodore, T.G., McGibbon, D.H., and Felder, R.P., 1995, Field guide for geology and ore deposits of the Battle Mountain Mining District, Humboldt and Lander Counties, Nevada: Reno, Nevada, Geological Society of Nevada and U.S. Geological Survey, Great Basin Symposium III, Geology and ore deposits of the American Cordillera, 92 p.

—1996, Field trip guidebook for Trip H—Geology and ore deposits of the Battle Mountain Mining District, in Green, S.M., and Struhsacker, Eric, eds., Field Trip Guidebook Compendium: Reno, Nevada, Geological Society of Nevada, U.S. Geological Survey, and Sociedad Geologica de Chile, Geology and ore deposits of the American Cordillera, p. 327–388.

Dohrenwend, J.C., and Moring, B.C., 1991, Reconnaissance photogeologic map of young faults in the Winnemucca 1° by 2° quadrangle, Nevada: U.S. Geological Survey Miscellaneous Field Studies Map MF–2175, 1 sheet, 1:250,000.

Dunham, R.J., 1962, Classification of carbonate rocks according to depositional texture, in Ham, W.E., ed., Classification of carbonate rocks: American Association of Petroleum Geologists Memoir 1, p. 108–121.

Einaudi, M.T., 1990, Zoning of gold and silver in central portions of porphyry copper districts [abs.], in Cuffney, Bob, ed., Geology and ore deposits of the Great Basin, Program with Abstracts: Reno, Nevada, Geological Society of Nevada, April 1–5, 1990, p. 59–60.

Elison, M.W., Speed, R.C., Kistler, R.W., 1990, Geologic and isotopic constraints on the crustal structure of the northern Great Basin: Geological Society of America Bulletin, v. 102, p. 1,077–1,092.

Elthon, Don, 1990, The petrogenesis of primary mid-ocean ridge basalts: Reviews in Aquatic Sciences, v. 2, no. 1, p. 27–53.

Emmons, D.L., and Eng, T.L., 1995, Geologic map of the McCoy Mining District, Lander County, Nevada: Nevada Bureau of Mines and Geology Map 103, 12 p.

Emsbo, Poul, Hofstra, A.H., Park, D., Zimmerman, J.M., and Snee, L., 1996, A mid-Tertiary age constraint on alteration and mineralization in igneous dikes on the Goldstrike property, Carlin trend, Nevada [abs.]: Geological Society of America Abstracts with Programs, v. 28, no. 7, p. A476.

Erickson, R.L., and Marsh, S.P., 1974, Geologic map of the Iron Point quadrangle, Humboldt County, Nevada: U.S. Geological Survey Geologic Quadrangle Map GQ-1175, 1:24,000, 1 sheet.

Evans, J.G., 1974a, Geologic map of the Welches Canyon quadrangle, Eureka County, Nevada: U.S. Geological Survey Geological Quadrangle Map GQ-1117 [scale 1:24,000].

—1974b, Geologic map of the Rodeo Creek NE quadrangle, Eureka County, Nevada: U.S. Geological Survey Geological Quadrangle Map GQ–1116 [scale 1:24,000].

Evans, J.G., and Cress, L.D., 1972, Preliminary geologic map of the Schroeder Mountain quadrangle, Nevada: U.S. Geological Survey Miscellaneous Field Studies Map MF–324, 1 sheet [1:24,000].

Evans, J.G., and Ketner, K.B., 1971, Geologic map of the Swales Mountain quadrangle and part of the Adobe Summit quadrangle, Elko County, Nevada: U.S. Geological Survey Miscellaneous Geologic Investigations MF-324.

Evans, J.G., and Theodore, T.G., 1978, Deformation of the Roberts Mountains allochthon in north-central Nevada: U.S. Geological Survey Professional Paper 1060, 18 p.

Farmer, G.L., and DePaolo, D.J., 1983, Origin of Mesozoic and Tertiary granite in the western United States and implications for pre-Mesozoic crustal structure: 1. Nd and Sr isotopic studies in the geocline of the northern Great Basin: Journal of Geophysical Research, v. 88, no. B4, p. 3,379–3,401.

—1984, Origin of Mesozoic and Tertiary granite in the western United States and implications for pre-Mesozoic crustal structure: 2. Nd and Sr isotopic studies of unmineralized and Cu– and Mo–mineralized granite in the Precambrian craton: Journal of Geophysical Research, v. 89, no. B12, p. 10,141–10,160.

Ferguson, H.G., and Cathcart, S.H., 1954, Geology of the Round Mountain quadrangle, Nevada: U.S. Geological Survey Geologic quadrangle Map GQ–40, 1 sheet.

Ferguson, H.G., Roberts, R.J., and Muller, S.W., 1952, Geologic map of the Golconda quadrangle, Nevada: U.S. Geological Survey Geologic Quadrangle Map GQ–15, 1 sheet.

Fitton, J.G., James, Dodie, Kempton, P.D., Ormerod, D.S., and Leeman, W.P., 1988, The role of lithospheric mantle in the generation of late Cenozoic basic magmas in the western United States: Journal of Petrology, Special Lithosphere Issue, p. 331–349.

Fitton, J.G., James, Dodie, and Leeman, W.P., 1991, Basic magmatism associated with late Cenozoic extension in the western United States: Compositional variations in space and time: Journal of Geophysical Research, v. 96, no. B8, p. 13,693–13,711.

Fleck, R.J., Theodore, T.G., Sarna-Wojcicki, Andrei, and Meyer, C.E., 1998, Age and possible source of air-fall tuffs of the Miocene Carlin Formation, northern Nevada, *in* Tosdal, R.M., ed., Contributions to the gold metallogeny of northern Nevada: U.S. Geological Survey Open-File Report 98–338, p. 176–192.

Folger, H.W., Snee, L.W., Mehnert, H.H., Hofstra, A.H., and Dahl, A.R., 1995, significance of K–Ar and ^{40}Ar/^{39}Ar dates from mica in Carlin-type gold deposits: Evidence from the Jerritt Canyon District, Nevada [abs.]: Geology and ore deposits of the American cordillera, Symposium, Reno/Sparks, Nevada, April 10–13, 1995, Program with Abstracts, p. A31.

Folk, R.L., 1968, Petrology of sedimentary rocks: Austin, Tex., Hemphill's Book Store, 170 p.

—1974, Petrology of sedimentary rocks: Austin, Tex., Hemphill's Book Store, 182 p.

Gabrielse, Hubert, Snyder, W.S., and Stewart, J.H., 1983, Sonoma orogeny and Permian to Triassic tectonism in western North America: Geology, v. 11, p. 484–486.

Garside, L.J., Bonham, H.F. Jr., Tingley, J.V., and McKee, E.H., 1993, Potassium-argon ages of igneous rocks and alteration minerals associated with mineral deposits, western and southern Nevada: Isochron/west, no. 59, p. 17–23.

Gehrels, G.E., Dickinson, W.R., and Smith, M.T., 1993, Detrital zircon provenance of Cambrian toTriassic miogeoclinal and eugeoclinal strata in Nevada [abs.]: Geological Society of America Abstracts with Programs, v. 25, no. 6, p. A–284.

Geist, D.J., McBirney, A.R., and Baker, B.H., 1989, MacGPP—A program package for creating and using geochemical data files: Clinton, New York, 33 p.

Gill, J.B., 1981, Orogenic andesites and plate tectonics: Berlin, Springer-Verlag, 390 p.

Gilluly, James, 1967, Geologic map of the Winnemucca quadrangle, Pershing and Humboldt Counties, Nevada: U.S. Geological Survey Geologic Quadrangle Map GQ–656, 1 sheet.

Gilluly, James, and Gates, O., 1965, Tectonic and igneous geology of the northern Shoshone Range, Nevada: U.S. Geological Survey Professional Paper 465, 153 p.

Gilluly, James, and Masursky, Harold, 1965, Geology of the Cortez quadrangle, Nevada, *with a section on* Gravity and aeromagnetic surveys by D.R. Mabey: U.S. Geological Survey Bulletin 1175, 117 p.

Girty, G.H., Reiland, D.N., and Wardlaw, M.S., 1985, Provenance of the Silurian Elder Sandstone, north-central Nevada: Geological Society of America Bulletin, v. 96, no. 7, p. 925–930.

Gostyayeva, Natalya, Theodore, T.G., and Lowenstern, J.B., 1996, Implications of fluid-inclusion relations in the Elder Creek porphyry copper system, Battle Mountain Mining District, Nevada: U.S. Geological Survey Open-File Report 96–268, 60 p.

Graney, J.R., and McGibbon, D.H., 1991, Geological setting and controls on gold mineralization in the Marigold Mine area, Nevada, *in* Raines, G.L., Lisle, R.E., Schafer, R.W., and Wilkinson, W.H., eds., Geology and ore deposits of the Great Basin, Symposium Proceedings: Reno, Nevada, Geological Society of Nevada, p. 865–874.

Graney, J.R., and Wallace, A.B., 1988, Stratigraphic and structural controls of gold mineralization in the Marigold project area, Nevada [abs.]: Geological Society of America Abstracts with Programs, v. 20, no. 7, p. A141.

Grauch, V.J.S., Jachens, R.C., and Blakely, R.J., 1995, Evidence for a basement feature related to the Cortez disseminated gold trend and implications for region exploration in Nevada: Economic Geology, v. 90, p. 203–207.

Grommé, C.S., McKee, E.H., and Blake, M.C., Jr., 1972, Paleomagnetic correlations and potassium-argon dating of middle Tertiary ash-flow sheets in the eastern Great Basin, Nevada and Utah: Geological Society of America Bulletin, v. 83, p. 1,619–1,638.

Hall, D.L., Sterner, S.M., and Bodnar, R.J., 1988, Freezing point depression of NaCl–KCl–H$_2$O solutions: Economic Geology, v. 83, p. 197–202.

Hammarstrom, J.M., and Zen, E-An, 1986, Aluminum in hornblende: an empirical igneous geobarometer: American Mineralogist, v. 71, p. 1,297–1,313.

Hanger, R.A., Strong, E.E., and Ashinhurst, R.T., 1994, Helicoprion sp. from the Pennsylvanian and Permian Antler Peak Limestone, Lander County, Nevada, *in* Theodore, T.G., Preliminary geologic map of the Snow Gulch quadrangle, Humboldt and Lander Counties, Nevada: U.S. Geological Survey Open-File Report 94–436, p. 24–26.

Harbaugh, D.W., and Dickinson, W.R., 1981, Depositional facies of Mississippian clastics, Antler foreland basin, central Diamond Mountains, Nevada: Journal of Sedimentary Petrology, v. 51, no. 4, p. 1,223–1,234.

Hayes, R.C., Jr., Bradley, M.A., and McCormack, J.K., 1994, Rift-related structural controls to gold mineralization in the Cortez-Gold Acres-Pipeline area, north-central Nevada, *in* Williams, Cindy, ed., Structural geology and mineral deposits in northeastern Nevada: Geological Society of Nevada, Elko Chapter, Symposium Handout, July 15–16, 1994, [p. 41–47].

Hedenquist, J.W., and Lowenstern, J.B., 1994, The role of magmas in the formation of hydrothermal ore deposits: Nature, v. 370, p. 519–527.

Henry, C.D., Boden, D.R., and Castor, S.B., 1998, Geology and mineralization of the Eocene Tuscarora volcanic field, Elko County, Nevada, *in* Tosdal, R.M., ed., Contributions to the gold metallogeny of northern Nevada: U.S. Geological Survey Open-File Report 98–338, p. 279–288.

Hess, P.C., 1989, Origins of igneous rocks: Cambridge, Massachusetts, Harvard University Press, 336 p.

Hofstra, A.H., Landis, G.P., Rye, R.O., Birak, D.J., Dahl, A.R., Daly, W.E., and Jones, M.B., 1989, Geology and origin of the Jerritt Canyon sediment-hosted disseminated gold deposits, Nevada [abs.], *in* Schindler, K.S., ed., USGS Research on Mineral Resources—1989, Programs and Abstracts: U.S. Geological Survey Circular 1035, p. 30–32.

Hofstra, A.H., Northrop, H.R., Rye, R.O., Landis, G.P., and Birak, D.J., 1988, Origin of sediment-hosted disseminated gold deposits by fluid mixing—Evidence from jasperoids in the Jerritt Canyon gold district, Nevada, U.S.A. [extended abs.], *in* Goode, A.D.T., and Bosma, L.I., eds., Bicentennial Gold 88, Extended Abstracts, Oral Programme: Geological Society of Australia, Abstract Series no. 22, p. 284–289.

Hofstra, A.H., and Rye, R.O, 1998, δD and δ¹⁸O data from Carlin-type gold deposits—Implications for genetic models, in Tosdal, R.M., ed., Contributions to the gold metallogeny of northern Nevada: U.S. Geological Survey Open-File Report 98–338, p. 202–210.

Hotz, P.E., and Willden, Ronald, 1964, Geology and mineral deposits of the Osgood Mountains quadrangle, Humboldt County, Nevada: U.S. Geological Survey Professional Paper 431, 128 p.

Howe, S.S., and Theodore, T.G., 1993, Sulfur isotopic composition of vein barite—A guide to the level of exposure of disseminated gold-porphyry copper systems in north-central Nevada? [abs.]: Geological Society of America, Abstracts with Programs, v. 25, no. 6, p. A162–A163.

Howe, S.S., Theodore, T.G., and Arehart, G.B., 1995, Sulfur and oxygen isotopic composition of vein barite from the Marigold mine and surrounding area, north-central Nevada: Applications to gold exploration [abs.]: Geology and ore deposits of the American cordillera, Symposium, Reno/Sparks, Nevada, April 10–13, 1995, Program with Abstracts, p. A39.

Hulen, J.B., Pinnell, M.L., Nielson, D.L., Cox, J.W., and Blake, Jason, 1994, The Yankee Mine oil occurrence, Alligator Ridge District, Nevada–an exhumed and oxidized, paleogeothermal oil reservoir, in

Schalla, R.A., and Johnson, E.H., eds., Oil fields of the Great Basin: Reno, Nevada Petroleum Society, p. 131–142.

Ingamells, C.O., 1970, Lithium metaborate flux in silicate analysis: Analytica Chimica Acta, v. 52, no. 2, p. 323–334.

Irvine, T.N., and Baragar, W.R.A., 1971, A guide to the chemical classification of the common volcanic rocks: Canadian Journal of Earth Sciences, v. 8, p. 523–548.

Ivosevic, S.W., and Theodore, T.G., 1995, Weakly developed porphyry system at upper Paiute Canyon, Battle Mountain Mining District, Nevada [abs.], *in* Geology and ore deposits of the American Cordillera, Program with Abstracts, Symposium, Reno/Sparks, Nevada, April 10–13, 1995, p. 40.

—1996, Weakly developed porphyry system at Upper Paiute Canyon, Battle Mountain Mining District, Nevada, *in* Coyner, A.R., and Fahey, P.L., eds., Geology and ore deposits of the American Cordillera, Symposium Proceedings, Geological Society of Nevada, Reno/Sparks, Nevada, April, 1995, v. 3, p. 1,573–1,594.

Jachens, R.C., Blakely, R.J., and Moring, B.C., 1989, Analysis of concealed mineral resources in Nevada: Constraints from gravity and magnetic studies [abs.], *in* Schindler, K.S., ed., USGS Research on mineral resources—1989: Program and Abstracts: U.S. Geological Survey Circular 1035, p. 35–37.

Jackson, L.L., Brown, F.W., and Neil, S.T., 1987, Major and minor elements requiring individual determination, classical whole rock analysis, and rapid rock analysis, in Baedecker, P.A., ed., Methods for geochemical analysis: U.S. Geological Survey Bulletin 1770, p. G1–G23.

Jansma, P.E., and Speed, R.C., 1993, Deformation, dewatering, and decollement development in the Antler foreland basin during the Antler orogeny: Geology, v. 21, p. 1,035–1,038.

Javoy, M., and Pineau, F., 1991, The volatiles record of a "popping" rock from the mid-Atlantic Ridge at 14 °N: chemical and isotopc composition of gas trapped in the vesicles: Earth and Planetary Science Letters, v. 107, p. 598–611.

Jensen, L.S., 1976, A new cation plot for classifying subalkalic volcanic rocks: Ontario Division of Mines, Miscellaneous Paper 66, 22 p.

Johnson, J.G., 1983, Comment on Mid-Paleozoic age of the Roberts thrust unsettled by new data from northern Nevada [by K.B. Ketner]: Geology, v. 11, no. 1, p. 60–61.

Johnson, J.G., and Visconti, Robert, 1992, Roberts Mountains thrust relationships in a critical area, northern Sulphur Spring Range, Nevada: Geological Society of America Bulletin, v. 104, no. 9, p. 1,208–1,220.

Johnson, M.C., and Rutherford, M.J., 1989, Experimental calibration of the aluminum-in-hornblende geobarometer with application to Long Valley caldera (California) volcanic rocks: Geology, v. 17, p. 837–841.

Johnson, M.G., 1973, Placer gold deposits of Nevada: U.S. Geological Survey Bulletin 1356, 118 p.

Johnson, R.G., and King, B.-S.L., 1987, Energy-dispersive X-ray flourescence spectrometry, *in* Baedecker, P.A., ed., Methods for geochemical analysis: U.S. Geological Survey Bulletin 1770, p. F1–F5.

Jones, B.K., 1992, Application of metal zoning to gold exploration in porphyry copper systems: Journal of Geochemical Exploration, v. 43, no. 2, p. 127–155.

Jones, A.E., 1991a, Upper Paleozoic rocks of the Osgood Mountains region, *in* Wallace, A.R., Jones, A.E., and Madden-McGuire, D.J., Geology and ore deposits of the Getchell trend and northern Nevada rift, northern Nevada, Field Trip 6, *in* Buffa, R.H., and Coyner, A.R., eds., Geology and ore deposits of the Great Basin, Field Trip Guidebook Compendium: Reno, Nevada, Geological Society of Nevada, v. 1, p. 372–381.

—1991b, Provenance of sandstones in the Golconda terrane, north-central Nevada [abs.]: American Association of Petroleum Geologists Bulletin, v. 75, no. 2, p. 368–369.

—1993, Northwest vergent folding in the Harmony Formation, north-central Nevada: Lower Paleozoic tectonics revisited [abs.]: Geological Society of America, Abstracts with Programs, Cordilleran and Rocky Mountain Sections Meeting, v. 25, no. 5, p. 59.

—1997, Geologic map of the Hot Springs Peak quadrangle and the southeastern part of the Little Poverty quadrangle, Nevada: Nevada Bureau of Mines and Geology Field Studies Map FS–14, 1 sheet.

Jones, D.L., Wrucke, C.T., Holdsworth, Brian, and Suczek, C.A., 1978, Revised ages of chert in the Roberts Mountains allochthon, northern Nevada [abs.]: Geological Society of America Abstracts with Programs, v. 10., no. 3, p. 111.

Kamali, C., and Norman, D.I., 1996, Mineralization at the Lone Tree deposit [abs.]: Society for Mining, Metallurgy, and Exploration, Annual Meeting, March 11–14, 1996, Phoenix, Arizona, Technical Program, p. 48.

Kamb, W.B., 1959, Ice petrofabric observations from Blue Glacier, Washington, in relation to theory and experiment: Journal of Geophysical Research, v. 64, p. 1,891–1,909.

Kay, Marshall, 1960, Paleozoic continental margin in central Nevada, western United States, *in* Rasmussen, L.B., and Larsen, Gunnar, eds., Regional paleogeography, International Geological Congress, Report of the Twenty-First Session, Norden, Part XII, p. 94–103.

Kelly, M.A., and Zangerl, Rainer, 1976, *Helicoprion* (Edestidae) in the Permian of west Texas: Journal of Paleontology, v. 50, no. 5, p. 992-994.

Kempton, P.D., Fitton, J.G., Hawkesworth, C.J., and Ormerod, D.S., 1991, Isotopic and trace element constraints on the composition and evolution of the lithosphere beneath the southwestern United States: Journal of Geophysical Research, v. 96, no. B8, p. 13,713–13,735.

Kesler, S.E., Haynes, P.S., Creech, M.Z., and Gorman, J.A., 1986, Application of fluid inclusion and rock-gas analysis in mineral exploration: Journal of Geochemical Exploration, v. 25, no. 1–2., p. 201–215.

Ketner, K.B., 1966, Comparison of Ordovician eugeosynclinal and miogeosynclinal quartzites of the Cordilleran geosyncline: U.S. Geological Survey Professional Paper 550–C, p. C54–C60.

—1977, Deposition and deformation of lower Paleozoic western facies rocks, northern Nevada, *in* Stewart, J.H., Stevens, C.H., and Fritsche, A.E., eds., Paleozoic paleogeography of the western United States: Pacific Coast Paleogeography Symposium 1: Los Angeles, Society of Economic Paleontologists and Mineralogists, Pacific Section, p. 251–258.

—1984, Recent studies indicate that major structures in northeastern Nevada and the Golconda thrust in north-central Nevada are of Jurassic or Cretaceous age: Geology, v. 12, p. 483–486.

—1991, Mt. Ichabod and Major Gulch, east of Wildhorse Reservoir [abs.], *in* Thorman, C.H., ed., Some current research in eastern Nevada and western Utah: U.S. Geological Survey Open-File Report 91–386, p. 28.

—1994, Paleozoic tectonisn in Nevada: New data from geologic mapping in key areas [abs.], Williams, Cindy, ed., Structural geology and mineral deposits in northeastern Nevada: Geological Society of Nevada, Elko Chapter, Symposium Handout, July 15–16, 1994, [p. 8–10].

Ketner, K.B., Ehman, K.D., Repetski, J.E., Stamm, R.G., and Wardlaw, B.R., 1993a, Paleozoic atratigraphy and tectonics in northernmost Nevada: Implications for the nature of the Antler orogeny [abs.]: Geological Society of America, Abstracts with Programs, Cordilleran and Rocky Mountain Sections Meeting, v. 25, no. 5, p. 62.

Ketner, K.B., Murchey, B.L., Stamm, R.G., and Wardlaw, B.R., 1993b, Paleozoic and Mesozoic rocks of Mount Ichabod and Dorsey Canyon, Elko County, Nevada—Evidence for post-Early Triassic emplacement of the Roberts Mountains and Golconda allochthons: U.S. Geological Survey Bulletin 1988–D, p. D1–D12.

Ketner, K.B., and Smith, J.F., Jr., 1982, Mid-Paleozoic age of the Roberts thrust unsettled by new data from northern Nevada: Geology, v. 10, p. 298–303.

Kistler, R.W., 1983, Isotope geochemistry of plutons in the northern Great Basin, in The role of heat in the development of energy and mineral resources in the northern Basin and Range Province, Special Report, Geothermal Resources Council, v. 13, p. 3–8.

—1991, Chemical and isotopic characteristics of plutons in the Great Basin, *in* Raines, G.L., Lisle, R.E., Schafer, R.W., and Wilkinson, W.H., eds., Geology and ore deposits of the Great Basin, Symposium Proceedings: Reno, Nevada, Geological Society of Nevada, p. 107–109.

Kistler, R.W., and Fleck, R.J., 1994, Field guide for a transect of the central Sierra Nevada, California; geochronology and isotope geology: U.S. Geological Survey Open-File Report 94–267, 50 p.

Kizis, J.A., Jr., Bruff, S.R., Crist, E.M., Mough, D.C., and Vaughan, R.G., 1997, Empirical geologic modeling in intrusion-related gold exploration: An example from the Buffalo Valley area, northern Nevada: SEG Newsletter, no. 30, p. 1, and p. 6–13.

Kotlyar, B.B., Theodore, T.G., and Jachens, R.C., 1995, Re-examination of rock geochemistry in the Copper Canyon area, Lander County, Nevada: U.S. Geological Survey Open-File Report 95–816, 47 p.

Kotlyar, B.B., Theodore, T.G., Singer, D.A., Moss, Ken, Campo, A.M., and Johnson, S.D., 1998, Geochemistry of the gold skarn environment at Copper Canyon, Nevada, *in* Lentz, D.R., ed., Mineralized intrusion-related skarn systems: Mineralogical Association of Canada Short Course Series, v. 26, p. 415–443.

Kretschmer, E.L., 1991, Geology of the Kramer Hill gold deposit, Humboldt County, Nevada, *in* Raines, G.L., Lisle, R.E., Schafer, R.W., and Wilkinson, W.H., eds., Geology and Ore Deposits of the Great Basin, Symposium Proceedings: Reno, Nevada, Geological society of Nevada, p. 857–863.

Lawton, T.F., Boyer, S.E., and Schmitt, J.G., 1994, Influence of inherited taper on structural variability and conglomerate distribution, Cordilleran fold and thrust belt, western United States: Geology, v. 22, no. 4, p. 339–342.

Le Bas, M.J., Le Maitre, R.W., Streckeisen, A., and Zanettin, B., 1986, A chemical classification of volcanic rocks based on the total alkali-silica diagram: Journal of Petrology, v. 27, p. 745–750.

Leeman, W.P., Oldow, J.S., and Hart, W.K., 1992, Lithosphere-scale thrusting in the western U.S. Cordillera as constrained by Sr and Nd isotopic transitions in Neogene volcanic rocks: Geology, v. 20, no. 1, p. 63–66.

LeMaitre, R.W., 1976, The chemical variability of some common igneous rocks: Journal of Petrology, v. 17, no. 4, p. 589–637.

Lichte, F.E., Golightly, D.W., and Lamothe, P.J., 1987, Inductively coupled plasma-atomic emission spectrometry, in Baedecker, P.A., ed., Methods for geochemical analysis: U.S. Geological Survey Bulletin 1770, p. B1–B10.

Lipman, P.W., Prostka, H.J., and Christiansen, R.L., 1972, Cenozoic volcanism and plate-tectonic evolution of the Western United States: Philosophical Transactions Royal Society of London, Series A, v. 271, p. 217–248.

Little, T.A., 1987, Stratigraphy and structure of metamorphosed upper Paleozoic rocks near Mountain City, Nevada: Geological Society of America Bulletin, v. 98, no. 1, p. 1–17.

Lofgren, G.E., 1983, Effect of heterogenous nucleation on basaltic textures: Journal of Petrology, v. 24, p. 229–255.

Loucks, T.A., and Johnson, C.A., 1992, Economic geology, in Theodore, T.G., Blake, D.W., Loucks, T.A., and Johnson, C.A., 1992, Geology of the Buckingham stockwork molybdenum deposit and surrounding area, Lander County, Nevada: U.S. Geological Survey Professional Paper 798–D, p. D101–D138.

Lovering, T.G., 1972, Jasperoid in the United States—Its characteristics, origin, and economic significance: U.S. Geological Survey Professional Paper 710, 164 p.

Lowe, D.R., 1982, Sediment gravity flows: II, Depositional models with special reference to the deposits of high-density turbidity currents: Journal of Sedimentary Petrology, v. 52, no. 1, p. 279–297.

MacDonald, A.J., and Spooner, E.T.C., 1981, Calibration of a Linkam TH600 programmable heating-cooling stage for microthermometric examination of fluid inclusions: Economic Geology, v. 76, p. 1,248–1,258.

Macdonald, G.A., 1967, Forms and structures of extrusive basaltic rocks, in Hess, H.H., and Poldervaart, Arie, eds., Basalts, the Poldervaart treatise on rocks of basaltic composition: New York, John Wiley & Sons, Ltd., p. 1–61.

Mack, G.H., 1984, Exceptions to the relationship between plate tectonics and sandstone composition: Journal of Sedimentary Petrology, v. 54, no. 1, p. 212–220.

Madden-McGuire, D.J., Hutter, T.J., and Suczek, C.A., 1991, Late Cambrian-Early Ordovician microfossils from the allochthonous Harmony Formation at its type locality, northern Sonoma Range, Humboldt County, Nevada [abs.]: Geological Society of America Abstracts with Programs, v. 23, no. 2, p. 75.

Madeisky, H.E., 1995, A lithogeochemical and radiometric study of hydrothermal alteration and metal zoning at the Cinola epithermal gold deposit, Queen charlotte Islands, British Columbia [abs.]: Reno, Nevada, Geological Society of Nevada and U.S. Geological Survey, Great Basin Symposium III, Geology and ore deposits of the American Cordillera, Program with Abstracts, p. A48.

Madrid, R.J., 1987, Stratigraphy of the Roberts Mountains allochthon in north-central Nevada: Stanford, California, Stanford University, Ph.D. dissertation, 341 p.

Madrid, R.J., and Bagby, W.C., 1986, Structural alignment of sediment-hosted gold deposits in north-central Nevada; An example of inherited fabrics [abs.]: Geological Society of America Abstracts with Programs, v. 18, no. 5, p. 393.

Madrid, R.J., and Roberts, R.J., 1991, Origin of gold belts in north-central Nevada, in Madrid, R.J., Roberts, R.J., and Mathewson, David, Stratigraphy and structure of the Battle Mountain gold belt and their relationship to gold deposits, Field Trip 15, in Buffa, R.H., and Coyner, A.R., eds., Geology and ore deposits of the Great Basin: Reno, Nevada, Geological Society of Nevada, Fieldtrip Guidebook Compendium, v. 2, p. 927–939.

Maher, B.J., Browne, Q.J., and McKee, E.H., 1993, Constraints on the age of gold mineralization and metallogenesis in the Battle Mountain-Eureka mineral belt: Economic Geology, v. 88, p. 469–478.

Mamet, B.L., Mortelmans, G., and Sartenaer, P., 1965, Réflexions à propos du Calcaire d'Etroeungt: Belgique Geological Society Bulletin, v. 74, p. 41–51.

Mamet, B.L., Skipp, Betty, Sando, W.J., and Mapel, W.J., 1971, Upper Mississippian and Carboniferous rocks, Idaho: American Association of Petroleum Geologists Bulletin, v. 55, no. 1, p. 20–33.

Mansfield, G.R., 1927, Geography, geology, and mineral resources of part of southeastern Idaho: U.S. Geological Survey Professional Paper 152, 409 p.

Mark, R.K., Lee Hu, C., Bowman, H.R., Asaro, F., McKee, E.H., and Coats, R.R., 1975, A high $^{87}Sr/^{86}Sr$, low alkali mantle source for MORB-like tholeite, northern Great Basin: Geochimica et Cosmochimica Acta, v. 39, p. 1,671–1,678.

Marsh, S.P., and Erickson, R.L., 1977, Geologic map of the Brooks Spring quadrangle, Humboldt County, Nevada: U.S. Geological Survey, Geologic Quadrangle Map GQ–1366, 1: 24,000.

McBride, E.F., 1963, A classification of common sandstones: Journal of Sedimentary Petrology, v. 33, no. 3, p. 664–669.

McCann, T., 1991, Petrological and geochemical determination of provenance in the southern Welch Basin, in Morton, A.C., Todd, S.P., and Haughton, P.D.W., eds., Developments in sedimentary provenance studies: Geological Society Special Publication, v. 57, p. 215–230.

McCollum, L.B., and McCollum, M.B., 1989, Carboniferous continental slope to ocean basin deposits adjacent to the Antler orogenic belt of western North America: XIe Congrès International de Stratigraphie et de Géologie du Carbonifère, Beijing 1987, Compte Rendu 4, p. 237–244.

McCollum, L.B., McCollum, M.B., Jones, D.L., and Repetski, J.E., 1987, The Scott Canyon Formation, Battle Mountain, Nevada: structural amalgamation of the Ordovician Valmy Formation and Devonian

Slaven Chert [abs.]: Geological Society of America Abstracts with Programs, v. 19, no. 17, p. 764.

McDonald, G.A., and Katsura, T., 1964, Chemical composition of Hawaiian lavas: Journal of Petrology, v. 5, p. 82–133.

McFarlane, D.N., 1991, Gold production on the Carlin trend, *in* Knutsen, G.C., Ekburg, and McFarlane, D.N., Geology and mineralization of the Carlin trend, Fieldtrip 12, *in* Buffa, R.H., and Coyner, A.R., eds., Geology and ore deposits of the Great Basin: Reno, Nevada, Geological Society of Nevada, Fieldtrip Guidebook Compendium, v. 2, p. 841–843.

McKee, E.H., 1970, Fish Creek Mountains Tuff and volcanic center: U.S. Geological Survey Professional Paper 681, 17 p.

—1971, Tertiary igneous chronology of the Great Basin of the western United States-implications for tectonic models: Geological Society of America Bulletin, v. 82, p. 3,497–3,502.

—1992, Potassium argon and $^{40}Ar/^{39}Ar$ geochronology of selected plutons in the Buckingham area, *in* Theodore, T.G., Blake, D.W., Loucks, T.A., and Johnson, C.A., Geology of the Buckingham stockwork molybdenum deposit and surrounding area, Lander County, Nevada: U.S. Geological Survey Professional Paper 798–D, p. D36–D40.

—1996, Cenozoic magmatism and mineralization in Nevada, *in* Coyner, A.R., and Fahey, P.L., eds., Geology and ore deposits of the American Cordillera, Symposium Proceedings, Geological Society of Nevada, Reno/Sparks, Nevada, April, 1995, v. 3, p. 581–588.

McKee, E.H., and Moring, B.C., 1996, Cenozoic mineral deposits and Cenozoic igneous rocks of Nevada, *in* Singer, D.A., Ed., Au Analysis of Nevada's metal-bearing mineral resources: Nevada Bureau of Mines and Geology Open-File Report 96–2, Chapter 6, 16 p., 1 sheet, scale 1:1,000,000.

McKee, E.H., Duffield, W.A., and Stern, R.J., 1983, Late Miocene and early Pliocene basaltic rocks and their implications for crustal structures, northeastern California and south-central Oregon: Geological Society of America Bulletin, v. 94, p. 292–304.

McKee, E.H., and Mark, R.K., 1971, Strontium isotopic composition of two basalts representative of the southern Snake River volcanic province: U.S. Geological Survey Professional Paper 750–B, p. B92–B95.

McKee, E.H., and Silberman, M.L., 1970, Geochronology of Tertiary igneous rocks in central Nevada: Geological Society of America Bulletin, v. 81, no. 8, p. 2,317–2,328.

McKelvey, V.E., 1967, Phosphate deposits: U.S. Geological Survey Bulletin 1252–D, p. D1–D21.

Meier, A.L., 1980, Flameless atomic absorption determination of gold in geologic materials: Journal of Geochemical Exploration, v. 13, p. 77–85.

Merriam, C.W., and Anderson, C.A., 1942, Reconnaissance survey of the Roberts Mountains, Nevada: Geological Society of America Bulletin, v. 53, p. 1,675–1,727.

Meschede, M., 1986, A method of discriminating between different types of mid-ocean ridge basalts and continental tholeiites with the Nb–Zr–Y diagram: Chemical Geology, v. 56, p. 207–218.

Middleton, G.V., 1960, Chemical composition os sandstones: Geological Society of America Bulletin, v. 71, no. 7, p. 1,011–1,026.

Miller, E.L., Holdsworth, B.K., Whiteford, W.B., and Rodgers, D., 1984, Stratigraphy and structure of the Schoonover sequence, northeastern Nevada: Implications for Paleozoic plate-margin tectonics: Geological Society of America Bulletin, v. 95, no. 9, p. 1,063–1,079.

Miller, E.L., Kanter, L.R., Larue, D.K., Turner, R.J., Murchey, B.L., and Jones, D.L., 1982, Structural fabric of the Paleozoic Golconda allochthon, Antler Peak quadrangle, Nevada: Progressive deformation of an oceanic sedimentary assemblage: Journal of Geophysical Research, v. 87, p. 3,795–3,804.

Miller, E. L., and Larue, D.K., 1983, Ordovician quartzite in the Roberts mountains allochthon, Nevada: Deep sea fan deposits derived from cratonal North America, *in* Stevens, C.H., ed., Pre-Jurassic rocks in western North American suspect terranes: Los Angeles, California, Society of Economic Paleontologists and Mineralogists, Pacific Section, Symposium Volume, p. 91–102.

Miller, E.L., Miller, M.M., Stevens, C. H., Wright, J.E., and Madrid, Raul, 1992, Late Paleozoic paleogeographic and tectonic evolution of the western U.S. Cordillera, *in* Burchfiel, B.C., Lipman, P.W., and Zoback, M.L., eds., The Cordilleran orogen: Conterminous U.S. : Geological Society of America, The Geology of North America, v. G–3, p. 57–106.

Moore, J.G., Batchelder, J.N., and Cunningham, C.G., 1977, CO_2 filled vesicles in mid-ocean basalts: Journal of Volcanology and Geothermal Resources, v. 2, p. 309–327.

Moore, T.J., and Murchey, B.L., 1998, Initial results of stratigraphic and structural studies in The Cedars quadrangle, southern Shoshone Range, *in* Tosdal, R.M., ed., Contributions to the gold metallogeny of northern Nevada: U.S. Geological Survey Open-File Report 98–338, p. 119–140.

Morimoto, N., Fabries, J., Ferguson, A.K., Ginzburg, I.V., Ross, M., Seifert, F.A., and Zussman, J., 1988, Nomenclature of pyroxenes: Mineralogy and Petrology, v. 39, no. 1, p. 55–76.

Motooka, J.M., 1988, An exploration geochemical technique for the determination of preconcentrated organometallic halides by ICP–AES: Applied Spectroscopy, v. 42, no. 7, p. 1,293–1,296.

Muffler, L.J.P., 1964, Geology of the Frenchie Creek quadrangle, north-central Nevada: U.S. Geological Survey Bulletin 1179, 99 p.

Muller, Daniel, and Groves, D.I., 1993, Direct and indirect associations between potassic igneous rocks, shoshonites and gold-copper deposits: Ore Geology Reviews, v. 8, p. 383–406.

Muller, Daniel, Rock, N.M.S., and Groves, D.I., 1992, Geochemical discrimination between shoshonitic and potassic volcanic rocks in different tectonic ssttings: A pilot study: Mineralogy and Petrology, v. 46, p. 259–289.

Murchey, B.L., 1990, Age and depositional setting of siliceous sediments in the upper Paleozoic Havallah sequence near Battle Mountain, Nevada: Implications for the paleogeography and structural evolution of the western margin of North America, *in* Harwood, D.S., and Miller, M.M., eds., Paleozoic and early Mesozoic paleogeographic relations; Sierra Nevada, Klamath Mountains, and related terranes: Geological Society of America Special Paper 255, p. 137–155.

—1994, Radiolarians in the Ordovician Valmy Formation and the Devonian Scott Canyon Formation, *in* Theodore, T.G., Preliminary geologic map of the Snow Gulch quadrangle: U.S. Geological Survey Open-File Report 94–436, 31 p..

Murchey, B.L., Theodore, T.G., and McGibbon, D.H., 1995, Regional implications of newly discovered relations of the Permian Edna Mountain Formation, north-central Nevada [abs.]: Geology and ore deposits of the American Cordillera, Symposium, Reno/Sparks, Nevada, April 10–13, 1995, p. A57–A58.

Murowchik, J.B., and Barnes, H.L., 1986, Marcasite precipitation from hydrothermal solutions: Geochemica and Cosmichemica Acta, v. 50, p. 2,615–2,629.

Mutti, Emiliano, 1974, Examples of ancient deep-sea fan deposits from circum-Mediterranean geosynclines, in Dott, R.H., Jr., and Shaver, R.H., eds., Modern and ancient geosynclinal sedimentation: Tulsa, Oklahoma, Society of Economic Paleontologists and Mineralogists Special Publication 19, p. 92–105.

Mutti, Emiliano, and Ricci Lucchi, F., 1978, Turbidites of the northern Appenines: Introduction to facies analysis: International Geological Review, v. 20, no. 2, p. 125–166 [translated from Mutti, Emiliano, and Ricci Lucchi, F., 1972, Le torbiditi dell' Appennino settentrionale: introduzione all' analisi di facies: Memorie della Societa Geologica Italiana, v. 11, p. 161–199].

Myers, G.L., 1994, Geology of the Copper Canyon-Fortitude skarn system, Battle Mountain, Nevada: Pullman, Washington, Washington State University, Ph.D. dissertation, 338 p.

Myers, G.L., and Meinert, L.D., 1991, Alteration, mineralization, and gold distribution in the Fortitude gold skarn, in Raines, G.L., Lisle, R.E., Schafer, R.W., and Wilkinson, W.H., eds., Geology and ore deposits of the Great Basin, Symposium Proceedings: Reno, Nevada, Geological Society of Nevada, p.407–417.

Nakamura, Noboru, 1974, Determination of REE, Ba, Fe, Mg, Na and K in carbonaceous and ordinary chondrites: Geochimica et Cosmochimica Acta, v. 38, p. 757–775.

Nash, J.T., 1976, Fluid-inclusion petrology-data from porphyry copper deposits and applications to exploration: U.S. Geological Survey Professional Paper 907–D, p. D1–D16.

Nash, J.T., and Theodore, T.G., 1971, Ore fluids in the porphyry copper deposit at Copper Canyon, Nevada: Economic Geology, v. 66, no. 3, p. 385–399.

Nichols, K.M., and Silberling, N.J., 1990, Delle phosphatic member: An anomalous phosphatic interval in the Mississippian shelf sequence of central Utah: Geology, v. 18, p. 46–49.

Nilsen, T.H., and Stewart, J.H., 1980, Penrose Conference Report; The Antler orogeny—mid-Paleozoic tectonism in western North America: Geology, v. 8, no. 6, p. 298–302.

Norman, D.I., Groff, John, Camalli, Cem, Musgrave, John, and Moore, J.N., 1996, Gaseous species in fluid inclusions: Indicators of magmatic input into ore-forming geothermal systems [abs.]: Geological Society of America, Abstracts with Programs, v. 28, no. 7, p. A–401.

Oakes, C.S., Bodnar, R.J., and Simonson, J.M., 1990, The system NaCl-$CaCl_2$-H_2O: 1. The ice liquidus at 1 atm total pressure: Geochimica et Cosmochimica Acta, v. 54, p. 603–610.

Oldow, J.S., Bally, A.W., Avé Lallemant, H.G., and Leeman, W.P., 1989, Phanerozoic evolution of the North Maerican Cordillera; United States and Canada, in Bally, A.W., and Palmer, A.R., eds., The geology of North America–An overview: Boulder, Colorado, Geological Society of America, The Geology of North America, v. A, p. 139–232.

Ormerod, D.S., Hawkesworth, C.J., Rogers, N.W., Leeman, W,P., and Menzies, M.A., 1988, Tectonic and magmatic transitions in the western Great Basin: Nature, v. 333, p. 349–353.

Osterberg , M.W., and Guilbert, J.M., 1991, Geology, wall-rock alteration, and new exploration techniques at the Chimney Creek sediment-hosted gold deposit, Humboldt County, in Raines, G.L., Lisle, R.E., Schafer, R.W., and Wilkinson, W.H., eds., Geology and ore deposits of the Great Basin, Symposium Proceedings: Reno, Nevada, Geological Society of Nevada, p. 805–819.

Palmer, A.R., 1971, The Cambrian of the Great Basin and adjoining areas, western United States, in Holland, C.H., ed., Cambrian of the New World: New York, Wiley-Interscience, p. 1–78.

Parker, R.L., 1967, Composition of the Earth's crust: U.S. Geological Survey Professional Paper 440–D, p. D1–D17.

Parratt, R.L., and Tapper, C.J., and Bloomstein, E.I.,1999, Geology and mineralization of The Rabbit Creek gold deposit [Extended abs.]: Geological Society of Nevada, Abstracts 1986 through 1998, 1999, p. 1,989–15 to 1,989–17.

Parsons, T., 1995, The Basin and Range province, in Olsen, K.H., ed., Continental rifts: evolution, structure, tectonics: Amsterdam, Elsevier, Developments in geotectonics 25, p. 277–322.

Pearce, J.A., and Cann, J.R., 1973, Tectonic setting of basic volcanic rocks determined using trace element analyses: Earth and Planetary Science Letters, v. 19, p. 290–300.

Peccerillo, A., and Taylor, S.R., 1976, Geochemistry of Eocene calc-alkaline volcanic rocks from the Kastamono area, northern Turkey: Contributions to Mineralogy and Petrology, v. 58, p. 63–81.

Peters, S.G., 1998, Evidence for the Crescent Valley-Independence lineament, north-central Nevada, in Tosdal, R.M., ed., Contributions to the gold metallogeny of northern Nevada: U.S. Geological Survey Open-File Report 98–338, p. 106–118.

Peters, S.G., Nash, J.T., John, D.A., Spanski, G.T., King, H.D., Connors, K.A., Moring, B.C., Doebrich, J.L., McGuire, D.J., Albino, G.V., Dunn, V.C., Theodore, T.G., and Ludington, Steve, 1996, Metallic mineral resources in the U.S. Bureau of Land Management's Winnemucca District and Surprise Resource Area, northwest Nevada and northeast California: U.S. Geological Survey Open-File Report 96–712, 147 p.

Pettijohn, F.J., 1963, Chemical composition of sandstones—excluding carbonate and volcanic sands, in Fleischer, Michael, ed., Data of geochemistry, sixth edition: U.S. Geological Survey Professional Paper 440–S, p. S1–S19.

Pettijohn, F.J., Potter, P.E., and Siever, Raymond, 1972, Sand and sandstone: New York, Springer-Verlag, 618 p.

Piper, D.Z., and Medrano, M.D., 1994, Geochemistry of the Phosphoria Formation at Montpelier Canyon, Idaho: Environment of deposition: U.S. Geological Survey Bulletin 2023, p. B1–B27.

Poole, F.G., Stewart, J.H., Palmer, A.R., Sandberg, C.A., Madrid, R.J., Ross, R.J., Jr., Hintze, L.F., Miller, M.M., and Wrucke, C.T., 1992, Latest Precambrian to latest Devonian time; Development of a continental margin, in Burchfiel, B.C., Lipman, P.W., and Zoback, M.L., eds., The Cordilleran orogen: Conterminous U.S.: Boulder, Colorado, Geological Society of America, The geology of North America, v. G–3, p.9–56.

Powers, M.C., 1953, A new roundness scale for sedimentary particles: Journal of Sedimentary Petrology, v. 23, no. 2, p. 117–119.

Pring, A., Birch, W.D., Dawe, J., Taylor, M., Deliens, M., and Walenta, K., 1995, Kintoreite, $PbFe_3(PO_4)_2(OH,H_2O)_6$, a new mineral of the jarosite-alunite family, and lusingite discredited: Mineralogical Magazine, v. 59, p. 143–148.

Ratajeski, Kent, 1995, Estimation of initial and saturation water concentrations for three granitic plutons in the north-central Great Basin, Nevada: College Park, Maryland, University of Maryland, M. S. thesis, 299 p.

Ressell, M.W., Noble, D.C., and Connors, K.A., 1998, Eocene dikes of the Carlin Trend, Nevada; magmatic As, Sb, Cs, Tl, CO_2 & excess Ar suggest a deep degassing model for gold mineralization [abs.]: Geological Society of America, Abstracts with Programs, v. 30, no. 7, p. 126.

Reynolds, R.C., 1980, Interstratified clay minerals, in Brindley, G.W. and Brown, G., eds., Crystal structures of clay minerals and their X–ray identification, Mineralogical Society of London, Chapter 4, p. 249–303.

Roberts, R.J., 1949, Geology of the Antler peak quadrangle, Humboldt and Lander Counties, Nevada: U.S. Geological Survey Open-File Report, 110 p.

—1964, Stratigraphy and structure of the Antler Peak quadrangle, Humboldt and Lander Counties, Nevada: U.S. Geological Survey Professional Paper 459–A, 93 p.

—1966, Metallogenic provinces and mineral belts in Nevada: Nevada Bureau of Mines Report 13, part A, p. 47–72.

Roberts, R.J., and Arnold, D.C., 1965, Ore deposits of the Antler Peak quadrangle, Humboldt and Lander Counties, Nevada: U.S. Geological Survey Professional Paper 459–B, 94 p.

Roberts, R.J., Hotz, P.E., Gilluly, James, and Ferguson, H.G., 1958, Paleozoic rocks in north-central Nevada: American Association of Petroleum Geologists Bulletin, v. 42, no. 12, p. 2,813–2,857.

Roberts, R.J., and Madrid, R.J., 1991, Relationship of the Roberts mountains thrust to the Antler orogeny, in Madrid, R.J., Roberts, R.J., and Mathewson, David, Stratigraphy and structure of the Battle Mountain gold belt and their relationship to gold deposits, Field Trip 15, in Buffa, R.H., and Coyner, A.R., eds., Geology and ore deposits of the Great Basin: Reno, Nevada, Geological Society of Nevada, Fieldtrip Guidebook Compendium, v. 2, p. 940–947.

Roberts, R.J., and Thomasson, M.R., 1964, Comparison of late Paleozoic depositional history of northern Nevada and central Idaho: U.S. Geological Survey Professional Paper 475–D. p. D1–D6.

Robinson, A.C., 1993, Regionally and compositionally coherent magma sources for Mesozoic and Tertiary granitoids in Nevada and eastern California north of 37 degrees latitude [abs.]: Geological Society of America Abstracts with Programs, v. 25, no. 5, p. 139.

—1994, Strontium in the Tuolumne intrusive suite, Sierra Nevada, CA; a guide to magma source, fractionation, and emplacement [abs.]: Eos, Transactions, American Geophysical Union, v. 75, p. 749.

Robinson, Peter, 1980, The composition of terrestrial pyroxenes-Internal and external limits, in Prewitt, C.T., ed., Pyroxenes: Mineralogi-

cal Society of America, Reviews in Mineralogy, v. 7, p. 419–494.

Rock, N.M.S., 1987, The need for standardization of normalized multi-element diagrams in geochemistry: A comment: Geochemical Journal, v. 21, p. 75–84.

Roedder, Edwin, 1984, Fluid inclusions: Mineralogical Society of America Reviews in Mineralogy, v. 12, 644 p.

Rogers, J.J.W., Burchfiel, B.C., Abbott, E.W., Anepohl, J.K., Ewing, A.H., Koehnken, P.J., Novitsky-Evans, P.J., and Talukdars, S.C., 1974, Paleozoic and lower Mesozoic volcanism and continental growth in the western United States: Geological Society of America Bulletin, v. 85, p. 1,913–1,924.

Root, D.H., Menzie, W.D., and Scott, W.A., 1992, Computer Monte Carlo simulation in quantitative resource estimation: Nonrenewable Resources, v. 1, no. 2, p. 125–138.

Roser, B.P., and Korsch, R.J., 1986, Determination of tectonic setting of sandstone-mudstone suites using SiO_2 content and K_2O/Na_2O ratio: Journal of Geology, v. 94, no. 5, p. 635–650.

Rowan, L.C., and Wetlaufer, P.H., 1981, Geologic evaluation of major Landsat lineaments in Nevada and their relationship to ore districts: U.S. Geological Survey Open-File Report, 79–544, 65 p.

Rowell, A.J., Rees, M.N., and Suzcek, C.A., 1979, Margin of the North American continent in Nevada during Late Cambrian time: American Journal of Science, v. 279, p. 1–18.

Rupke, N.A., 1978, Deep clastic seas, in Reading, H.G., ed., Sedimentary environments and facies: Oxford, Blackwell Scientific Publications, p. 372–415.

Rye, R.O., Bethke, P.M., and Wasserman, M.D., 1992, The stable isotope geochemistry of acid sulfate alteration: Economic Geology, v. 87, no. 2, p. 225–262.

Rytuba, J.J., 1985, Development of disseminated gold deposits of Cortez, Horse Canyon, and Gold Acres, Nevada, at the end stage of the caldera-related volcanism [abs.], in Krafft, Kathleen, ed., USGS Research on Mineral Resources—1985, Program and Abstracts: U.S. Geological Survey Circular 949, p. 47.

Saller, A.H., and Dickinson, W.R., 1982, Alluvial to marine facies transition in the Antler overlap sequence, Pennsylvanian and Permian of north-central Nevada: Journal of Sedimentary Petrology, v. 52, no. 3, p. 925–940.

Sandberg, C.A., Gutschick, R.C., Johnson, J.G., Poole, F.G., and Sando, W.J., 1982, Middle Devonian to Late Mississippian geologic history of the overthrust belt region, western United States, in Powers, R.B., ed., Geologic studies of the Cordilleran thrust belt, volume II: Denver, Colorado, Rocky Mountain Association of Geologists. p. 691–719.

Sando, W.J., 1993, Coralliferous carbonate shelves of Mississippian age, west side of Antler orogen, central Nevada: U.S. Geological Survey Bulletin 1988–F, p. F1–F29.

Sanford, R.F., Pierson, C.T., and Crovelli, R.A., 1993, An objective replacement method for censored geochemical data: Mathematical Geology, v. 25, no. 1, p. 59–80.

Sarda, Philippe, and Graham, David, 1990, Mid-ocean ridge popping rocks: implications for degassing at ridge crests: Earth and Planetary Science Letters, v. 97, p. 268–289.

Sasada, Masakatsu, Sawaki, Takayuki, and Takeno, Naoto, 1992, Analysis of fluid inclusion gases from geothermal systems using a rapid-scanning quadrupole mass spectrometer: European Journal of Mineralogy, v. 4, p. 895–906.

Saunders, A.D., 1984, The rare earth element characteristics of igneous rocks from the ocen basins, *in* Henderson, P., ed., Rare earth element geochemistry: Amsterdam, Elsevier, p. 206–236.

Saunders, J.A., 1993, Supergene oxidation of bonanza Au–Ag veins at the Sleeper deposit, Nevada, USA: implications for hydrogeochemical exploration in the Great Basin: Journal of Geochemical Exploration, v. 47, nos. 1–3, p. 359–375.

Sawlan, M.G., 1991, Magmatic evolution of the Gulf of California rift, *in* Dauphin, J.P., and Simoneit, B.R.T., eds., The Gulf and Peninsular Province of the Californias: American Association of Petroleum Geologists Memoir, v. 47, p. 301–369.

—1992, Concordant geometry of map-scale folds and plutons: Implications for the structural framework of the Sierra Nevada, California [abs.]: AAPG-SEPM-SEG-EMD Pacific Section Meeting, April 27–May 1, 1992, Sacramento, California, American Association Petroleum Geologists Bulletin, v. 76, no. 3, p. 430.

Schmidt, K.W., Wotruba, P.R., and Johnson, S.D., 1988, Gold-copper skarn and related mineralization at Copper Basin, Nevada: Reno, Nevada, Geological Society of Nevada, Fall (1988) Field-trip Guidebook, 6 p.

Schmidt, M.W., 1992, Amphibole composition in tonalite as a function of pressure: an experimental calibration of the Al-in-hornblende barometer: Contributions to Mineralogy and Petrology, v. 110, p. 304–310.

Seedorff, Eric, 1991, Magmatism, extension, and ore deposits of Eocene to Holocene age in the Great Basin—mutual effects and preliminary proposed genetic relationships, *in* Raines, G.L., Lisle, R.E., Schafer, R.W., and Wilkinson, W.H., eds., Geology and ore deposits of the Great Basin, Symposium Proceedings: Reno, Nevada, Geological Society of Nevada, p. 133–178.

Seedorff, Eric, Bailey, C.R.G., Kelley, David, and Parks, Wright, 1990, Buffalo Valley Mine: A porphyry-related gold deposit, Lander County, Nevada, *in* Madrid, R. J., Roberts, R.J., and Mathewson, D., eds., Structure and stratigraphy of the Battle Mountain gold belt: Their relation to active and prospective mines and exploration for gold: Field Trip Guidebook, Trip 15, Geological Society of Nevada, Symposium, Geology and ore deposits of the Great Basin, Reno/Sparks, 1990, p. 118–145.

Shand, S.J., 1947, Eruptive rocks, their genesis, composition, classification, and their relatio to ore deposits, with a chapter on meteorites: London, Thomas Murby, 488 p.

Sharp, Z.D., Essene, E.J., and Kelly, W.C., 1985, A re-examination of the arsenopyrite geothermometer: pressure considerations and applications to natural assemblages: Canadian Mineralogist, v. 23, p. 517–534.

Shawe, D.R., 1988, Introduction to geology and resources of gold, and geochemistry of gold, *in* Shawe, D.R., Ashley, R.P., and Carter, L.M.H., eds., Geology and resources of gold in the United States: U.S. Geological Survey Bulletin 1857, Chapter A, p. A1–A8.

Shawe, D.R., and Stewart, J.H., 1976, Ore deposits as related to tectonics and magmatism, Nevada and Utah: Society of Mining Engineers of AIME Transactions, v. 260, p. 225–231.

Shepherd, T.J., 1981, Temperature programmable heating-freezzing stage for micro-thermometric analysis of fluid inclusions: Economic Geology, v. 76, p. 1,244–1,247.

Siddeley, G. and Araneda, R., 1986, The El Indio-Tambo gold deposits, Chile, *in* MacDonald, A.J., ed., Proceedings of Gold '86: Toronto, Canada, International Symposium on the Geology of Gold, p. 445–456.

Silberling, N.J., 1973, Geologic events during Permian-Triassic time along the Pacific margin of the United States, *in* Logan, A., and Hills, L.V., eds., The Permian and Triassic Systems and their mutual boundary: Calgary, Alberta, Alberta Society Petroleum Geology, p. 345–362.

—1986, Pre-Tertiary stratified rocks of the Tonopah 1° x 2° quadrangle [abs.], *in* D.H. Whitebread, compiler, Abstracts of the symposium on the geology and mineral deposits of the Tonopah 1° x 2° quadrangle, Nevada: U.S. Geological Survey Open-File Report 86–467, p. 2–3.

Silberling, N.J., Jones, D.L., Blake, M.C., Jr., and Howell, D.G., 1984, Lithotectonic terrane map of the western conterminous United States, Part C of Silberling, N.J. and Jones, D.L., eds., Lithotectonic maps of the North American Cordillera: U.S. Geological Survey Open-File Report 84–523, 43 p.

—1987, Lithotectonic terrane map of the western conterminus United States: U.S. Geological Survey Miscellaneous Field studies Map MF–1874–C, 20 p.

Silberling, N.J., and Roberts, R.J., 1962, Pre-Tertiary stratigraphy and structure of northwestern Nevada: Geological Society of America Special Paper 72, 58 p.

Silberman, M.L., and McKee, E.H., 1971, K–Ar ages of granitic plutons in north-central Nevada: Isochron/West, no. 71–1, p. 15–32.

Sillitoe, R.H., 1991, Intrusion-related gold deposits, in Foster, R.P., ed., Gold metallogeny and exploration: Glasgow and London, Blackie and Son, Ltd., p. 165–209.

—1994, Indonesian mineral deposits—introductory comments, comparisons, and speculations, *in* Van Leeuwen, T.M., Hedenquist, J.W., James, L.P., and Dow, J.A.S., eds., Mineral deposits in Indonesia-Discoveries of the past 25 years: Journal of Geochemical Exploration, v. 50, nos. 1–3, p. 1–11.

Sillitoe, R.H., and Bonham, H.F., Jr., 1990, Sediment-hosted gold deposits: Distal products of magmatic-hydrothermal systems: Geology, v. 18, no. 2, p. 157–161.

Simmons, G.L., 1992, Development of low temperature pressure oxidation at Lone Tree gold mine: Littleton, Colorado , Society of Mining, Metallurgy and Exploration, Inc., Preprint no. 93–86.

Smith, A.G., Hurley, A.M., and Briden, J.C., 1981, Phanerozoic paleocontinental world maps: Cambridge, Cambridge University Press, 102 p.

Smith, J.F., Jr., and Ketner, K.B., 1978, Geologic map of the Carlin-Piñon range area, Elko and Eureka Counties, Nevada: U.S. Geological Survey Miscellaneous Series Investigations Map I–1028, 2 sheets [1:62,500].

Smith, J.V., 1974, Feldspar minerals, chemical and textural properties, volume 2: New York, Springer-Verlag, 690 p.

Smith, M.T., Dickinson, W.R., and Gehrels, G.E., 1993, Contractural nature of Devonian-Mississippian Antler tectonism along the North American continental margin: Geology, v. 21, no. 1, p. 21–24.

Smith, M.T., and Gehrels, G.E., 1992, Detrital zircon constraints on the provenance of the Harmony and Valmy Formations, Roberts Mountains allochthon, Nevada [abs.]: Geological Society of America, Abstracts with Programs, v. 24, no. 5, p. 82.

—1994, Detrital zircon geochronology and the provenance of the Harmony and Valmy Formations, Roberts Mountains allochthon, Nevada: Geological Society of America Bulletin, v. 106, no. 7, p. 968–979.

Snyder, W.S., and Brueckner, H.K., 1983, Tectonic evolution of the Golconda allochthon, Nevada: Problems and Perspectives, *in* Stevens, C.H., ed., Pre-Jurassic rocks in western North American suspect terranes: Los Angeles, California, Society of Economic Paleontologists and Mineralogists, Pacific Section, Symposium Volume, p. 103–123.

Solomon, G.C., and Taylor, H.P., Jr., 1989, Isotopic evidence for the origin of Mesozoic and Cenozoic plutons in the northern Great basin: Geology, v. 17, p. 591–594.

Southern Pacific Company, 1964, Minerals for industry, northern Nevada & northwestern Utah: San Francisco, Calif., Southern Pacific Company, Summary of Geological Survey of 1955–1961, v. 1, 188 p.

Speed, R.C., 1963, Layered picrite-anorthositic gabbro, West Humboldt Range, Nevada: Mineralogical Society of America Special Paper 1, p. 69–77.

—1977, Island arc and other paleogeographic terranes of late Paleozoic age in the western Great Basin, *in* Stewart, J.H., Stevens, C.H., and Fritsche, A.E., eds., Paleozoic paleogeography of the western United States: Pacific Coast Paleography Symposium 1, The Pacific Section: Los Angeles, Society of Economic Paleontologists and Mineralogists, p. 349–362.

Speed, R.C., and Sleep, N.H., 1982, Antler orogeny and foreland basin: A model: Geologic Society of America Bulletin, v. 93, p. 815–828.

Spurr, J.E., 1898, Geology of the Aspen mining district, Colorado, with atlas: U.S. Geological Survey Monograph 31, 260 p., and atlas of 30 folio sheets.

Stacey, J.S., Sherrill, N.D., Dalrymple, G.B., Lanphere, M.A., and Carpenter, N.V., 1981, A five collector system for the simultaneous measurement of argon isotopic ratios in a static mass spectrometer: International journal of Mass Spectrometry and Ion Physics, v. 39, p. 167–180.

Stanley, C.R., and Madeisky, H.E., 1994, Lithogeochemical exploration for hydrothermal ore deposits using Pearce element analysis, *in* Lentz, D.R., ed., Alteration and alteration processes associated with ore-forming systems: Geological Association of Canada, Short Course Notes, v. 11, p. 193–211.

Steiger, R.H., and Jäger, E., 1977, Subcomission on geochronology—convention on the use of decay constants in geo- and cosmochronology: Earth and Planetary Science Letters, v. 36, p. 359–362.

Stevens, C.H., 1977, Permian depositional provinces and tectonics, western United States, *in* Stewart, J.H., Stevens, C.H., and Fritsche, A.E., eds., Paleozoic paleogeography of the western United States: Pacific Coast Paleography Symposium 1, The Pacific Section: Los Angeles, Society of Economic Paleontologists and Mineralogists, p. 113–135.

—1987, Affinities of Permian fusulinid faunas in the Golconda allochthon and northern Sierra Nevada [abs.]: Geological Society of America Abstracts with Programs, v. 19, no. 6, p. 455.

Stewart, J.H., 1980, Geology of Nevada—A discussion to accompany the Geologic Map of Nevada: Nevada Bureau of Mines and Geology Special Publication 4, 136 p.

—1985, East-trending dextral faults in the western Great Basin: An explanation for anomalous trends of pre-Cenozoic strata and Cenozoic faults: Tectonics, v. 4, no. 6, p. 547–564.

Stewart, J.H., MacMillan, J.R., Nichols, K.M., Stevens, C.H., 1977, Deep-water upper Paleozoic rocks in north-central Nevada—a study of the type area of the Havallah Formation, *in* Stewart, J.H., Stevens, C.H., and Fritsche, A.E., eds., Paleozoic paleogeography of the western United States: Society of Economic Paleontologists and Mineralogists, Pacific Coast Paleogeography Symposium 1, p. 337–348.

Stewart, J.H., and McKee, E.H., 1977, Geology, Part I, *in* Geology and mineral deposits of Lander County, Nevada: Nevada Bureau of Mines and Geology Bulletin 88, p. 1–59.

Stewart, J.H., Moore, W.J., and Zietz, Isadore, 1977, East-west patterns of Cenozoic igneous rocks, aeromagnetic anomalies, and mineral deposits, Nevada and Utah: Geological Society of America Bulletin, v. 88, no. 1, p. 67–77.

Stewart, J.H., Murchey, B.L., Jones D.L., and Wardlaw, B.R., 1986, Paleontologic evidence for complex tectonic interlayering of Mississippian to Permian deep-water rocks of the Golconda allochthon in Tobin Range, north-central Nevada: Geological Society of America Bulletin, v. 97, no. 9, p. 1,122–1,132.

Stewart, J.H., and Poole, F.G., 1974, Lower Paleozoic and uppermost Precambrian Cordilleran miogeocline, Great Basin, Western United States, *in* Dickinson, W.R., ed., Tectonics and sedimentation: Society of Economic Paleontologists and Mineralogists Special Publication 22, p. 28–57.

Stewart, J.H., and Suczek, C.A., 1977, Cambrian and latest Precambrian paleogeography and tectonics in the western United States, *in* Stewart, J.H., Stevens, C.H., and Fritsche, A.E., eds., Paleozoic paleogeography of the western United States: Pacific Coast Paleogeography Symposium 1: Los Angeles, Society of Economic Paleontologists and Mineralogists, Pacific Section, p. 1–17.

Stoffregen, R.E. and Alpers, C.N. , 1992, Observations on the unit cell dimensions, water contents and ∂D values of natural and synthetic alunites: American Mineralogist, v. 77, p. 1,092–1,099.

Stow, D.A.V., 1985, Deep-sea clastics: Where are we and where are we going?, *in* Brenchley, P.J., and Williams, B.P.J., eds., Sedimentology: Recent developments and applied aspects: Oxford, Blackwell Scientific Publications, Geological Society Special Publication, p. 67–93.

Streckeisen, A.L., chairman, 1973, Plutonic rocks: Classification and nomenclature recommended by the IUGS Subcommission on the

Systematics of Igneous Rocks: Geotimes, v. 18, no. 10, p. 26–30.

Struhsacker, E.N., 1980, The geology of the Beowawe geothermal system, Eureka and Lander Counties, Nevada: University of Utah Research Institute, Earth Science Laboratory Report ESL–37, U.S. Department of Energy 12079–7, 78 p.

Suczek, C.A., 1977, Tectonic relations of the Harmony Formation, northern Nevada: Stanford, Calif., Stanford University, Ph.D. thesis, 96 p.

Teal, Lewis, and Jackson, Mac, 1997, Geologic overview of the Carlin trend gold deposits and descriptions of recent deep discoveries, in Vikre, Peter, Thompson, T.B., Bettles, Keith, Christensen, Odin, and Parratt, Ron, eds., Carlin-type gold deposits field conference: Society of Economic Geology Guidebook Series, v. 28, p. 3–37.

Theodore, T.G., 1991a, Preliminary geologic map of the North Peak quadrangle, Humboldt and Lander Counties, Nevada: U.S. Geological Survey Open-File Report 91–429, 10 p.

—1991b, Preliminary geologic map of the Valmy quadrangle, Humboldt County, Nevada: U.S. Geological Survey Open-File Report 91–430, 11 p.

—1991c, Geology and ore deposits of the Valmy and North Peak quadrangles, Battle Mountain Mining District, Nevada [abs.], in Thorman, C.H., ed., Some current research in eastern Nevada and western Utah: U.S. Geological Survey Open-File Report 91–386, p. 4–6.

—1994, Preliminary geologic map of the Snow Gulch quadrangle, Humboldt and Lander Counties, Nevada, with a section on Radiolarians in the Ordovician Valmy Formation and Devonian Scott Canyon Formation, by B.L. Murchey, and a section on Helicoprion sp. from the Pennsylvanian and Permian Antler Peak Limestone, Lander County, Nevada, by R.A. Hanger, E.E. Strong, and R.T. Ashinhurst: U.S. Geological Survey Open-File Report 94–436, 31 p.

—1996, Geology and implications of silver/gold ratios of the Elder Creek porphyry copper system, Battle Mountain Mining District, Nevada, in Coyner, A.R., and Fahey, P.L., eds., Geology and ore deposits of the American Cordillera, Symposium Proceedings, Geological Society of Nevada, Reno/Sparks, Nevada, April, 1995, p. 1,557–1,571.

Theodore, T.G., Armstrong, A.K., Harris, A.G., Stevens, C.H., and Tosdal, R.M., 1998, Geology of the northern terminus of the Carlin trend, Nevada: Links between crustal shortening during the late Paleozoic Humboldt orogeny and northeast-striking faults, in Tosdal, R.M., ed., Contributions to the gold metallogeny of northern Nevada: U.S. Geological Survey Open-File Report 98–338, p. 69–105.

Theodore, T.G., and Blake, D.W., 1975, Geology and geochemistry of the Copper Canyon porphyry copper deposit and surrounding area, Lander County, Nevada: U.S. Geological Survey Professional Paper 798–B, p. B1–B86.

—1978, Geology and geochemistry of the West ore body and associated skarns, Copper Canyon porphyry copper deposits, Lander County, Nevada, with a section on Electron microprobe analyses of andradite and diopside by N.G. Banks: U.S. Geological Survey Professional Paper 798–C, p. C1–C85.

Theodore, T.G., Blake, D.W., Loucks, T.A., and Johnson, C.A., 1992, Geology of the Buckingham stockwork molybdenum deposit and surrounding area, Lander County, Nevada, with a section on: Potassium-argon and $^{40}Ar/^{39}Ar$ geochronology of selected plutons in the Buckingham area by E.H. McKee, and a section on Economic geology by T.A. Loucks and C.A. Johnson, and a section on Supergene copper deposits at Copper Basin by D.W. Blake, and a section on Mineral chemistry of Late Cretaceous and Tertiary skarns by J.M. Hammarstrom: U.S. Geological Survey Professional Paper 798–D, p. D1–D307.

Theodore, T.G., and Hammarstrom, J.M., 1991, Petrochemistry and fluid-inclusion study of skarns from the northern Battle Mountain mining district, Nevada, in Aksyuk, A.M., and others, eds., Skarns—their genesis and metallogeny: Athens, Greece, Theophrastus Publications, p. 497–554.

Theodore, T.G., Howe, S.S., Blake, D.W., and Wotruba, P.R., 1986, Geochemical and fluid zonation in the skarn environment at the Tomboy-Minnie gold deposits, Lander County, Nevada: Journal of Geochemical Exploration, v. 25, p. 99–128.

Theodore, T.G., and Jones, G.M., 1992, Geochemistry and geology of gold in jasperoid, Elephant Head area, Lander County, Nevada: U.S. Geological Survey Bulletin 2009, 53 p.

Theodore, T.G., and McKee, E.H., 1983, Geochronology and tectonics of the Buckingham porphyry molybdenum deposit, Lander County, Nevada [abs.]: Geological Society of America Abstracts with Programs, May 2–4, 1983, Salt Lake City, Utah, v. 15, no. 5, p. 275.

Theodore, T.G., and Menzie, W.D., 1984, Fluorine-deficient porphyry molybdenum deposits in the cordillera of western North America, in Janelidze, T.V., and Tvalchrelidze, A.G., eds., International Association on the Genesis of Ore Deposits Quadrennial Symposium, 6th, Tbilisi, U.S.S.R., 1982, Proceedings: Stuttgart, E. Schweizerbart'sche Verlagsbuchhandlung, v. 1, p. 463–470.

Theodore, T.G., and Nash, J.T., 1973, Geochemical and fluid zonation at Copper Canyon, Lander County, Nevada: Economic Geology, v. 68, no. 4, p. 565–570.

Theodore, T.G., Orris, G.J., Hammarstrom, J.M., and Bliss, J.D., 1991, Gold-bearing skarns: U.S. Geological Survey Bulletin 1930, 61 p.

Theodore, T.G., and Roberts, R.J., 1971, Geochemistry and geology of deep drill holes at Iron Canyon, Lander County, Nevada, with a section on Geophysical logs of drill hole DDH–2 by C.J. Zablocki: U.S. Geological Survey Bulletin 1318, 32 p.

Theodore, T.G., Silberman, M.L., and Blake, D.W., 1973, Geochemistry and K–Ar ages of plutonic rocks in the Battle Mountain mining district, Lander County, Nevada: U.S. Geological Survey Professional Paper 798–A, 24 p.

Thomas, A.V., and Spooner, E.T.C., 1992, The volatile chemistry of magmatic H_2O-CO_2 fluid inclusions from the Tanco zoned granitic pegmatite, southeastern Manitoba, Canada: Geochemica at Cosmochimica Acta, v. 56, no. 1, p. 49–65.

Thorman, C.H., and Brooks, W.E., 1994, Field trip guide to the southern East Humboldt Range and northern Currie Hills, northeast Nevada: Age and style of attenuation faults in Permian and Triassic rocks: U.S. Geological Survey Open-File Report 94–439, 10 p.

Thorman, C.H., Ketner, K.B., and Peterson, Peter, 1990, The Elko orogeny—Late Jurassic orogenesis in the Cordilleran miogeocline [abs.]: Geological Society of America, Abstracts with Programs, v. 22, no. 3, p. 88.

Thornton, C.P., and Tuttle, O.F., 1960, Chemistry of igneous rocks I.Differentiation index: American Journal of Science, v. 258, no. 9, p. 664–684.

Tingley, J.V., 1992, Mining districts of Nevada: Nevada Bureau of Mines and Geology Report 47, 124 p.

Tomlinson, A.J., 1990, Biostratigraphy, stratigraphy, sedimentary petrology, and structural geology of the upper Paleozoic Golconda allochthon, north-central Nevada: Stanford, Calif., Stanford University, Ph.D. dissertation, 492 p.

Trexler, J.H., Jr., and Nitchman, S.P., 1990, Sequence stratigraphy and evolution of the Antler foreland basin, east-central Nevada: Geology, v. 18, p. 422–425.

Turekian, K.K., and Wedepohl, K.H., 1961, Distribution of the elements in some major units of the earth's crust: Geological Society of America Bulletin, v. 72, p. 175–192.

Turner, R.J., 1985, A late Devonian stratiform Pb–Zn and stratiform barite metallogenic event in the North American Cordillera: Marginlong extension and implications for the Antler orogeny [abs.]: Geological Society of America Abstracts with Programs, v. 17, no. 6, p. 414.

Turner, R.J.W., Madrid, R.J., and Miller, E.L., 1989, Roberts Mountains allochthon: Stratigraphic comparison with lower Paleozoic outer continental margin strata of the northern Canadian Cordillera: Geology, v. 17, no. 4, p. 341–344.

Verville, G.J., Sanderson, G.A., and Drowley, D.D., 1986, Wolfcampian fusulinids from the Antler Peak Limestone, Battle Mountain, Lander County, Nevada: Journal of Foraminiferal Research, v. 16, no. 4, p. 353–362.

Vieten, K., and Hamm, H.M., 1978, Additional notes on the calculation of the crystal chemical formula of clinopyroxenes and their contents of Fe^{3+} from microprobe analyses: Neues-Jahrb., -Mineral., Monatsh., v. 2, p. 71–83.

Vikre, P.G., 1981, Silver mineralization in the Rochester District, Pershing County, Nevada: Economic Geology, v. 76, no. 3, p. 580–609.

Volk, J.A., and Lauha, E.A., 1993, Structural controls on mineralization at the Goldstrike property, Elko and Eureka Counties, Nevada [abs.]: Geological Society of America Abstracts with Programs, Cordilleran and Rocky Mountain Section Meeting, v. 25, no. 5, p. 159.

Wallace, A.R., 1991, Effect of late Miocene Extension on the exposure of gold deposits in north-central Nevada, in Raines, G.L., Lisle, R.E., Schafer, R.W., and Wilkinson, W.H., eds., Geology and ore deposits of the Great Basin, Symposium Proceedings: Reno, Nevada, Geological Society of Nevada, p. 179–183.

Wallace, A.R., and Hruska, D.C., 1991, Miocene volcanic rocks and gold deposits, Gold Circle (Midas) district and Ivanhoe gold deposit, in Geology and ore deposits of the Getchell trend and northern Nevada rift, northern Nevada, in Buffa, R.H., and Coyner, A.R., eds., Geology and ore deposits of the Great Basin, Field Trip Guidebook Compendium, v. 1: Reno, Nevada, Geological Society of Nevada, p. 382–388.

Wallace, A.R., and John, D.A., 1998, New studies on Tertiary volcanic rocks and mineral deposits, northern Nevada rift, in Tosdal, R.M., ed., Contributions to the gold metallogeny of northern Nevada: U.S. Geological Survey Open-File Report 98–338, p. 264–278.

Wallace, A.R., and McKee, E.H., 1994, Implications of Eocene through Miocene ages for volcanic rocks, Snowstorm Mountains and vicinity, northern Nevada, in Berger, B.R., ed., Advances in research on mineral resources, 1994: U.S. Geological Survey Bulletin 2081, p. 13–18.

Wallin, E.T., 1990, Provenance of selected lower Paleozoic siliciclastic rocks in the Roberts Mountains allochthon, Nevada, in Harwood, D.S., and Miller, M.M., eds., Paleozoic and early Mesozoic paleogeographic relations; Sierra Nevada, Klamath Mountains, and related terranes: Geological Society of America Special Paper 255, p. 17–32.

Watkins, Rodney, and Browne, Q.J., 1989, An Ordovician continental-margin sequence of turbidite and seamount deposits in the Roberts Mountains allochthon, Independence Range, Nevada: Geological Society of America Bulletin, v. 101, p. 731–741.

Westra, Gerhard, and Keith, S.B., 1981, Classification and genesis of stockwork molybdenum deposits: Economic Geology, v. 76, no. 4, p. 844–873.

Wheeler, H.E., 1939, *Helicoprion* in the Anthracolithic (late Paleozoic) of Nevada and California, and its stratigraphic significance: Journal of Paleontology, v. 13, no. 1, p. 103–114.

White, W.H., Bookstrom, A.A., Kamilli, R.J., Ganster, M.W., Smith, R.P., Ranta, D.E., and Steininger, R.C., 1981, Character and origin of Climax-type molybdenum deposits, in Skinner, B.J., ed., Economic Geology, seventy-fifth anniversary volume, 1905–1980: New Haven, Conn., Economic Geology Publishing Co., p. 270–316.

Whitebread, D.H., 1994, Geologic map of the Dun Glen quadrangle, Pershing County, Nevada: U.S. Geological Survey Miscellaneous Investigations Series Map I–1209, 1 sheet (scale 1:48,000).

Whiteford, W.B., 1990, Paleogeographic setting of the Schoonover sequence, Nevada, and implications for the late Paleozoic margin of western north America, in Harwood, D.S., and Miller, M.M., eds., Paleozoic and early Mesozoic paleogeographic relations; Sierra Nevada, Klamath Mountains,and related terranes: Geological Society of America Special Paper 255, p. 115–136.

Whiteford, W.B., Little, T.A., Miller, E.L., and Holdsworth, B.K., 1983, The nature of the Antler orogeny: View from north-central Nevada [abs.]: Geological Society of America Abstracts with Programs, v. 15, no. 5, p. 382.

Whitehill, H.R., 1873, Biennial report of the State Mineralogist of the State of Nevada for the years 1871 and 1872: Carson City, Nevada, 191 p.

Wiebe, R.A., 1993, The Pleasant Bay layered gabbro-diorite, coastal Maine: Ponding and crystallization of basaltic injections into a silicic magma chamber: Journal of Petrology, v. 34, p. 461–489.

Willden, Ronald, 1963, Geology and mineral deposits of Humboldt County, Nevada: Nevada Bureau of Mines and Geology Bulletin 59, 154 p.

—1979, Ruby orogeny—A major early Paleozoic tectonic event, in Newman, G.W., and Goode, H.D., eds., Basin and Range Symposium and Great Basin Field Conference: Denver, Colorado, Rocky Mountain Association of Geologists, p. 55–73.

Williams, J.S., and Dunkle, D.H., 1948, *Helicoprion*-like fossils in the Phosphoria Formation (western U.S.) [abs]: Geological Society of

America Bulletin, v. 59, no. 12, p. 1,362.

Wilson, S.A., Kane, J.S., Crock, J.G., and Hatfield, D.B., 1987, Chemical methods of separation for optical emission, atomic absorption spectrometry, and colorimetry, *in* Baedecker, P.A., ed., Methods for geochemical analysis: U.S. Geological Survey Bulletin 1770, p. D1–D14.

Wood, D.A., 1980, The application of a Th–Hf–Ta diagram to problems of tectonomagmatic classification and to establishing the nature of crustal contamination of basaltic lavas of the British tertiary volcanic province: Earth and Planetary Science Letters, v. 50, p. 11–30.

Wood, D.A., Joron, J.-L., and Treuil, M., 1979, A reappraisal of the use of trace elements to classify and discriminate between magma series erupted in different tectonic settings: Earth and Planetary Sciences Letters, v. 45, no. 2, p. 326–336.

Wooden, J.L., Kistler, R.W., and Tosdal, R.M., 1998, Pb isotopic mapping of crustal structure in the northern Great Basin and relationships to Au deposit trends, *in* Tosdal, R.M., ed., Contributions to the gold metallogeny of northern Nevada: U.S. Geological Survey Open-File Report 98–338, p. 20–33.

Wotruba, P.R., Benson, R.G., and Schmidt, K.W., 1986, Battle Mountain describes the geology of its Fortitude gold-silver deposit at Copper Canyon: Mining Engineering, July 1986, v. 38, no. 7, p. 495–499.

Wright, J.E., and Wooden, J.L., 1991, New Sr, Nd, and Pb isotopic data from plutons in the northern Great Basin: Implications for crustal structure and granite petrogenesis in the hinterland of the Sevier thrust belt: Geology, v. 19, no. 5, p. 457–460.

Wrucke, C.T., and Armbrustmacher, T.J., 1975, Geochemical and geologic relations of gold and other elements at the Gold Acres open-pit mine, Lander County, Nevada: U.S. Geological Survey Professional Paper 860, 27 p.

Wrucke, C.T., Churkin, Michael, Jr., and Heropoulos, Chris, 1978, Deep-sea origin of Ordovician pillow basalt and associated sedimentary rocks, northern Nevada: Geological Society of America Bulletin, v. 89, no. 8, p. 1,272–1,280.

Zoback, M.L., McKee, E.H., Blakely, R.J., and Thompson, G.A., 1994, The northern Nevada rift: Regional tectono-magmatic relations and middle Miocene stress direction: Geological Society of America Bulletin, v. 106, no. 3, p. 371–382.

Zoback, M.L., and Thompson, G.A., 1978, Basin and Range rifting in northern Nevada; clues from a mid-Miocene rift and its subsequent offsets: Geology, v. 6, p. 111–116.

Zuffa, G.G., 1980, Hybrid arenites: Their composition and classification: Journal of Sedimentary Petrology, v. 50, no. 1, p. 21–29.

—1985, Optical analysis of arenites: Influence of methodology on compositional results, *in* Zuffa, G.G., ed., Provenance of arenites: Dordrecht, Reidel, p. 165–190.